ANALYTICAL ADVANCES FOR HYDROCARBON RESEARCH

MODERN ANALYTICAL CHEMISTRY

A Continuation Order Plan is available for this series. A continuation order will bring delivery of each new volume immediately upon publication. Volumes are billed only upon actual shipment. For further information please contact the publisher.

ANALYTICAL ADVANCES FOR HYDROCARBON RESEARCH

Edited by

CHANG SAMUEL HSU

ExxonMobil Research and Engineering Co.
Baton Rouge, Louisiana

KLUWER ACADEMIC / PLENUM PUBLISHERS
New York, Boston, Dordrecht, London, Moscow

Library of Congress Cataloging-in-Publication Data

Hsu, Chang Samuel.
    Analytical advances for hydrocarbon research/edited by Chang Samuel Hsu.
        p.   cm. — (Modern analytical chemistry)
    ISBN 0-306-47476-X
        1. Hydrocarbons—Research.   2. Petroleum—Analysis.   I. Title.   II. Series.

TP691 .H77 2003
665.5—dc21

                                                                    2002040700

ISBN 0-306-47476-X

©2003 Kluwer Academic/Plenum Publishers, New York
233 Spring Street, New York, New York 10013

http://www.wkap.nl/

10 9 8 7 6 5 4 3 2 1

A C.I.P. record for this book is available from the Library of Congress

ac

Contributors

Bruzual, Jenny
Analytical Chemistry Department, PDVSA Intevep,
P.O. Box 76343 Caracas 1070-A Venezuela

Cameron, A. S.
Research Department, Products and Chemicals Division
Imperial Oil, Sarnia, Ontario, Canada, N7T 8C8

Carbognani, Lante
PDVSA-INTEVEP. P.O. Box 76343. Caracas 1070 A.
Venezuela

Cebolla, V. L.
Instituto de Carboquímica, Consejo Superior de Investigaciones
Científicas (CSIC), P.O. Box 589, 50080 Zaragoza, Spain

Chawla, Birbal
ExxonMobil Research & Engineering, Paulsboro Technical
Center, Paulsboro, New Jersey 08066

Colaiocco, Silvia
PDVSA-INTEVEP. P.O. Box 76343. Caracas 1070 A.
Venezuela

Di Sanzo, Frank P.
ExxonMobil Research & Engineering Co.
Annandale, New Jersey USA

Domingo, M. P.
Instituto de Carboquímica, Consejo Superior de Investigaciones
Científicas (CSIC), P.O. Box 589, 50080 Zaragoza, Spain

Espidel, Joussef
PDVSA-INTEVEP. P.O. Box 76343. Caracas 1070 A.
Venezuela

Fedora, J. W.
Research Department, Products and Chemicals Division
Imperial Oil, Sarnia, Ontario, Canada, N7T 8C8

Fitzgerald, W. P.
Research Department, Products and Chemicals Division
Imperial Oil, Sarnia, Ontario, Canada, N7T 8C8

Gálvez, E. M.
Instituto de Carboquímica, Consejo Superior de Investigaciones
Científicas (CSIC), P.O. Box 589, 50080 Zaragoza, Spain

Hormes, Josef	Center for Advanced Microstructures and Devices (CAMD) Louisiana State University, Baton Rouge, LA 70806
Hsu, Chang Samuel	ExxonMobil Research and Engineering Co., Baton Rouge, LA 07821
Isaksen, Gary H.	ExxonMobil Upstream Research Co., Houston, TX 77252
Kennedy, Gordon J.	ExxonMobil Research and Engineering Company, 1545 Route 22 East, Annandale, NJ 08801
Lai, Wei-Chuan	Department of Energy & Geo-Environmental Engineering, The Pennsylvania State University, University Park, PA 16802
Limbach, Patrick A.	Department of Chemistry, University of Cincinnati, Cincinnati, OH 45221
Macha, Stephen F.	Department of Chemistry, University of Cincinnati, Cincinnati, OH 45221
Matt, M.	Instituto de Carboquímica, Consejo Superior de Investigaciones Científicas (CSIC), P.O. Box 589, 50080 Zaragoza, Spain
Membrado, L.	Instituto de Carboquímica, Consejo Superior de Investigaciones Científicas (CSIC), P.O. Box 589, 50080 Zaragoza, Spain
Mendez, Aaron	Analytical Chemistry Department, PDVSA Intevep, P.O. Box 76343 Caracas 1070-A Venezuela
Modrow, Hartwig	Institute of Physics, Bonn University, Nussallee 12, D-53115 Bonn, Germany
Nadkarni, R. A. Kishore	Millennium Analytics, Inc., East Brunswick, NJ 08816
Nero, Vincent P.	ChevronTexaco, 3901 Briarpark, Houston, Texas 77042
Peters, Kenneth E.	ExxonMobil Upstream Research Co., Houston, TX 77252
Proulx, R.	Research Department, Products and Chemicals Division Imperial Oil, Sarnia, Ontario, Canada, N7T 8C8
Reddy, K. Madhusudan	Department of Energy & Geo-Environmental Engineering, The Pennsylvania State University, University Park, PA 16802
Riazi, M. R.	Department of Chemical Engineering, Kuwait University, P.O.Box 5969, Safat 13060, Kuwait
Robins, Chad L.	Department of Chemistry, University of Cincinnati, Cincinnati, OH 45221
Roussis, S. G.	Research Department, Products and Chemicals Division Imperial Oil, Sarnia, Ontario, Canada, N7T 8C8
Rudzinski, Walter E.	Department of Chemistry, Southwest Texas State University, San Marcos, Texas 78666
Schaps, M.	ExxonMobil Upstream Research Co., Houston, TX 77252

Song, Chunshan Department of Energy & Geo-Environmental Engineering, The Pennsylvania State University, University Park, PA 16802

Vandell, Victor E. Analytical Science Division, Hercules Incorporated Wilmington, DE. 19808

Walters, Clifford C. ExxonMobil Research and Engineering Co. Annandale, NJ 08801

Wei, Boli Department of Energy & Geo-Environmental Engineering, The Pennsylvania State University, University Park, PA 16802

Preface

Petroleum and fossil fuels (coal, oil shale, etc.) consist of complex hydrocarbon mixtures. Hydrocarbon research has been playing an important role in all phases of the petroleum business. In the upstream, hydrocarbon fingerprints can provide clues for oil/gas potential in basin assessments. The distribution of some specific hydrocarbons, so called biomarkers, can be used for the assessment of source input, age, maturity and alteration of oils. In the downstream, the studies of molecular transformation of hydrocarbons (including heteroatom-containing) in various processes would help the design and improvement of petroleum refining and utilization. The compositional analysis of the products provides important information on quality and performance. In the environmental area, the understanding of hydrocarbon degradation in nature would help the clean-up effort after accidental release of petroleum and its refined products into the environment. The properties and processibility of petroleum and its fractions are closely related to the composition. Hence, in hydrocarbon research analytical characterization has been playing important roles in the determination of composition and the understanding of compositional changes during refining and other chemical processes.

There have always been constant needs for new and improved analytical technology for a better analysis of petroleum composition. At the 220th American Chemical Society (ACS) National Meeting in Washington, DC on August 20-24, 2000, there was a symposium in the Division of Petroleum Chemistry dedicated to analytical characterization of petroleum and its fractions. This book is based on the material given at the symposium and extends its scope to broadly cover recent advances in the analytical arena. We are fortunate to have many experts who agree to contribute articles in their respective areas of expertise. Each article will give brief review, recent developments and future perspectives of the area covered.

To determine the composition of complex petroleum mixtures, it usually requires a battery of analytical techniques that detect and measure specific features of the molecules, such as mass, boiling point, vibrational frequencies, nuclear magnetic resonance frequencies, etc. This book is intended to broadly cover modern analytical techniques that are commonly used to determine the petroleum composition for upstream, downstream, petrochemical and environmental applications.

The analytical areas covered include physical property measurement and estimation, elemental analysis, chromatography, mass spectrometry (MS), nuclear magnetic resonance (NMR) spectroscopy, and x-ray absorption spectroscopy. In recent years many of these analytical techniques are combined, or "hyphenated", to take advantages of the unique features of each technique.

Chapter 1 presents characterization methods and estimation of thermodynamic and physical properties of hydrocarbon and petroleum fractions through the use some easily measurable properties. Chapter 2 overviews elemental composition analysis by a variety of instruments. Many of ASTM methods based on instrumental elemental analysis are described.

Prior separations of petroleum into its fractions by boiling point or polarity facilitate subsequent analysis by spectroscopies. Lower boiling petroleum fractions are normally separated by gas chromatography (GC), while higher boiling fractions by liquid chromatography (LC) or supercritical fluid chromatography (SFC). In petroleum, sulfur and nitrogen-containing compounds are important because of processing and environmental concerns. Chapter 3 describes sulfur- and nitrogen-specific detection when a hydrocarbon mixture or a petroleum fraction is analyzed directly by GC. Complex hydrocarbon mixtures can be analyzed by high- and low-resolution MS for compound type distributions. This is described in Chapter 4. This chapter also describes recent developments in combined gas chromatography-mass spectrometry (GC-MS) for hydrocarbon analysis. An ingenious method of using thin-layer chromatography (TLC) for hydrocarbon type analysis of middle distillates is reported in Chapter 5, particularly the use of berberine-impregnated silica plate for the detection of saturated hydrocarbons by fluorescence.

Chapter 6 describes the analyses of fuels using GC, SFC, high performance liquid chromatography (HPLC) and TLC. The use of 2-dimensional GC for fuel analysis is briefly introduced. Chapter 7 selectively reviews temperature-programmed retention indices for GC and GC-MS analysis of distillate fuels, and high temperature simulated distillation GC analysis of residual oils and their upgraded products. Chapter 8 focuses on the analysis of sulfur species, including elemental sulfur, using MS, tandem mass spectrometry (MS-MS), and GC-MS.

Chromatography and mass spectrometry have also been widely applied in upstream research. Chapter 9 describes the applications of GC-MS and GC-MS-MS for the characterization of petroleum biomarkers. A chemometric approach has been used to retrieve important geochemical information from the data set. Chapter 10 gives a comprehensive review of the light hydrocarbons in petroleum geochemistry. Their molecular and isotropic compositions have been applied to thermal maturity determination, oil-condensate correlation and thermal sulfate reduction.

There are many new analytical techniques developed during the last decade to characterize high boiling and heavy hydrocarbons and petroleum fractions. Chapter 11 describes the coupling of liquid chromatography and mass spectrometry (LC-MS) for molecular-level characterization of saturates, aromatics, polars and resids. Chapter 12 illustrates the uses of advanced mass spectrometric methods for solving challenging analytical problems in refineries and chemical plants of the petroleum industry. Chapter 13 discusses the fractionation of crude oils, particularly nitrogen-, sulfur- and oxygen-containing (NSO) compounds that are analyzed by atmospheric pressure chemical ionization (APCI), an method for LC-MS characterization of high boiling and heavy petroleum fractions. Chapter 14 reviews and discusses the characterization schemes and strategies for heavy crude oils and heavy fractions (heavy ends) of crude oils, including fractionation, chromatographic separations, and molecular, microscopic and x-ray spectroscopic analyses. Chapter 15 highlights some of the recent improvements in nuclear magnetic resonance (NMR) instrumentation that impacts on hydrocarbon characterization, such as higher magnetic fields, facile implementation of pulsed gradient and on-line coupling with chromatography (such as HPLC-NMR, SFC-NMR, etc.)

Chapters 16 and 17 broadly cover the use of matrix-assisted laser desorption/ionization (MALDI) for heavy hydrocarbons and hydrocarbon polymers. For analyzing non-polar hydrocarbon materials, applications of non-polar matrices with cationization agents have been developed that are different than conventional MALDI in ionization mechanisms. More discussions of laser desorption/ionization and the utilization of ultrahigh-resolution Fourier-transform ion cyclotron resonance mass spectrometers (FT/ICR/MS) are introduced in Chapter 17. Chapter 18 reviews the applications of x-ray absorption spectroscopy, particularly x-ray absorption fine structure (XAFS), extended XAFS (EXAFS) and x-ray near edge structure (XANES), for the characterization of heavy hydrocarbon matrices, polymers/rubbers and hydrocarbon synthesis catalysts.

I would like to thank all of the authors for their contributions. I am also grateful to the reviewers for their valuable comments and for their precious time. I sincerely hope this book will become a useful reference to not only experienced researchers, but also new comers and graduate students who are active in hydrocarbon research. I wish to thank ExxonMobil Research and Engineering for supporting my research. Last but not least, to my wife Grace Miao-miao Chen, for her understanding and support of my time and effort in editing this book.

Chang Samuel Hsu
Editor

CONTENTS

3. Selective Detection of Sulfur and Nitrogen Compounds in Low Boiling Petroleum Streams by Gas Chromatography
Birbal Chawla

4. Molecular Characterization of Petroleum and Its Fractions by Mass Spectrometry
Aaron Mendez and Jenny Bruzual

5. Thin-Layer Chromatography for Hydrocarbon Characterization in Petroleum Middle Distillates
V. L. Cebolla, L. Membrado, M. Matt, E. M. Galvez and M. P. Domingo

11. Coupling Mass Spectrometry with Liquid Chromatography for Hydrocarbon Research
Chang Samuel Hsu

12. Advanced Molecular Characterization by Mass Spectrometry: Applications for Petroleum and Petrochemicals
S. G. Roussis, J. W. Fedora, W. P. Fitzgerald, A. S. Cameron and R. Proulx

13. Chromatographic Separation and Atmospheric Pressure Ionization/Mass Spectrometric Analysis of Nitrogen, Sulfur and Oxygen Containing Compounds in Crude Oils
Walter E. Rudzinski

14. Characterization of Heavy Oils and Heavy Ends
Lante Carbognani, Joussef Espidel and Silvia Colaiocco

15. Advances in NMR Techniques for Hydrocarbon Characterization
Gordon J. Kennedy

18. X-Ray Absorption Spectroscopy for the Analysis of Hydrocarbons and Their Chemistry
Josef Hormes and Hartwig Modrow

ANALYTICAL ADVANCES FOR HYDROCARBON RESEARCH

Chapter 1

ESTIMATION OF PHYSICAL PROPERTIES AND COMPOSITION OF HYDROCARBON MIXTURES

M. R. Riazi
Department of Chemical Engineering
Kuwait University
P.O.Box 5969
Safat 13060, Kuwait
E-mail: riazi@kuc01.kuniv.edu.kw

1.   INTRODUCTION

Economical and optimum design and operation of refinery units and related industry largely depend on accurate values of thermodynamic and physical properties of related fluid mixtures used in process simulations. Thermodynamic and physical properties are also known as thermophysical or simply *physical properties* and include but not limited to PVT (density), phase equilibrium (equilibrium ratios), thermal properties (i.e., enthalpy, heat capacity), and transport properties (i.e., viscosity, thermal conductivity, diffusion coefficients and interfacial tension). In petroleum production, reservoir engineers use these properties in reservoir simulation to calculate rate of oil production. Physical properties are state functions and for every fluid depend on temperature, pressure and composition of the mixture. As experimental determination of these properties for every hydrocarbon mixture under different conditions is prohibitive in both time and cost, prediction and estimation of these properties become increasingly important.

Thermophysical properties of fluids are calculated through equations of state (EOS) or generalized correlations. These methods generally require critical temperature and pressure (T_c, P_c) and a third parameter such as acentric factor, ω.[1] For a pure compound, critical temperature and pressure are the highest temperature and pressure at which vapor and liquid phases become indistinguishable and identical. The acentric factor is defined in terms of vapor pressure, P^{vap} at 0.7 T_c as: $\omega = - [\log(P^{vap}/P_c)] - 1.$[2] In addition molecular weight, M, is needed to convert molar properties to

1

weight basis properties. Experimental values of critical properties for pure hydrocarbons are reported up to C_{18} [1,3] while for heavier hydrocarbons due to thermal cracking of hydrocarbon bonds such data are not available. Reported data for the critical properties of heavier hydrocarbons by some researchers are not yet confirmed.[4] For mixtures pseudocritical properties are used in EOS or generalized correlations which are calculated rather than measured properties. For pure compounds, pseudocritical and true critical properties are the same. Reservoir fluids, crude oils and petroleum fractions are all mixtures of various hydrocarbon compounds. Usually laboratories measure and report a few properties for such mixtures. For example for petroleum fractions properties such as specific gravity (SG) or API gravity, distillation data (boiling point, T_b), refractive index (n), kinematic viscosity (v) at a reference temperature and in some cases composition or elemental composition are reported. Characterization of hydrocarbons and petroleum fractions involve the estimation of basic parameters such as critical properties from easily measurable parameters. For petroleum products such as kerosene and gasoline, the quality of these mixtures largely depends on their composition and amount of heteroatoms (i.e., sulfur content). Estimation of these quantities from available laboratory data is also a part of characterization scheme of petroleum fractions.

Characterization methods play an important role in economic design and operation of various units in petroleum related industries. It has been shown that a small error in the critical temperature can cause a much larger errors in other properties such as enthalpy, viscosity, density and vapor pressure for toluene when calculated through an equation of state.[5] Similar results apply to other properties for other compounds through other correlations. Effect of error in critical properties of several hydrocarbons on the error in calculated vapor pressure through Lee-Kesler correlation is demonstrated in Figure 1.[6] In this figure when actual critical temperature (0% deviation) is used error in calculation of vapor pressure for different compounds is nearly zero. However, when critical temperature is underestimated by 5% (-5% deviation in critical temperature) error in calculation of vapor pressure increases to +60%. Errors associated with such physical properties can significantly affect unit design and operation. For example a -5% error on the relative volatility can cause a 100% error on the calculated minimum number of trays for a distillation column.[7] Relative volatility itself is calculated through equilibrium ratios and vapor pressure.[2] A 100% error on the number of trays means that design calculation for the length of a distillation column is twice of the actual length. This error can cause doubling the initial cost of plant and subsequent problems in its operation. Such example can be extended to other units and signifies the role of characterization in the unit design and operation.

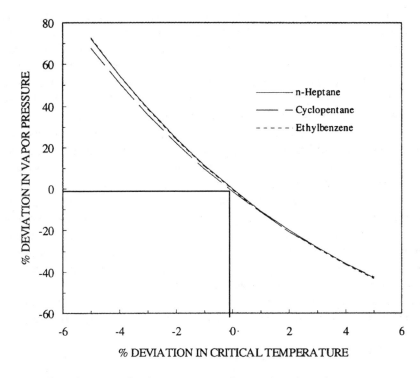

Figure 1. Effect of error in critical temperature on the error in estimated vapor pressure

Development of methods of characterization of hydrocarbons began in the 30's by Watson et al. [8] and has continued to the present time [9]. As advanced analytical and computer tools have become available, recently developed methods are more accurate and general than the old methods. In addition, methods developed until the 60's are mainly graphical [10] while the new methods are in the form of equations and correlations suitable for computer applications. Analytical form of the Winn nomogram for M, T_c and P_c are given by Sim-Daubert [11]. Methods of estimation of critical properties that are widely used in the industry include API-TDB [2], Kesler-Lee [12], Winn[10], Cavett [13], Riazi-Daubert [14] and Twu [15]. These methods use normal boiling point (T_b) and specific gravity (SG) as the input parameters, which are available from laboratory measurements. Most of these correlations are empirically developed with various degree of sophistication using pure compound databases and are included in many process simulators for the characterization of hydrocarbons. However, if the limitations for various methods are not considered, significant errors may be associated with the calculated physical property of a hydrocarbon system. Most of these methods are applicable to light fractions.

Petroleum fluids are mixtures of many compounds from different families, especially paraffins, olefins, naphthenes and aromatics. Since olefins are generally unstable, most petroleum products are olefin free and contain the other three hydrocarbon groups. Properties of a petroleum fluid depend on the composition of the mixture. Basically there are two approaches to estimate the composition of petroleum fractions. The API-TDB method uses v, n and density (d) to estimate the fractions of paraffins (x_P), naphthenes (x_N) and aromatics (x_A).[2] The n-d-M method uses refractive index (n), density (d) and molecular weight (M) to estimate % of carbon atoms in the three hydrocarbon groups.[16] Recently a method is proposed to estimate sulfur weight % of petroleum fractions using n, d, SG and M.[17]

Characterization of crude oils and reservoir fluids is based on representation of the mixture by a number of pure and pseudo components with known basic parameters such as M, T_b and SG.[18] In this chapter various methods of characterization of hydrocarbons and petroleum fractions are presented. Different characterization parameters are reviewed and minimum laboratory data for various petroleum mixtures are introduced. The advantage and disadvantages of different methods and the basis for selection of a characterization method or development of new methods are discussed.

2.   PURE HYDROCARBONS

Properties of pure hydrocarbons are known for many compounds and are tabulated in sources such as API-TDB[2], TRC[19] and DIPPR[20]. Measured critical properties of pure hydrocarbons are reported up to C_{18}, recently reported data for heavier hydrocarbons (up to C_{36})[21] have been reported but not confirmed[3]. However, experimental data on most other properties such as T_b, d, n or freezing point are available for heavier compounds. Properties of pure hydrocarbons with known structure are usually calculated through group contribution methods.[1,2] But these methods cannot be applied to petroleum mixtures containing many different hydrocarbons. For this reason, methods of estimation of properties of hydrocarbons through use of some easily measurable properties are useful. Properties that are used as the input parameters to estimate other physical properties for a hydrocarbon mixture with unknown composition should be easily and directly measurable. As mentioned in the Introduction, most of predictive methods are empirically developed and use T_b and SG as the input parameters. However, the generalized correlation proposed by Riazi and Daubert[14,22] are developed based on EOS parameters and theory of intermolecular forces as discussed in Section 2.1.

2.1 Generalized Correlations for Physical Properties

Properties of a substance depend on the intermolecular forces existing between the molecules. These forces are determined through potential energy, Γ as discussed by Prausnitz et al.[23] Low and medium molecular weight hydrocarbons are generally non-polar substances and their potential energy can be presented by two parameter relations such as Lennard-Jones potential.[23]

$$\Gamma = 4\varepsilon \left[\left(\frac{\sigma}{r}\right)^{12} - \left(\frac{\sigma}{r}\right)^{6} \right] \tag{1}$$

where r is the distance between molecules, ε is the characteristic parameter representing molecular energy and σ is another parameter representing molecular size. For fluids that follow such potential energy relations there exists a universal two-parameter equation of state in the following form.[24]

$$V = f(T, P, \theta_1, \theta_2) \tag{2}$$

where θ_1 and θ_2 are two parameters in a two-parameter potential energy relation that should be able to characterize molecular size and energy. At the critical point there is a criteria of: $\partial P / \partial V = \partial^2 P / \partial V^2 = 0$.[7] Applying these equations at $T=T_c$ and $P=P_c$ to Equation (2) yields

$$T_c, P_c, V_c = g(\theta_1, \theta_2) \tag{3}$$

where g is a universal function for all compounds that follow the same two-parameter potential energy relation. A simple analysis of two-parameter EOSs, such as Redlich-Kwong and van der Waals,[1] shows that Equation (3) can be generalized as

$$T_c, P_c, V_c = a\theta_1^b \theta_2^c \tag{4}$$

where a,b,c are constants specific for each property but same for all compounds. The most suitable parameter for θ_1 is T_b, and for θ_2 is SG. For these input parameters constants in Equation (4) have been determined from the critical properties of pure hydrocarbons from all different groups and are reported by Riazi and Daubert.[14] It was shown that Equation (4) could be applied to other properties such as M or V at a specific temperature. Equation (4) was also applied to estimate thermal conductivity of liquid and vapor hydrocarbons at various temperature through correlating a, b and c to temperature.[25]

For non-polar compounds the main intermolecular force is the London force characterized by polarizability, α, defined as [23]

$$\alpha = (\frac{3}{4\pi N_A}) \times (\frac{M}{d}) \times (\frac{n^2 - 1}{n^2 + 2}) \qquad (5)$$

where N_A is the Avogadro's number, M is the molecular weight, d, is the absolute density and n is the refractive index. Polarizability is proportional to molar refraction, R_m, defined as

$$R_m = (\frac{M}{d}) \times (\frac{n^2 - 1}{n^2 + 2}) = VI \qquad (6)$$

$$V = \frac{M}{d} \qquad (7)$$

$$I = \frac{n^2 - 1}{n^2 + 2} \qquad (8)$$

where V is the molar volume and I is a characterization parameter that was first used by Huang.[26] Through the above relations, parameter I can be presented as the ratio of two volumes

$$I = \frac{R_m}{V} = \frac{\text{actual molar volume of molecules}}{\text{apparent molar volume of molecules}} \qquad (9)$$

where R_m is the molar refraction which represents the actual molar volume of molecules but V represents the apparent molar volume and parameter I is proportional to the fraction of total volume occupied by molecules. Therefore, parameter I can be used as a size parameter. Both parameters "I" and "V" are functions of temperature, but their ratio is nearly independent of temperature.[27] Reference temperature of 20 °C is usually used to report d and n. Refractive index is defined as the ratio of velocity of light in the vacuum to the velocity of light in a substance. Thus n is greater than unity and parameter I is between 0 and 1. Since the velocity of light depends on its wavelength, sodium D line is usually used to measure the refractive index.

Equation (4) was later modified and was applied to various pairs of input parameters in the following form:

$$\theta = a[\exp(b\theta_1 + c\theta_2 + d\theta_1\theta_2)]\theta_1^e\theta_2^f \qquad (10)$$

where a-f are numerical constants for each property θ and each pair of input parameters (θ_1, θ_2). As mentioned earlier, any pair of parameters which is capable of characterizing molecular energy and molecular size can be used as the input parameters. For example kinematic viscosity at a fixed reference temperature (i.e. 100 °F), v_{100}, may be used as a parameter to characterize molecular energy while carbon-to-hydrogen weight ratio, C/H, is a suitable parameter to characterize molecular size. For the same carbon number, aromatics have higher C/H values than paraffins.

Properties of hydrocarbons vary with carbon number within a single hydrocarbon group and from one hydrocarbon group to another for the same carbon number. This is demonstrated in Table 1 for hydrocarbons C_5 and C_6 through the percent differences in various parameters within adjacent hydrocarbons in n-paraffins or for C_6 from paraffinic and aromatic groups. For higher carbon number compounds the difference in properties on percentage basis decreases. Two input parameters chosen for (θ_1, θ_2) should be a pair that one characterizes carbon number and the other hydrocarbon molecular type. Parameters suitable for θ_1 to characterize carbon number within a hydrocarbon group are T_b, M and v_{100} and for θ_2 appropriate parameters are SG, I and C/H as they characterize hydrocarbon type. When there is a choice, the characterization power of these parameters can be written in the order of: T_b > M > v_{100} for θ_1 and SG > I > C/H for θ_2. Therefore, a pair of (T_b, SG) is more suitable than (M, C/H) as input parameters.

Table 1. Comparison of properties of adjacent members of paraffin family and two families of C_6 hydrocarbons

Hydrocarbon	T_b, K	M	v_{100}	SG	I	C/H
Paraffin Family						
C_5H_{12} (n-pentane)	309.2	72.2	0.209	0.6311	0.219	4.96
C_6H_{14} (n-hexane)	341.9	86.2	0.414	0.6638	0.229	5.11
% Difference in property	+10.6	+19.4	+98	+5.2	+4.5	+3.0
Two Families (C_6)						
Paraffin (n-hexane)	341.9	86.2	0.414	0.6638	0.229	5.11
Aromatic (benzene)	353.2	78.1	0.587	0.8829	0.295	11.92
% Difference in property	+3.3	-9.4	41.8	+33.0	+28.8	+133.3

Parameters a-f in Equation (10) for various physical properties in terms of different pairs of input parameters are given in Reference 22. For the critical properties, T_b, M, I and C/H for the molecular weight range of 70-300 (C_5-C_{22}) these constants are given in Table 2. Equation (10) for prediction of M, T_c, P_c, and I in terms of T_b and SG is also included in the API-TDB-

1987.[2] Constants in Equations (10) for various properties in terms of M and SG are also given in reference 22. It should be noted that with the exception of M, Equation (10) with constants given in Table 2 is not suitable for heavy hydrocarbons (> C_{25}). Accuracy of Equation (10) for T_c is about 1% and for M and P_c is about 3%.

Table 2. Constants in Equation (10) for various properties with θ_2=SG.

	θ_1	a	b	c	d	e	f
M	T_b	42.965	2.097×10^{-4}	-7.78712	2.08476×10^{-3}	1.26007	4.98308
T_c	T_b	9.5233	-9.314×10^{-4}	-0.544442	6.4791×10^{-4}	0.81067	0.53691
P_c	T_b	3.1958×10^{-4}	-8.505×10^{-3}	-4.8014	5.749×10^{-3}	-0.4844	4.0846
V_c	T_b	6.049×10^{-2}	-2.6422×10^{-3}	-0.26404	1.971×10^{-3}	0.7506	-1.2028
I	T_b	2.3435×10^{-2}	7.029×10^{-4}	2.468	-1.0267×10^{-3}	0.0572	-0.72
C/H	T_b	3.4707	1.485×10^{-2}	16.94	-1.2492×10^{-2}	-2.725	-6.798
T_b	M	3.76587	3.7741×10^{-3}	2.98404	-4.25288×10^{-3}	0.40167	-1.58262

2.2   Properties of Heavy Hydrocarbons

Lee and Kesler [5,12] developed correlations for T_c, P_c and ω in the following forms:

$$T_c = 189.8 + 450.6 \text{ SG} + (0.4244 + 0.1174 \text{ SG})T_b$$
$$+ (0.1441 - 1.0069 \text{ SG})10^5 / T_b \tag{11}$$

$$\ln P_c = 5.689 - 0.0566 / \text{SG} - (0.43639 + 4.1216 / \text{SG} + 0.21343 / \text{SG}^2) \times 10^{-3} T_b$$
$$+ (0.47579 + 1.182 / \text{SG} + 0.15302 / \text{SG}^2) \times 10^{-6} / T_b^2 \tag{12}$$
$$- (2.4505 + 9.9099 / \text{SG}^2) \times 10^{-10} / T_b^3$$

$$\omega = \frac{-\ln \dfrac{P_c}{1.01325} - 5.97214 + 6.09648 / T_{br} + 1.28862 \ln T_{br} - 0.169347 T_{br}^6}{15.2518 - 15.6875 / T_{br} - 13.4721 \ln T_{br} + 0.43577 T_{br}^6} \tag{13}$$

where T_{br} is the reduced boiling point (T_b/T_c) and P_c is in bar. These equations for T_c and P_c can be applied to hydrocarbons from C_5 to C_{45}. Critical properties of heavy hydrocarbons used to develop the above relations were back calculated from other properties. In addition $T_c=T_b$, $P_c = 1.01325$ bar have been imposed as an internal constraint. Therefore, for hydrocarbons heavier than C_{22} critical properties may be estimated from Equations (11) and (12). Although a special relation is proposed for the estimation of acentric factor of heavy compounds,[12] Equation (13) can be safely used up to C_{45} if accurate input parameters for T_c and P_c are used. The Edmister equation for acentric factor is simpler than Equation (13) but it is less accurate.[1] Lee et al.[28] used perturbation theory to correlate the difference between critical properties of n-alkanes and other hydrocarbons with the SG. Twu [15] later used the same approach with data for critical properties of heavy hydrocarbons calculated from vapor pressure to develop correlations for M, T_c, P_c, and V_c. The correlations contain many constants and are inter-correlated to each other. In phase behavior calculations for heavy reservoir fluids, both Lee-Kesler and Twu correlations are recommended.[29] However for fractions with M of less than 300 (< C_{22}), both Equations (4) and (10) in terms of T_b and SG are recommended for the critical properties and molecular weight in various references [2,29-33] and are included in some simulators [34-37]. Equation (13) predicts acentric factor of pure hydrocarbons with average deviation of about 1.5%.

To correlate physical properties of heavy hydrocarbons and more complex compounds, two parameter correlations are not suitable and a third parameter is needed. However, properties of homologous hydrocarbons within a single family can be correlated to only one characterization parameter such as carbon number (N_C), T_b or M. For the three groups of n-alkanes, n-alkylcyclopentanes and n-alkylbenzenes, various properties are related to M through the following relation with constants given in Table 3.[38]

$$\theta = \theta_\infty - \exp(a - bM^c) \tag{14}$$

where, in this relation θ_∞ is the value of θ for extremely large molecules (M→ ∞).

Table 3. Constants of Equation 14 for various hydrocarbon groups

θ	Carbon No. range	Constants in eq. 14			
		θ_∞	a	b	C
n-Alkanes					
T_b	C_5-C_{40}	1070	6.982 91	0.020 13	2/3
$T_{br} = T_b/T_c$	C_5-C_{20}	1.15	-0.419 66	0.024 36	0.58
-P_c	C_5-C_{20}	0	4.657 57	0.134 23	0.5
-ω	C_5-C_{20}	0.3	-3.068 26	-1.049 87	0.2
n-Alkylcyclopentanes					
T_b	C_6-C_{41}	1028	6.956 49	0.022 39	2/3
$T_{br} = T_b/T_c$	C_5-C_{18}	1.2	0.067 65	0.137 63	0.35
-P_c	C_6-C_{18}	0	7.258 57	1.131 39	0.26
-ω	C_6-C_{20}	0.3	-8.256 82	-5.339 34	0.08
n-Alkylbenzenes					
T_b	C_6-C_{42}	1015	6.910 62	0.022 47	2/3
$T_{br} = T_b/T_c$	C_6-C_{20}	1.03	-0.298 75	0.068 14	0.5
-P_c	C_6-C_{20}	0	9.779 68	3.075 55	0.15
-ω	C_6-C_{20}	0	-14.97	-9.483 45	0.08

In obtaining the constants for Equation (14) consistency criteria between T_c, P_c and T_b was imposed so that for a heavy hydrocarbon in each family when P_c = 1.0133 bar, T_c becomes equal to T_b. Equation (14) predicts T_b and T_c with errors of about 0.3%, P_c with an error of about 0.8% and acentric factor with an error of about 1.2%. Figure 2 shows prediction of critical pressure for naphthenic compounds of n-alkylcyclopentanes from Equations (4), (10), (12) and (14) and comparison with data given in the API-TDB-1997.[2] Values reported by the API for alkylcyclopentane compounds with C_N > 10 are predicted by a group contribution method. It clearly shows that values of P_c for compounds with N_C greater than 22 reported by the API-TDB-1997 are not accurate. Values of P_C predicted by Equation (12) are lower than API values. Equations of (4) and (10) predict values of P_c very close to the API values for compounds up to C_{20}. Values predicted by Equation (14) are nearly identical to the API values up to C_{20}. Equation (14) is particularly useful to estimate properties of heavy petroleum fraction through the pseudocomponent method discussed in the next section.

3.   PROPERTIES OF PETROLEUM FRACTIONS

Petroleum fractions are mixtures of different hydrocarbons within a specified boiling point range. The most accurate method of estimating properties of petroleum fractions is through knowledge of exact composition of all compounds existing in the mixture and use of appropriate mixing rule to estimate mixture properties. However, as such a complete analysis is usually not available, bulk properties of the fraction may be used to estimate the properties. Some of the bulk properties that may be available are: T_b, SG, M, v and the composition. Although in many cases only a few of these parameters are available.

Contrary to pure compounds where T_b is a single value of temperature, petroleum fractions have a range of boiling points from the initial boiling point (IBP) to the final boiling point (FBP). Boiling points are measured at certain volume percents (or volume fractions) distilled from 0 to 100% which are usually referred to as distillation data or distillation curve. ASTM D86 is the simplest form of distillation where the temperatures are reported at 10,30,50,70 and 90% volume vaporized.[39] Nowadays distillation data are obtained by gas chromatography called simulated distillation (SD) which can be measured by ASTM D2887 test method and it is more accurate than ASTM D86. However, the actual boiling point is represented by true boiling point (TBP) and methods of conversion of various distillation data are given in API-TDB[2] and reference 40. Average boiling points are generally defined in terms of ASTM D86,[2] however, for narrow boiling range fractions the average boiling point is near the ASTM D86 temperature at 50%. Narrow

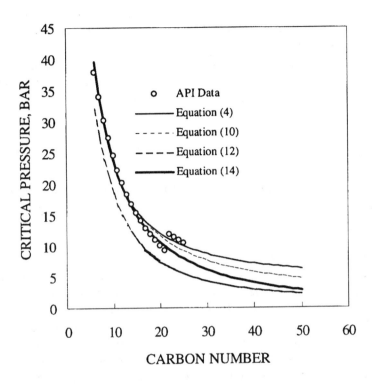

Figure 2. Prediction of critical pressure of n-alkylcyclopentanes from different methods.

fractions are considered as those fractions which the difference between ASTM temperature at 90% and 10% is less than 60 °C (~110 °F) although this is just an approximate range. Another approximate value to represent average boiling point of narrow petroleum cuts is the mid
true boiling point temperature. Usually boiling point information are available for light fractions (M<300) and for heavy fractions, kinematic viscosity at 100 or 210 °F (38 or 99 °C) are measured as boiling point cannot be measured due to cracking of heavy hydrocarbons. Specific gravity or API gravity (API = 141.5/SG – 131.5) is generally available for all types of fractions.

A convenient method of estimating properties of narrow fractions is to consider the mixture as one pseudocomponent having the same properties as the mixture. Properties of this pseudocomponent may be estimated from the correlations presented in Section 2 for pure compounds. For light fractions the most suitable and available parameters are boiling point and specific gravity. Therefore, Equations (4) and (10)-(13) can be applied directly to a petroleum fraction using the mid boiling point and SG. For heavy fractions that the boiling point may not be available, correlations in terms of parameters such as M and SG can be used.[22] Heavy petroleum fractions contain complex compounds, which are not really non-polar, and a two-parameter potential energy cannot describe molecular forces. For such compounds a third parameter is required to characterize the complexity of molecules. As an example, the following relation was developed to estimate M of heavy petroleum fractions using kinematic viscosities at 100 and 210 °F and specific gravity as input parameters.[42]

$$M = 223.56 \, [v_{100}^{(-1.2435+1.1228SG)} \, v_{210}^{(3.4758-3.038SG)}] \, SG^{-0.6665} \qquad (15)$$

This equation can be used to estimate M of heavy fractions in the range of 200-800 when boiling point is not available but viscosity is available. This relation is also recommended by the API-TDB.[2]

A more accurate approach to estimate properties of petroleum fractions is to consider a fraction as a mixture of several pseudocomponents. As shown in Table 1, properties of hydrocarbons vary both by carbon number and molecular type. T_b mainly varies with carbon number and for narrow boiling range fractions the variation of carbon number in the fraction is limited. For such fractions, all compounds within each family may be grouped together as a single pseudocomponent. For fractions with paraffinic (P), naphthenic (N) and aromatic (A) groups, three model pseudocomponents can represent the mixture with PNA composition in terms of x_P, x_N and x_A. If properties of the three pseudocomponents are known, then the mixture property can be estimated through the following simple mixing rule:

$$\theta = x_P \theta_P + x_N \theta_N + x_A \theta_A \qquad (16)$$

where θ is a physical property for the mixture and θ_P, θ_N, θ_A are the values of θ for the model pseudocomponents from the three groups. In this equation the composition should be in mole fraction, however, since the difference between molecular weights of different groups having the same boiling point is not significant the weight fractions may also be used in Equation (16). This method is called pseudocomponent method and was used by Huang to estimate enthalpy of petroleum fractions.[41] The three model compounds are taken from n-alkane, n-alkylcyclopentane and n-alkylbenzene groups which have one characterization parameter same as the fraction. This characteristic parameter could be T_b, M or v. If distillation data are available, 50% ASTM or mid true boiling point can be used as the characteristic parameter to select three pseudocomponents from the three model groups having boiling point same as that of the fraction. Equation (14) with constants in Table 3 is suitable for this purpose. The following example shows application of both approaches.

   Example 1: A heavy gasoline has TBP cut of 80-180 °C, specific gravity of 0.742 and PNA composition of 73, 15 and 12 in weight %.[43] Estimate molecular weight of this fraction using bulk properties of T_b and SG and compare with the value estimated from the pseudocomponent method.

   Solution: The mid boiling point of this fraction is [(80+180)/2]+273=403 K. SG =0.742. Equation (10) with constants given in Table 2 is used to estimate the molecular weight.

$$M = 42.9654 \,[\exp(2.097 \times 10^{-4} \times 403 - 7.787 \times 0.742 + 2.0848 \times 10^{-3} \times 403 \times 0.742)]$$
$$\times [(403)^{1.26007}] \times [(0.742)^{4.98308}] = 107.5$$

Constants for M in Table 2, have been obtained from data on both pure compounds and petroleum fractions with molecular weights from 70 to 700. For light fractions, an equation derived from Equation (4) with constants given in Reference 14 is more suitable:

$$M = 1.6607 \times 10^{-4} \, T_b^{2.1962} SG^{-1.0164} \tag{17}$$

From this equation M =118.3. To use the pseudocomponent method, Equation (16) is applied to molecular weight.

$$M = x_P M_P + x_N M_N + x_A M_A$$

Molecular weight of paraffinic, naphthenic and aromatic compounds having boiling point the same as of the fraction can be estimated from Equation (14) when it is solved to get M from T_b for each group. For example M_P is calculated from Equation (14) and constants in Table 3 as:

$$M_p = \left\{ \frac{1}{0.02013}[6.98291 - \ln(1070 - 403)] \right\}^{3/2} = 116.5$$

Similarly, M_N = 111.5 and M_A = 103.0. The PNA composition is given as: x_P = 0.73, x_N = 0.15 x_A = 0.12. Therefore, M = 0.73 × 116.5 + 0.15 × 111.5 + 0.12 × 103 = 114.1. A more accurate value can be obtained if the composition is converted to mole fraction through molecular weight for each group. Mole fraction of paraffinic compounds is calculated as x_P^m = (0.73/116.5)/(0.73/116.5 + 0.15/111.5 + 0.12/103) = 0.714. Similarly, x_N^m = 0.153 and x_A^m = 0.133. Based on the molar composition the mixture molecular weight is 113.9 which is very close to the value of 114.1 calculated based on the weight fraction. Similarly if volume fraction is used in Equation (16) there is no significant change in the results and the difference is within the experimental uncertainty for the composition. Although an experimental value is not reported, the value of 114.1 generally is considered as the most accurate estimated value. Similarly, other properties such as T_c and P_c may be estimated through Equation (16). Values of M_P, M_n, and M_A estimated from T_b should be used in Equation (14) when it is used for the estimation of other properties from boiling point as the input parameter.

4.   COMPOSITION OF PETROLEUM FRACTIONS

As mentioned earlier, petroleum fractions are mixtures of many compounds from various hydrocarbon groups that vary in both carbon number and molecular type. Prediction of complete composition of all compounds existing in a mixture is nearly impossible. Distribution of compounds in terms of carbon number (or molecular weight) is determined through TBP distillation curves. For example all compounds with 6 carbon atoms have a limited boiling range of 50-90 °C. Therefore from distillation curve, the difference between cumulative volume % vaporized at 90 and 50 °C gives the volume percent of C_6 compounds in the fraction. But for narrow boiling range petroleum fractions composition is generally presented in terms of fraction of paraffins, naphthenes and aromatics (x_P, x_N and x_A) and it is referred as PNA composition. Experimental data are generally reported in terms of weight fractions, but through SG they are conveniently converted to volume fractions or by molecular weight to mole fractions. As mentioned in Example 1, the difference between various forms of composition is within the experimental uncertainty in the measurement of composition. Since carbon is the dominant element in a petroleum mixture on the weight basis, the percentages of paraffins, naphthenes and aromatics (P%, N%, A%) in a petroleum fraction are approximately proportional to the distribution of carbon atoms in these hydrocarbon groups expressed as CP%, CN% and CA%.

Composition of petroleum fractions is measured by gas chromatography and mass spectrometry (GC-MS) which identify volatile compounds by their boiling point and molar mass. PIONA analyses determine the amounts of n-paraffins, isoparaffins, olefins, naphthenes and aromatics. Naphthenic and aromatic compounds may also be divided according to their number of ring structures. For example alkylcyclopentanes have only one cyclic structure and are referred as monocycloparaffins. Naphthenic compounds having more than one ring structure are called polycycloparaffins. Aromatics with one benzene rings are called monoaromatics or mononuclear aromatics while aromatics having more than one benzene ring structure are called polyaromatics or polynuclear aromatics. A crude oil when distilled is converted to various fractions with different boiling points. As a general rule when boiling point of fractions increases the amounts of polycyclic compounts from both naphthenic and aromatic groups also increase while the amount of paraffins decreases as shown in Figure 3.

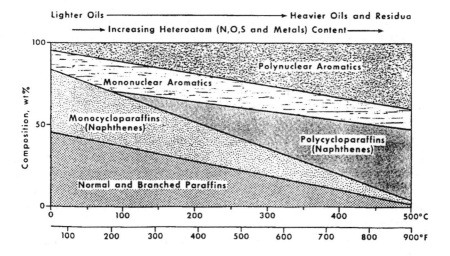

Figure 3. Variation of composition of petroleum fractions with boiling point (Taken from Reference 44 with permission).

4.1   Characterization Parameters for Molecular Type

There are several parameters that are defined to characterize molecular type of petroleum fractions. The most useful parameters are: Watson or UOP characterization factor (K_W), viscosity gravity constant (VGC), SG (or API gravity), refractivity intercept (R_i) and refractive index-molecular weight parameter (m) defined by the following equations.[45,46]

$$K_W = \frac{(1.8T_b)^{1/3}}{SG} \tag{18}$$

$$VGC = \frac{10SG - 1.0752\log(V_{100} - 38)}{10 - \log(V_{100} - 38)} \tag{19}$$

$$R_i = n - d/2 \tag{20}$$

$$m = M(1.475 - n) \tag{21}$$

where T_b is in Kelvin, SG is the specific gravity at 15.5 °C (60 °F), V_{100} is the viscosity at 100 °F (38 °C) in Saybolt universal seconds (SUS), n and d are liquid refractive index and density both at 20 °C and M is the molecular weight. VGC could also be defined in terms of viscosity at 210 °F (99 °C). Variation of these parameters for the three hydrocarbon groups commonly found in petroleum fractions is shown in Figure 4 and for parameter m is given in Table 4.[44,45] K_W was originally defined to identify hydrocarbon groups but as seen in Figure 4, there is an overlap between K values from aromatic and naphthenic groups. Similar overlap exists for parameter "I" for paraffinic and naphthenic groups. From Figure 4, one can see that parameters R_i and VGC separate hydrocarbon groups better than other parameters. A good parameter to characterize hydrocarbon type is a parameter whose variation within a group is minimal but varies significantly from one group to another and there is no overlap from one group to another so that a single value can be assigned for each group. As an example, a pure hydrocarbon compound whose R_i is 1.1 has to be aromatic and cannot be from other groups. Definition of R_i is based on this observation that variation of n with density is linear for each hydrocarbon group but with different intercept.[47] Parameter "m" defined by Equation (21) is based on this observation that plot of n versus 1/M is a straight line for each hydrocarbon group and m represents slope of that line for each specific group. Parameter m identifies hydrocarbon types even better than R_I and VGC. In addition it also characterizes various aromatic groups. For example if a pure hydrocarbon has m value of 2.6 it should be a monoaromatic compound.

Table 4. Values of parameter m for different types of hydrocarbons

Hydrocarbon Type	m
Paraffins	-8.79
Cyclopentanes	-5.41
Cyclohexanes	-4.43
Benzenes	2.64
Naphthenes	19.5
Condensed Tricyclics	43.6

One of the problems with VGC as defined by Equation (19) is that it cannot be applied to fractions with viscosities at 100 °F of less than 38 SUS (~3.6 cSt). Therefore, for light fractions another parameter equivalent to VGC is required. This parameter is defined by Riazi and Daubert [45] and is called viscosity gravity function (VGF):

hydrocarbon type	M	R_i	value range VGC	K	I
paraffin	337–535	1.048–1.05	0.74–0.75	13.1–13.5	0.267–0.273
naphthene	248–429	1.03–1.046	0.89–0.94	10.5–13.2	0.278–0.308
aromatic	180–395	1.07–1.105	0.95–1.13	9.5–12.53	0.298–0.362

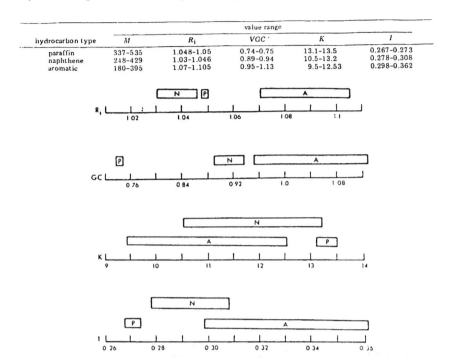

Figure 4. Comparison of various characterization parameters for hydrocarbon molecular type (Taken from Reference 45 with permission).

$$\text{VGF} = -1.816 + 3.484\text{SG} - 0.1156\ln v_{100} \qquad (22)$$

where v_{100} is kinematic viscosity at 100 °F in cSt. There is a similar relation for estimation of VGF in terms of kinematic viscosity at 210 °F.[45] Equation (22) is basically defined for fractions with molecular weights of less than 200. This equation is defined based on this observation that for each hydrocarbon group, SG varies linearly with viscosity as shown in Figure 5.

Another useful parameter capable of characterizing molecular type is the carbon to hydrogen weight ratio, C/H. Aromatic oils have higher C/H values than the paraffinic oils. Higher C/H values correspond to lower H/C atomic ratios as follows:

$$\text{C/H(weight ratio)} = \frac{11.9147}{\text{H/C (atomic ratio)}} \qquad (23)$$

in which H/C is the hydrogen to carbon atomic ratio. Petroleum fractions with higher H/C (lower C/H) have higher heating values and better quality. Hydrogen as a perfect fuel has C/H value of 0 and carbon as a worse fuel has C/H value of ∞ (H/C =0).

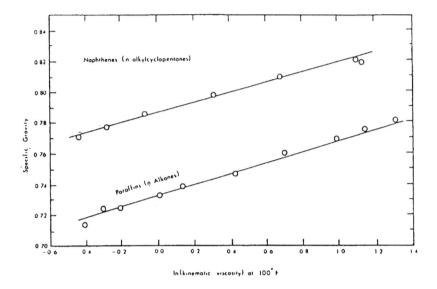

Figure 5. Viscosity-gravity relationship (Taken from Reference 45 with permission).

4.2 Development of Predictive Methods

Generally VGF or VGC are low for paraffinic fractions and high for aromatic oils. Based on the properties of pure hydrocarbons from the API RP-42, average values of 1.0482, 1.0138 and 1.081 were obtained for the refractivity intercept, R_i, of paraffins, naphthenes and aromatics, respectively.[48] Applying Equation (16) to R_i gives:[45]

$$R_i = 1.0482x_P + 1.038x_N + 1.081x_A \tag{24}$$

Based on 33 defined hydrocarbon mixtures Equation (24) was slightly modified as:

$$R_i = 1.0486x_P + 1.022x_N + 1.11x_A \tag{25}$$

A similar relation was obtained for the VGC of hydrocarbon mixtures as:

$$VGC = 0.7426x_P + 0.9x_N + 1.112x_A \tag{26}$$

For olefin free fractions we have:

$$x_P + x_N + x_A = 1 \tag{27}$$

By simultaneous solution of Equations (25)-(27) and modifying the constants with additional data the following relations were obtained for fractions with molecular weight greater than 200.[46]

$$x_P = 2.5737 + 1.0133R_i - 3.573VGC \qquad (28)$$

$$x_N = 2.464 - 3.6701R_i + 1.96312VGC \qquad (29)$$

$$x_A = -4.0377 + 2.6568R_i + 1.60988VGC \qquad (30)$$

Similarly based on the composition of light fractions with M<200, the following relations in terms of R_i and VGF were obtained.[46]

$$x_P = -13.359 + 14.4591R_i - 1.41344VGF \qquad (31)$$

$$x_N = 23.9825 - 23.333R_i + 0.81517VGF \qquad (32)$$

$$x_A = -9.6235 + 8.8739R_i + 0.59827VGF \qquad (33)$$

These equations generally estimate the composition of petroleum fractions within errors of 3-5% based on evaluations made with experimental data on composition of over 150 petroleum fractions.[46] In cases that viscosity is not available, alternative relations have been developed in terms of other characterization parameters, e.g., C/H, SG, R_i and m.[46]
For fractions with M<200:

$$x_P = 2.57 - 2.877SG + 0.02876C/H \qquad (34)$$

$$x_N = 0.52641 - 0.7494x_P - 0.02181m \qquad (35)$$

$$x_A = 1 - (x_P + x_N) \qquad (36)$$

For fractions with M>200:

$$x_P = 1.9842 - 0.27722R_i - 0.15643C/H \qquad (37)$$

$$x_N = 0.5977 - 0.761745R_i + 0.068048C/H \qquad (38)$$

in which x_A can be estimated from Equation (36). For highly aromatic fractions such as coal liquids, the following relations have been proposed to estimate monoaromatics and polyaromatics for fractions with M<250.[46]

$$x_{MA} = -62.8245 + 59.90816R_i - 0.0248335m \qquad (39)$$

$$x_{PA} = 11.88175 - 11.2213R_i + 0.023745m \qquad (40)$$

$$x_A = x_{MA} + x_{PA} \tag{41}$$

These equations estimate fractions of x_P, x_N, x_{MA}, x_{PA}, x_A with average deviations of about 5 to 6% for 70 to 80 fractions.[46] Equations (28)-(33) and (39)-(41) are also included in the API-TDB-87.[2] More recently, based on a broader data set, [49] the following set of equations were developed for fractions with M<250 [27].

$$x_P = 3.2574 - 3.48148SG + 0.011666m \tag{42}$$

$$x_N = -1.9571 + 2.63853SG - 0.03992m \tag{43}$$

where x_A is again calculated from Equation (36). The C/H value may be estimated from Equation (10). Other physical properties needed to estimate the composition are M, n and d (at 20 °C). For fractions with molecular weights from 70 to 300, M may be estimated from Equation (17) and parameter I can be estimated from a correlation derived from Equation (4) and reported in Reference 14 as follows:

$$I = 0.3773T_b^{-0.02269}SG^{0.9182} \tag{44}$$

where n is calculated from definition of parameter I, Equation (8) as :

$$n = \sqrt{\frac{1 + 2I}{1 - I}} \tag{45}$$

Density at 20 °C can be estimated from specific gravity at 15.5 °C [50]

$$d = SG - 4.5 \times 10^{-3}(2.34 - 1.9SG) \tag{46}$$

Average errors for these equations are about 0.5% for the refractive index and 0.1% for the density when evaluated with data for pure hydrocarbons.[2]

4.3   Prediction of Sulfur Content and Carbon Residue

Sulfur is one of the most important hetroatoms in petroleum. Its presence in a crude oil or petroleum product reduces quality and heating value of the fuel. It promotes corrosion and reduces oil resistance to oxidation. Environmental and governmental authorities in developed countries have imposed regulations of very low sulfur content of 0.01 wt.% or less to most petroleum products. Sulfur content of a crude oil varies from 0.05 to 6%,[44] but when the crude oil is refined the sulfur is distributed within heavy products. Sulfur is mainly associated with heavy aromatic compounds and for this reason the sulfur content also increases with increase in the boiling point or specific gravity of a petroleum fraction. Based on the assumption that the sulfur is mainly associated with cyclic compounds especially aromatics, the following relations were developed for the estimation of sulfur content of a fraction in terms of parameters R_i, m and SG:[17]

For fractions with M < 250

$$S\% = 177.448 - 170.946\ R_i + 0.2258\ m + 4.054\ SG \qquad (47)$$

For fractions with M > 200

$$S\% = -58.02 + 38.463\ R_i - 0.023\ m + 22.4\ SG \qquad (48)$$

where R_i and m are defined by Equations (20) and (21). For fractions with molecular weights between 200 and 250 both Equations (47) and (48) may be used. For light fractions in which Equation (47) may estimate a negative value, S% is set equal to zero. The average absolute errors for these equations is about 0.25%. Estimation of sulfur content and the PNA composition of a petroleum fraction is demonstrated in Example 2.

Another parameter that represents the quality of a petroleum fraction is carbon residue which quantifies the non-distillable heavy constituents of a fraction. ASTM D 189 test method describes the method to determine carbon residue known as Conradson Carbon Residue (CCR) in weight %. It is possible to estimate CCR% from H/C atomic ratio through the following relation:[51]

$$CCR\% = 148.7 - 86.96\ H/C \qquad (49)$$

if H/C ≥ 1.71 (CCR<0), set CCR%=0.0 and if H/C < 0.5 (CCR>100), set CCR%=100. H/C ratio can be estimated from Equation (23). Generally heavier oils (lower API gravity) have higher boiling points, lower hydrogen contents, higher aromatic contents, higher carbon residue, higher sulfur, nitrogen and metallic contents. If H/C is calculated from Equation (23), then CCR% can be estimated from Equation (49).

Example 2: For the gasoline sample of Example 1, estimate: P%, N%, A%, CCR% and S% of the fuel.

Solution: From Example 1 we have: T_b = 403 K, SG = 0.742. Since M <300, Equations (17) and (44) can be used to estimate M, I while C/H can be estimated from Equation (10) and constants in Table 2: M = 118.5, I = 0.25,

C/H = 5.723. From Equation (45), n can be calculated as: n = 1.415. From Equation (46), d = 0.7378 g/cm^3. From Equations (20) and (21) parameters R_i and m are calculated as: R_i = 1.0462, m = -7.117. Using C/H, SG and m we calculate x_P, x_N and x_A from Equations (34), (35) and (36), respectively which give: P%=60, N%=23 and A%=17. The experimental values given in Example 1 are 73, 15 and 12% for P%, N%, and A%, respectively. In this case the PNA prediction is not very accurate mainly because all input parameters have been estimated. If the experimental value of viscosity is available, Equations (31)-(33) should give better prediction of the composition. Since M<250, S% can be estimated from Equation (47): S%=177.448-170.946×x1.0462+0.2258×(-7.117)+4.054×0.742=0.01.   The experimental value for the S% is 0.035%.[43] Therefore the absolute error is 0.035%-0.01% = 0.025%. To estimate CCR%, H/C ratio is calculated from Equation (23) and C/H value. H/C=11.9147/5.723=2.08. Substituting H/C value into Equation (49) gives CCR=-32.2%. Since CCR<0 and H/C>1.71, therefore, CCR%=0. That was expected as gasoline is a light petroleum fraction and has no carbon residue.

It should be noted that in experimental determination of composition of heavy fractions non-volatile compounds cannot be separated by GC analysis. Composition and analysis of heavy petroleum fractions are discussed by Altgelt and Boduszynski.[51]

5.   SUMMARY

In this chapter characterization and methods of estimation of properties of hydrocarbons and petroleum fractions are discussed. For light fractions (M<300; N_C<22, T_b<350°C), at least two parameters are needed for a petroleum fraction to estimate its quality and physical properties. The most suitable parameters for characterization of light fractions according to their characterization power are listed below:

- Boiling point (i.e, simulated distillation)
- Specific gravity
- Molecular weight
- Refractive index at 20 °C
- Viscosity at 38 and/or at 99 °C
- PNA composition
- Elemental analysis (C, H, S)

Minimum data required for property prediction are two parameters that could be any pair as: (T_b,SG), (T_b,I), (M,SG), (M,I), (T_b,C/H), (v,SG). The quality of a petroleum product can be determined through prediction of PNA composition, sulfur content and carbon residue. For heavy petroleum fractions in which distillation data may not be available viscosity at two temperatures together with specific gravity would be useful. Measurement of molecular weight especially for heavy fraction is very useful for better prediction of properties and quality of these fractions.

Methods presented in this chapter are recommended to be applied to relatively narrow petroleum fractions. To estimate properties of crude oils and reservoir fluids, it is required that the mixture be split into several narrow boiling range pseudocomponents with known characterization parameters such as M and SG.[18] For each pseudocomponent, properties can be calculated through methods discussed in this chapter and through an appropriate mixing rule, the properties of crudes can then be calculated.[17] Further details on the methods of characterization and estimation of properties of hydrocarbon systems are provided in Reference 52.

6. NOMENCLATURE

API	API Gravity (141.5/SG − 131.5)
a,b,...i	Correlation constants in various equations
C/H	Carbon-to-hydrogen weight ratio
CCR%	Conradson Carbon Residue, Weight %
d	Liquid density of liquid at 20 °C and 1 atm, g/cm^3
H/C	Hydrogen-to-carbon atomic ratio
K_W	Watson (UOP) K factor defined in Equation (18)
I	Refractive index parameter defined in Equation (8)
M	Molecular weight, g/mol [kg/kmol]
m	Refractive index - molecular weight parameter defined in Equation (21)
n	Sodium D line refractive index of liquid at 20 °C and 1atm, dimensionless
N_A	Avogadro's number
N_C	Carbon number (number of carbon atoms in a hydrocarbon molecule)
P_c	Critical pressure, bar
R	Universal gas constant, 8.314 J/mol.K
R_i	Refractivity intercept defined in Equation (20)
R_m	Molar refraction defined in Equation (6), cm^3/mol
r	Distance between molecules
SG	Specific gravity of liquid substance at 15.5 °C (60 °F) defined in Equation (2.2), dimensionless
S%	Sulfur content, Weight%
T_b	Boiling point, K
T_c	Critical temperature, K

V	Saybolt universal viscosity, SUS (or molar volume)
V_c	Critical volume, cm^3/g
VGC	Viscosity gravity constant defined in Equations (19)
VGF	Viscosity gravity function defined in Equations (22)
x_P	Fraction of paraffins in a hydrocarbon mixture

Greek Letters

Γ	Potential energy
α	Polarizability defined in Equation (5), cm^3/mol
ε	Energy parameter in a potential energy relation
μ	Absolute (dynamic) viscosity, cp [mPa.s]
ν	Kinematic viscosity (absolute viscosity/density), cSt [mm^2/s]
θ	A property of hydrocarbon such as: M, T_c, P_c, I, C/H, T_b, ...
σ	Size parameter in potential energy relation
ω	Acentric factor, dimensionless

Subscripts

A	Aromatic
N	Naphthenic
P	Paraffinic
T	Value of a property at temperature T
∞	Value of a property at M $\rightarrow \infty$
20	Value of a property at 20 °C
100	Value of kinematic viscosity at (100 °F) 39 °C

Acronyms

| API-TDB | American Petroleum Institute – Technical Data Book |
| EOS | Equation of State |

7.   REFERENCES

1. Poling, B. E.; Prausnitz, J. M.; O'Connell, J. P., *Properties of Gases & Liquids,* 5th ed., McGraw Hill: New York, 2000.
2. Smith, J. M.; Van Ness, H. C.; Abbott, M. M. *Introduction to Chemical Engineering Thermodynamics.* 5th ed., McGraw-Hill: New York, 1996.
3. *API Technical Data Book - Petroleum Refining* T. E. Daubert; Danner, R. P. (Eds.), 6th ed., American Petroleum Institute (API): Washington, D.C., 1997.
4. Firoozabadi, A. *Thermodynamics of Hydrocarbon Reservoirs,* McGraw Hill: New York, 1999.
5. Brule, M. R.; Kumar, K. H.; Watanasiri, S. "Characterization Methods Improve Phase Behavior Predictions", *Oil & Gas J.* **1985**, 87-93.
6. Lee, B. I.; Kesler, M. G. "A Generalized Thermodynamic Correlation Based on Three-Parameter Corresponding States", *AIChE J.* **1975**, 21, 510-527.
7. Dohrn, R.; Pfohl, O. "Thermophysical Properties – Industrial Direction", Paper presented at the Ninth International Conference on Properties and Phase Equilibria for Product and Process Design (PPEPPD 2001), Kurashiki, Japan, May 20-25, 2001.

8. Watson, K. M.; Nelson, E. F.; Murphy, G. B. "Characterization of Petroleum Fractions", *Ind. & Eng. Chem.* **1935**, 27, 1460.
9. Korsten, H. "Characterization of Hydrocarbon Systems by DBE Concept", *AIChE J.* **1997**, 43(6), 1559-1568.
10. Winn, F. W. "Physical Properties by Nomogram", *Petroleum Refiners* **1957**, 36(21), 157.
11. Sim, W. J.; Daubert, T. E. "Predcition of Vapor-Liquid Equilibria of Undefined Mixtures", *I & EC Process Des. & Dev.* **1980**, 19, 386-393.
12. Kesler, M. G.; Lee, B. I. "Improve Prediction of Enthalpy of Fractions", *Hydrocarbon Processing,* **1976**, 153-158.
13. Cavett, R. H. "Physical Data for Distillation Calculations, Vapor-Liquid Equilibria", *Proceeding of 27th API Meeting*, API Division of Refining, **1962**, 42(3), 351- 366.
14. Riazi, M. R.; Daubert, T. E. "Simplify Property Predictions", *Hydrocarbon Processing* **1980**, 59(3), 115-116.
15. Twu, C. H. "An Internally Consistent Correlation for Predicting the Critical Properties and Molecular Weights of Petroleum and Coal-Tar Liquids", *Fluid Phase Equilbria* **1984**, 16, 137-150.
16. Van Nes, K.; van Westen, H. A. *Aspects of the Constitution of Mineral Oils.* Elsevier Publishing: New York, 1951.
17. Riazi, M. R.; Nasimi, N.; Roomi, Y. "Estimating Sulfur Content of Petroleum Products and Crude Oils", *I & EC Res.* **1999**, 38(11), 4507-4512.
18. Riazi, M. R. "A Distribution Model for C_{7+} Fractions Characterization of Petroleum Fluids", *I & EC Res.* **1997**, 36, 4299-4307.
19. *TRC Thermodynamic Tables – Hydrocarbons*, K. R. Hall (Ed.) Thermodynamic Research Center, The Texas A&M University System, College Station, Texas (1993).
20. Daubert, T. E.; Danner, R. P.; Sibul, H. M.; Stebbins, C. C. *Physical and Thermodynamic Properties of Pure Compounds: Data Compilation,* DIPPR-AIChE, Taylor & Francis: Bristol, PA, 1994. (www.aiche.org/dippr)
21. Nikitin, E. D.; Pavlov, P. A.; Popov, A. P. "Vapor-Liquid Critical Temperatures and Pressures of Normal Alkanes from 19 to 36 Carbon Atoms, Naphthalene and m-Terphenyl Determined by the Pulse-Heating Technique", *Fluid Phase Equilibria* **1997**, 141, 155.
22. Riazi, M. R.; Daubert, T. E. "Characterization Parameters for Petroleum Fractions", *I & EC Res.* **1987**, 26, 755-759; Corrections, 1268.
23. Prausnitz, J. M.; Lichtenthaler, R. N.; de Azevedo, E.G., *Molecular Thermodynamics of Fluid Phase Equilibria.* 3rd ed., Prentice-Hall: New Jersey, 1999.
24. Hill, T. L. *"An Introduction to Statistical Thermodynmaics,"* Wesley: Addison, 1960.
25. Riazi, M. R.; Faghri, A. "Thermal Conductivity of Liquid and Vapor Hydrocarbon Systems: Pentanes and Heavier at Low Pressures", *I & EC Process Des. & Dev.* **1985**, 24(2), 398-401.
26. Huang, P. K. "Characterization and Thermodynamic Correlations for Undefined Hydrocarbon Mixtures", Ph.D. Dissertation, Pennsylvania State University, University Park, Pennsylvania 1977.
27. Riazi, M. R.; Roomi, Y. "Use of the Refractive Index in the Estimation of Thermophysical Properties of Hydrocarbons and Their Mixtures", *I & EC Res.* **2001**, 40(8), 1975-1984.
28. Kesler, M. G.; Lee, B. I.; Sandler, S. I. "A Third Parameter for Use in Generalized Thermodynamic Correlations", *I & EC Fund.* **1979**, 18, 49-54.
29. Whitson, C. H.; Brule, M. R. *"Phase Behavior",* Monograph Volume 20, Society of Petroleum Engineers Inc.: Richardson, Texas, 2000.
30. Ahmed, T. *Reservoir Engineering Handbook,* Gulf Publishing Company: Houston, 2000.
31. Ahmed, T. *Hydrocarbon Phase Behavior,* Gulf Publishing Company: Houston, 1989.

32. Edmister, W. C.; Lee, B. I. *Applied Hydrocarbon Thermodynamics*, 2nd ed., Gulf Publishing Company: Houston, 1985.
33. Soreide, I. "Improved Phase Behavior Predictions of Petroleum Reservoir Fluids from Cubic Equation of State", Dr. Ing. Dissertation, Norwegian Institute of Technology, Trondheim, Norway, 1989.
34. Aspen Plus, *Introductory Manual Software Version,* Aspen Technology, Inc.: Cambridge, Massachusetts, December 1986.
35. HYSYS, "Reference Volume 1., Version 1.1", HYSYS Reference Manual for Computer Software, HYSYS Conceptual Design, Hyprotech Ltd., Calgary, Alberta, Canada, 1996.
36. Stange, E.; Johannesen, S.O. "HYPO*S , A Program for Heavy End Characterization", An Internal Program from the Norsk Hydro Oil and Gas Group (Norway), Paper presented at the 65th Annual Convection, Gas Processors Association, San Antonio, Texas, March 10, 1986.
37. PRO/ II, *Keyword Manual,* Simulation Sciences Inc.: Fullerton, California, October 1992.
38. Riazi, M. R.; Al-Sahhaf, T. "Physical Properties of n-Alkanes and n-Alkyl Hydrocarbons: Application to Petroleum Mixtures", *I & EC Res.,* **1995**, 34, 4145-4148.
39. *ASTM Annual Book of Standards,* American Society for Testing and Materials: Philadelphia, 1995.
40. Riazi, M. R.; Daubert, T. E. "Analytical Correlations Interconvert Distillation Curve Types", *Oil & Gas J.* **1986**, 50-57.
41. Daubert, T. E. "Property Predictions", *Hydrocarbon Processing* **1980**, 59(3), 107-112.
42. Riazi, M. R.; Daubert, T. E. "Molecular Weight of Heavy Fractions from Viscosity", *Oil and Gas J.* **1987**, 58(52), 110-113.
43. Wauquier, J.-P. *Petroleum Refining. Vol. 1 Crude Oil. Petroleum Products. Process Flowsheets*, Editions Technip: Paris, 1995.
44. Speight , J. G. *The Chemistry and Technology of Petroleum*, 3rd ed., Marcel Dekker: New York, 1998.
45. Riazi M. R.; Daubert T. E. "Prediction of the Composition of Petroleum Fractions", *I & EC Process Des. & Dev.* **1980**, 19(2), 289-294.
46. Riazi M. R.; Daubert T. E. "Prediction of Molecular Type Analysis of Petroleum Fractions and Coal Liquids", *I & EC Process Des. & Dev.* **1986**, 25(4), 1009-1015.
47. Kurtz Jr., S. S.; Ward, A. L. "The Refractivity Intercept and the Specific Refraction Equation of Newton. I. Development of the Refractivity Intercept and Composition with Specific Refraction Equations", *J. of Franklin Institute* **1936**, 222, 563.
48. *API Research Project 42: Properties of Hydrocarbons of High Molecular Weight.* American Petroleum Institute: New York, 1962.
49. *Oil and Gas Journal Data Book.* PennWell: Tulsa, Oklahoma, 2000; p. 315.
50. Denis, J.; Briant, J.; Hipeaux, J. C. *Lubricant Properties Analysis & Testing.* Translated to English by G. Dobson, Editions TECHNIP, Paris, France, 1997.
51. Altgelt, K. H.; Boduszynski, M. M. *Composition and Analysis of Heavy Petroleum Fractions.* Marcel Dekker: New York, 1994.
52. Riazi, M. R., *Characterization and Properties of Petroleum Fractions.* ASTM: Philadelphia (in preparation).

Chapter 2

ADVANCES IN ELEMENTAL ANALYSIS OF HYDROCARBON PRODUCTS

R. A. Kishore Nadkarni
Millennium Analytics, Inc., East Brunswick, NJ 08816

1.   INTRODUCTION

Probably about half of the elements in the periodic table are found in petroleum products varying in concentration from major percentages of carbon/hydrogen, minor percentages of nitrogen/sulfur, to mg/kg and sub-mg/kg amounts of several transition metals. Much of the presence of inorganic elements in crude oil originate from marine animal and vegetative materials deposited with sediment in sea and coastal waters millions of years ago. Many transition metals such as vanadium, nickel, and iron are of porphyrin origin. Many transition metal elements were incorporated into hydrocarbon skeletons from minerals in the depositional environment of ancient organisms during paleotransformation. For example, vanadium and nickel porphyrins derived from chlorophyll through transmetallation of magnesium. Through refining process much of these inorganic elements are eliminated from the final end products of refining such as gasoline, jet fuel, and diesel. Only small amounts of sulfur etc. are left in these end products. Several other inorganic metals remain in other refinery products such as fuel oil and heating oil.

There is extensive data available on the elemental composition of crude oils. Some of them are summarized by Nadkarni.[1] On the other hand, specifically, some alkaline earth and other elements are added in the form of organometallic compounds to finished additives, lubricating oils, gear oils, etc. to enhance their performance in internal combustion engines. These additives act as dispersants, detergents, anti-oxidants, anti-wear agents, friction modifiers, pour point depressants, extreme pressure agents, corrosion inhibitors, anti-foamants, viscosity index improvers, and metal deactivators.[2-4] Several such additive elements and the reasons they are added to oils are listed in Table 1. However, most widely used inorganic elements for this purpose include boron, calcium, copper, magnesium, phosphorus, and zinc. Elemental compositional analysis is also important and often critical for characterization of the catalysts used in hydrocarbon, i.e. petroleum processing.

Table 1. Lubricant and Additive Materials

METAL	COMPOUNDS	PERFORMANCE CHARACTERISTICS
ANTIMONY	Dialkyldithiocarbamates, Dialkylphosphorodithionates	Anti-wear, Extreme Pressure, Antioxidant
BARIUM	Sulfonates, Phenates, Diorganodiphosphhates, Phosphonates, Thiophosphonates	Detergent Inhibitors, Corrosion Inhibitors, Detergents, Rust Inhibitors, ATF, Greases
BORON	Borax and Esters	Anti-wear Agents, Antioxidant, Deodorant Cutting Oils, Greases, Brake Fluids
CADMIUM	Dithiophosphates	Steam Turbine Oils
CALCIUM	Sulfonates, Phenates	Detergent Inhibitors, Dispersants
CHROMIUM	Salts	Grease Additive
LEAD	Naphthenate	Extreme Pressure Additive, Greases, Gear Oils
MAGNESIUM	Sulfonates, Phenates	Detergent Inhibitors
MERCURY	Organic Compounds	Bactericide in Cutting Oil Emulsions
MOLYBDENUM	Disulfide, Phosphate, Dibutyldithiocarbamate	Greases, Extreme Pressure Additives
NICKEL	Cyclopentadienyl Complexes	Anti-wear Agents, Carbon Deposit Reduction, Improved Lubrication and Combustion
PHOSPHROUS	Metaldialkyldithiophosphates	Detergents, Anti-rusting Agents
SELENIUM	Selenides	Oxidation and Bearing Corrosion Inhibitors
SILICON	Silicone Polymers	Foam Inhibitors
TIN	Organo Compounds	Anti-scuffing Additives, Metal Deactivators
ZINC	Dialkyldithiophosphates, Dithiocarbamates, Phnolates	Anti-oxidant, Corrosion Inhibitor, Anti-wear Additive, Detergents, Extreme Pressure Additives, Crankcase Oils, Hypoid Gear Lubricants, Greases, Aircraft Piston Engine Oils, ATF, Railroad Diesel Engine Oils, Brake Lubricants

Excerpted from Reference 1.

Another fertile area of elemental analysis is used oils and in-service oils for monitoring purposes. Trace metals in such samples originate from mechanical wear from the engine components in contact with oil as well as from contamination from air, fuel, and liquid coolant. The presence of specific metals in used lubricating oils is attributed to wear of specific metal parts of an engine. A sudden increase in one or more metals in the in-service oils is indicative of excessive wear of some engine component. The analyses of in-service engine oil is used to estimate when the oil should be changed based on among other measurements, the metal analysis. The analysis of used oils is also important for reprocessing the oils for environmental protection. Table 2 summarizes the common wear metals encountered and the engine wear they indicate.

Table 2. Wear Metals in Used Lubricating Oils

METAL	WEAR INDICATION
ALUMINUM	Piston and bearings wear, push rods, air cooler, pump housings, oil pumps, gear castings, box castings
ANTIMONY	Crankshaft and camshaft bearings
BARIUM	Detergent additive, grease additive
BORON	Coolant leakage in system
CADMIUM	Bearings
CALCIUM	Hard water, detergent additive, oxidation inhibitor, road salt
CHLORINE	Anti-wear additive, extreme pressure additive
CHROMIUM	Ring wear, cooling system leakage, chromium plated parts in aircraft engines, cylinder liners, seal rings
COPPER	Wear in bushings, injector shields, coolant core tubes, thrust washers, valve guides, connecting rods, piston rings, bearings, sleeves, bearing cages
IRON	Wear from engine block, cylinders, gears, cylinder liners, valve guides, wrist pins, rings, camshaft, oil pump, crankshaft, ball and roller bearings, rust
LEAD	Bearings, fuel blowby in leaded fuels, thrust bearings, bearing cages, bearing retainers
MAGNESIUM	Cylinder liner, gear box housings in aircraft engines
MOLYBDENUM	Wear in bearing alloys and in oil coolers, various Mo alloyed components in aircraft engines, piston rings
NICKEL	Bearings, valves, gear platings
PHOSPHORUS	Anti-wear additive, extreme pressure additive
POTASSIUM	Coolant additive leakage
SILICON	Dirt intrusion from improper air cleaner, seal materials, anti-foamant Leakage
SILVER	Wrist pin bearings in railroad and auto engines, silver plated spline lubricating pump
SODIUM	Anti-freeze leakage
TIN	Bearings and coatings of connecting rods and iron pistons
TITANIUM	Various titanium alloyed components in aircraft engines
TUNGSTON	Bearings
VANADIUM	Surface coatings on piston rings, turbine impeller beds, valves
ZINC	Neoprene seals, galvanized piping

In part from Reference 1.

In the early part of the twentieth century, the determination of metals in petroleum products was essentially carried out using wet chemistry techniques. Elaborate separations of metal species and subsequent determination, usually using gravimetric, titrimetric, or colorimetric finishes was the standard practice. With the introduction of modern instrumental methods of elemental analysis all such wet chemistry methods have fallen by the wayside except perhaps in rare cases for some specialized analysis in a particular matrix.

Since the mid-twentieth century as newer and newer instrumental techniques were developed, each of them has been widely used for the elemental analysis of petroleum products and lubricants. Many of these techniques are now quite mature, and continue to be used throughout the world in research and commercial analytical laboratories in the oil industry. A large body of literature exists regarding the elemental analysis of various types of petroleum products. It is beyond the scope of this chapter to include references to all such articles. Only typical examples of such applications of each technique used in this field will be given.

Committee D-02 on Petroleum Products and Lubricants of American Society for Testing and Materials (ASTM) through its Sub-Committee 3 on Elemental Analysis has published numerous standard test methods for elemental analysis of petroleum products and lubricants. Most available analytical techniques have been utilized for this work. Standardization of test methods through ASTM has helped all laboratories involved in such analysis of petroleum products throughout the world, and has helped commerce between buyers and sellers of petroleum products to resolve quality complaints regarding elemental concentration levels between product specifications and certificates of analyses of such products. Nearly 35 instrumental methods out of over 65 test methods on elemental analysis published by ASTM D-02 Committee are listed in Table 3.[5] See reference 6 for brief summaries and precision data on these methods. Similar to ASTM, other standard developing organizations such as Institute of Petroleum (IP), International Organization for Standardizations (ISO), Japan Institute of Standards (JIS), Association Francaise de Normalisation (AFNOR), Deutsche Institut fur Normung (DIN), etc. have also issued similar elemental analysis standards. Many of them are based on the original ASTM standards. See Table 4 for a list of such international standards equivalent to each other in the elemental analysis area. This table is excerpted from a much larger table of equivalent test methods in the petroleum products and lubricants area.[6]

TABLE 3. Instrumental Elemental Analysis Methods Issued by ASTM D-02 Committee

DESIGNATION-YEAR   SUBJECT

D 1552 – 01	Sulfur by High Temperature Method
D 2622 – 98	Sulfur by Wavelength Dispersive X-Ray Fluorescence
D 3120 – 96	Trace Sulfur by Oxidative Micro-Coulometry

D 3237 – 00	Lead by Atomic Absorption Spectrometry
D 3605 – 00	Trace Metals by Atomic Absorption and Flame Emission Spectrometry
D 3701 – 99a	Hydrogen by Low Resolution NMR
D 3831 – 98	Manganese by Atomic Absorption Spectrometry
D 4045 – 99	Sulfur by Hydrogenolysis and Rateometric colorimetry
D 4294 – 98	Sulfur by Energy Dispersive X-Ray Fluorescence
D 4628 – 97	Additive elements by Atomic Absorption Spectrometry
D 4629 – 96	Nitrogen by Combustion – Chemiluminescence
D 4808 – 98	Hydrogen by Low Resolution NMR
D 4927 – 96	Additive Elements by Wavelength Dispersive X-Ray Fluorescence Spectrometry
D 4951 – 00	Additive Elements by ICP/AES
D 5056 – 96	Trace Metals in Coke by Atomic Absorption Spectrometry
D 5059 – 98	Lead by X-Ray Spectroscopy
D 5184 – 00	Aluminum and Silicon by ICP/AES and AAS
D 5185 – 97	Additive Elements by ICP/AES
D 5291 – 96	Carbon/Hydrogen/Nitrogen by Instrumental Techniques
D 5453 – 00	Sulfur by Combustion – UV Fluorescence
D 5600 – 98	Trace Metals in Coke by ICP/AES
D 5622 – 01	Oxygen by Reductive Pyrolysis
D 5708 – 00	Metals by ICP/AES
D 5762 – 98	Nitrogen by Combustion-Chemiluminescence
D 5863 – 00	Metals in Crude Oils by Atomic Absorption Spectrometry
D 6334 – 98	Sulfur by Wavelength Dispersive X-Ray Fluorescence
D 6443 – 99	Additive Metals by Wavelength Dispersive X-Ray Fluorescence Spectrometry
D 6445 – 99	Sulfur by Energy Dispersive X-Ray Fluorescence
D 6481 – 99	Additive Metals by Energy Dispersive X-Ray Fluorescence
D 6595 – 00	Wear Metals in Lubes by Rotrode Spectroscopy
D 6667 – 01	Sulfur in LPG by UV-Fluorescence
D 6728 – 01	Wear Metals in Fuels by Rotrode Spectroscopy
D 6732 – 01	Copper in Jet Fuels by GF/AAS

Excerpted from Reference 5.

Table 4. Equivalent Test Methods for Elemental Analysis

ANALYSIS	ASTM	IP	ISO	DIN	JIS	AFNOR
S, BOMB METHOD	D 129	61		51577		T60-109
S, LAMP METHOD	D 1266	107				M07-031
V IN FUEL OILS	D 1548					M07-027
S, HIGH TEMP.	D 1552					M07-025
Pb, VOLUMETRIC	D 2547	248	2083			M07-014
S, WDXRF	D 2622	.		51400T6	K2541	
S, WICKBOLD	D 2785	243	4260			
S, COULOMETRY	D 3120		16591		K2276	M07-022
Pb BY AAS	D 3237	428				
S, COULOMETRY	D 3246	373				M07-052
Pb, ICl METHOD	D 3341		3830			
METALS BY AAS	D 3605	413	8691	51790T3		
P, COLORIMETRY	D 4047	149	4265			

S, NDXRF	D 4294	336	8754		M07-053
METALS BY AAS	D 4628	308		51391T1	
N, CHEMILUMIN.	D 4629	379			M07-058
METALS, WDXRF	D 4927	407		51391T2	
Al, Si IN FUELS	D 5184	377	10478	51416	

Excerpted from reference 6.

The most widely used instrumental techniques are discussed below. Brief technical summary of the technique and a few typical applications in the petroleum products and lubricant area are described. The instrumental techniques discussed here include atomic absorption spectrometry, graphite furnace atomic absorption spectrometry, inductively coupled plasma atomic emission spectrometry, inductively coupled plasma-mass spectrometry, ion chromatography, micro-elemental analysis, neutron activation analysis, and X-ray fluorescence. Given the importance of the topics, separate discussions are included on the analysis of used engine oils and determination of trace amounts of sulfur in these products.

2.   ATOMIC ABSORPTION SPECTROMETRY (AAS)

The most widely used technique for metals analysis of petroleum products throughout the world is atomic absorption spectrometry (AAS). Since the technique was first introduced in mid-fifties, tens of thousands of AAS instruments have been sold, and thousands of publications have appeared describing the fundamentals or the applications of this technique to every imaginable matrix, petroleum products being no exception.

In AAS, the sample solution, whether aqueous or non-aqueous, is vaporized in a flame, and the elements are atomized. The elemental concentration is determined by absorption of the analyte atoms of a characteristic wavelength emitted from a light source, typically a hollow cathode lamp which consists of a tungsten anode and a cylindrical cathode made of the analyte metal, encased in a gas-tight chamber. The detector is usually a photomultiplier tube. A monochromator separates the elemental lines.

Usually, the AAS instrument uses flame as the atomization source. An air-acetylene flame is used for most elements; the nitrous oxide-acetylene flame reaches higher temperature (2300°C for air-C_2H_2 versus 2955°C for N_2O-C_2H_2), and is used for atomizing the more refractory oxide forming metals.

Over 70 elements can be determined by AAS usually with a precision of 1-3 % and with detection limits of the order of sub-mg/kg levels, and with little or no atomic spectral interference. However, there could be molecular spectral, ionization, chemical, and matrix interferences.

Petroleum products can be wet ashed with mineral acids and brought into aqueous solution for AAS measurements. Alternatively, many liquid petroleum products can be simply diluted with organic solvents such as kerosene, xylenes, or methyl isobutyl ketone (MIBK), etc. and direct AAS measurements can be made. Aqueous metallic standards are used in the first instance, and organometallic standards in organic solvents are used in the latter case for instrument calibration.

There are hundreds of publications on the applications of AAS in the petroleum products area. See references 7 – 11 for comprehensive reviews of this subject. AAS is truly a workhorse of the metal analysis of petroleum products to the point that "cookbook" methods are available for almost all elements that are amenable to AAS determination. With tongue in cheek, Hieftje predicted that AAS instruments will be removed from the market place, based on the trends of AAS publications from 1960 to 2000.[12] Obviously, this dire prediction has not come to pass, but there is little doubt that that's where the future lies. Inherent limitations of AAS such as sample preparation and limited dynamic range can be overcome on a technique based peristaltic pumps performing on-line multiple calibration from a single bulk standard and fast automatic on-line dilution of over-range samples.[13] Simultaneous multi-element flame AAS has been used to determine cadmium, lead, and nickel in burned and unburned Venezuelan crude oils. Better than 1 % precision could be obtained.[14]

About a dozen test methods have been written by ASTM D-02 Committee. Of these seven remain as active methods. In D 3237-97 test method, the gasoline sample is diluted with MIBK and the alkyl lead compounds are stabilized by reaction with iodine and a quaternary ammonium salt, followed by AAS determination of lead. Determination of trace metals such as Pb, Na, Ca, and V in gas turbine fuels is the subject of D 3605-00 test method. Certain organo-metallic compounds (e.g. methylcyclopentadienyl manganese tricarbonyl) are added as anti-knock agents to gasoline. The test method D 3831-98 describes the determination of manganese in such gasoline samples by AAS after reacting the sample with bromine and diluting with MIBK. A widely used test method for the determination of additive elements - Ba, Ca, Mg, and Zn – in lubricating oils is described in the test method D 4628-97. Oils which contain viscosity index (VI) improvers may give low results unless calibration standards also contain VI improvers.

The results by this method are found equivalent to those obtained by using D 4927 X-ray fluorescence or D 4951 inductively coupled plasma atomic emission spectrometry test methods. Trace metals in petroleum coke are determined by the test method D 5056-96 after ashing the sample, fusion of the ash with lithium borate, dissolution of the melt in HCl, and AAS measurement of Al, Ca, Fe, Ni, Si, Na, and V. Aluminum and silicon in fuel oils are determined by the test method D 5184-00 following the same procedure as in D 5056 above. The method also alternatively allows to use inductively coupled plasma for this determination. Crude oils and residual fuels are analyzed for Ni, V, Fe, and Na content by either ashing or direct dilution with an organic solvent, and AAS determination in the test method D 5863-00.

2.1 Graphite Furnace Atomic Absorption Spectrometry (GF/AAS)

Detection limits of AAS can be extended by using graphite furnace atomic absorption spectrometry (GF/AAS), although somewhat at the expense of precision and speed. It is principally used for trace elements at sub-mg/kg or μg/kg levels. Although higher levels of elements can certainly be determined by this technique by appropriate dilutions, errors introduced by dilutions makes it less than an ideal technique for the determination of major or minor elements in a matrix.

In GF/AAS, a heated graphite tube replaces the flame as the atomization source. After depositing a small volume of the sample solution in the graphite cuvette, it is heated with electric current through three stages of evaporation of the solvent, ashing of the sample, and finally, atomization of the sample at a temperature of about 3000°C.

Measurement time of GF/AAS is somewhat longer than that of AAS. Also, there could be serious matrix interference in GF/AAS that is negligible in AAS. These matrix interferences can be taken care of by adding matrix modifiers such as nitrates of Ca, Mg, La, Ni, or Pd. Overall, the precision of GF/AAS ranges between 5 to 10 % versus about 1-2 % for AAS.

Because of the extra effort needed to perform GF/AAS compared to AAS, this technique has been used more commonly in research than in routine analysis of petroleum products.

Usually oil samples can be directly analyzed with simple dilution with an organic solvent. Some of the applications of GF/AAS in the petroleum area include the determination of wear metals in aircraft engine oils, vanadium and nickel in crude oils, and lead and several refractory transition metals in gasoline. See references 8, 11, and 15 for reviews of the use of GF/AAS in petroleum products area.

3.   INDUCTIVELY COUPLED PLASMA ATOMIC EMISSION SPECTROMETRY (ICP/AES)

Without a doubt the most significant milestone in the field of atomic spectroscopy in last thirty years has been the development and immense popularity of the technique of inductively coupled plasma atomic emission spectrometry (ICP/AES). This technique has revolutionized the rapid and sensitive multi-element determination of metals in a wide variety of matrices including petroleum products and lubricants.

In ICP/AES, the liquid sample (most often in case of petroleum products and lubricants, the sample is used straight or diluted with an organic solvent) is aspirated through a nebulizer into the argon plasma where it is vaporized, atomized, and/or ionized, and electronically excited. A plasma is an electrically neutral highly ionized gas consisting of ions, electrons, and neutral particles. Most commonly used gas is argon that is ionized by a strong electric field by radiofrequency (RF) discharge at 27.12 MHz frequency. This causes the flowing argon gas to become electrically conductive and reach a temperature up to 7000 to 9000°C. Nebulizer is a critical part of the overall ICP/AES instrument. Commonly used nebulizers are glass concentric pneumatic, cross-flow or grid pneumatic, more advanced ultrasonic, and Babington high solids nebulizers.

The excited state ions and atoms emit photons of wavelengths characteristic of an element on returning to their ground states. A polychromator in a simultaneous spectrometer or a single photomultiplier in a scanning or sequential spectrometer measures the emitted light. By comparing the intensity of emission from a sample to that of a calibration standard containing a known amount of the analyte, quantification is achieved.

Noble [16] has prepared a critical review of various ICP/AES instrumentation commercially available.

One of the most important advantages of ICP/AES over other atomic spectroscopic techniques such as AAS or GF/AAS, is its linear dynamic range of over five orders of magnitude for many elements. This permits simultaneous determination of trace, minor, and major elements in a sample without extensive or multiple dilutions. Because of the high temperature of the plasma, chemical or matrix interferences are minimal; also, refractory elements can be determined without any difficulty.

Detection limits of ICP/AES are usually not as low as those of GF/AAS. But for most practical applications, this does not pose a serious problem. Spectral interference is a critical disadvantage in ICP/AES, particularly when a trace element may emit at a wavelength same as or close to that emitted by an element present at much higher levels. Hence, usually careful consideration is given in choosing interference free wavelengths for quantitation. High-resolution spectrometers are used for separating the overlapping spectral lines and correcting for background interferences.

A major development in the ICP/AES area is the incorporation of array detectors allowing to view large portions of the spectrum at once, and collect multiple lines per element on a flexible basis.

ICP/AES is probably the most widely used metal analysis technique in the oil industry. As far as is known, the first ICP/AES instrument was installed in the Baytown, TX laboratory of Exxon Research and Engineering Company.[17] There are about 11 companies in U.S. and five others worldwide who offer more than 30 different ICP/AES instruments. It is estimated that about 1000 instruments are sold each year.[16] Totally there must be thousands of these instruments in every research and plant-refinery laboratories of every oil company in the world. Hundreds of publications are available in this area, several of which have been reviewed by Nadkarni earlier.[1] Some of the more recent work is summarized below. Hausler and Carlson have reviewed the applications of ICP/AES as used at Phillips Petroleum Company, which is pretty typical of other oil companies also.[18] Similarly, Botto has reviewed the work done in this area in Exxon Research and Engineering Company.[19]

Petroleum products and lubricant samples are usually (a) wet-ashed with concentrated mineral acids and brought into aqueous solution, (b) directly aspirated, or (c) diluted with organic solvents such as xylenes, kerosene, MIBK, MEK, etc. Aqueous calibration standards are used in the first case; organometallic standards in the latter two.

Botto has determined over 15 trace elements in National Institute of Standards and Technology (NIST) fuel oil SRM 1634a in excellent agreement with the certified values. This sample was routinely used as a laboratory quality control sample and was analyzed by ICP/AES after sulfated ash fusion.[19]

Ultrasonic nebulizers (USN) offer some advantages over conventional pneumatic cross-flow type nebulizers in ICP/AES systems. USN have high efficiency of aerosol production resulting in improved sensitivity and lower detection limits. However, a chief drawback of USN is extremely long rinsing time needed for some elements. Botto has shown applications of USN equipped ICP/AES to the analysis of NIST fuel oil SRM 1634b, reference fuel for lead SRM 2715, and lead in aviation fuel. The samples were prepared both by acid digestion in a microwave oven and by direct dilution with toluene.[20] Using USN coupled with a microporous membrane desolvator (MMD) for direct ICP/AES, with a single element universal calibration, Botto and Zhu have determined a variety of trace elements in a number of organic solvents, petroleum naphthas, and gasoline samples.[21] Detection limits are similar to aqueous USN limits, and provided, the analyte species are not volatile, accurate and precise results can be obtained. Utilizing universal calibration for organic solutions, it should be possible to determine trace metals at ng/mL concentrations in naphtha feedstocks which are fed into catalytic units and steam crackers.

Gonzales and Lynch have determined 14 elements in crude oils by using scandium as an internal standard and diluting with kerosene to overcome drift, sample transport delivery and nebulization problems associated with different sample viscosities and surface tension variations. It also obviates the need for sample ashing and associated loss of volatile analytes.[22]

The sample viscosity and the presence of viscosity improver modifiers in oils play an important part in precise and accurate determination of metals by ICP/AES. Brown determined 21 elements in new and used oils after xylene dilution.[23] The solution viscosity caused large errors, which were significantly reduced by adding neutral base oil to the calibration standards. In the additive samples diluted with xylene, the presence of detergents in lube oils was not found to have any effect on the analysis.[24] Bansal and McElroy [25] note that particularly for the multi-grade oils errors of as much as 20 % when viscosity modifiers are present in the lube oils. It was found that viscosity modifier polymers interfere with the formation of aerosol, a critical step in the analysis of ICP/AES analysis procedure, thus affecting the sample delivery to the plasma torch. By using both an internal standard (such as cobalt) and adequate sample dilution, this undesirable effect can be eliminated. Based on this research, ASTM test methods D 4951 and D 5185 now include both these steps in the standard procedure. Similar technique of using an internal standard has been used in on-line dilution and ICP/AES analysis of lube oils to eliminate the effect of the presence of viscosity index improvers in oils.[26]

Although metal determination continues to be the *forte* of ICP/AES applications, some non-metals such as sulfur and halogens have also been determined by this technique. Usually a vacuum spectrometer and measurement in ultra-violet region are used for these determinations. Of these, by far the widely determined element is sulfur. A number of cross checks have shown equivalent results for sulfur in oils using ICP/AES or X-ray fluorescence, the primary technique used for this determination. Richter, et al.[27] have shown examples of non-metal determination in waste oils, particularly for the analysis of polychlorinated biphenyls by diluting with kerosene after a simplex optimization. They give following detection limits:

ELEMENT	LINE (nm)	DETECTION LIMITS (mg/kg)
Chlorine	134.72	1
Bromine	163.34	1.4
Iodine	161.76	0.47
Phosphorus	177.5	0.04
Sulfur	180.73	0.07

Several ICP/AES test methods have been issued by ASTM D-02 Committee for applications in the oil analysis. The first such test method to be issued was D 4951-00 for the determination of additive elements - B, Ba, Ca, Cu, Mg, P, S, and Zn - in fresh lubricating oils at 1 to 5 m% concentration. The samples are diluted with xylenes or other solvents. The use of an internal standard is mandatory to overcome the bias caused by the presence of viscosity index improvers in multi-grade lubricating oils. A similar test method D 5185-97 is used for the determination of additive elements, wear metals, and contaminants in used as well as base oils. Several more elements, in addition to those done by D 4951 test method, are measured by this test method.

Aluminum and silicon in fuel oils are contaminants from catalyst fines and can cause abnormal engine wear. These elements are measured by the test method D 5184a-00 after ashing the sample, fusion with alkali borate, and analysis of the aqueous solution by ICP/AES. The presence and concentration of various metals in petroleum coke are major factors in determining the suitability of coke for various end uses. Such metals in petroleum coke are determined by the test method D 5600-98 in which the ash from the coke sample is fused with lithium borate. The melt is dissolved in dilute mineral acid and the metals are determined by ICP/AES. The test method D 5708-00 covers the determination of Ni, V, and Fe in crude oils and residual fuels by ICP/AES. Two sample preparation approaches are used: dilution with an organic solvent or decomposition with concentrated sulfuric acid.

4.   INDUCTIVELY COUPLED PLASMA/MASS SPECTROMETRY (ICP/MS)

The ICP/MS technique combines the power of ICP for atomizing and ionizing the sample with the sensitivity and selectivity of MS detection. Also, by coupling ICP/MS with high performance liquid chromatography (HPLC) or ion Chromatography (IC), metal speciation can be achieved. The technique is capable of simultaneously determining over 70 elements in a sample, has a wide dynamic range, and detection limits of parts per trillion (ppt) for most elements.

Although Sciex introduced the first commercial ICP/MS instrument in 1983, in nearly two decades, less than 4000 systems have been installed worldwide. By contrast, the first commercial ICP/AES instrument was introduced in 1974, and in about 20 years since then, over 9000 units were sold. In the same time frame as ICP/AES availability (1983 to-date), over 17,000 ICP/AES systems have been installed.[28] A major reason for this lack of sales is the high price of an ICP/MS system (about 200 K$ versus about 100

K$ for an ICP/AES instrument). The operational complexity of an ICP/MS system is also a drawback compared to the ease of operating an ICP/AES instrument or certainly an AAS instrument. Specifically in the petrochemical field, ICP/MS has been mostly confined to research laboratories, partly because in field or commerce the level of trace elements to be analyzed in the petrochemical products can be easily determined by ICP/AES or other more conventional techniques without resorting to ICP/MS. Newman has reviewed the field of commercial ICP/MS instrumentation.[29]

The other major drawback to ICP/MS is that the first row transition elements have oxide or other polyatomic mass interferences; ICP/AES does not. In effect ICP/MS is most useful with higher atomic number elements. This is in contrast to ICP/AES where there are more emission line interferences for the higher atomic number elements. High-resolution mass spectrometers can decrease this interference, but preference still would be to use ICP/AES. In reality this is one of the major reasons why ICP/MS has not replaced ICP/AES technology.

ICP/MS uses ICP as the ion source and a MS as the ion analyzer. Liquid samples are vaporized in a nebulizer and enter the ICP. An inert gas such as argon ionizes the sample, and the ions are drawn into the vacuum of the interface region, and are focused by the ion lenses, and analyzed in the quadrupole according to their mass-to-charge ratios. An electron multiplier detects the resultant separated ions. ICP/MS can simultaneously determine about 75 elements in one analysis. The sensitivity is as good as or better than that of GF/AAS. Precisions of 1-2 % relative standard deviation can be routinely obtained in solutions. However, a major source of inaccuracy is the spectral overlap resulting from polyatomic species formed in the plasma torch. Measurement of ultra-trace concentration of elements can be obscured by the formation of large amounts of other species such as argon compounds, e.g., The measurement of ultratrace iron with typical qudrupole ICP/MS is almost impossible with the ArO interference. This problem can be sometimes overcome by using higher resolution instrument or using different and, hopefully, interference-free elemental isotopes for quantitation. Because of such interferences, the claimed parts per trillion detection limits in pure aqueous solutions may not be achieved in complex matrices. Also, while ICP/AES can tolerate up to 2 % solids in the solution, ICP/MS cannot handle > 0.1 % solids. Given the ppt sensitivity of the technique, extreme care must be exercised in controlling the blank contamination of solutions. For ultra-trace elemental determination, memory effect is also a problem. Repeated analyses of solutions with somewhat higher levels of analytes will produce lingering memory effects that will need prolonged and thorough rinsing of the nebulizer with ultra-clean water.

To overcome the limitations of quadrupole based ICP/MS instruments such as limited capability of simultaneous detection of ions, relatively slower speed of analysis, and interferences from elements in higher concentrations, time-of-flight ICP/MS has been developed. Here, all ions that contribute to the mass spectrum are sampled through the interface, and groups of ions are electrostatically injected into a flight tube at exactly the same time. The mass measurement is related to the time taken to reach the ion detector. Time-of-flight systems can collect a full mass spectrum significantly faster than by a quadrupole. Ultra-trace direct analysis of petrochemical solvents used in the semiconductor industry can be achieved using high resolution ICP/MS.

Analyses at such ultra-trace levels require clean room facilities to minimize or eliminate trace contamination. In practice, such need is perhaps more critical in the semiconductor industry or environmental analysis than in the petrochemical field.

Al-Swaidan [30] determined sub-mg/kg levels of several elements in Saudi Arabian crude oils and gasoline products by diluting the sample in xylene and extracting in 40 % nitric acid. Williams [31] has analyzed diluted waste oils for the determination of several trace elements as a screening tool to decide whether the used oil can be recycled or disposed off as a hazardous waste. While many ICP/MS applications use aqueous liquids, McElroy et al [32] have demonstrated injecting organic samples directly into an ICP/MS. Direct organic liquid injection has several advantages over alternate sample preparation steps. For example, in the determination of vanadium and arsenic in oils, direct analysis can be completed in less than an hour with no isotopic interferences, while a sample prepared by sulfated ash (ASTM D 874 test method) needs about two days and results in interference from a host of polyatomic species. It also avoids the loss of volatile arsenic species during ashing. By using standard addition and a refined specific elemental equation, phosphorus could be determined overcoming the interference from several nitrogen and carbon species. These authors have also shown the determination of nickel and vanadium in crude oils (an important analysis from the viewpoint of petroleum processing) by developing suitable interference correction factors. There was excellent agreement between the results obtained by SASH-ICP/AES technique and direct ICP/MS technique. [32]

A unique advantage of ICP/MS is its ability to determine elemental species when coupled with IC or HPLC. Often this is an important analysis in the petroleum industry given that some elemental species are toxic to environment and to petroleum processing units, while other species of the same element may be benign. McElroy et al [32] have shown separation of nickel and vanadium octaethylporphyrin in crude oil distillate cuts by this technique. In other work, McElroy [33] has shown the feasibility of determining ultra-trace levels (parts-per-billion or ppb) of selenite, selenate, and selenocyanate in oil samples by linking IC with ICP/MS.

5. OVERVIEW OF ATOMIC SPECTROSCOPIC METHODS

The principal atomic spectroscopic methods have been reviewed regarding their principles of operation along with selected examples of their applications in the petrochemical field. Clearly AAS and ICP/AES are most widely used techniques at present for metals determination, and this is unlikely to change in the near future. Other than in specific instances where detection limit is of utmost importance, GF/AAS is used infrequently because of the relatively long time it takes for determination of multi-elements. High cost and complexity of instrumentation has limited the applications of ICP/MS generally only in the research laboratories.

Table 5 compares the detection limits of these four popular atomic spectroscopic techniques for some selected elements, excerpted from a review by Parsons et al.[34]

Table 5 Detection Limits of Atomic Spectroscopic Methods (ng/mL except for ICP/MS in ppt)

ELEMENT	AAS	GF/AAS	ICP/AES	ICP/MS
As	0.02	0.08	2	1
B	700	15	0.1	10
Ba	8	0.04	0.01	0.1
Ca	0.5	0.01	0.0001	10
Cd	0.5	0.0002	0.07	1
Cr	2	0.004	0.08	1
Cu	1	0.005	0.04	1
Fe	3	0.01	0.09	1
K	1	0.004	30	10
Mg	0.1	0.0002	0.003	1
Mn	0.8	0.0005	0.01	1
Mo	10	0.02	0.2	0.1
Na	0.2	0.004	0.1	1
Ni	2	0.05	0.2	1
P	-	0.3	15	0.1
Pb	10	0.007	1	0.1
S	-	10	30	1 ppb
Se	0.02	0.05	1	10
Si	20	0.005	2	0.1 ppb
Ti	10	0.3	0.03	1
V	20	0.1	0.06	1
Zn	0.8	0.0006	0.1	1

Data for AAS, GF/AAS, and ICP/AES excerpted from Parsons et al[34]; ICP/MS data from reference 40.

These authors correctly point out that:
(a) Quantitative determination cannot be made at the detection limit.
(b) Relative precision at this level will be ± 30 to 50 % relative standard deviation.
(c) For quantitation, 5 to 10 times the level of detection limits are necessary for simple matrices, and 20 to 100 times for complex matrices.

Table 6 gives a summary of comparison of advantages and limitations of the atomic spectroscopic techniques. Several reviews have covered these aspects.[35 – 39]

Table 6   *Comparison of Atomic Spectroscopic Methods*

PARAMETER	AAS	GF/AAS	ICP/AES	ICP/MS	XRF
Instrument Cost	$ 20 K	$ 50 K	$ 100 K	$ 200 + K	$ 100-200 K
Maintenance	Minimal	Minimal	Moderate	High	Minimal
Elements	Single	Single	Multi	Multi	Multi
No. of Elements	`70	`70	`70	`70	`65
Analysis Time	Few Min.	> AAS	Few Min.	Few Min.	Few Min.
Detection Limits	Mg/kg	Sub-mg/kg	Sub-mg/kg	µg/kg to ng/kg	mg/kg
Dynamic Range (orders of magnitude)	1 – 3	1 – 2	5 – 6	5 – 8	3 – 6
Interferences	Chemical: Ionization	Chemical; Matrix: Physical: Molecular Absorption	Spectral; Matrix	Spectral; Chemical	Matrix
Sample State	Liquid	Liquid	Liquid	Liquid	Liq. or Solid
Sample Volume	Few mL	Microliter	Few mL	Few mL	10 – 15 mL
Precision, %	< 1	1 – 3	0.5 – 2	0.5 – 3	0.2 – 2
Applications	Extensive	Limited	Extensive	Limited	Extensive
ASTM Test methods	7	1	5	None	8

6.   ION CHROMATOGRAPHY (IC)

Until the advent of ion chromatography (IC) in the seventies, anions such as halides, sulfate, nitrate, etc. had to be determined with elaborate and time intensive laborious wet chemistry methods. IC dramatically altered that situation usually simultaneously determining a number of anions sequentially in a small volume of liquid sample. IC is a combination of ion exchange chromatography, eluent suppression, and at least in the early years conductimetric detection of eluted anions. The anions are eluted through the separating columns in the background of carbonate-bicarbonate eluent, and detected by electrical conductivity. Quantitation is achieved by use of anionic standards and comparing the peak areas of sample and standard for specific anions. In about 10 minutes several anions such as fluoride, chloride, nitrate, sulfate, etc. can be determined. The technique has a large dynamic range over four decades of concentration, and the detection limits for most common anions are about 0.1 mg/kg.

Even though IC was originally developed for anions, several alkaline earth and transition metals were later determined using conductimetric and UV-visible detection. Some drawbacks of IC technique are that the ions with similar retention times are coeluted; determination of smaller quantities of anions is difficult in presence of an ion in very large concentration; the conductimetric determination is not suitable for anions with pKa of greater than 7. Although the technique has been extended to some transition metal determinations, its strong suite remains the determination of anions.

Since petroleum products are non-aqueous materials, they have to be converted to aqueous phase before subjecting to the IC analysis. This is usually achieved by combusting the sample in oxygen under pressure in an enclosed vessel. This technique has been used by several workers for the determination of sulfur, fluorine, chlorine, bromine, phosphorus, and nitrogen in fuel oil, gasoline, diesel fuel, waste oil, etc.[41-48]

7.   MICROELEMENTAL ANALYSIS

Although not as glamorous as ICP/AES or ICP/MS techniques, several elements in petroleum products are determined by diverse techniques, usually initiating the procedure with sample combustion and completing it with some form of detection and quantitation.

The "micro" in this context of analysis refers to the elemental concentration as well as the fact that these methods utilize very small quantities of sample for analysis. The elements such as carbon, hydrogen, nitrogen, chlorine, oxygen, and sulfur are usually included in this group. Determination of sulfur will be discussed separately later given the importance of this element in petrochemistry. Others are briefly discussed below.

CARBON-HYDROGEN-NITROGEN (CHN) – Several instruments based on the principle of combustion of the sample to form carbon dioxide, water, and nitrogen oxides are available. These gases formed are separated using a series of chemical absorbers or gas chromatography. Final quantitation of the separated gaseous compounds is done employing thermal conductivity or infra red detector. Usually a complete analysis takes less than 15 minutes; uses < 10 mg of sample; and has a precision of about 2 to 5 %. A standard test method comprising a number of such instruments and based on a round robin among 33 laboratories using 14 petroleum products is issued by ASTM as D 5291.

NITROGEN – The classical Kjeldahl method of nitrogen determination has been replaced in most of the oil company laboratories with ASTM test methods D 4629 or D 5762, both based on pyrolysis – combustion of the sample and the reaction of the gaseous products (NO) with ozone to produce metastable nitrous oxide. When this unstable compound returns to normal state it emits the radiation at 650 to 900 nm. A photomultiplier measures the intensity of this light emission (or chemiluminescence). The technique is extremely sensitive and can determine ppb levels of nitrogen with 2 to 4 % relative standard deviation.

HALOGENS – Trace quantities of halogens present in some petroleum products can be determined using IC, NAA, or XRF (see elsewhere in this chapter). Additionally, some instruments are available which can directly determine chlorine by pyrolysis – combustion of the sample and coulometric determination of resultant chloride and oxychloride ions produced. The detection limit is as low as 200 ppb chlorine.

OXYGEN – Although not widely determined in industry, oxygen can be determined in petroleum products using fast neutron activation analysis with 14 MeV neutrons in a Cockroft-Walton generator. The technique is extremely fast, on the order of less than a minute, and is essentially interference free. The moisture, if any, in the sample will also contribute to total oxygen determined by this technique. The detection limit is in low ppm range.

ASTM test method D 5622 employs reductive pyrolysis to quantitatively convert oxygen into carbon monoxide which after separation is determined using infrared or thermal conductivity detector. Better than 1 % precision can be obtained.

8.   NEUTRON ACTIVATION ANALYSIS (NAA)

Neutron activation analysis (NAA) methods are comparable in precision and accuracy to X-ray fluorescence or atomic spectroscopic methods for most elements, particularly when the analytical matrix is a "clean" one such as gasoline. Particularly in national research laboratories and universities NAA was widely used for the determination of trace elements in petroleum products. The use of this technique, however, has been limited in the oil industry laboratories because of the need to have access to a nuclear reactor for most sensitive determinations. See review by Nadkarni [49] for a detailed discussion and literature citations of applications of NAA methods for the analysis of fossil fuels.

Although there are several versions of NAA available, the largest body of work in the application of NAA to fossil fuels is using reactor thermal neutron irradiation followed by direct instrumental gamma ray counting or appropriate radiochemical separations followed again by gamma ray counting. High resolution Ge(Li) detectors have made instrumental NAA (INAA) a reliable multi-element analytical tool. INAA is a non-destructive method that can measure about 40 elements in fossil fuels.

In practice about 100 mg of a sample in sealed quartz vials along with elemental standards are irradiated for varying periods of time in a nuclear reactor in a thermal neutron flux of 1×10^{12} n.cm^{-2}.sec^{-1}. During irradiation, stable isotopes of elements are excited to radioactive isotopes that emit characteristic gamma rays during their decay. Post-irradiation, these gamma rays are counted on a high resolution Ge(Li) or hyper purity Ge detectors coupled to a multi-channel detector for varying periods of time. Since different isotopes have different half-lives, judicious selection of measurement time enables one to measure a number of elements over a period of few weeks. The elements with shorter half-life isotopes are measured soon after the irradiation, and those with longer half-life isotopes are measured at a later date after the short half-life isotopes have decayed to an insignificant level. A typical plan for sequential irradiation and measurement is given in Table 7.

Table 7   *Sequential Thermal INAA Scheme*

IRRADIATION TIME	DECAY TIME	ELEMENTS MEASURED
2 min.	2- 4 min.	Al, Cu, S, V, Ti
	10 min.	Ca, Mg
	26 min.	Cl, I
	30 – 60 min.	Ba, Cu, Mn, Na, Ni
8 – 24 hrs.	2 days	Cu, K, Na
	3 – 4 days	As, Br
	7 days	Ca, Sb
	15 days	Ba, Cr
	25 days	Co, Fe, Ni, Sb, Se, Sr, Zn

Since the petroleum products are essentially a carbon-hydrogen-oxygen matrix that does not become radioactive by thermal neutron irradiation, usually background gamma ray interference is minimal thus facilitating the determination of trace elements present in the sample.

8.1 Radiochemical NAA

In spite of the high resolution capability of modern gamma ray detectors, there are several trace elements which cannot be easily determined by INAA due to their low concentration level in the sample, their poor activation parameters, and/or interferences from matrix elements. In such cases, radiochemical separations are employed to isolate the radionuclides of interest from other high intensity gamma ray emitters or matrix. These separations may involve solvent extractions, ion exchange, gravimetric precipitations, etc.

9.   X-RAY FLUORESCENCE (XRF)

A technique widely used in the oil industry laboratories and rivaling the capabilities of AAS or ICP/AES is X-ray fluorescence spectrometry (XRF). Similar to AAS or ICP/AES, XRF has become a workhorse for the determination of trace, minor, and major levels of metals and some non-metals in petroleum products and lubricants. The technique has advantages of capability of non-destructive simultaneous multi-element determination with excellent precision and accuracy. It can also analyze solids or liquids. The cost of instrumentation is generally high, particularly for large multi-element capability instruments. Non-expensive instruments for dedicated single element determinations are also widely used for routine analysis. See Table 6 for a comparison of XRF technique versus other atomic spectroscopic methods.

In XRF technique a sample is bombarded with X-rays resulting in ejection of electrons from the inner shells of the target atoms. In the process, X-rays of discrete characteristic energy are emitted as electrons from the outer shells and replace the ejected electrons. Every element produces a unique secondary X-ray spectrum whose intensity is proportional to the element concentration in the sample. XRF has been typically applied to all elements above the atomic number 10 (i.e. neon). The basic components of X-ray spectrometers are a source of excitation consisting of a high power X-ray tube (usually comprised of a tungsten, molybdenum, chromium, rhodium, or scandium target), a wavelength or energy dispersive spectrometer for selecting the characteristic fluorescence signals, and integrating or counting circuits for measurements.

Due to numerous inter-element matrix effects, standards and samples should have similar compositions and corrections made for matrix element absorption and enhancement of the analyte element. Particle size and shape of solids can affect the accuracy of the analysis. Many of these interferences and matrix effects can be compensated by corrective techniques such as standard addition, internal standards, matrix dilution, thin film method, and mathematical corrections.

Speed, convenience, minimal sample preparations, and often non-destructive nature are the main advantages of the XRF technique. The method rivals the accuracy of wet chemistry techniques for the analysis of major constituents provided interferences and matrix effects are accounted for.

Two main varieties of XRF instrumentation are wavelength dispersive XRF (WDXRF) and energy dispersive XRF (EDXRF). In WDXRF, primary X-rays irradiate the sample and generate fluorescent X-rays that are then diffracted by a crystal. A goniometer selects the geometry between the crystal and the detector. The geometry controls the detection of X-rays from the elements of interest. Different crystals can have different d-spacings and cover different wavelength ranges, and are used depending on the X-ray wavelength of the analyte element. Many commercial WDXRF instruments have two detectors and up to seven crystals in order to optimize the instrumental conditions for each element.

The EDXRF instruments generally use a lower power X-ray tube. The emitted X-ray radiation from the sample impinges directly on a detector, typically a Si(Li) , which generates a change proportional in magnitude to the energy of the incident X-rays. The series of changes are sorted and counted by a multi-channel analyzer. Simultaneous determination of all elements, with atomic numbers greater than magnesium, is possible. Optimization of the specific elements is accomplished through the use of the secondary targets and filters. Radioisotope sources can be used in place of the X-ray tubes in instruments designed for limited element applications.

Sensitivity in XRF analysis is highly dependent on the sample matrix. Detection limits are usually in the mg/kg range, with the highest sensitivity observed for the transition metals. Sensitivity declines for the lower atomic number elements. Resolution and sensitivity of EDXRF is typically an order of magnitude worse than for WDXRF. However, the cost of an EDXRF spectrometer, especially the single element model, can be an order of magnitude less than for a WDXRF instrument. Newman [50] and MacRae [51] have reviewed the currently available instrumentation both for ED- and WD-XRF analyzers. The technique has also been used on-line for metal and sulfur determination in many refineries for monitoring the process streams. The pre-1990 literature on XRF applications to petrochemical products has been reviewed by Nadkarni earlier.[1]

To overcome the matrix effects in XRF analysis, the standards and samples need to be closely matched in their density. Often, an internal standard is used to overcome the density and viscosity differences. Background scattering is usually compensated for by using a pulse height analyzer that discriminates between the different wavelength pulses. The inter-element effects from other elements present in the sample are corrected by either using the internal standard or the inter-element correction factors (often called alphas) incorporated in the system software.

Over the years, one drawback of XRF technique has been the difficulty of measuring low atomic number (< 10) elements. The main reasons for this limitation have been fluorescent yields, dispersion, the sample window required for liquids, and sample absorption. With improvement in detector technology, associated electronics, detector windows, and availability of multi-layers, both qualitative and quantitative determination of low atomic number elements is now feasible with XRF.[52]

Shay and Woodward [53] determined V, Ni, and Fe in petroleum and petroleum residue by EDXRF, first ashing the samples with sulfuric acid and using cobalt as an internal standard to eliminate the matrix effects for low detection limits with high precision and accuracy. Several NIST SRMs were analyzed by this method. The EDXRF results were comparable to those obtained by flame AAS. Wheeler [54] has shown application of EDXFR for the determination of several elements in finished lubricating oils. Mackey, et al.[55, 56] have compared the performance of XRF and ICP/AES techniques, latter with and without robotic sample preparation, for the determination of several additive elements in lubricating oils. Except for magnesium, they found the precision of XRF and ICP/AES comparable. Precision improved using a robot for sample dilutions.

Sieber et al [57] have similarly demonstrated the equivalence of XRF technique to AAS and ICP/AES for the determination of several additive elements in lubricating oils. Sieber and Van Driessche [58] have empirically determined the alphas used for the determination of 13 additive and trace elements in oils and compared the alphas obtained with theoretically calculated ones. They found agreement between the two sets of alphas quite good. The empirical alphas correcting for magnesium and sodium absorbency of all other elements were about twice the magnitude of the theoretically calculated alphas.

Overman et al [59] have shown the usefulness of EDXRF for the determination of S, Pb, Ni, V, and Cl in crude oil, diesel, gasoline, waste oil, reformulated gasoline, jet fuel, gas oil, heating oil, and kerosene. In lubricating or cutting oils, P, S, Cl, Ca, and Zn, and chlorine in waste oils can be determined with EDXRF with results comparable to those obtained by WDXRF. Because of the simplicity and the ruggedness of the EDXRF analyzers they can be used on-line for analyses of gasoline for sulfur with detection limit as low as 9 ± 3 mg/kg.[59]

A number of methods have been issued by the ASTM D-02 Committee on Petroleum Products and Lubricants for the analysis of such products using WD- and ED-XRF techniques. Earliest of these is D 2622-98 which determines total sulfur in diesel, jet fuel, kerosene, naphtha, residual oil, lubricating base oil, hydraulic oil, crude oil, unleaded gasoline, M-85 and M-100 gasohols. The method is applicable in the range of 20 mg/kg to 5.3 m%. The standard and the sample matrix must be well matched to avoid poor results obtained by different carbon to hydrogen ratio difference between the samples and the standards. Oxygenated gasolines can lead to significant absorption of sulfur K alpha radiation. Correction factors or standards matching the samples must be used to overcome this interference. Test method D 2622 is the method of choice used by U.S. Environmental Protection Agency for compliance purposes.

D 4294-98 test method uses EDXRF for the determination of sulfur in hydrocarbons such as diesel, naphtha, kerosene, residuals, lubricating base oil, hydraulic oil, jet fuel, crude oil, unleaded gasoline, and gasohol. The applicable concentration range is from 0.015 to 5.0 m% sulfur. Spectral, matrix, and high oxygen levels interferences are corrected by appropriate software and matrix matching.

A widely used test method for the analysis of lubricating oils is D 4927-96 which determines Ba, Ca, P, S, and Zn in 0.03 to 2.0 m% range using WDXRF. Both internal standard or mathematical correction factors can be used for correcting the interference effects. An extension of this method to also include the determination of Cl, Cu, and Mg is designated as D 6443-99.

Lead in gasoline in the range 0.01 to 5 g/U.S. gallon can be determined using D 5059-98 test method. This WDXRF method allows three options: bismuth internal standard at high lead concentrations, scattered tungsten radiation, and bismuth internal standard at low lead concentrations.

D 6334-98 test method also is used for the determination of sulfur in the 15 to 940 mg/kg range in gasoline and oxygenated blends by WDXRF. Usual interferences are corrected in same fashion as described for other XRF methods. An equivalent method using EDXRF is D 6445-99 for the determination of sulfur in gasoline and oxygenated blends in the 48 to 1000 mg/kg range.

Finally, D 6428-99 test method uses EDXRF for the determination of Ca, P, S, and Zn in lubricating oils, generally in the concentration range 0.01 to 1.0 m%.

10.   ANALYSIS OF USED OILS

An area of particular interest in oil analysis is the analysis of used oils, also called in-service engine oils. Engine oils analyzed on a regular basis help predict likely occurrence of engine failure, thus reducing the high cost of major overhaul and repairs of costly engines, be they in heavy duty trucks, railroad engines, bus fleets, or air planes. Several articles describe the role engine oil analyses play in such preventive maintenance.[60 - 62]

Usually, an used oil analysis consists of determining several parameters : base number, water, sediment, FT-IR, particle count, ferrography, metals, etc. In spite of this array of testing done on a sample the cost of such multi-parameter analysis remains at about $ 20 per sample! There are hundreds of commercial laboratories that offer such services. The users are spending between $ 250M to $ 300M annually on the analysis of perhaps 50 million samples from some 15 million machines. U. S. military alone spends at least $ 40M a year on such analyses.[63]

Trace metal analysis of such oils helps indicate mechanical wear from oil-wetted components of an engine or as a contaminant from air, fuel, and liquid coolant. Generally, these metals are present as particulates rather than in true solutions. The presence of specific metals in used oils is indicative of wearing of specific metal components in the engine (See Table 2). A sudden increase in the metal content of the oil sampled from an engine on a regular frequency indicates failure or excessive wear of an engine part. Such early diagnosis helps in preventing catastrophic engine failure.

Atomic spectroscopy is the main tool used for metals determination in used oils. Although for decades, AAS was the main tool for such analyses, because of their simultaneous multi-element capability, ICP/AES and rotating disc electrode emission spectrometry are now carrying out the bulk of such analyses.

As mentioned above, because the metals are present as discrete particles of different sizes, sampling is of paramount importance to obtain accurate results. The particles typically above 3 to 10 microns do not reach the torch. A direct AAS or ICP/AES analysis will tend to give metal results biased low since larger metal particles will not be aspirated. Wet ashing with mineral acids as well as dry ashing has been used before converting the used oil sample into aqueous solution. However, given the extremely low price that is charged for such analyses, it is inconceivable that such sample preparations are used in commercial laboratories. The dilution of samples with an organic solvent such as xylenes, MIBK, kerosene, etc. is most commonly used.

Since the used oil analyses are used mainly to follow trends of oil degradation, it is not of utmost importance to obtain absolutely accurate and precise results, only to look for sudden increase in an analyte. Recently ASTM D-02 Committee has initiated an in-service diesel oil analysis inter-laboratory cross check program for several analytes normally determined in the used oils. Since many/most commercial laboratories do not strictly follow the ASTM standard test methods or the required calibration and quality control practices, it would be interesting to see how well different laboratories agree with each other, or how well does the precision obtained in these cross checks compares with those given in the ASTM test methods.

As part of the Joint Oil Analysis Program (JOAP), U. S. armed forces use rotating disk electrode atomic emission spectrometry to determine Fe, Ag, Al, Cr, Cu, Mg, Ni, Si, and Ti in mg/kg concentrations to diagnose an aircraft engine system including turbine engine, compressor, transmission, and gear box. The used oil samples, of course, have to be sent back to the home bases from aircraft deployed elsewhere. To be able to perform this analysis in field, a portable GF/AAS analyzer has been developed which determines the above nine trace metals. A special sample introduction device, an air cooled furnace, compact furnace power supply, and an multi-element scheme were developed to successfully use this instrument in the field by U.S. air force and navy.[64]

To overcome the drawback of single element measurement in AAS, a new sequential AAS capable of rapid multi-element measurement for up to 24 elements per sample in a single run was developed. The instrument can change the hollow cathode lamps and other instrumental parameters automatically within 2 seconds, thus allowing ten elements to be determined per minute. A sample is completely analyzed for all elements before the next sample is aspirated.[65]

Three sample preparation methods were compared for wear metals analysis by AAS: dry ashing, ashing with mineral acids and silica gel, and direct dilution with an organic solvent. The use of porous silica improves the dry ashing preparation and shortens the time required.[66]

Nygaard et al [67] and Lukas and Anderson [68] have demonstrated the application of rotating disc electrode (rotrode) AES for wear metal analysis of used oils. Nygaard et al [67] found inconsistencies between ICP/AES and rotrode results attributable to differences of sample viscosities and particulates size in the used oils.

Almeida [69] has described a high throughput high performance ICP/AES system that can do an analysis every 30 seconds. A V-groove nebulizer made of inert polymer resistant to the solvents used and designed to handle the large particles present in the used oil samples without clogging was employed. Garavaglia et al. [70] used butanol as a diluent for the determination of boron and phosphorus in used lubricating oils by ICP/AES. Ekanem et al.[71] used sulfanilic acid as an ashing agent in the sample preparation for wear metals analysis by AAS. Humphrey [72] used EDXRF to measure wear metals in the lubricant of the F104 engine on board F-18 weapon system. Prabhakaran and Jagga [73] have compared ferrography, ICP/AES, scanning electron microscopy (SEM), and EDXRF for their capability to study the wear particles in the lubricant of a steam turbine generator.

Two standard test methods have been issued by the ASTM D-02 Committee for such work. The test method D 5185-97 uses ICP/AES for the determination of a large number of metals in used (as well as fresh) lube oils. The oils are diluted ten fold with an organic solvent. An optional internal standard may be used to compensate for the variations in the sample introduction efficiency. The test method D 6595-00 uses rotrode AES to determine a number of wear metals and contaminants in the used lubricating oils or hydraulic fluids. A similar test method for analyzing fuels is expected to be issued in the fall of 2001.

11.   SULFUR

Given the importance of sulfur in petrochemical industry both from the processing and the environmental pollution viewpoints a closer look at its determination is worthwhile. A number of organo-sulfur compounds present in the crude oil adversely affect during petroleum processing and poisoning the process catalysts. The sulfur present in gasoline, diesel, or heating oil converts into sulfur oxides on combustion. Reaction with atmospheric water forms sulfuric acid resulting in acid rain polluting the waterways and affecting plant-animal welfare. Recently U. S. Environmental Protection Agency (EPA) has mandated < 20 mg/kg sulfur levels in gasoline and diesel fuels by the year 2006.

Several multi-elemental techniques (ICP/AES, XRF, IC) as well as specific instrumental methods are employed for the determination of sulfur at concentration levels ranging from mg/kg to mass percent in petroleum products. ASTM has published about 17 standard test methods for this determination employing wet chemistry, ICP/AES, XRF, and other specific instrumental techniques.[6,74] Of these, U.S. EPA at one time chose D 2622 (modified), WDXRF technique, as a mandatory method for regulatory compliance for sulfur in gasoline. Based on the data from a number of ASTM proficiency testing programs, Nadkarni [75,76] has shown that this method is inadequate at low levels (about 20 mg/kg) of sulfur present in future gasoline and diesel. Since then U.S. EPA has mandated D 6428, combustion-electrochemical detection technique, as the compliance method for sulfur in diesel fuels of the future. At present, work is being conducted to study the precision of this method for low sulfur levels in gasoline and diesel samples. It is expected that this inter-laboratory study may be completed by the end of the year 2001.

Kohl [77] did an in-depth study of fitness for use of three most widely used methods in the oil industry for the determination of low levels of sulfur in gasoline. These methods were D 2622 (WDXRF), D 4294 (EDXRF), and D 5453 (UV-fluorescence). All three were found to be equivalent in the 150 to 500 mg/kg sulfur range. Test methods D 5453 and D 2622 were equivalent down to 20 mg/kg. Of these, D 5453 can be fit for use in multi-laboratory situations down to 1 mg/kg sulfur level. The study also included analysis of 24 commonly occurring organo-sulfur compounds diluted in iso-octane matrix. No bias was found and the accuracy was within the precision limits for each of the three methods studied. The superior low level precision was found for D 2622 and D 5453. Of these, D 5453 generated better data in the low (< 50 mg/kg) sulfur range.

In a large study completed in Europe, 69 laboratories from nine countries determined sulfur in 5 to 500 mg/kg range in eight gasoline and seven diesel samples.[78] Principally five test methods were used: ISO versions of D 2622 (WDXRF), D 4294 (EDXRF), D 5453 (UV-Fluorescence), D 3120 (micro-coulometry), and D 1266 (Wickbold combustion). Although at sulfur concentration levels tested all five test methods gave essentially equivalent results, the precision of different methods varied considerably. Similar to the SWRI study mentioned above [77], CEN study also found D 2622 and D 5453 to have the best reproducibility. Of these, only D 5453 was considered suitable for determination of sulfur at < 10 mg/kg content required in future European fuels.

Finally, the ASTM inter-laboratory cross check programs (ILCP) being conducted over last decade confirm the conclusions of above two studies, i.e. for gasoline types of samples at < 10 mg/kg sulfur level, D 2622 has extremely poor precision. At around 30 mg/kg level, D 2622 reproducibility improves, but D 5453 still has superior reproducibility. At sulfur levels equal to or above 50 mg/kg both D 2622 and D 5453 produce equivalent results in reformulated gasoline, gasoline, diesel, and jet fuels. Thus, overall for very low level sulfur samples, D 5453 is far superior among the test methods available.[74]

12. CONCLUDING REMARKS

In this chapter we have reviewed the most widely used modern elemental analysis techniques for the analysis of hydrocarbons, petroleum products, and lubricants. Most of these are mature techniques, and without a paradigm shift in analytical technology (such as from wet chemistry to AAS to ICP/AES or XRF in the past), future developments can be expected to be only incremental. Elemental analysis can also be accomplished via hyphenated techniques such as gas chromatography where element specific atomic emission detectors (AED) are used for elemental determination. Most commonly used AEDs are for silicon, phosphorus, etc and Hall electrolytic detectors for halogens.

Although not covered in this chapter, an associated area of hydrocarbon analysis involves characterization of catalysts used in petroleum processing. X-ray diffraction technique is extensively used for structural determination combined with quantitative elemental determination by classical gravimetric or colorimetric methods in order to obtain very precise results. X-ray fluorescence, being a reliable and precise technique, is also used for this work. See McElroy and Mulhall [79] for an example of such work for the precious metal assay of fresh reforming catalysts.

Hieftje [12] lists a number of criteria for choosing an ideal analytical technique: high sensitivity, broad linear dynamic range, high precision, no matrix interference, inexpensive to acquire and to operate, applicable to all elements, simple to operate, compact, needing minimal sample preparation, micro-sampling capability, rapid, nondestructive, etc. Based on these criteria, Nadkarni [1] compared the leading elemental analysis techniques against the "ideal". Alas, no one single technique can meet each and every one of these requirements! Few, if any, techniques can even meet most of these criteria. However, many techniques meet many of the criteria, thus, enabling an analyst to make an intelligent choice of the best available technique or a combination of more than one technique to satisfy the analytical requirements of a specific analysis in a specific matrix.

13.   REFERENCES

1. Nadkarni, R. A. In *Modern Instrumental Methods of Elemental Analysis of Petroleum Products and Lubricants*, R. A. Nadkarni, Ed. ASTM 1991; pp. 5-51, STP 1109.
2. Benfaremo, N.; Liu, C. S. *Lubrication* **1990**, 76(1), 1.
3. Vartanian, P. F. *J. Chem. Educ.* **1991**, 68(12), 1015.
4. Liston, T. V. *J. STLE*, 389 (May 1992).
5. Nadkarni, R. A. In *ASTM D-02 Manual on Fuels, Lubricants, and Standards: Application and Interpretation,* Totten, G. E. (Ed.), ASTM, MNL 37, 2001; Chapter 27.
6. Nadkarni, R. A. *Guide to ASTM Test Methods for the Analysis of Petroleum Products and Lubricants*, ASTM, MNL 44, 2000.
7. Hofstader, R. A.; Milner, O. I.; Runnels, J. H. *Analysis of Petroleum for Trace Metals*, Adv. Chem. Series No. 156, Washington, D.C.: American Chemical Society, 1976.
8. Sychra, V.; Lang, I.; Sebor, G. *Progress in Anal. At. Spectr.* **1981**, 4, 341.
9. de al Guardia, M. ; Salvador, A. *At. Spectrosc.* **1984**, 5, 150.
10. Pradhan, N. K. *Coal and Petroleum Analysis by Atomic Absorption Spectroscopy*, Springvale: Varian Techtron, 1976.
11. Van Loon, J. C. *Analytical Atomic Spectroscopy: Selected Methods*" Academic Press: New York, 1980.
12. Hieftje, G. M. *J. Anal. At. Spectrom.* **1989**, 4, 117.
13. Vanclay, E. *Today's Chemist at Work*, 14 (January 1996).
14. Sheddon, J.; Lee, Y. I.; Hammond, J. L.; Noble, C. O.; Beck, J. N.; Proffit, C. E. *Amer. Environ. Lab.* **1998**, 10(6), 4.
15. Meyer, G. A. *Anal. Chem.* **1987**, 59, 1345A.
16. Noble, D. *Anal. Chem.* **1994**, 66(2), 105A.
17. Botto, R. I. *Jarrell Ash Plasma Newsletter* **1981**, 4(2), 7.
18. Hausler, D.; Carleson, R. *Spectrochim. Acta Rev.* **1991**, 14, 125.
19. Botto, R. I *Spectrochim. Acta Rev.* **1991**, 14, 141.
20. Botto, R. I *J. Anal. At. Spectrom.* **1993**, 8, 51.
21. Botto, R. I; Zhu, J. J. *J. Anal. At. Spectrom.* **1996**, 11, 675.
22. Gonzales, M.; Lynch, A. W. In *Modern Instrumental Methods of Elemental Analysis of Petroleum Products and Lubricants*, Nadkarni, R. A. (Ed.), ASTM 1991; pp. 62.
23. Brown, R. J. *Spectrochim. Acta* **1983**, 38B, 283.
24. Merryfield, R. N.; Loyd, R. C. *Anal. Chem.* **1979**, 51, 1965.
25. Bansal, J. G.; McElroy, F. C. *SAE Techn. Paper* **1993**; 932694.
26. Jansen, E. B. M., Knipscheer, J. H., Magtegaal, M. *J. Anal. At. Spectro.* **1992**, 7, 127.
27. Richter, U.; Krengel–Rothensee, K.; Heitland, P. *Amer. Lab.*, 170 (Feb. 1999).
28. Thomas, R. *Spectroscopy* **2001**, 16(4), 38.
29. Newman, A. *Anal. Chem.* **1996**, 68, 46A.
30. Al-Swaidan, H. M. *Anal. Letters* **1998**, 21, 1487.
31. Williams, M. C. In *Modern Instrumental Methods of Elemental Analysis of Petroleum Products and Lubricants*, Nadkarni, R. A. (Ed.), ASTM 1991; pp. 96.
32. McElroy F. C., Mennito, A., Debrah, E., Thomas, R. *Spectroscopy* **1998**, 13(2), 42.
33. McElroy, F. C. Unpublished Work.
34. Parsons, M. L.; Majors, S.; Forster, A. R. *Appl. Spectro.* **1983**, 5, 411.
35. Routh, M. W. *Spectroscopy* **1987**, 2(2), 45.
36. Slavin, W. *Spectroscopy* **1991**, 6(8), 16.

37. Beauchemin, D. *Spectroscopy* **1992**, 7(7), 12.
38. Thomas, R. *Today's Chemist at Work* **1999**, 8(10), 42.
39. idem – *ibid*, 9(9), 19 (2000).
40. Supplement to *Spectroscopy* **2001**, 16(4).
41. Mizsin, C. S.; Kuivinen, D. E.; Otterson, D. A. In *Ion Chromatographic Analysis of Environmental Pollutants*, Vol. 2, Mullik, J. D.; Sawicki, E. (Eds.), Ann Arbor Science: Ann Arbor, 1979; pp. 129.
42. Butler, F. E.; Toth, F. J.; Driscoll, D. J.; Hein, J. N.; Jungers, R. H. In *Ion Chromatographic Analysis of Environmental Pollutants*, Vol. 2, Mullik, J. D.; Sawicki, E. (Eds.), Ann Arbor: Ann Arbor Science, 1979; pp. 185.
43. Koch, W. F. *J. Ressearch NBS* **1979**, 84, 241.
44. McCormick, M. J. *Anal. Chim. Acta* **1980**, 121, 233.
45. Saitoh, H.; Oikawa, K. *Bunseki Kagaku* **1982**, 31, E375.
46. Vishwanadham, P.; Smick, D. R.; Pisney, J. J.; Dilworth, W. F. *Anal. Chem.* **1982**, 54, 2431.
47. Dionex Application Note, No. 15 (1979).
48. Wetzel, R.; Smith, F. C.; Cathers, E. *Industrial Res. Devlop. News* **1981**, 23(1), 152.
49. Nadkarni, R. A. *J. Radioanal. Nucl. Chem.* **1984**, 84, 67.
50. Newman, A. *Anal. Chem.* **1997**, 69, 493A.
51. MacRae, M. *Spectroscopy* **1997**, 12(6), 26.
52. Vrebos, B. *Spectroscopy* **1997**, 12(6), 54.
53. Shay, J. Y.; Woodward, P. W. In *Modern Instrumental Methods of Elemental Analysis of Petroleum Products and Lubricants*, Nadkarni, R. A. (Ed.), ASTM 1991; pp. 128.
54. Wheeler, B. D. In *Modern Instrumental Methods of Elemental Analysis of Petroleum Products and Lubricants*, Nadkarni, R. A. (Ed.), ASTM 1991; pp. 136.
55. Mackey, J. R.; Meunier, C. A.; Windsor, B. K. Winter Conf. On Plasma Spectrochemistry, San Diego, CA, Jan. 1988.
56. Mackey, J. R.; Watt, S. T.; Cardy, C. A.; Smith, S. I.; Meunier, C. A. In *Modern Instrumental Methods of Elemental Analysis of Petroleum Products and Lubricants*, Nadkarni, R. A. (Ed.), ASTM 1991; pp. 52.
57. Sieber, J. R.; Salmon, S. G.; Williams, M. C. In *Modern Instrumental Methods of Elemental Analysis of Petroleum Products and Lubricants*, Nadkarni, R. A. (Ed.), ASTM 1991; pp. 118.
58. Van Driessche, M.; Sieber, J. R. *Adv. X-Ray Anal.* **1999**, 41, 770.
59. Overman, A. R.; Comtois, R. R.; Fess, S. B. *Amer. Lab. News Ed.* **1994**, 26, 14.
60. Eisentraut, K. J.; Newman, R. W.; Saba, C. C.; Kauffman, R. E.; Rhine, W. E. *Anal. Chem.* **1984**, 56, 1086A.
61. Leugner, L. *Lubes'N'Greases;* pp. 20 (May 1997).
62. Thomas, R. J. *Today's Chemist at Work* **2000**, 9(10), 53.
63. Carnes, K. *Lubricants World*; p. 13 (June 2001).
64. Niu, W.; Haring, R.; Newman, R. *Amer. Lab.* **1987**, 19(11), 40.
65. Carter, J. M.; Battie, W.; Bernhard, A. E. In *Modern Instrumental Methods of Elemental Analysis of Petroleum Products and Lubricants*, Nadkarni, R. A. (Ed.), ASTM 1991; pp. 70.
66. Barbooti, M. M.; Zaki, N. S.; Baha-Uddin, S. S.; Hassan, E. B. *Analyst* **1991**, 115, 1059.
67. Nygaard, D.; Bulman, F.; Alavosys, T. – in Reference 1, 77 (1991).
68. Lukas, M.; Anderson, D. P. In *Modern Instrumental Methods of Elemental Analysis of Petroleum Products and Lubricants*, Nadkarni, R. A. (Ed.), ASTM 1991; pp.83.
69. Almeida, M. *Amer. Lab.* **2000**, 32(16), 52.
70. Garvaglia, R. N.; Rodriguez, R. E.; Batistoni, D. A. *Fresenius' J. Anal. Chem.* **1998**, 360, 683.

71. Ekanem, E. J.; Lori, J. A.; Thomas, S. A. *Talanta* **1997**, 44, 2103.
72. Humphrey, G. R. *Lubr. Engr.* **1999**, 55(10), 19.
73. Prabhakaran, A.; Jagga, C. R. *Tribol. Int.* **1999**, 32(3), 145.
74. Nadkarni, R. A. *Amer. Lab.* **2000**, 32(22), 16.
75. Nadkarni, R. A. *World Refining* June 2000; 10(5), S-14.
76. Nadkarni, R. A. *Today's Refinery* August 2000.
77. Kohl, K. *ASTM Research Report* **1999**; D-02-1456.
78. Tittarelli, P. *Round Robin Exercise for Sulfur Test Methods for EN 228 and EN 590 Fuel Specifications*, CEN/TC 19/WG 27, April 2000.
79. McElroy, F. C.; Mulhall, J. M. In *Modern Instrumental Methods of Elemental Analysis of Petroleum Products and Lubricants*, Nadkarni, R. A. (Ed.), ASTM 1991; pp. 105.

Chapter 3

SELECTIVE DETECTION OF SULFUR AND NITROGEN COMPOUNDS IN LOW BOILING PETROLEUM STREAMS BY GAS CHROMATOGRAPHY

Birbal Chawla
ExxonMobil Research & Engineering Co.
Paulsboro Technical Center
Paulsboro, New Jersey 08066

1.   BACKGROUND

A significant number of organic molecules in petroleum crudes and their products contain one or more than one heteroatoms. Amongst them, sulfur and nitrogen are the most important heteroatom constituents of petroleum.

Although petroleum product specifications are not new, the requirements are getting tighter and are, therefore, becoming increasingly difficult to meet. In order to meet the requirements of changing environmental regulations/legislation regarding the specification of lower sulfur content in diesel and gasoline streams, either rigorous desulfurization processes need to be optimized/developed or the feed composition needs to be revised.

Though it is generally believed that nitrogen compounds cause color, gum formation and are responsible for catalyst poisoning during catalytic hydrotreating/hydrocracking processes, the refiners have not been very enthusiastic about nitrogen speciation. This is probably due to relatively small quantities of nitrogen compounds being present in the conventional feed stocks and also partially due to the fact that there has been no specification for total nitrogen in the final product. However, the situation has been changing rapidly because of the growing need to process relatively inexpensive, heavy, and low quality feed stocks containing large quantities of nitrogen and other heteroatom compounds.

In order to obtain the desired product quality and also to troubleshoot and understand the effect of sulfur and/or nitrogen compounds on the product upgrading processes, it is essential to identify and quantify the sulfur and/or nitrogen containing organic molecules present in various petroleum process streams. Of particular interest are the light streams (gasolines and diesels) where tight specifications are in place.

2.   SULFUR COMPOUNDS IN LIGHT STREAMS

Over the past decade, a number of analytical methods for speciation of sulfur compounds present in the process feeds and their products have been developed by petroleum scientists and utilized by refinery engineers for optimization of desulfurization processes.

The sulfur compounds are often speciated by gas chromatography (GC) using a variety of commercially available detectors, e.g. flame photometric detector (FPD), electrolytic conductivity (Hall) detector, atomic emission detector (AED), electron capture sulfur detector (ECD) and an "universal" sulfur chemiluminescence detector (SCD). About four decades ago, the FPD was introduced by Brody and Chaney[1] and it continues to have widespread use[2-5] today. Although the FPD is widely used, it has several major drawbacks. The FPD is based on the S_2^* chemiluminescence emission bands at 384 and 394 nm.[3] The electronically excited S_2^* are formed by the sulfur atoms produced in a hydrogen rich flame. The intensity (I) of this emission, which is due to molecular band, is of the form $I = S^n$. The exponent "n" is theoretically 2, but the response to sulfur ranges between first and second order, depending on the heteroatom environment.[6] The FPD response is also affected by co-eluting water or hydrocarbons, which can quench the chemiluminescence to a significant extent.[7,8] Although dual-flame photometric detectors have been developed to overcome some of these problems, they generally do not have the low sulfur detection limits that are required for process development. The dual flame photometric detector can also exhibit some hydrocarbon interference particularly at low sulfur concentration. The pulsed flame photometric detector (PFPD) developed recently is considered to overcome most of the drawbacks of FPD and is also supposed to respond to the different type of sulfur compounds at the equimolar level.[9] However, the PFPD detector response is linear only over a narrow range of sulfur content.

The atomic emission detector (AED) has also been used for the identification and quantitation of a variety of elements including sulfur and nitrogen,[2,10-13] but has many limitations. The AED is relatively costly, requires time-consuming, laborious calibrations and is not very rugged. Additionally, it requires a highly trained professional to perform routine maintenance, run the system, and analyze the data.

Other sulfur detectors, such as electrolytic conductivity [2,4,14] and electron capture detectors [2,10,11] are rarely used.

About fifteen years ago a "universal" sulfur-selective chemiluminescence detector (SCD) was developed and subsequently became commercially available.[16-19] Benner and Stedman[20] demonstrated that the *SCD response to individual sulfur compounds was equal on a weight sulfur basis* and hydrocarbon interference was minimal. Furthermore, there was no loss in the detector response or any interference from carbon dioxide and water vapors. It was also shown that the SCD was very sensitive and provided relatively low detection limits that are at least 10-15 times lower than that of the FPD. Over the past ten years, several improvements have been made to the operation of the SCD. These improvements have enhanced the SCD's performance, simplified maintenance, and minimized downtime. The American Society of Testing and Materials (ASTM) has also adopted a SCD based standard method (ASTM D 5623-94) for analysis of sulfur compounds in light petroleum streams. The operational characteristics (linear response, sensitivity, and its long-term stability) of the SCD detectors are discussed in detail elsewhere.[21-23]

The GC based methods for identification and quantification of a series of sulfur components such as mercaptans (RSH), sulfides (RSR'), disulfides (RSSR'), thiophenes (T), benzothiophenes (BT), and dibenzothiophenes (DBT) present in the various diesel and gasoline process streams are outlined below. The studies discussed in the preceding sections on chemiluminescence detection cover a variety of the light petroleum streams (catalytically cracked gasolines, kerosines, and diesels) containing a wide range of sulfur content (approx. 0.01 to 4 wt%).[21,22]

2.1   Instrumentation

Gas chromatography (GC) is an essential first step for speciation of sulfur compounds. Once the sulfur compounds are separated on a GC column, they are detected by sulfur selective detection either directly or immediately after their flame ionization detection. In the direct detection system, the effluent from the GC column is fed directly into the sulfur detection system where the sulfur compounds are quantitatively converted to sulfur dioxide and traced. In this arrangement only the sulfur signal is recorded. This arrangement is preferred if the sulfur compounds are present in trace amount. In the second arrangement, the GC column effluent goes first to a flame ionization detector (FID) and then to the sulfur detection system. The carbon and sulfur traces are recorded simultaneously from a single injection. The second arrangement is generally preferred if possible. A schematic diagram of the major components of typical instrumentation for the second arrangement, where a sulfur chemiluminescence detector (SCD) is coupled to a gas chromatograph, is shown in Figure 1. The combustion products formed in the flame are transferred under a reduced pressure of about 9 ± 2 torrs to the chemiluminescence reaction chamber with the help of a high capacity vacuum pump. To minimize oil loss, a large coalescing oil return/recovery filter is fit to the pump exhaust outlet to trap vaporized oil. The pump can remain in continuous use (24 hr a day) for about 2-3 months without requiring an oil change or any replacement of the oil return/recovery filter. In order to remove unreacted ozone, NO_x and other potentially oil destructive gases, a trap containing Hopcalite™ (Callery Chemical Co., Pittsburgh, PA) and sodalime in a 7 to 3 ratio by volume is placed between the reaction cell and the vacuum pump.

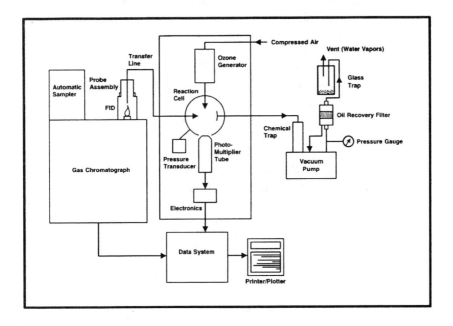

Figure 1. Schematic Diagram of the Major Components of a Sulfur Chemiluminescence Detection System.

2.2   Sulfur Chemiluminescence Detection System

Organo-sulfur compounds, emerging from a GC column and entering the flame of the FID, produce sulfur oxide (SO) which emits a strong blue chemiluminescence of SO_2^* when it reacts with ozone (O_3) in the reaction chamber of the SCD.[24-26] The SCD is believed to operate according to the following series of reactions:

$$R - S - R' + O_2 \quad \text{-----}> \quad SO + \text{Other Products}$$
$$SO + O_3 \quad \text{-----}> \quad SO_2^* + O_2$$

$$SO_2^* \quad \text{------>} \quad SO_2 + h\nu$$

The light is emitted in the wavelength range of 260 to 480 nm.[27] Since a hydrogen-rich flame is used in the SCD operation, a portion of the observed light may also be due to the reaction between the H_2S produced in the flame and O_3 ($H_2S + O_3$ ----> $HSO^* + HO_2$). HSO emission occurs in the wavelength range of 360 to 380 nm. However, there is evidence that a single collision reaction between H_2S and ozone results in SO_2^*.[28] The single collision reaction was postulated as

$$H_2S + O_3 \quad \text{------>} \quad SO_2^* + H_2O$$

At present, there are two commercially available SCD's based upon the same chemiluminescence chemistry. The manufacturers are: Sievers (Sievers Research Inc., Boulder, Colorado) and Antek (Antek Instruments, Inc., Houston, Texas). Both the SCD's (Sievers & Antek) have been used for more than a decade for speciating sulfur compounds in the light petroleum streams. Based upon our experience, the Sievers detector is more rugged than the Antek and is, therefore, preferred. Also, the Sievers SCD is available in two options. In one of the options, the SCD is coupled directly to the flame housing of a flame ionization detector (FID) and the carbon and sulfur traces are recorded simultaneously from a single injection. Whereas in the second arrangement, the exit end of a GC column is inserted deep into a hot burner (800°C) where sulfur compounds are quantitatively converted to sulfur dioxide and traced. In the second arrangement only the sulfur signal is recorded. The Antek SCD is available only in the second option where sulfur traces (and no carbon) are obtained if the GC column effluent is not split at the exit. If the column effluent is split, a significant discrimination between the low and high boiling materials may occur. Additionally, the detector furnace (approx. 15" tall), which operates at 1000°C, requires more frequent maintenance and is not operator friendly. Recently, Antek has come up with a "two-in-one" detector for obtaining nitrogen and sulfur traces from a single injection. However, the sensitivity for the sulfur signal suffers by a factor of ten or greater in the two-in-one Antek detector.

2.3   Gas Chromatography

In principle, the SCD can be coupled to any commercially available GC after some minor modifications. Most commonly used GC's are: Hewlett Packard, Varian, and Perkin Elmer. If the SCD (Sievers only) is coupled through a flame ionization detector, a special adapter is required. While the GC's equipped with the split/splitless- and/or direct/cool-on-column injection-ports have been used successfully, the direct-on-column injector port is preferred in order to avoid sample discrimination.

The light petroleum streams are most often analyzed by three separate methods operated under different GC conditions and columns. The three GC methods used are based upon the type of the streams to be analyzed. One is for the light gases, the second is for the gasolines, and the third is for the diesels. The separation of light gases is not à subject matter of this chapter. In order to obtain the desired baseline separations of sulfur compounds, the GC column selection is very crucial. Only a few fused silica and/or silicosteel capillary columns of 30-100 meter length with 0.10-4.0 micron film thickness have been tested for gasoline and diesel streams.[17] Most commonly used columns are: 30 or 60 meter x 0.2 µm (Diesels) and 30 or 60 meter x 3 µm (gasolines) with a methyl-silicone stationary phase (e.g. DB-1, SPB-1, HP-1, MXT-1 etc.). Most of the time, the neat samples (except very high sulfur content samples where samples are diluted) of gasolines or diesels are analyzed as received. Some typical GC traces of diesel and gasoline streams are shown in Figures 2-5.

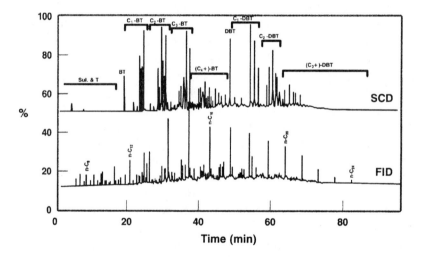

Figure 2. Comparison of SCD and FID Chromatograms Obtained from a Single Injection of a Light Petroleum Stream (diesel, S = 1.7%, Ref. 21).

2.4   Identification of Sulfur Compounds

The sulfur containing compounds are most often identified using the GC retention time data for the reference compounds and by GC-mass spectrometry (GC-MS). The retention time data for the selected reference compounds[21] and the sulfur compounds identified in the GC traces of the light streams[21,22] are shown in Figures 2-6 and in Table 1.

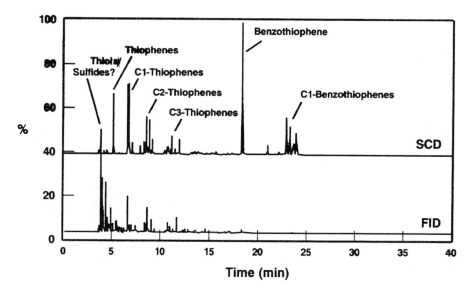

Figure 3. Comparison of SCD and FID Chromatograms Obtained from a Single Injection of an FCC Gasoline (Ref. 21).

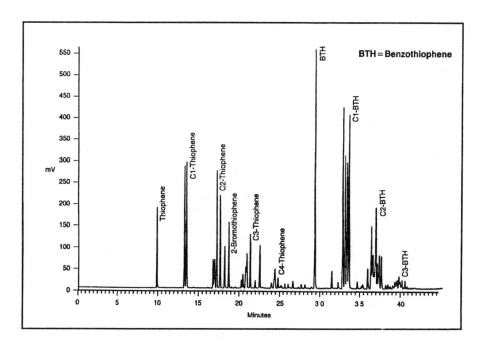

Figure 4. SCD Chromatogram of a Diluted Fluid Catalytic Cracking (FCC) Naphtha (Ref. 22).

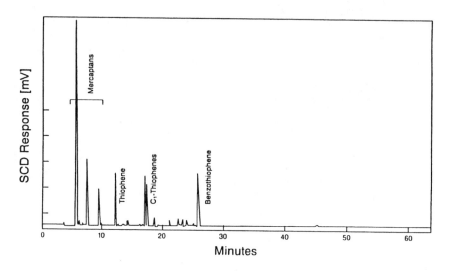

Figure 5. SCD Chromatogram of a Gasoline Range Sample Containing Mercaptans and Thiophenes Ref. 22).

Table 1. Retention Time Data for Selected Sulfur Compounds (Ref.21 & 22)

Compound	RT (min)	Compound	RT (min)
Reference # 21 (See Figure 2)		Reference # 22 (See Figure 4)	
Thiophene	5.26	H$_2$S	1.75
2-Methyl	6.71	C$_2$ Mercaptan	4.79
3-Methyl	6.81	Dimethyl Sulfide	5.44
2,5-Dimethyl	8.56	CS$_2$	6.00
Benzothiophene	18.43	1-Propyl Mercaptan	6.45
3-Methyl	24.08	**Thiophene**	9.75
Dibenzothiophene	48.25	Diethyl Sulfide	10.71
4-Methyl	53.71	n-Butyl Mercaptan	11.11
3-Methyl	54.77	2-Methyl Thiophene	13.19
2-Methyl	54.74	3-Methyl Thiophene	13.42
1-Methyl	55.89	n-Pentyl Mercaptan	14.79
4,6-Dimethyl	59.01	2-Ethyl Thiophene	16.68
2-Ethyl	59.59	C$_2$ Thiophene	16.91
2,6-Dimethyl	60.04	C$_2$ Thiophene	17.18
1-Ethyl	60.14	C$_2$ Thiophene	17.41
3,6-Dimethyl	60.15	n-Propyl Sulfide	17.52
3-Ethyl	60.18	C$_2$ Thiophene	18.17
3,7-Dimethyl	60.91	2-Bromo Thiophene (Std.)	18.61
2,8-Dimethyl	60.94	n-Hexyl mercaptan	18.82
3,8-Dimethyl	60.98	C$_3$ Thiophene	20.19
1,4-Dimethyl	61.17	n-Propyl Thiophene	20.47
1,6-Dimethyl	61.25	C$_3$ Thiophenes	20.80-22.53
1,8-Dimethyl	61.4	n-Heptyl Mercaptan	22.77
1,3-Dimethyl	61.82	C$_4$ Thiophenes	24.01-24.45
3,4-Dimethyl	61.95	C$_7$ Mercaptan	24.58
1,9-Dimethyl	62.04	C$_4$ Thiophenes	24.84-25.17

2,4-Dimethyl	62.07	n-Butyl Sulfide	25.31
1,7-Dimethyl	62.08	C_4 Thiophenes	25.70-26.12
1,2-Dimethyl	62.79	n-C_8 Mercaptan	26.63
2,3-Dimethyl		C_4 Thiophenes	26.73-28.07
		Benzothiophene	29.42
		C_9 Mercaptan	30.53
		Methylbenzothiophene	32.97
		Methylbenzothiophene	33.27
		Methylbenzothiophene	33.5
		Methylbenzothiophene	33.74
		C_2 Benzothiophenes	36.01- 37.14

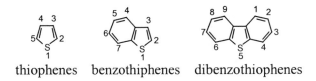

thiophenes   benzothiphenes   dibenzothiophenes

2.5 Quantitation of Sulfur Compounds

It has been reported in the literature that the SCD has been shown to have a linear response, which is independent of the sulfur compound type.[21,22,29,30] It is also reported that in addition to sulfur speciation, total sulfur can simultaneously be determined with reasonable accuracy.[21] Once the total sulfur is determined, the quantitation of individual components is obtained by multiplying its peak-area with the total sulfur divided by the total peak-area under all the peaks of the SCD trace.

The total SCD peak-area can also be converted into total sulfur using either an internal standard or by obtaining a correlation between the total-peak area and the total sulfur determined independently for the light petroleum streams. The later technique seems to work well and is reported by Chawla and Di Sanzo[21]. The SCD results reported on ten samples with various sulfur contents from about 300 to 5000 ppm were shown to be in excellent agreement (mostly better than 10%) with the elemental sulfur determinations.[21]

Use of an internal standard for quantitation of individual sulfur components or of total sulfur has also been reported.[22,29-31] Two internal standards, namely 2-bromothiophene [22] and 3-chlorothiophene, have been used for the SCD analyses whereas 2-fluorodibenzothiophenehas been used with the AED traces [31].

For sulfur quantitation of the low-level sulfur samples (ppm level), it is important that the integration performed is carefully scrutinized and performed manually when necessary. Each peak must be evaluated for the detection saturation limit of the SCD. If the GC run contains peak(s) close to the saturation limit, the sample must be rerun with a smaller injection volume or with a diluted sample. Also, if a significant peak tailing occurs, the quantification will not be accurate and the sample should be rerun with a smaller injection volume.

Total sulfur determination and individual component quantification may be obtained as follows:

$$S_{total} \text{ (ppm)} = [(A_{total} - A_{IS})/ A_{IS}] \text{ x } S_{IS} \text{ (ppm)}$$

And

$$S_{IC} \text{ (ppm)} = [A_{IC}/A_{IS}] \text{ x } S_{IS} \text{ (ppm)}$$

Where S_{total}, S_{IS}, and S_{IC} are the amounts of total sulfur, internal standard sulfur, and individual component sulfur respectively. The A_{total}, A_{IS}, and A_{IC} are the total area under all of the peaks including the internal standard peak, the area under internal standard peak, and the area under any individual peak being quantified, respectively.

3.   NITROGEN COMPOUNDS IN LIGHT STREAMS

Although nitrogen compounds effect product quality, only a few attempts have been made to identify and quantify some of the nitrogen compounds in the petroleum streams.[32-37] In these studies, nitrogen compounds were analyzed after being concentrated using different laborious techniques such as acid-treatment extraction, ion-exchange chromatography, etc. Results reported in these studies on quantitation of nitrogen compounds are questionable because it is possible that some of the nitrogen compounds might not have been extracted or might have been lost during the extraction. In these studies, the concentrated fractions were analyzed by GC and/or GC-MS.

The most commonly used nitrogen-detectors are Hall detectors and nitrogen-phosphorus-detector (NPD). Hall detectors are selective for halogens, sulfur, nitrogen, phosphorus and oxygen[38], and NPD is selective for nitrogen, phosphorus, sulfur, halogens, argon and lead[39]. Relatively recently developed nitrogen chemiluminescence detectors (NCD) are selective for only nitrogen and have greater than 10^6 nitrogen-to-carbon selectivity under optimum conditions, whereas Hall and NPD have only within 10^4 to 10^6 nitrogen-to-carbon selectivity.[35] In the last few years, the NCD has been used for a wide range of applications,[35-45] including qualitative analyses of light cycle oil and crude oil.[32,44,45]

3.1   Instrumentation

The instrumentation for the nitrogen speciation and quantification by NCD is very similar to the one discussed in section 2.1 for the SCD (see Figure 1). But the gases' flows are very different for the two detectors (NCD & SCD). The NCD needs oxygen for the ozone generator instead of air.

At present, there are two commercially available NCD's: Antek NCD and Sievers NCD. Since the Antek NCD was developed and marketed earlier than the Sievers, there is more published data on speciation of nitrogen compounds using the Antek detector. Antek detector optimization and performance are discussed in more detail in the published literature.[32,42-44] Although the Antek detector may be more sensitive and selective for nitrogen compounds, the Sievers detector is more rugged and user friendly.

3.2   Principle of Nitrogen Chemiluminescence Detection

The principle of operation is illustrated in the following chemical equations. The emerging components from the GC column are combined with oxygen in a furnace maintained at high temperature to produce oxidation products, which include NO, CO_2, H_2O, SO_2, etc. The NO (produced from all chemically bound nitrogen compounds) combines with ozone (O_3) to form nitrogen dioxide in the excited state (NO_2^*). The NO_2^* emits light on its decay to the ground state. The emitted light is detected at specific wavelengths by a photomultiplier tube. This chemiluminescent emission is specific for nitrogen and is directly proportional to the amount of nitrogen in the sample. Although atmospheric diatomic nitrogen is not detected, it gives a relatively high background if air is used instead of oxygen for generating ozone.

$$=N- \ + \ -X- \ + \ R\text{-}H \ + \ O_2 \ ------> \ NO \ + \ \text{Other Combustion Products}$$
$$NO \ + \ O_3 \ ------> \ NO_2^* \ + \ O_2$$
$$NO_2^* \ ------> \ NO_2 \ + \ h\nu$$

Where =N- is a chemically bound nitrogen compound, -X- is any other chemically bound heteroatom compound and R-H is any hydrocarbon.

3.3   Gas Chromatography

The NCD can be coupled to any commercially available GC after some modifications similar to that for the SCD discussed in section 2.2 of this chapter.

Speciation and quantitation of the nitrogen compounds, which are polar and somewhat basic in nature, require careful selection of an appropriate GC column and rigorous optimization of the gas chromatographic conditions. Seven different capillary columns from five different manufacturers have been tested using a light cycle oil (LCO) sample.[32] A direct on-column injection technique is preferred in order to prevent boiling point discrimination and adsorption due to strong interactions of the basic nitrogen compounds on active surfaces in the injection port. A DB-1 type column has been reported to be acceptable for the good separation of nitrogen compounds present in the light petroleum stream.[32,44] The NCD separation data from Reference 32 is reproduced below.

Figures 6 and 7 represent the typical chromatograms of LCO and gasoline samples with major components identified, respectively. As shown in Figures 6 and 7, the major nitrogen components identified in the LCO are carbazoles whereas the gasoline sample contained mostly anilines.

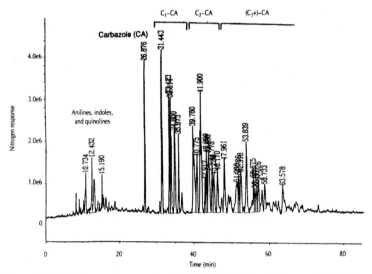

Figure 6. NCD Chromatogram of a Light Cycle Oil (N = 596 ppm, Ref. 32)

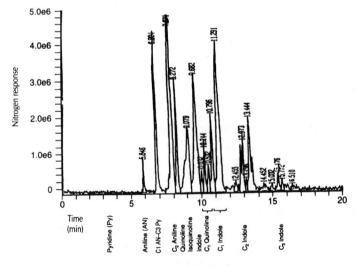

Figure 7. NCD Chromatogram of a Heavy Naphtha (N = 180 ppm, Ref. 32)

Table 2 lists the retention times for 62 reference nitrogen compounds of different classes such as pyridines, piperidine, pyrroles, anilines, indoles, carbazoles, and other miscellaneous nitrogen compounds including aliphatic amines. The model compounds were chromatographed at a concentration level of 4-5 mg/20 ml of toluene. Structures of selected classes of nitrogen compounds are provided in Table 2.

Table 2. Retention Time Data for Selected Classes of Nitrogen Compounds (Ref. 32)

Compound	B.P. (°F)	Retention Time (min.)
Pyridines		
Pyridine	240	3.65
Hexahydropyridine (piperidine)	223	3.82
4-Methylpyridine	294	4.72
2,4-Dimethylpyridine	318	5.36
Pyrroles		
1-Methylpyrrole	235	3.53
Pyrrole	268	4.07
2,5-Dimethylpyrrole	333	5.35
1,2,5-Trimethylpyrrole	343	7.7
Anilines		
Aniline	363	5.91
N-Methylaniline	385	6.74
N,N-Dimethylaniline	379	6.79
2,6-Dimethylaniline	418	7.78
2,4-Dimethylaniline	424	7.76
2,5-Dimethylaniline	424	7.82
2,3-Dimethylaniline	430	7.83
3,5-Dimethylaniline	432	7.84
3,4-Dimethylaniline	439	8.1
2-Propylaniline	435	8.61
4-Octylaniline	590	24.25
4-Decylaniline	~670	37.37
4-Dodecylaniline	-	60.39
4-Tetradecylaniline	-	Retained
Indoles		
Tetrahydroindole	-	8.23
2-Methylindoline	-	8.47
1-Methylindole	468	9.15
Indole	489	9.44
2,3-Dimethylindoline	482	9.5
7-Methylindole	511	10.66
3-Methylindole	509	11.05
2-Methylindole	523	11.06
6-Methylindole	501	11.14
5-Methylindole	-	11.14
4-Methylindole	513	11.2
1,2-Dimethylndole	-	11.85
2,5-Dimethylindole	-	13.3
2,3-Dimethylindole	545	13.5

Compound	B.P. (°F)	Retention Time (min.)
2-Phenylindole	770	42.46
Quinolines		
Quinoline	459	8.81
Isoquinoline	468	9.04
8-Methylquinoline	482	9.9
7-Methylquinoline	496	10.5
6-Methylquinoline	498	10.52
2,8-Dimethylquinoline	500	10.92
2,6-Dimethylquinoline	-	11.85
2,7-Dimethylquinoline	-	11.85
7,8-Benzoquinoline	650	24.86
Acridine	655	25.43
Carbazoles		
Tetrahydrocarbazole	617	25.42
9-Methylcarbazole	-	26.39
Carbazole	671	26.97
9-Ethylcarbazole	-	27.87
9-Phenylcarbazole	-	63.52
Miscellaneous		
Propylamine	118	1.71
1-Phenylpiperazine	547	14,21
Diphenylamine	576	17.51
Dibenzylamine	~600	21.61
4,4'-Trimethylene-bis (p-methylpiperdine)	~610	25.9
Hexadecylamine	626	27.88
Phenoxazine	-	27.93
Phenothiazine	~700	41.53
Octadecylamine	-	44.45
N-Phenyl-1-naphthaylamine	~700	52.32

pyrrole   indole   carbazole

aniline   pyridine   quninoline

3.4 Quantitation of Nitrogen Compounds

In order to avoid any error due to drift in the detector response, the use of an internal standard is recommended for accurate quantitation of nitrogen compounds.[32] Because of the NCD's equimolar response for the different classes of nitrogen compounds (see Reference 32 for details), the absolute nitrogen concentration of each compound can be obtained by using the peak areas and the internal standard concentration (see Section 2.4 for details). 4-Decylaniline has been used as an internal standard for the nitrogen speciation work reported in Reference 32. The NCD can also be used for determination of total nitrogen by utilizing the total peak area from the NCD chromatogram. Details of total nitrogen determination are the same as that of total sulfur determination and are discussed in Section 2.4 of this chapter and also in Reference 32.

4. FUTURE WORK

In order to be cost effective, the two routinely used sulfur speciation methods for the gasoline and diesel streams should be combined into a single method. This could be achieved by using a thick film long GC column packed with the methyl silicone stationary phase.

The sulfur compounds in the gasoline range streams are almost completely identifiable, whereas the sulfur compounds in the diesel range streams are only partially identifiable. The C_2-benzothiophenes and the higher homologues are hard to separate to a level where all of the sulfur compounds could be identified and hence speciated. The same holds true for the substituted carbazoles. The development of new GC columns is warranted.

Since the SCD and NCD work on similar chemiluminescence principles, it will be cost effective to combine the two techniques into one. It should also have the capability to be coupled (without any column effluent split) with an FID if so desired. With this kind of arrangement, the FID, sulfur, and nitrogen traces, with almost no changes in the peak retention times, could be obtained from a single injection with minimum sample discrimination. Also, the three traces obtained from a single injection could be used to obtain the compositional distributions of the sulfur and nitrogen compounds with respect to their boiling points.

5. REFERENCES

1. Brody, S. S.; Chaney, J. E. *J. Gas Chromatogr.* **1966**, 4, 42.
2. Dressler, M. in *Selective Gas Chromatographic Detectors*, Elsevier: Amsterdam, 1986. Chapters 7-9.
3. Farwell, S. O.; Barinaga, C. J. *J. Chromatogr. Sci.* **1986**, 24, 483.
4. Drushel, H. V. *J. Chromatogr. Sci.* **1983**, 21, 375.
5. Kim, B. H.; Kim, H. Y.; Kim, T. S.; Park, D. H. *Fuel Processing Technology* **1995**, 43, 87.

6. McGaughey, J. F.; Gangwal, S. K. *Anal. Chem.* **1980**, 52, 2079.
7. Pearson, C. D.; Hines, W. J. *Anal. Chem.* **1977**, 49, 123.
8. Berthou, F.; Dreano, Y.; Sandra, P. *HRC CC, J. High Resolut. Chromatogr. Chromatogr. Commun.* **1984**, 7, 679.
9. Tzanani, N.; Amirav, A. *Anal. Chem.* **1995**, 67, 167 and the references cited therein
10. Uden, P. C.; Young, Y.; Wang, T.; Cheng, Z. *J. Chromatogr.* **1989**, 468, 319.
11. Zakaria, M.; Gonnord, M. F.; Guiochon, G. J. *J. Chromatogr.* **1983**, 271, 127.
12. Quimby, B. D.; Glarrocco, V.; Sullivan, J. J. *J High Resolut. Chromatogr.* **1992**, 15, 705 and the references cited therein.
13. McCormack, A. J.; Sudhakar, C.; Levine, S. A.; Patel, M. S.; McCann, J. M. *LC-GC* **1994**. 12, 30.
14. Hall, R. C. *J. Chromatogr. Sc.* **1974**, 12, 152.
15. Johnson, J. E.; Lovelock, J. E. *Anal. Chem.* **1983**, 60, 812.
16. Nelson, J. K.; Getty, R. H.; Birks, J. W. *Anal. Chem.* **1985**, 55, 1767.
17. Nyarady, S. A.; Barkley, R. M.; Sievers, R. E. *Anal. Chem.***1985**, 57, 2074.
18. Gaffney, J. S.; Spandau, D. J.; Kelly, T. J.; Tanner, R. *J. Chromatogr.* **1985**, 347, 121.
19. Turnipseed, A. A.; Birks, J. W. in *Chemiluminescence and Photochemical Reaction Detection in Chromatography*, J. W. Birks, Ed., VCH Publishers: New York, 1989.
20. Benner, R. L.; Stedman, D. H. *Anal. Chem.* **1989**, 61, 1268.
21. Chawla, B.; Di Sanzo, F. P. *J. Chromatogr.* **1992**, 589, 271 and the references cited therein.
22. Di Sanzo, F. P.; Bray, W.; Chawla, B. *J. High Resolut. Chromatogr.* **1997**, 17, 255 and the refernces cited therein.
23. Shearer, R. L. *Anal. Chem.* **1992**, 64, 2192.
24. Zacharia, M. R.; Smith, O. I. *Combust. Flames* **1987**, 69, 125.
25. Muller, C. H.; Schofield, K.; Steinberg, M.; Brolda, H. P. *Int. Symp. Combust.* **1979**, 17, 867.
26. Halstead, C. J.; Thrush, B. A. *Proc. R. Soc. London* **1966**, 295, 380.
27. Kenner, R. D.; Ogryzlo, E. A. in *Chemiluminescence and Bioluminescence*, J. G. Burr, Ed., Marcel Dekker: New York, 1985;,pp. 139.
28. Glinski, R. J.; Sedarski, J. A.; Dixon, D. A. *J. Amer. Chem. Soc.* **1982**, 104, 1126.
29. Shearer, R. L.; O'Neil, D. L.; Rios, R.; Baker, M. D. *J. Chromatogr. Sci.* **1990**, 28, 24.
30. Hutte, R. S.; Johansen, N. G.; Legier, M. F. *J. High Resolut. Chromatogr.* **1990**, 13, 421.
31. Schade, T.; Roberz, B.; Andersson, J. T. *Polycyclic Aromatic Compounds* **2002** (in press)
32. Chawla, B. *J. Chromatogr. Sci.* **1997**, 35, 97.
33. Li, M.; Larten, S. R.; Stoddart, D. *Anal. Chem.* **1992**, 64, 1337.
34. Dorbon, M.; Bernasconi, C. *Fuel* **1989**, 68, 1067.
35. Drushel, H. V. *J. Chromatogr. Sci.* **1983**, 21, 375.
36. Xie-qing, W.; Xin-fen, H.; Cheng-min, R.; Qing-en, Z. *Proc. Sino-West Ger. Symp. Chromatogr.* **1983**, 52WMAL, 275.
37. Di Sanzo, F. P. *J. HRC & CC* **1981**, 4, 649.
38. Dressler, M., *Selective Gas Chromatography Detectors*, J. Chromatogr. Library, 36, Elsevier: New York, 1986.
39. Onuska, F. I.; Karasek, F. W. *Open Tubular Column Gas Chromatography in Environment Sciences*, Plenum: New York, 1984.
40. Britten, A. J. *Research & Development* **1989**, 31(9), 77.
41. Parks, R. E.; Marietta, R. L. U. S. Patent 4018 562, 1975.
42. Fujinari, E. M.; Courthaudon, L. O. *J. Chromatog.* **1992**, 592, 209.
43. Young, R. J.; Fujinari, E. M. *Am. Lab.* **1994**, 26(15), 38.
44. Courthaudon, L. O.; Fujinari, E. M. *LC-GC* **1991**, 9(10), 732.
45. Tourres, D.; Langellier, C.; Leborgne, D. *Analusis Megazine* **1995**, 23, 29.

Chapter 4

MOLECULAR CHARACTERIZATION OF PETROLEUM AND ITS FRACTIONS BY MASS SPECTROMETRY

Aaron Mendez and Jenny Bruzual
Analytical Chemistry Department
PDVSA Intevep
P.O. Box 76343 Caracas 1070-A Venezuela

"I feel sure that there are many problems in chemistry which could be solved with far greater ease by this than by any other method. The method is surprisingly sensitive- more so even than that of spectrum analysis- requires an infinitesimal amount of material and does not require this to be specially purified...".
J.J. Thomson. Rays of positive electricity and their application to chemical analysis. Longmans London, 1913.

1. INTRODUCTION

Energy consumption is foreseen to have a sustained demand in industrialized countries and an increasing demand in the developing world. The industrialized countries are engaged in extensive programs to improve the fuel formulations to gain efficiency, especially in transportation and energy generation.

Despite of many energy sources, such as solar, hydraulic, thermal, electrochemical and fuel cells, petroleum is still the primary source of energy in the industrialized world. However the oil industry has to face many challenges of economical and political nature.

The economical challenges involve both producing (upstream) and refining (downstream) with efficiency to ensure the profitability to survive in highly competitive global markets. The characteristic low profit refining margins calls for continuous improvements to current practices through basic and applied research.

Political challenges encompass the concerns for the environment as well as the government regulations on both refining processes and generated products. Another political challenge that must be considered is the continuous consumers' demand for fuels to be safer, less polluting, less expensive and higher performance. As in the economical challenges, the political ones can only be met by improving the processes and reformulating the products through systematic and considerable investment in research and development programs.

These aspects make the oil industry very dynamic in the sense that it has to be alert to incorporate and generate cost-effective technologies partly through its own research and mainly through partnerships with government and academia. These will bring the refinery the flexibility to produce wider range of fuels and materials from the processing of crude oils of continuously lower qualities and to process non-conventional feedstock that would render products of required characteristics while at the same time to reduce the amounts of byproducts outside specifications, for compliance with environmental regulations. Therefore it is expected a variety of non-conventional refining processes will appear, based on totally different principles such as biotechnology, and new catalysts to produce new generation fuels, lubricant oils and petrochemicals.

All these reflect that not only state-of-the-art technologies are essential but the demand for better methods of analysis in terms of precision, specificity and sensitivity are of paramount importance to support the oil industry core business. In essence it should be emphasized is the fact that petroleum compositional analysis is far from being a dead issue.

Mass spectrometry (MS) is an analytical technique whose outburst in to the scientific and industrial scenes due to its early applications to quantitative analysis of petroleum fractions in the early fifties. It would definitely play a decisive role in the forthcoming developments of modern petroleum products.

2.   LOW RESOLUTION/HIGH IONIZING VOLTAGE MASS SPECTROMETRIC ANALYSIS

Analysis of oil samples by mass spectrometry has been taking place for more than five decades. This application in fact is considered to be the first analytical application rapidly growing into other industries, such as environmental, agriculture, forensic, medicine and fundamental scientific research. Mass spectrometry is based on the ionization of neutral molecules with energy by various means. The most common ionization technique is the use of high-energy (50-70 eV) electron beams. Ionization of molecules occurs mainly in gas phase through thermodynamic processes that causes the molecules to loss, valence electrons. When this process takes place, molecules yield a particular pattern of molecular and fragment ions characteristic to the molecules of interest. The high-energy electron-impact ionization (EI) mass spectra are highly reproducible by keeping the same experimental conditions. This phenomenon was first applied in a petroleum laboratory by O. L. Roberts at the Atlantic Refining Company in October 1942.[1] Several researchers also applied the technique to analyze light hydrocarbon samples and low boiling organic compounds. However the first standard procedure on gasoline and naphtha fractions was introduced by Brown,[2] providing a quantitative report by grouping the constituents of complex mixtures in families or compound types by summation and normalization of characteristic ions.

The various methods reported, covering the whole distillation range, are generally accurate within ± 5% relative standard deviation. They are based on the grouping of the characteristic molecular and fragment ions presented in Table 1. Because the masses of molecular ions and their corresponding fragment ions are identical for olefins and naphthenes of the same carbon number, the olefinic content in the sample should not exceed the maximum concentration recommended by the particular method being used.

Table 1 Molecular and fragment ion masses of various hydrocarbon types

Alkanes	43 = 43 + 57 + 71 + 85 + 99 + ...
Noncondensed cycloalkanes	55 = 55 +69 + 83 + 97 + 98 + 111 + 112 + ...
Condensed dicycloalkanes	67 = 67 + 81 + 95 + 96 + 109 + 110 + 123 + 124 +...
Condensed tricycloalkanes	93 = 93 + 107 + 121 + 135 + 149 + ...
	+ 94 + 108 + 122 + 136 + 150 + 164 + ...
Alkyl benzenes	77 = 77 + 91 + 105 + 119 + 133 + ...
Indanes – Tetralins	104 = 104 + 117 + 118 + 131 + 132 + 145 + ...
Dinaphthenebenzenes	157 = 157 + 158 + 171 + 172 + 185 + 186 + ...
Alkyl naphthalenes	127 = 127 + 128 + 141 + 142 + 155 + 156 + 169 + ...
Fluorenes	165 = 166 + 180 + 194 +...+ 165 + 179 + 193 +...
Acenaphthenes, Dibenzofurans	153 = 154 + 168 + 182 + ...+ 167 + 181 + 195 +...
Phenanthrenes	177 = 178 + 192 + 206 +...+ 177 + 191 + 205 + ...
Naphthenephenanthrenes	217 = 217 + 218 + 231 + 232 + 245 + 246 + ...
Chrysenes	241 = 241 + 255 + 269 + ...+ 242 + 256 + 268 + ...
Biphenyls	153 = 154 + 168 + 182+...+ 153 + 167 + 181+...
Dibenzanthracenes	277 = 278 + 292 + 306 +...+ 277 + 291 + 305 +...
Benzothiophenes	133 = 133 + 147 + 161 +...+ 134 + 148 + 162 +...
Naphthobenzothiophenes	247 = 248 + 262 + 276 +...+ 247 + 261 + 275 +...
Dibenzothiophenes	183 = 184 + 198 + 212 +...+ 183 + 197 + 211 +...

This procedure works well on total samples only for gasoline range hydrocarbons. For high boiling fractions the possibility of having mass overlap among different compound types become greater that would require preparative liquid chromatographic separation of saturate and aromatic hydrocarbons prior to mass analysis. For example, paraffins overlap with alkyl naphthalenes above C_9 and with dibenzothiophenes above C_{13}. Naphthenes overlap with biphenyls above 154 daltons (Da). Dinaphthenes overlap with fluorenes and dibenzanthracenes above 278 Da. Tricycloparaffins overlap with phenanthrenes and naphthobenzothiophenes above mass 178. From the ions shown in Table 1 a series of inverse sensitivity matrices are constructed for quantitative analysis after corrections for isotopic contribution in the mass spectra. Based on this methodology, various ASTM standard methods of analysis were developed: D-2650 for gases, D-2789 for gasolines, D-2425 for kerosenes, D-2786 for gas oil saturates and D-3239 for gas oil aromatics.[3,4]

Procedures in a similar approach are published for the different boiling point ranges by the oil industry.[5-8] M. J. O'Neal [5] set the basis of calculating the hydrocarbon type distribution for middle distillates up to C_{40}. In this approach an All Glass Heated Inlet System (AGHIS) was used for the first time to ensure constant vapor pressure for reproducible results. Efforts were on the identification of sulfur compounds, especially for the thiophenic types, and condensed aromatics. At the same time the expansion of the number of naphthenic condensed rings in the calculations and the increase of boiling range extend the methods to lubricating oil fractions.

The applications of mass spectrometry to other than virgin hydrocarbon samples came immediately.[9] Gordon et al.[9] applied low ionizing voltage EIMS instead of the conventional 70 eV EIMS and ultraviolet spectrophotometry to cracked gas oils to resolve ambiguities in the determination of aromatic compounds. At 10 – 20 eV ionizing energy only aromatic molecular ions are produced with practically no fragmentation. In this circumstance information based on fragment ions are no longer available. Since spectra were recorded at low resolution the measured mass values correspond to nominal masses. As a result the concept of Z-series to represent hydrogen deficiency of the compound series was developed. The report is given according to the general formula:

$$C_nH_{2n+z} \qquad\qquad (1)$$

Pioneering work on the molecular characterization of catalytic stocks was performed at low resolution, low- and high-voltage mass spectrometry.[10,11] The combination of MS with other techniques such as nuclear magnetic resonance spectroscopy (NMR), ultraviolet (UV) and infrared (IR) was established as a basic approach to gain information on the composition of very complex hydrocarbon mixtures of different nature.

Efforts continued on expanding the application of mass spectrometry towards higher boiling ranges. Special attention was paid to sulfur and nitrogen compounds. Lumpkin[8] and Hood and O'Neal[12] presented a method for middle distillates based on ASTM D-2425. In their works matrices were constructed for lineal paraffins and up to six naphthene condensed rings. On the same track Snyder et al.[13], developed a method for the direct analysis of heavy reformates and hydrodealkylated products boiling in the kerosene range. These authors reported nine compound types: alkanes, noncondensed and condensed cycloalkanes, alkyl benzenes, indanes, tetralins and alkyl naphthalenes, aromatic hydrocarbon of the –14 and –16 Z series as well as benzothiophenes. One very interesting feature of Snyder's method is that even at the low mass resolution and the high ionizing voltage, it did not require the prior separation of the saturates and aromatic fractions.

Following the expansion of mass spectrometry, it was rapidly recognized the convenience of avoiding preparative liquid chromatographic separations. This is particularly meaningful as it reduces the time of analysis and avoids sources of errors. Chromatographic separations lower the reproducibility and repeatability of final results due to solvent stripping especially in the low boiling range, i.e. diesels, jet fuels and kerosenes. Gallegos et al. [14] applied high resolution MS to analyze whole samples without chromatographic separation, which will be discussed later. Robinson [15] developed a low resolution procedure applicable to a wide boiling range, 200 – 1100°F. A baseline technique artificially separates the aromatics from saturates as well as the contribution of side chain substitution of aromatics to the saturate signal. The procedure determines four saturate and twelve aromatic hydrocarbon types, including three sulfur compound types as reported in ASTM D-3239.

The vast majority of contributions to group type analysis by mass spectrometry was developed in the magnetic mass spectrometers of the Consolidated Electrodynamics Co. CEC 100 series. With the advent of new instrumentation discrepancies in the correlation of Robinson procedure with the ASTM methods started to appear. Piemonti and Hazos [16] introduced a modified Robinson procedure to overcome this deficiency. The base line separation was corrected using a Kratos MS- 25Q mass spectrometer that allowed reasonable correlation with the standard ASTM procedures.

This concern was brought up by Ashe [17], who pointed out on the scarce demand of expensive double focusing sectors instrument for routine refinery operations. As a replacement for the traditional magnetic sector instruments he attached a quadrupole mass filter to a conventional All Glass Heated Inlet System (AGHIS). After some experimental considerations derived from the different philosophy in the construction of bench top quadrupole mass analyzers it was possible to reach very reproducible results and a good correlation with historical data from sector instruments.

3. HIGH RESOLUTION MASS SPECTROMETRY

Mass spectrometry has been an important analytical tool for analyzing petroleum products since the 1950s. The spread of the technique is mainly due to its success in providing detailed information on the composition of very complex mixtures and its capability in giving qualitative and quantitative information of the components. It requires only very small amount of sample and can be applied to organic and inorganic substances. However, there are some disadvantages such as the complexity of the spectra, the difficulty associated with isomer identification, and the requirement of vaporization of samples which limits its application to samples with high boiling points and poor thermal stability. The loss of sensitivity as slits have to be significantly narrowed in order to achieve the ultimate mass resolution also limits its range of quantitation.

The application of mass spectrometry in the petroleum industry started with the analysis of gas and low boiling liquid samples. Significant breakthroughs in its evolution include the introduction of heated inlet systems for the analysis of liquid and volatile solid compounds, low voltage EI techniques, high resolution instrumentation capable of measuring accurate masses and separating multiplets of the same nominal mass, and computerized data acquisition and handling systems. However, the composition analysis of hydrocarbons and aromatics compounds containing one or more of sulfur, nitrogen, and oxygen in distillates has become very important in refining of crude oils and in the storage and use of refined products. Moreover, sulfur- and nitrogen-containing species may poison catalysts and must be removed prior to certain oil refinery processes.

The use of high-resolution mass spectrometry eliminates most of the interferences between saturate and aromatic hydrocarbons and, further, remove some of the interferences of sulfur compounds and overlapping hydrocarbons. Resolution and resolving power (RP) are of vital importance for high-resolution mass spectrometric methods that determine elemental composition in terms of homologs series of compounds. Molecular ions containing one or more less-abundant isotopes of its constituent elements (e.g. ^{13}C, ^{34}S) can contribute significantly to complicate the mass spectra of samples of interest. In addition, low-voltage electron ionization can produce non-negligible quantities of fragment ions. Different combination of atoms can be nearly the same mass, e.g., $^{12}C_3$ vs. $^{32}SH_4$, ^{13}C vs. CH, CH_2 vs. ^{14}N, and CH_4 vs. O. Thus the presence of these overlapping ions can complicate the identification and quantitation of various compound types present in a given sample, such as aromatic and polar crude oil fractions boiling up to 1050°F. Consequently, resolution of isobaric ions in these distillates requires mass resolution considerably in excess of 10,000. The resolving power needed for the separation of the most common doublets found in hydrocarbon samples is illustrated in Table 2.

Table 2. Resolution required for Mass Doublets in Petroleum and Synthetic Samples

Doublet	ΔMass, u	Required Mass Resolution @ m/z 400
H_{16}/O	0.1303	3,100
O_2H_{16}/C_4	0.1150	3,500
H_{12}/C	0.0939	4,300
C_2H_8/S	0.0905	4,400

Doublet	ΔMass, u	Required Mass Resolution @ m/z 400
C_2H_8/O_2	0.0728	5,500
N_2H_8/C_3	0.0687	5,800
OH_8/C	0.0575	7,000
CH_4O/S	0.0541	7,400
N_2H_4/O_2	0.0476	8,400
O_2H_8/CN_2	0.0463	8,600
CH_4/O	0.0364	11,000
C_2H_4/N_2	0.0252	15,900
O_2H_4/C_3	0.0211	19,000
O_2/S	0.0178	22,500
$N_2H/^{13}CO$	0.0157	25,500
C_4/O_3	0.0153	26,100
CH_2/N	0.0126	31,700
N_2/CO	0.0112	35,700
H_4O_3/C_2N_2	0.0099	40,400
$^{13}CH/N$	0.0081	49,400
SH_4/C_3	0.0034	117,600

In the high-resolution mass spectrometric methods, composition is determined in terms of amounts of homologs of various hydrocarbon and heteroatom-containing compound types. When the samples become more complex the grouping in hydrocarbon types is not straightforward, therefore results are expressed in terms of hydrogen deficiency Z- series. For high-resolution mass spectrometers, precise standardization of operating conditions is even more important for quantitative measurements than for low-resolution mass spectrometers. Generally, the higher the resolution, the weaker the signal and the longer the data acquisition time is needed. Special mass standards that do not interfere with sample ions, mostly chlorinated hydrocarbon petroleum fractions, are employed for mass calibration. The chloro compounds are preferred because of their higher sensitivities at the low ionization voltage compared to those of the fluorinated standards common in regular EI-MS.

A technique was developed by McMurray et al.[18] to obtain high resolution mass spectra from fast magnetic scans of gas chromatographic effluents.

The use of low- and high-resolution mass spectrometry for quantitative analysis based upon group type calculation of saturate and aromatic fractions has been reported.[19,20] Table 3 lists a number of series that can be identified in one high resolution mass scan. As indicated in this table, it is possible to obtain 27 different hydrocarbon series and a number of single sulfur-, oxygen-containing series. So in one high-resolution scan, it's possible to determine over 200 series and to have quantitative measurement of not only the Z series but also carbon number in the series. In this case the range can be expanded to gasoil fractions boiling up to 1200°F and eventually applicable to crudes oils, coal liquids, shale oils and refinery streams. These methods would be facilitated by the separation of the hydrocarbon mixture into saturate and aromatic fractions. This fact affects the efficiency of the technique in terms of turn-around times.

Table 3. Common compound-types in petroleum.

Class	Compound Types Range in General Formulas	Number of Types
Hydrocarbons	C_nH_{2n} to C_nH_{2n-52}	27
Sulfur Compounds	$C_nH_{2n+2}S$ to $C_nH_{2n-42}S$	23
	$C_nH_{2n-12}S_2$ to $C_nH_{2n-30}S_2$	11
Sulfur-Oxygen Compounds	$C_nH_{2n-10}SO$ to $C_nH_{2n-30}SO$	11
	$C_nH_{2n-10}SO_2$ to $C_nH_{2n-18}SO_2$	5
	$C_nH_{2n-10}SO_3$ to $C_nH_{2n-20}SO_3$	6
Nitrogen Compounds	$C_nH_{2n-3}N$ to $C_nH_{2n-51}N$	25
	$C_nH_{2n-12}N_2$ to $C_nH_{2n-26}N_2$	8
	$C_nH_{2n-5}N_3$ to $C_nH_{2n-11}N_3$	4
Oxygen Compounds	$C_nH_{2n-2}O$ to $C_nH_{2n-50}O$	25
	$C_nH_{2n-6}O_2$ to $C_nH_{2n-46}O_2$	19
	$C_nH_{2n-6}O_3$ to $C_nH_{2n-32}O_3$	14
	$C_nH_{2n-10}O_4$ to $C_nH_{2n-20}O_4$	7
Nitrogen-Oxygen Compounds	$C_nH_{2n-7}NO$ to $C_nH_{2n-41}NO$	18
		203

Gallegos et al.[21] reported the first multicomponent group-type analysis using high resolution mass spectrometry to analyze high boiling petroleum fractions without silica gel separation. Their method made possible to get quantitative data for seven saturate hydrocarbon compound classes from zero to six naphthene rings, nine aromatic compound classes, and three aromatic sulfur compound types covering the range of 500 to 950°F. However, this method does not account for all of the sulfur compounds that are present. Nevertheless, the agreement was very good when the weight per cent sulfur by mass spectrometry was compared with that determined by x-ray fluorescence.

Joly [22] used a 30 x 30 matrix at 10,000 mass resolution to obtain quantitative data for three saturate compound classes, thirteen aromatics and fourteen sulfur compound classes.

Early applications of high-resolution Fourier-transform ion cyclotron resonance mass spectrometric (FT/ICR/MS) measurements for hydrocarbon characterization were described by Hsu and coworkers.[23] They demonstrated the potential use of FT/ICR/MS in the analysis of complex hydrocarbon mixtures with resolving power beyond sector instruments. The high stability of accurate mass measurement even under different ion source conditions eliminated the need of mass reference standards. However, ultrahigh-resolution mass measurement was limited to rather narrow mass ranges to avoid space-charge effects. Nevertheless, these narrow mass ranges can be acquired and reconstructed into a wide mass range spectrum. For the determination of compound type distribution, it required the development of software capable of linking high-resolution quantitation results of all individual mass segments.

Marshall et al.[24] have published a report in which FT/ICR/MS was used to provide the best available mass resolution and mass accuracy for molecular ion type analysis of aromatic neutral fraction of gas oils. Their results resolved most common doublets found in hydrocarbon samples, such as C_3/SH_4, $^{13}C/CH$, and $^{13}CH_3S/C_4$. Additional application of this technique was reported for the compositional analysis of processed and unprocessed diesel fuels.[25]   By comparing the relative abundances of mass-resolved sulfur-containing species before and after hydrotreating, they found complete removal of the sulfur-containing species except for dimethyl-dibenzothiophene and higher alkyl-substituted dibenzothiophenes.

4.   GAS CHROMATOGRAPHY-MASS SPECTROMETRY (GC-MS)

Instrumentation in mass spectrometry has changed dramatically, partly pressed by the need to reduce costs. Additionally, mass spectrometry as it was before, required highly qualified and experienced personnel to perform the analysis.

With the advent of GC-MS, its application to the oil industry and particularly for group type analysis was encouraged by the ability of the gas chromatograph to calibrate and quantify the separated peaks followed by the unambiguous mass spectrometric identification based on the fragmentation pattern library search.[18,26] The distribution of aromatics by carbon numbers and boiling points can be enhanced by carrying out low voltage EI experiments.[27]   With the miniaturization and automation of GC-MS apparatus and the reduction of costs, this instrumentation became highly disseminated in the scientific and industrial communities. The application of GC-MS in the hydrocarbon analysis passed through a explosive period very much similar to mass spectrometry in the 1950's.[28-30]

Dzidic et al.[29] and Wadsworth and Villalanti [30] used a Townsend discharge source and nitric oxide (NO) as a chemical reagent to ionize hydrocarbons mixtures. The way NO^+ reacts with the different hydrocarbon types allows to perform PNA (paraffins, naphthenes and aromatics) analysis and sulfur compounds. In this reaction aliphatic hydrocarbons will give only $(M-1)^+$ ions whereas aromatic hydrocarbons react to produce exclusively $M^{+\cdot}$ ions which reflects the possibility to analyze total samples avoiding the preparative liquid chromatography separation step. The method also determines the distribution of hydrocarbons according to carbon number and hydrogen deficiency Z-series and can be applied to the whole distillation range up to VGO of approximately 1000°F.

Combining gas chromatography with mass spectrometry also enables the PIONA (n-paraffins, iso-paraffins, naphthenes, and aromatics) analysis of gasolines to be performed in a very systematic way.[31] The procedure takes advantage of the resolving power of the chromatograph and the ability of the mass spectrometer to detect eluting compound at very low levels and to identify them unambiguously. The special software by means of pure compound response factors allows for the resolution of overlapping compounds.

Several ionization modes have been used apart from electron impact at 70 eV. We have already described low voltage eV and NO chemical ionization. Malhotra et al.[32] used field ionization (FI), which is a "soft" ionization technique, especially amenable to volatile compounds, yielding mainly molecular ions with little or no fragmentation. Using a special ion source design known as a volcano source on a HP 5971A mass selective detector, it was possible to produce a PNA and benzothiophenes report on several fuel samples such as gasolines, jet fuels and an Arabian sweet crude oil. The gas chromatograph in this case provided the separation for the mass overlapping compounds (see Table 1). The quantitation was performed in a very straightforward way: peak integration normalized by applying compound specific sensitivity factors.

As it was noted by several authors, there is a need for having a GC-MS system with an universal soft ionization mode.[17] This point was expanded by Schoemakers et al.[33] who also presented an interesting comparison of two dimensional GC (GCxGC) and GC-MS. In search of this ideal soft ionization technique, important conclusions are made on the still unreachable potential of the technique. Lately, FI-TOF (time-of-flight) mass spectrometers combined with gas and liquid chromatography become commercially available.[34] A very useful feature of this type of instrumentation is that TOF allows for the determination of accurate mass at low resolution. Normally high resolving power is reached on sector instruments at the expenses of sensitivity. Therefore, keeping the full sensitivity of the technique and the tremendous separation capabilities of chromatography in general will certainly increase enormously the possibility of measuring low concentrations of components and determining compound type distribution of hydrocarbons.

A very novel application of GC-MS that expands the use of the technique well beyond the separation capabilities of a chromatographic column has been reported by several authors.[35-39] Applying spectra deconvolution techniques,[27] the relative concentration of four saturate and four aromatic compound types were reported on hydroconversion products of deasphalted vacuum residua by avoiding the need for separation. Ashe [36] and Roussis [37,38] patented and developed methods to predict physical and chemical properties of crude oils by means of simulated distillation mass spectrometry. The crude oils distillation profiles are converted to both weight and volume true boiling point curves allowing the characterization of fractions of interest without the need for obtaining the physical cuts.

Mendez et al.[39] similarly obtained the boiling point profile of light crude oils samples and applied an algorithm [16] based on C. J. Robinson Approach,[15] to get a quantitative report for both the total distillable fraction of the crude oil and the lubricant fractions of interest. Good correlations were obtained when the fractions of interest calculated from the total ion chromatogram (TIC) of whole crude oil compared with several lube-oil basestocks. Table 3 show a quantitative group type report of three different lubricant basestocks and Table 4 represent the same fractions taken out from the whole crude chromatogram applying the algorithm in the same intervals for every one of the fractions. Figure 1 represents a typical paraffinic whole crude oil chromatogram using a high-temperature metallic capillary GC column. In this figure, the carbon number and boiling point are calibrated against retention time.

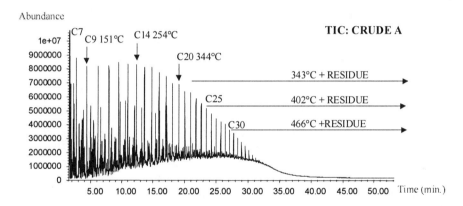

Figure 1. Fingerprint representation of whole crude oil

Table 3. Hydrocarbon type analysis of Lub-oil basestocks

	SPO[a]	LMO[b]	MMO[c]
Total Saturates, %p	*75.3*	*69.5*	*53.6*
Paraffins	39.7	27.2	17.4
Monocycloparaffins	13.1	14.2	13.4
Dicycloparaffins	7.2	12.6	10.3
Tricycloparaffins+	15.4	15.4	12.6
Total Aromatics, %p	*24.7*	*30.5*	*46.4*
Monoaromatics	**10.5**	**12.9**	**21.0**
Benzenes	3.3	4.0	7.0
Naphthenebenzenes	3.4	4.0	6.5
Dinaphthenebenzenes	3.7	4.9	7.4
Diaromatics	**5.2**	**3.5**	**6.9**
Naphthalenes	0.0	0.0	0.0
Acenaphthenes, Dibenzofurans	2.5	0.8	2.0
Fluorenes	2.7	2.7	4.9

	SPO[a]	LMO[b]	MMO[c]
Triaromatics	1.9	2.2	3.0
Phenanthrenes	1.1	1.0	1.6
Naphthenophenanthrenes	0.9	1.2	1.4
Tetraaromatics	2.1	4.3	3.4
Pyrenes	1.6	2.7	1.8
Chrysenes	0.5	1.6	1.6
Pentaaromatics	0.6	0.5	1.3
Perylenes	0.6	0.4	1.1
Dibenzanthracenes	0.0	0.1	0.2
Thiophenes	4.2	6.8	6.8
Benzothiophenes	2.1	2.3	2.4
Dibenzothiophenes	2.1	2.9	3.2
Naphthobenzothiophenes	0.0	1.6	1.2
Unidentified Aromatics	0.0	0.3	4.1
Class II	0.0	0.0	1.1
Class III	0.0	0.0	0.4
Class IV	0.0	0.3	2.2
Class V	0.0	0.0	0.0
Class VI	0.0	0.0	0.1
Class VII	0.0	0.0	0.3

a. Spindle Oil
b. Light Machine Oil
c. Medium Machine Oil

Table 4. Hydrocarbon type analysis from whole crude

	SPO[a] Time interval 21-33 min.	LMO[b] Time interval 24-42 min.	MMO[c] Time interval 28-54 min.
Total Saturates, %p	*76.0*	*69.2*	*54.2*
Paraffins	42.7	33.5	20.2
Monocycloparaffins	14.3	13.8	12.2
Dicycloparaffins	10.4	10.9	9.4
Tricycloparaffins+	8.6	11.0	12.3
Total Aromatics, %p	*24.0*	*30.8*	*45.8*
Monoaromatics	10.8	11.9	20.7
Benzenes	3.8	3.8	7.1
Naphthenebenzenes	3.2	3.7	6.3
Dinaphthenebenzenes	3.9	4.4	7.2
Diaromatics	4.0	5.4	7.6
Naphthalenes	0.0	0.0	0.0
Acenaphthenes, Dibenzofurans	1.4	1.9	1.8
Fluorenes	2.6	3.5	5.7
Triaromatics	2.7	3.8	1.7
Phenanthrenes	2.0	2.4	1.2
Naphthenophenanthrenes	0.7	1.4	0.6
Tetraaromatics	1.7	3.1	4.7
Pyrenes	1.4	2.2	2.7
Chrysenes	0.3	1.0	1.9
Pentaaromatics	0.0	0.3	0.8
Perylenes	0.0	0.2	0.5

	SPO[a] Time interval 21-33 min.	LMO[b] Time interval 24-42 min.	MMO[c] Time interval 28-54 min.
Dibenzanthracenes	0.0	0.1	0.3
Thiophenes	**4.7**	**6.2**	**7.3**
Benzothiophenes	2.4	2.5	2.6
Dibenzothiophenes	2.1	3.2	3.6
Naphthobenzothiophenes	0.2	0.5	1.1
Unidentified Aromatics	**0.0**	**0.1**	**3.1**
Class II	0.0	0.1	0.3
Class III	0.0	0.0	0.3
Class IV	0.0	0.0	2.3
Class V	0.0	0.0	0.0
Class VI	0.0	0.0	0.0
Class VII	0.0	0.0	0.1

a. Spindle Oil
b. Light Machine Oil
c. Medium Machine Oil

The boiling points and carbon numbers were performed by injecting on the same instrumental conditions a standard normal paraffin mixture. All of these contributions not only expand the use of a GC-MS system but provide for a method to rapidly screen a crude oil to asses the quality and yield of distillable fractions of interest

5. LIQUID CHROMATOGRAPHY-MASS SPECTROMETRY (LC-MS)

Mixtures of hydrocarbons, such as those resulting from oil-refining processes, are extremely complex.[40,41] The properties of petroleum and its products are closely related to their chemical compositions. This information is relevant for decision making. To analyze these complex mixtures, coupling chromatography and mass spectrometry becomes highly desirable. GC-MS has been widely used for hydrocarbon analysis of light fractions in oil industry,[42-46] but the requirement of vaporization of samples limits its application. Liquid chromatography-mass spectrometry (LC-MS) is another approach to characterize molecular structures in relation to their polarities and extends the capabilities for analyzing mixtures of high boiling, low thermostability and polar compounds.

Boduszynski combined FIMS with off-line LC for the molecular characterization of heavy petroleum.[47] The vacuum residua was characterized by Malhotra and coworkers using off-line HPLC and FIMS.[48] Nevertheless, the on-line LC-MS combination has several advantages over the separated applications of these techniques. It reduces analysis procedures as it automates solvent evaporation and minimizes its amount. Another unique characteristic of LC-MS, is that it allows for the differentiation between overlapping compound series, such as alkylaromatics from naphthenoaromatics, and aromatic hydrocarbons from sulfur compounds. This combination eliminates the high-resolution mass measurement for quantitation of these compounds. It makes possible to analyze the basic and neutral polars.[49]

Hsu et al.[50] described the application of on-line LC-MS to the study of heavy hydrocarbons. The sample was separated by HPLC into saturate, 1-ring aromatic, 2-ring aromatic, 3-ring aromatic, 4-ring aromatic and polar fractions. By incorporating low-voltage electron-impact ionization/high resolution MS with moving belt LC-MS, the differentiation of naphthenoaromatics and alkylaromatics could be made and also was possible to distinguish alkylaromatics from thiophenes.

Due to complexity of hydrocarbon mixture, such high resolution mass spectral data analyses are complicated and time consuming. This necessity is more critical when on-line liquid chromatography-mass spectrometry is used. Hsu and coworkers [51] has developed a data reduction procedure based on the Kendrick mass scale. Compounds are grouped together by multiple sorting in a homologous series according to their Kendrick mass defects (KMDs). The general formula of each group was determined from the average KMD of the whole group. As a result of this procedure, the accuracy of compound type identification was increased and a simpler computer program was developed.

The advantage of on-line LC-MS with low-voltage EI/medium-resolution MS are evident in a paper by Hsu and Qian[52]. This combined technique has been applied to study molecular transformation of high boiling petroleum fractions upon hydrotreating. The results clearly delineate the effects of the treatment in the composition of each compound class, the carbon distribution of homologous series, and in terms of compound series. The data indicated that most of hydroaromatic hydrocarbons formed were in the monoaromatic and diaromatic regions. Disulfur and oxygenated compounds were almost removed, while certain amounts of monosulfur and nitrogen compounds survive the hydrotreating.

Thermospray (TSP) was used in the hydrocarbon characterization of high boiling petroleum fractions by on-line liquid chromatography-mass spectrometry.[53] TSP is effected by gas phase ion/molecule reactions where proton transfer is the predominant ionization process. In consequence, the aromatic hydrocarbons are ionized selectively because of their high proton affinity. This technique can be used to determine molecular weight and compound type distributions of aromatic hydrocarbons.

Hsu, demonstrated that petroleum resids and asphaltenes can be characterized by using thermospray ionization technique (TSP) in on-line LC-MS instrument.[54] It is shown the compound type distribution as a function of the z-number of equivalent hydrocarbon series. The average of molecular weights of a Middle East crude oil 1120-1305°F distillate are about 850 for monoaromatics, 820 for diaromatics, 790 for triaromatics, and 730 for tetraromatics, respectively. This result indicates that in a finite boiling range compounds with higher aromaticity have lower average molecular weights.

Liang and Hsu[55] have successfully reconfigured a commercial field ionization (FI) source to adapt to a moving belt interface for on-line liquid chromatography-mass spectrometric (LC-MS) operations. This new LC-MS approach allows the characterization of saturates with maximum chromatographic resolution and minimal sample preparation. This combination produces only molecular ions for normal paraffins and naphthenes but essentially fragment ions for isoparaffins. By using the phenomenon the split between normal paraffins and isoparaffins can be directly determined, avoiding inaccurate results obtained from two different analytical techniques, i.e., GC and MS. The comparison of these results with the obtained from ASTM standard methods were favorable.

6.   FUTURE TRENDS

Since the early stages of mass spectrometry method development, mass spectroscopists are confronted with the dilemma of how to analyze high boiling range fractions and to resolve overlapping compounds. The problem is especially acute in the cases of beyond the separation capabilities of a chromatographic column and beyond the mass resolving power of the commercial instruments. As it was exposed in the previous sections, preparative liquid chromatographic separations still lack of the selectivity required for performing truly compound class separations. Moreover LC-MS interfaces for neutral and low polar compounds are not sufficiently developed. Although potentially high resolving power instruments,[56,24] capable of deconvoluting overlapping mass peaks, have been reported, their costs are excessively high as to have them in routine operations. This fact becomes pathetically strong if considerations of the low profit margins of modern refineries are taken into account.

Technical solutions of these limitations point to the development of artificial rather than instrumental ultra high mass resolution.[57] Lately due to the development of electronics, time-of-flight mass spectrometers (TOF) have revived and their sensitivity and ruggedness has been shown.[57,58] These instruments are specially amenable to exact mass measurements at low resolution.[34] This combination offers the sensitivity needed to detect compounds at very low concentrations enhanced by the capability of measuring accurately their mass so their elemental composition can be determined.

This approach converges in a more powerful effect combined with mass spectrometers of high sensitivity. The need for high sensitivity instruments derives from the fact that in very complex mixtures of high boiling range the vast majority of compound classes and individual species are in ppm or even lower concentration ranges.

Time of flight mass spectrometry is not new; however, recent technological advances in reflectron design and electronics have made possible to measure the traveling time and its corresponding mass of a particular ion very accurately. The flight time of a particular ion of mass m (most of the ions are singly charged) is given by equation 2.

$$T_m = T_{max} \times (m/m_{max})^{1/2} \qquad (2)$$

Where :

T_m is the time in μs for an ion of mass m to reach the detector

T_{max} is the time in μs of traveling of the maximum mass range, m_{max}.

A decisive development in TOFMS responsible for the high sensitivity of this instrumentation is the orthogonal acceleration minimizes the initial energy spread of ions as illustrated in Figure 2.

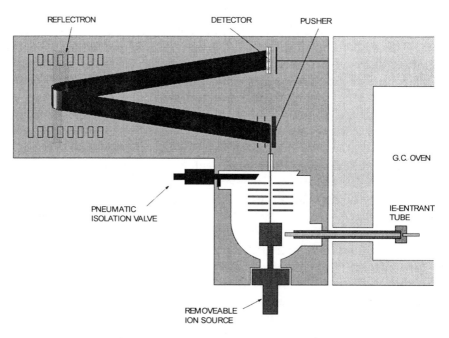

Figure 2. Schematic diagram of the orthogonal acceleration Time-of-Fight mass spectrometer (oa-TOF)

With the advent of orthogonal arrangement TOF and the capability to accurately measure the molecular masses it is obvious that soft techniques such as field ionization (FI) and field desorption (FD) are strongly needed to improve the quality of the analysis of high boiling hydrocarbons.

In FI or FD molecules are introduced into the ion source either in vapor phase (in FI) or deposited on to a carbon activated tungsten wire emitter (in FD). After applying a strong electric field (5-10 kV) which creates a high field gradient on the order of 10^7-10^8 V/cm an electron is removed from the neutral molecule by quantum mechanical tunneling effect. This means that ions are produced with very little or no excess energy, explaining the lack of fragment ions. It is obvious that for a hydrocarbon complex mixture it is highly desirable if the molecular ion masses can be measured accurately. Another excellent feature that favors FI over low ionizing voltage electron impact mass spectrometry is that FI ionizes both saturates and aromatics, while low voltage EI only ionizes aromatics and highly condensed naphthenes.

Efforts are needed however to produce rugged FI/FD sources that could be easily interfaced to gas or liquid chromatography systems for automated operations.

In this regard Hsu and Green [58] excellently combined FI, a truly soft ionization technique, literally producing fragment free spectra, with an orthogonal acceleration TOF analyzer and coupled to a capillary GC. They applied it to a middle distillate containing 0.135 mass % sulfur previously determined by x-ray fluorescence. They demonstrated the capabilities of GC-TOF-FIMS for resolving overlapping compound series of paraffins, naphthenes and dibenzothiophenes, and for obtaining isomer distributions of each series. Their results, clearly illustrated the benefits of the GC-FIMS system acting as a powerful separating and analyzing system with the features of a double focussing high resolution mass spectrometer but with increased full sensitivity of a low resolution conventional mass spectrometer.

Another recent development that has great potential of application to increase specificity and accuracy of analysis is the Metastable Atom Bombardment ion source.[59,60] By means of this technique it is possible to vary the ionizing energy and hence to enhance particular species depending upon the noble gas used as a reagent gas. The Penning ionization process is illustrated by the following reactions:

$$A^* + BC \rightarrow A + BC^+ + e^- \qquad (3)$$

$$A^* + BC \rightarrow A + B^+ + C + e^- \qquad (4)$$

In equations (3) and (4), A^* represents an energized metastable particle, usually a rare gas metastable atom and BC is the analyte organic molecule.

The metastable source produces the rare gas beam by means of a corona discharge over rare gases the source is energy tunable depending on the rare gas employed. Reaction (4) only takes place if the energy given by the interactions with the metastable atom is high enough to surpass the appearance potential of a particular fragment. In this fashion it is possible to control the ionization so as to enhance certain signals of interest. Because the metastable atom beam is produced externally typical spectra are clean free of the common interference of discharge ionization sources coming from energetic electrons, metastable species and ions formed in the process. The cleanliness of the spectra is also due to the fact that the interacting energy is controlled over a wide range depending on the rare gas. When working at low energies high enough to produce only molecular ions free of the normal contribution of pseudo molecular species common to another soft ionization technique like chemical ionization (CI). This is a good reason why CI is not amenable to accurate mass group type analysis.

This aspect brings up the possibility to analyze olefins and performed truly PIONA analysis in samples beyond the use of gas chromatography. However, the MAB sources available are designed for double focusing sector instruments. This implies that associated costs are relatively high as to consider it plausible to find large applications in the oil industry with the economical intricacies of nowadays. However, very interesting correlation to the conventional EI mass spectrometric methods has been already made which indeed shows the potential of this powerful technique.[61]

7.   REFERENCES

1.   Kurtz, S. S. Jr. "The development of hydrocarbon analysis", Prepr. Symp. - Am. Chem. Soc., Division of Petroleum Chemistry, New York, August 27-September 1, 1972.
2.   Brown, R.A. *Anal. Chem.* **1951**, 23, 430-437.
3.   Annual Book of ASTM Standards 2001, vol. 05.02.
4.   Annual Book of ASTM Standards 2001, vol. 05.03.
5.   O'Neal, M. J.; Wier, T. P. "Mass Spectrometry of Heavy Hydrocarbons",*Anal. Chem.* **1951**, 23, 830-843.
6.   Lumpkin, H. E.; Johnson, B. H. "Identification of Compound Types in a Heavy Petroleum Gas Oil", *Anal. Chem.* **1954**, 26, 1719-1722.
7.   Hastings, S. H.; Johnson, B. H.; Lumpkin, H. E. "Analysis of the Aromatic Fraction of Virgin Gas Oils by Mass Spectrometer", *Anal. Chem.* **1956**, 28, 1243-1247.
8.   Lumpkin, H. E. "Determination of Saturated Hydrocarbons in Heavy Petroleum Fractions by Mass Spectrometry", *Anal. Chem.* **1956**, 28, 1946-1948.
9.   Gordon, R. J.; Moore, R. J.; Mueller, C. E. "Aromatic Types in Heavily Cracked Gas Oil Fraction. Combined use of Ultraviolet and Mass Spectrometry", *Anal. Chem.* **1958**, 30, 1221-1224.
10. Bartz, K. W.; Aczel, T.; Lumpkin, H. E.; Stehling, F. C. "Characterization of Aromatics from Light Catalytic Cycle Stocks by Spectrometric Techniques. Compounds Types of the General Formula C_nH_{2n-16} and C_nH_{2n-18}", *Anal. Chem.* **1962**, 34, 1814-1828.
11. Aczel, T.; Bartz, K. W.; Lumpkin, H. E.; Stehling, F. C. "Characterization of Aromatics in Light Catalytic Cycle Stock by Spectrometric Techniques. Compound Types of the General Formula C_nH_{2n-12} and C_nH_{2n-14}", *Anal. Chem.* **1962**, 34, 1821-1828.
12. Hood, A.; O'Neal, M. J. *Advances in Mass Spectrometry*, AMSPA, Waldron, 1959.

13. Snyder, L. R.; Howard, H. E.; Ferguson, W. C. "Direct Mass Spectrometric Analysis of Petroleum Samples Boiling in the Kerosene Range", *Anal. Chem.* **1963,** 35, 1676-1679.
14. Gallegos, E. J.; Green, J. W.; Lindeman, L. P.; LeTourneau, R. L.; Teeter, R. M. "Petroleum group type analysis by high resolution mass spectrometry", *Anal. Chem.* **1967,** 29, 1833-1838.
15. Robinson, C. J. "Low–resolution mass spectrometric determination of aromatics and saturates in petroleum fractions", *Anal. Chem.* **1971,** 43, 1425-1434.
16. Piemonti, C.; Hazos, M. "Aromatic and Saturate Analysis by Low Resolution Mass Spectrometry", *Prepr. Symp. - Am. Chem. Soc., Division of Petroleum Chemistry,* **1992,** 37, 1521-1532. Washington D.C., August 23 - 28, 1992.
17. Ashe, T. R.; Colgrove, S. G. "Petroleum mass spectral hydrocarbon compound type analysis", *Energy Fuels* **1991,** 5, 356-360.
18. McMurray, W. J.; Green, B.,N.; Lipsky, S. R. "Fast scan high resolution mass spectrometry, operating parameters and its tandem use with gas chromatography", *Anal. Chem.* **1966,** 38, 1194-1204.
19. Aczel, T.; Lumpkin, H. E. "Detailed characterization of gas oil by high and low resolution mass spectrometry", Symp. - Advances in Analysis of Petroleum and its Products. American Chemical Society, New York, August 27 - September 1, 1972.
20. Chasey, K. L.; Aczel, T. "Polycyclic aromatic structure distribution by high-resolution mass spectrometry", *Energy Fuels* **1991,** 5, 386-394.
21. Gallegos, E. J.; Green, J. W.; Lideman, L. P.; LeTourneau, R. L.; Teeter, R. M. "Petroleum group-type analysis by high-resolution mass spectrometry", *Anal. Chem.* **1967,** 39, 1833-1204.
22. Joly, D. "Analyse directe des coupes pétrolierés lourdes par spectrométrie do masse", Proceed Int. Symp. Charactc. Heavy Oils Pet. Resid., Technip, Paris 1984; pp. 416-420.
23. Hsu, C. S.; Liang, Z.; Campana, J. E. "Hydrocarbon characterization by ultrahigh resolution fourier transform ion cyclotron resonance mass spectrometry", *Anal. Chem.* **1994,** 66, 850-855.
24. Guan, S.; Marshall, A. G.; Scheppele, S. E. "Resolution and chemical formula identification of aromatic hydrocarbons and aromatic compounds containing sulfur, nitrogen or oxygen in petroleum distillates and refinery streams", *Anal. Chem.* **1996,** 68, 46-71.
25. Rodgers, R. P.; White, F. M.; Hendrickson, C. L.; Marshall, A. G.; Andersen, K. V. "Resolution, elemental composition, and simultaneous monitoring by fourier transform ion cyclotron resonance mass spectrometry of organosulfur species before and after diesel fuel processing", *Anal Chem.* **1998,** 70, 4743-4750.
26. Hsu, C. S.; Drinkwater, D. "Gas chromatography-mass spectrometry for the analysis of complex hydrocarbon mixtures", *Chromatogr. Sci. Ser.* **2001,** 86, 55-94.
27. Kuras, M.; Hala, S. "The use of a gas chromatograph-mass spectrometer for the analysis of complex hydrocarbon mixtures", *J. Chromatog.* **1970,** 51, 45-57.
28. Aczel, T.; Hsu, C. S. "Recent Advances in the Low Voltage Mass Spectrometric Analysis of Fossil Fuel Distillates", *Int. J. Mass Spectrom. Ion Processes* **1989,** 92, 1-7.
29. Dzidic, I.; Petersen, H. A.; Wadsworth, P. A.; Hart, H. V. "Towsend Discharge Nitric Oxide Chemical Ionization Gas Chromatography/Mass Spectrometry for Hydrocarbon Analysis of the Middle Distillates", *Anal. Chem.* **1992,** 64, 2227-2232.
30. Wadsworth, P. A.; Villalanti, D. C. "Pinpoint Hydrocarbon Types. New analytical methods helps in processing clean fuels", *Hydrocarbon Processing* **1992,** 71, 109-112.
31. Teng, S. T.; Ragsdale, J.; Urdal, K. "SI-PIONA on the INCOS™ XL. A Novel GC-MS Approach to Gasoline Analysis", Application Report No. 225 Finnigan MAT.
32. Malhotra, R.; Coggiola, M. A.; Young, S. E.; Sprindt, C. A. "GC-FIMS Analysis of Transportation Fuels", Symp. - Advanced Testing for Fuel Quality and Performance. Division of Petroleum Chemistry, 212th American Chemical Society National Meeting, Orlando, Florida, August 25-29, 1996.

33. Schoemakers, P. J.; Oomen, J. L. L. M.; Blomberg, J.; Genuit, W.; Van Velzen, G. "Comparison of Comprehensive Two-dimensional Gas Chromatography-Mass Spectrometry for the Characterization of Complex Hydrocarbon Mixtures", *J. Chromatogr. A* **2000**, 892, 29-46.

34. Bateman, R.; Bordoli, R.; Gilbert, A.; Hoyes, J. "An investigation into the accuracy of mass measurement on a Q-TOF mass spectrometer", *Adv. Mass Spectrom.* **1998**, 14, 1-10.

35. Bacaud, R.; Rouleau, L. "Coupled Simulated Distillation-Mass Spectrometry for the Evaluation of Hydroconverted Petroleum Residues", *J. Chromatogr. A* **1996**, 750, 97-104.

36. Ashe, T. R.; Roussis, S. G.; Fedora, J. W.; Felsky, G.; Fitzgerald, P. "Method for predicting chemical or physical properties of crude oils", U.S. Patent No. 5,699,269, December 16, 1997.

37. Roussis, S. G.; Fedora, J. W.; Fitzgerald, W. P. "Direct method for determination of true boiling point distillation profiles of crude oils by gas chromatography/mass spectrometry", U.S. Patent No. 5,808,180, September 15, 1998.

38. Roussis, S. G.; Fitzgerald, W. P. "Gas Chromatographic Simulated Distillation-Mass Spectrometry for the Determination of the Boiling Point Distribution of Crude Oils", *Anal. Chem.* **2000**, 72, 1400-1409.

39. Méndez, A.; Piemonti, C.; Dassori, C. G.; Fernández, N. "Expanding the use of GC-MS beyond the separation capabilities of a chromatographic column. A system for crude oil assay and process control", XV International Mass Spectrometry Conference, Barcelona, Spain,August 25 - September 1, 2000.

40. Boduszynski, M. M. "Composition of heavy petroleum. 2. Molecular characterization", *Energy Fuels* **1988**, 2, 597-613.

41. Altgelt, K. H.; Boduszynski, M. M. *Composition and analysis of heavy petroleum fractions,* Marcel Dekker:New York, 1994.

42. Petrakis, L.; Allen, D. T.; Gates, B. C. "Analysis of synthetic fuels for functional group determination", *Anal. Chem.* **1983**, 55, 1557-1564.

43. Disanzo, F. P.; Giarroco, V. J. "Analysis of pressurized gasoline-range liquid hydrocarbon samples by capillary column and PIONA analyzer gas chromatography", *J. Chromatogr. Sci.* **1988**, 26, 258-266.

44. Matisova, E.; Juranyiova, E. "Analysis of multi-component mixtures by high resolution capillary gas chromatography and combined gas chromatography-mass spectrometry", *J. Chromatogr. A* **1991**, 552, 301-312.

45. Beardslay, J. D. "Fuels, gaseous and liquid", *Anal. Chem.* **1992**, 57, 195R

46. Sazonova, M. L.; Luskii, M. K. "Gas Chromatography determination of the composition of unfractionated natural hydrocarbon mixtures", *J. Chromatogr.* **1986**, 364, 267-298.

47. Boduszynski, M. M. "Composition of heavy petroleum. 1. Molecular weight, hydrogen deficiency, and heteroatom concentration as a function of atmospheric equivalent boiling point up to 1400°F (760°C)", *Energy Fuels* **1987**, 1, 2-11.

48. Malhotra, R.; McMillen, D. F.; Tse, D. S.; St. John, G. A.; Coggiola, M. L.; Matsui, H. "An approach to chemical characterization of vacuum residues", *Prepr.- Am. Chem. Soc., Div. Pet. Chem.* **1989**, 34, 330-338.

49. McLean, M. A.; Hsu C. S. "Combination of moving belt LC/MS and low voltage electron impact ionization for the characterization of heavy petroleum streams", Proc. 38th ASMS Conf. Mass Spectrom. Allied Top., Tucson AZ, June 3 - 8, 1990.

50. Hsu, C. S.; Qian, K.; Chen, Y. C. "An innovative approach to data analysis in hydrocarbon characterization by on-line liquid chromatography-mass spectrometry", *Anal. Chim. Acta* **1992**, 264, 79-89.

51. Hsu, C. S.; McLean, M. A.; Qian, K.; Aczel, T.; Blum, S. C.; Olmstead, W. N.; Kaplan, L. H.; Robbins, W. K.; Schulz, W. W. "On-line liquid chromatography/mass spectrometry for heavy hydrocarbon characterization", *Energy Fuels* **1991**, 5, 395-398.

52. Hsu, C. S.; Qian K. "Molecular transformation in hydrotreating processes studied by on-line liquid chromatography/mass spectrometry", *Anal. Chem.* **1992**, 64, 2377-2333.
53. Hsu, C. S.; Qian, K. "High boiling aromatic hydrocarbons characterized by liquid chromatography-thermospray-mass spectrometry", *Energy Fuels* **1993,** 7, 268-272.
54. Hsu, C. S. "Novel characterization of petroleum resids by liquid chromatography coupled with mass spectrometry", *prep. ACS Div. Fuel Chem.* **1997**, 42, 390-393.
55. Liang, Z.; Hsu, C. S. "Molecular speciation of saturates by online liquid chromatography-field ionization mass spectrometry", *Energy Fuels* **1998**, 12, 637-643.
56. Bower, M. T.; Marshall, A. G.; McLafferty, F. W. "Mass spectrometry: recent advances and future directions", *J. Phys. Chem.* **1996,** 100, 12897-12910
57. Little, D. "Exact mass measurements by flow injection ESMS with orthogonal acceleration-TOF mass detector", Application Brief AB1. Micromass U.K.
58. Hsu, C. S.; Green, M. "Fragment–Free accurate mass measurement of complex mixtures components by gas chromatography – field ionization – orthogonal acceleration time – of – flight mass spectrometry (GC/FI/oa-TOFMS): an unprecedent capability for mixture analysis", *Rapid Commun. Mass Spectrom.* **2001**, 15, 236-239.
59. Roussis, S. "Automated tandem mass spectrometry by orthogonal acceleration TOF data acquisition and simultaneous magnet scanning for the characterization of petroleum mixtures", *Anal. Chem.* **2001,** 73, 3611-3623.
60. Faubert, D.; Paul, G. J.; Giroux, J.; Bertrand, M. J. "Selective fragmentation and ionization of organic compounds using an energy-tunable rare-gas metastable beam source", *J. Int. Mass Spectrom. Ion Process* **1993**, 124, 69-77.
61. Faubert, D.; Mousselmal, M.; Roussis, S. G.; Bertrand, M. J. "Comparison of MAB and EI for petroleum mass spectrometry", Proc. 44th ASMS Conf. Mass Spectrom. Allied Top., Portland, OR, May 12 - 16, 1996.

Chapter 5

THIN-LAYER CHROMATOGRAPHY FOR HYDROCARBON CHARACTERIZATION IN PETROLEUM MIDDLE DISTILLATES

V. L. Cebolla, L. Membrado, M. Matt, E. M. Gálvez and M. P. Domingo
Instituto de Carboquímica
Consejo Superior de Investigaciones Científicas (CSIC)
P.O. Box 589, 50080 Zaragoza, Spain

1.   ANALYSIS OF PETROLEUM MIDDLE DISTILLATES

Petroleum is fractionated according to its boiling point by distillation methods. Depending on the boiling point range of the sample, fractionated products fall into three main categories: light, middle and heavy distillates. Middle distillates (from approximately 150°C to 400°C, which approximately corresponds to C_9-C_{24}) are mostly used for heating and transportation fuels (e.g., diesel, jet fuel, gas oil, aviation turbine oils, kerosene).

Middle distillates, like the other petroleum-derived products, are complex mixtures of hydrocarbons and related compounds. These compounds vary in molecular structure, size, polarity and functionality. Given this complexity, accurate characterization of petroleum products is a challenge for the analytical chemist.[1]

In the near future, development of new analytical techniques will play an important role in the petroleum industry for the choice of process conditions and the evaluation of fuel quality. Owing to legislation calling for the reduction of aromatic and sulfur content in transportation fuels, a progressive reduction of the polyaromatics content (< 5 %), sulfur (50 ppm) and cetane number has been proposed in Europe for 2005. This last parameter is related to the cycloalkanes content (also called naphthenes) in fuels.

An analysis usually required in the industry either at analytical or semi-preparative scale is the determination of hydrocarbon types. In effect, the thermal or catalytic behavior of a hydrocarbon- containing product is usually better defined in terms of principal groups or hydrocarbon families than performing an extensive separation of all the components in such complex mixture. A complete separation of individual compounds would be impossible.

For middle distillates, hydrocarbon type analyses (HTA) of interest include:

- Separation (and subsequent quantification) of hydrocarbon types into saturates including n-alkanes, isoalkanes, and cycloalkanes (naphthenes), olefins (if any), aromatics, and polars.
- Separation according to number of aromatic rings (mono-, di-, polyaromatics) and quantification of these aromatic families.
- Separation (and quantification) into aromatics and sulfur heterocyclic compounds, and for the latter, separation (and quantification) into families according to the ring number (e.g., benzothiophenes, dibenzothiophenes)

Liquid Chromatographic (LC) techniques, mostly High Performance Liquid Chromatography (HPLC), are used for these purposes.[2]

Thin-Layer Chromatography (TLC) has been clearly under-utilized in petrochemical applications. Its degree of technical maturity has now reached a stage that allows the development of accurate, sensitive, and quantitative analytical methods of interest for petroleum products, as will be justified later.

In this chapter, original methods based on TLC-ultraviolet (UV) and fluorescence scanning densitometry are presented for the separation and quantification of gas oils into alkanes, naphthenes and aromatics with adequate precision and sensitivity.

One of the most important problems in applying TLC-densitometry to petrochemical analysis was the lack of UV or fluorescence spectra of saturates (alkanes and naphthenes). Special attention is given to this question in this chapter. Thus, the developed methods include the detection and quantitative analysis of saturated compounds using berberine-induced fluorescence scanning densitometry. This type of detection is explained. Moreover, examples of application of developed methods to middle distillates and other petroleum products are also presented using TLC or High-Performance TLC methods.

2. INTRODUCTION TO MODERN THIN-LAYER CHROMATOGRAPHY (TLC)

TLC is usually perceived as a simple, rapid, robust and inexpensive technique although not sufficiently quantitative, efficient or sensitive. This view is somewhat erroneous. At present, and due to the improvements in instrumentation over the last decade, modern TLC is a mature, useful technique, which complements HPLC and, for certain separations, is the technique of choice.[3,4]

Modern TLC is the result of improvements in the quality and nature of stationary phases and consistency in plate manufacture, as well as the use of optimized techniques and equipment for sample application, plate development with eluants, and detection systems.[3]

Currently, automated instruments are available for most individual TLC steps (e.g., sample application, eluant development, detection). Despite the fact that the TLC plate must be manually handled between steps, the higher sample throughput of this technique makes it more rapid (in terms of the number of samples processed) than HPLC even given the total automation of the latter.[3]

A modern TLC system is a powerful tool that can be used as an analytical or a preparative technique. It can separate (and identify, using appropriate detection systems) a number of compounds with adequate resolution. Its detection limit is similar to that obtained by HPLC, usually in the nanogram (ng, 10^{-9} g) or picogram (pg, 10^{-12} g) range.

The choice of sample application and eluant development systems usually depends on the type of sample to be analyzed. Development systems determine the number of resolved compounds in a sample.[5,6] It should be pointed out that only conventional TLC elution systems have been used for analysis of petrochemicals. For HTA of petrochemicals, in which the separation of 4-6 peaks is considered, capillary-based, high performance TLC (HPTLC) elution techniques are adequate.

With regard to detection sensitivity as well as quantification, many systems such as scanning densitometry, FID, video densitometry, mass spectrometry interfaces, drift interfaces and laser Raman techniques have been developed. At present, in-situ densitometry (UV and fluorescence scanning) is the most convenient, accurate and precise approach.[3] Densitometry also allows the separated peaks to be identified by their UV spectra.

2.1   Advantages of TLC for the Analysis of Complex Mixtures

TLC is especially well suited for complex or dirty samples, and has advantages when compared to the use of column chromatographic techniques (e.g., HPLC). Sherma [7,10], Fried [8,9] and other authors have clearly explained and detailed the reasons for this. Here is a summary of the most important ones with regard to the analysis of complex mixtures of hydrocarbons.

*   Sample cleanup or preparation (e.g., deasphalting in petroleum products) is not necessary in TLC, and in contrast to HPLC, all components in a sample (even those that do not migrate) contribute to the resulting chromatogram by effective scanning of these peaks. In HPLC, some heavy and polar compounds could be irreversibly adsorbed onto the stationary phase, producing column deterioration and quantitative errors due to their non-elution. Moreover, TLC plates are disposable and not reused.

*   The development and detection steps are much more flexible and versatile than with HPLC. In the case of the development step, the availability of continuous, multiple, circular, and anticircular development methods and wide range of available mobile phases allow the TLC system to be optimized for separation of only the compounds of interest from mixtures containing different solutes.[3] The rest of the sample can be left at the origin or moved near the solvent front, away from the center region of maximum resolution. This leads to a considerable time saving when compared to HPLC, in which the most strongly sorbed materials have to be completely eluted for each sample.

In detection, the mobile phase never interferes because it is always removed first. Moreover, compounds are detected in TLC in a static way whilst they are detected in a dynamic, time-dependent fashion in HPLC. This means that different kinds of detection measurements (use of different scanning beam sizes, optimization of detection and/or excitation wavelength, use of different compatible visualization reagents, etc.) can be performed on the same separated sample without time constraints. This static detection also implies that selected parts of the plate can be detected at will. Since the plates act as a physical storage medium, measurements can often be repeated later after having been analyzed for the first time.

*   With regard to sensitivity, highly retained compounds (low R_f, with R_f=Distance traveled by a solute/Distance traveled by solvent front) in TLC form the narrowest peaks and are detected with the highest sensitivity. In the other column techniques, highly retained compounds form the widest peaks and are the most poorly resolved and detected (e.g., polyaromatics in fuels).

* TLC can work with a variety of compounds usually by pre- or post-impregnation or spraying. This treatment allows either the efficiency of separation to be improved (by modification of the retention of compounds) or "invisible" compounds to be detected through the use of visualizing reagents.

Despite these advantages, TLC is usually thought to be limited when compared to HPLC with regard to the number of compounds to be separated. Although this is often true, the number of compounds to be separated in TLC is strongly influenced by the development technique used and, in certain cases, this number can be even higher than in HPLC.[6,11]

2.2 Previous Research Done on TLC of Petroleum Products

Since the seventies, TLC has been overshadowed by the development of gas chromatography (GC), HPLC and supercritical fluid chromatography (SFC). As a result, relatively few works have been published on TLC-densitometry of hydrocarbons and petrochemical samples. These works have been almost exclusively devoted to the use of TLC either for collection of sample fractions for further analysis using other analytical techniques, or for qualitative or semi-quantitative analyses of compound classes.[1,12-16] As previously mentioned, the possibilities of efficient application, eluant development, and detection systems have scarcely been exploited.

There have been a large number of petrochemical studies involving quantitative hydrocarbon type determination of heavy petroleum distillates by TLC on thin rods coupled with Flame Ionization Detection (TLC-FID).[17-19] With this system, called Iatroscan, separations are carried out, not on planar plates, by using thin quartz rods (Chromarods) sintered with micrometer-sized silica or alumina particles. Owing to this, the term "Planar Chromatography" is often used to distinguish separations carried out on planar surfaces.

In comparison with Planar Chromatography-densitometry systems, Iatroscan presents several disadvantages: first of all, efficiency of separations is very limited in the case of TLC-FID chromarods because they are carried out on rods 11 cm in length. Efficiency is usually higher with TLC plates because the additional possibility of performing different elution modes is not available in the case of Iatroscan (e.g., horizontal elution, bidimensional elution, etc.). Secondly, TLC-FID does not give any structural information on the separated peaks whereas modern densitometers provide their UV spectra. Moreover, Iatroscan is a destructive technique and densitometry is not. Finally, a partial volatilization of some semi-volatile samples by the TLC-FID flame cannot be ruled out. These reasons illustrate the fact that TLC-densitometry has a higher potential usefulness than TLC-FID and could be applied to a wider range of samples.

The application of TLC-UV and fluorescence scanning densitometry to HTA of middle distillates is presented in the next sections.

3.   MATERIALS, METHODS AND TLC SYSTEMS USED IN THIS RESEACH

3.1   Samples Analyzed

G-1 gas oil was provided by Institut de Recherches sur la Catalyse (CNRS, Villeurbanne, France). The vis-breaking fuel was provided by CEPSA (Compañía Española de Petróleos, S.A., Madrid, Spain). Details on physico-chemical properties of these products can be found elsewhere.[20]

3.2   Stationary Phases

Conventional silica gel TLC plates (aluminium sheets, 20 x 20 cm; 5-25 μm particle size; 60 Å pore size; 0.2 mm thick layer) were obtained from Merck (Darmstadt, Germany), and Macherey-Nagel (Düren, Germany). High-Performance silica gel TLC plates (HPTLC plates, on glass, 10 x 10 cm; 3-10 μm particle size; 60 Å pore size; 0.2 mm thick layer) and preparative silicagel plates (glass, 60 Å pore diameter, 5-25 μm particle size, 2 mm thick layer) were also purchased from the above companies.

3.3   Preparation of Berberine-Impregnated Silica Gel Plates

Silicagel plates (described above) were impregnated with berberine sulfate (Across Chimica, Geel, Belgium) before use for the detection of saturates (alkanes and naphthenes), using the following method: a solution of berberine sulfate in methanol (2-6 mg x 200 mL^{-1}) was used for impregnation during 20 s. Plates were subsequently dried overnight at 40°C.

3.4   Application of samples

3.4.1   Automatic Sample Spotter

Samples were spotted, at least in duplicate, onto the plates using an SES 3202/IS-02 automatic spotter (Bechenheim-Alzey, Germany). Samples were dissolved in dichloromethane (DCM). Volumes typically applied were 0.4 μL. Effective sample loads usually ranged from 1 to 20 μg.

3.4.2   Band-sprayer Sample Applicator

Samples were dissolved in dichloromethane (DCM), and applied, at least in duplicate, onto TLC plates using a band-sprayer Linomat IV sample applicator (from Camag, Muttenz, Switzerland), which can apply the sample uniformly (at constant density) by spraying it, using N_2 as impulsion gas (3 bar). Therefore, solvent is removed by nebulizing before application.

In analytical TLC experiments, the effective masses of gas oil were 1 μg to quantify alkanes, and 10 μg for aromatics. The effective masses of pure compounds used were 0.2-0.8 μg in the case of alkanes, and 2-10 μg in the case of aromatics. Samples were always applied as 2 mm-bands. The space between adjacent bands was 8 mm.

3.5   Elution of Samples

3.5.1   Conventional Vertical Elution

After application, samples were eluted using a sequence of eluants. A standard TLC tank (22 cm x 25 cm) was used for developing. The tank was left to stand for 5 min to allow the inside atmosphere to be saturated with solvent vapor and to obtain equilibrium before the plate was inserted.

·The normal elution scheme for gasoil was: n-hexane (9 min.), and dichloromethane (DCM) (4.5 min.). Both sides of the plates were often used. Plates were always dried at room temperature.

3.5.2   Horizontal Developing Chamber

A horizontal development chamber (Camag, Muttenz, Switzerland) was used which takes 10 x 10-cm HPTLC plates. Development was carried out either from one end to the other (maximizing resolution), or from both ends into the middle (increasing sample capacity). When the latter configuration was used, 10 samples could be applied 2 mm apart on a 10 x 10 cm-HPTLC plate. This chamber provided an enhanced efficiency because the plate is horizontal (solvent flow is not against gravity) and free volume around the plate is minimum (only 2 ml of solvent is necessary). The elution on berberine-impregnated HPTLC plates was done as follows: 4.5 min. with n-hexane to separate alkanes and naphthenes. Aromatics were further refocused in silica gel HPTLC plates as a single peak using acetone or ethanol after elution with n-hexane.

3.6 Detection by Densitometry

A CS9301 TLC scanner (Shimadzu, Japan) was used in the fluorescence and ultraviolet modes to provide detection of peaks. Fluorescence scanning densitometry was used in the case of berberine-impregnated silicagel plates, usually under the following conditions: excitation wavelength, λ_{exc} = 365 nm; detection in the visible zone, between 450 and 550 nm; and linear scanning using a 1x1 mm beam size.

UV scanning densitometry was used in the cases of both silicagel and berberine-impregnated silicagel plates. The wavelength used was 250 nm. Linear or zigzag scanning and different beam sizes were used in order to obtain an adequate sensitivity.

Peak area data were taken from the densitometer and were collected, displayed and stored using Shimadzu CS9310 PC software. The Kubelka-Munk equation was applied by means of this software to each data point to linearize the area-mass response.[21] An SX=3 factor, recommended for silicagel plates (0.2 mm thickness), was used.

3.7 TLC Systems Used

3.7.1 Conventional TLC System

The combined use of conventional TLC plates (silica gel and berberine-impregnated ones), the automatic sample spotter, and the vertical conventional tank will hereafter be referred to as the conventional TLC system.

3.7.2 High-Efficiency TLC System

The combined use of HPTLC plates (silica gel and berberine-impregnated ones), the Linomat IV, and the horizontal developing chamber will hereafter be referred to as high-efficiency TLC system.

3.8 Quantification

Quantitative determination by UV and fluorescence scanning densitometry was carried out, in each case, by external calibration using fractions obtained from the corresponding fuel itself through preparative TLC.

3.8.1   Preparative TLC

Preparative TLC for isolating alkanes and cycloalkanes from G-1 was carried out using glass silicagel TLC plates (see section III.2). G-1 (500 mg) was applied as a 180 mm-band on a preparative layer and eluted with n-hexane (18 cm) in a conventional vertical development tank. The alkanes and naphthenes were detected by fluorescence, using a slice of the plate which was previously impregnated with berberine to exclusively monitor the elution of saturates. The cutting point between alkanes and naphthenes was fixed by detection of the latter by UV at 210 nm. Note that alkanes do not give response on UV. The distance of migration of the aromatic fraction was visualized by UV at 250 nm.

Once the different fractions were located, zones were scratched and, subsequently, extracted with dichloromethane from the silicagel in a Soxtec apparatus (Tecator, Hogänäs, Sweden) under mild conditions,[22] and dried under N_2 until constant weight was reached. Purity of fractions was monitored by the previously cited TLC/HPTLC runs at analytical scale with fluorescence or UV scanning densitometry.

3.9   Validation of Results

The content in aromatics of G-1 was provided by CEPSA (Madrid, Spain) according to the IP391 standard that is based on HPLC (aminocyano-silica gel) associated to Refractive Index (RI) detection.[23]

4.   APPLICATION OF TLC TO CHARACTERIZATION OF MIDDLE DISTILLATES

4.1   Phenomenon of Fluorescence Induced by Berberine in TLC

Saturated hydrocarbons have traditionally been considered inert molecules. Their detection with spectroscopic techniques has scarcely been investigated because they do not have ultraviolet or fluorescence spectra under analytical working conditions. However, we have found that an organic, readily available cation (berberine: a fluorescent, aromatic and heterocyclic alkaloid derivative) is a fluorescent chemosensor which allows saturated hydrocarbons and other low-polarity molecules (e.g., lipids) to be detected and quantified with great sensitivity. In other words, our technique converts the above-mentioned molecules into fluorescent entities.[24]

In this section, we report this phenomenon and its application to a TLC system, explain the nature of the produced fluorescent signal, and discuss the application of this detection to other petroleum-related compounds, before detailing its application to HTA of middle distillates.

The phenomenon takes place as follows: when a TLC-silicagel plate is impregnated with a methanolic solution containing berberine cation, and an alkane is applied onto the thin-layer, an enhancement of the fluorescence signal of the berberine in the visible region (between 450 and 550 nm) occurs when the system is irradiated with long-wave monochromatic UV light. (e.g., λ=365 nm). No emission signal is produced without berberine impregnation. This phenomenon will be hereafter referred to as berberine-induced fluorescence detection (Figure 1).

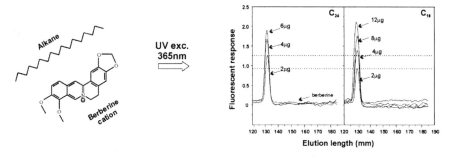

Figure 1. Phenomenon of berberine-induced fluorescence in TLC

For a given wavelength and concentration of berberine in the plate, the fluorescent response is proportional to the mass of injected alkane, and also depends on the alkane structure. The addition of each new -CH$_2$- group gives an increase in the fluorescent response.[24] Therefore, berberine sulfate can be seen as a sensitive fluorescent sensor for saturated hydrocarbons.

Although this phenomenon also takes place in solution, the use of silicagel plates provides a rigid support that enhances the fluorescent response, and also a chromatographic system to separate the analyte of interest from other components in complex mixtures. Moreover, the signal can be easily quantified using a fluorescence scanning densitometer after chromatographic separation.

For analytical purposes, the sensitivity of the determinations can be tailored through the choice of adequate impregnation conditions.[20] The limit of detection for pure alkanes is in the range of nanograms.

Since saturated hydrocarbons are supposed to be inert under normal working conditions, we were challenged to try to understand the nature of this fluorescent chemosensing. For this reason, we took on this task, basing our study on Computational Chemistry. Our results have been detailed elsewhere.[24,25]

According to computational results, explanations of the fluorescent emission based upon conformational changes of berberine were ruled out. Computational calculations have demonstrated that this increase in fluorescence signal involves an ion-induced dipole interaction between the corresponding alkane and the electron π-deficient system of berberine cation. This interaction can accurately model the experimentally obtained fluorescent response. Although weak, this interaction is useful from an analytical point of view.

The mechanisms of this fluorescent emission have been determined in detail.[25] The enhancement of intensity observed in the berberine-alkane systems can be explained by considering two factors. On the one hand, the intensity of the emitted signal is proportional to the Einstein coefficient of spontaneous emission A, which depends inversely on the dielectric constant of the medium. A lipophylic compound (e.g., an alkane) that interacts with berberine and surrounds it creates an apolar micro-environment which, in turn, lowers the dielectric permitivity of the silicagel-berberine medium, thus enhancing the intensity of the fluorescence signal. On the other hand, the alkane provides an apolar environment for the excited berberine cation that hinders alternative relaxation mechanisms and favors fluorescence emission.

According to the proposed theory, other compounds, which have poor ultraviolet or fluorescent responses, have been experimentally confirmed to interact with berberine cation and provide fluorescence response. As in the case of alkanes, cycloalkanes (naphthenes) also give a strong response.

In the case of aromatics in petroleum products, the fluorescent response seems to be due to the presence of alkylic chains of alkyl-aromatics. However, an intrinsic fluorescent response of some aromatic moieties cannot be discarded. Therefore, the particular characteristics of alkylic chains and aromatic compounds existing in a given product will greatly influence the limit of detection for aromatics in petroleum products using this technique.

4.2   HTA of Middle Distillates Using Conventional TLC System

Berberine-impregnated silica gel plates can be used to determine hydrocarbon types in petroleum products. The chromatogram of a heavy petroleum product includes determination of saturates, aromatics, polars and uneluted (Figure 2A).[20] In the case of middle distillates, saturates and aromatics represent the quasi-totality of sample (Figure 2B). The use of berberine as impregnating agent does not modify the separation of compounds with regard to silica gel plates, but only allows visualization of saturated compounds.

In the case of gas oil, saturates are eluted for 9 min with n-hexane, and aromatics for 4.5 min using dichloromethane (Figure 2). Saturates include alkanes and naphthenes under the same peak. This elution is carried out on berberine-impregnated plates.

The use of TLC-Densitometry offers different possibilities with regard to this analysis. When only determination of the saturates content is required, fluorescence scanning is used after eluting with n-hexane. When a complete HTA is to be performed, two different working methodologies can be used. The first alternative is to detect saturates using fluorescence scanning, and the other peaks using UV scanning. In this case, UV scanning can be performed either on the same berberine-impregnated plate or on a silica gel plate. Thus, the sample is scanned twice although the duration of each scan is only 30 s. Scanning can be performed either after complete elution or between eluants (with previous drying of the plates at room temperature). The other alternative is to use fluorescence scanning densitometry for simultaneous determination of saturates and aromatics on the same berberine-impregnated plate. However, this alternative presents problems related to sensitivity. As previously mentioned, the response of aromatics varies depending on the particular aromatic composition of fuels. Likewise, the response factors of saturates are much higher that those of aromatics. This occurs especially in middle distillates in which condensed aromatics structures are relatively scarce. In many cases this precludes the possibility of determining saturates and aromatics simultaneously on the same chromatogram.

Accuracy in saturates determination has been detailed elsewhere.[20] It was determined by comparing results from TLC densitometry with those from other external techniques (e.g. TLC-FID for heavy distillates, HPLC-RI for gas oil). Percentages of Relative Standard Deviation ranged from 2.2 to 6.8 for different heavy distillates. In the case of the gas oil studied, TLC-Fluorescence provided 69.8 mass percent whilst the use of IP391 standard (HPLC-RI) gave 70.8 mass percent. It is necessary to point out that our fluorescence method thus allows the determination of saturates to be performed directly. Table 1 shows quantitative results in aromatics determination, obtained by the application of TLC (conventional and high-efficiency TLC systems)-densitometry.

Evaporative losses in TLC runs were followed by SIMDIS and GC-FID of the whole gas oil and the fraction corresponding to saturates, which was previously isolated by preparative TLC, as explained elsewhere.[26] Chromatograms showed that C_{11} and C_{12} alkanes were evaporated during the runs. However, according to SIMDIS data, this material only represents 1 mass percent of the gas oil.

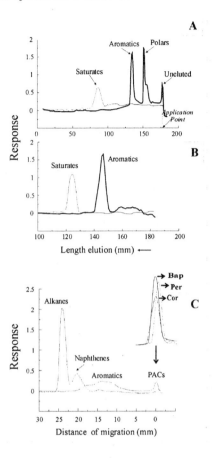

Figure 2. Hydrocarbon type determination of a vis-breaking fuel (A), G-1 gas oil (B) by conventional TLC, and G-1 gas oil by HPTLC (C, elution only with n-hexane). In red: saturates detected by berberine-induced fluorescence. Other: groups detected by UV. (Bap:benzo[a]pyrene; Per: perylene, Cor: coronene)

Table 1. Quantitative data of TLC and HPTLC methods in determination of aromatics

Sample	TLC system	Elution	Ref. Values (mass%) [1]	C Aromatics (mass%) [2]
G-1	Conventional	1) n-hexane (9 min)	29.2	27.0
		2) toluene (4.5 min)		
Synthetic G-1 [3]	High-efficiency	1) n-hexane (4.5 min)	16.8	15.3
		2) Acetone, to merge aromatics in one peak	22.2	22.7
			32.2	29.8
			52.3	47.1

[1] HPLC-RI according to IP391 standard
[2] Detection by UV at 250 nm
[3] By spiking a G-1 fraction of saturates with known amounts of G-1 aromatic fraction

It is emphasized that our TLC method based on berberine-induced fluorescence detection does allow the determination of saturates to be performed directly, without the technical problems associated with the use of Refractive Index detector in the HPLC-based standard.

4.3   HTA of Gas Oils Using High-Efficiency TLC System

The combined use of a band-sprayer sample applicator, HPTLC plates, and a horizontal development chamber leads to a notable improvement in separation efficiency and detection sensitivity. This is illustrated in Figure 2C. The application of such a system by eluting with only 2 mL of n-hexane (4.5 min) provides separation between alkanes and a smaller, well-separated peak which was referred to as naphthenes. These peaks were detected by berberine-induced fluorescence.

The attribution of naphthenic nature to the peak eluted after alkanes and before aromatics is justified by its R_f (cycloalkanes are slightly more polar than alkanes), its fluorescent response in the berberine system; its characteristic response in UV (no response at 254 and response at 210 nm) (Figure 3); its ^1H NMR spectra showing peak multiplicities characteristic of cyclic -CH_2-; and its GC-FID chromatograms.[26] Although none of these aspects alone would justify to identify them as naphtenic products, this is the most reasonable assumption when considered together.

As a consequence of n-hexane elution, the aromatics are like a wide peak on the layer (detected by UV), and a certain amount of heavy polyaromatics (with more than four rings) remains at the application point (Figure 2C). The attribution of polyaromatic nature to the peak at the application point after n-hexane elution, is justified by the migratory properties of some heavy pure polynuclear aromatic hydrocarbons (PAHs) used as standards (Figure 2C).

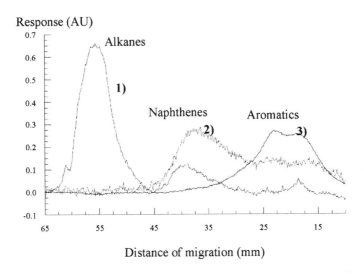

Figure 3. Hydrocarbon type determination of G-1.Detection by berberine-induced fluorescence (1), by UV at 200 nm (2)and by UV at 254 (3)

In order to determine total aromatics, it is necessary to merge these wide peaks (aromatics+polyaromatics) into one narrow, Gaussian peak. Dichloromethane is not able to elute heavy polynuclear aromatic compounds (PACs) which may remain at the application point. Acetone can be used instead but it also would elute polar compounds. In our case, as the gas-oil studied has not a significant amount of N-O polar compounds, acetone can be used with no undesired effect whatsoever. However, Figure 4 shows the complete elution of aromatics of G-1 using acetone, after n-hexane elution. No peaks remain at the sample application point. Thus, this allows the total aromatics to be quantified. This analysis must be performed on a non-impregnated HPTLC silica gel plate because berberine is highly soluble in acetone.

Alkanes, naphthenes and total aromatics are quantitatively determined by calibration using the fractions previously separated by preparative TLC as external calibration standards.

The results of total aromatics determination using the high-efficiency TLC system agree quite well with the reference values which were obtained by spiking a G-1 fraction of saturates with known amounts of a G-1 aromatic fraction (Table 1). These fractions were previously isolated by preparative TLC. Table 2 shows the linearity ranges for hydrocarbon type determinations of G-1, under the conditions described in this work, using both TLC systems.

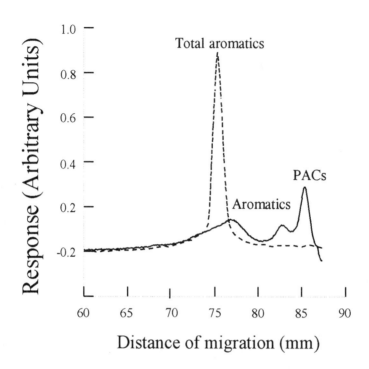

Figure 4. G-1 aromatics in HPTLC by UV (254 nm). Continuous line: elution of G-1 with n-hexane; dashed line: elution of G-1 with n-hexane and then acetone to merge all the aromatics into one peak

Table 2. Linearity ranges for hydrocarbon type determinations[1] of G-1

TLC system	Standard	Linearity ranges (µg)
Conventional	Saturates	4-16
	Aromatics	3-16
High-efficiency	Alkanes	0.05-15
	Naphthenes	0.6-2.4
	Total aromatics	0.1-2

[1] Under the conditions specified in the text

In short, the adaptation of this analysis to a high-efficiency TLC system has led to: i) a higher sensitivity; ii) a shorter analysis time; iii) the additional separation and determination of alkanes and cycloalkanes; and iv) determination of the total aromatics content.

5.   CONCLUSIONS AND FUTURE TRENDS

There is a growing need for accurate, and yet cost-effective and easy-to-use chromatographic techniques and methods that are capable of providing valuable information on petrochemical samples. In this context, the potential of TLC for petrochemical analysis has not been fully realized. TLC (high-efficiency TLC) with UV and fluorescence scanning densitometry is a powerful tool with more degrees of freedom than column techniques, and is applicable to a wide range of petroleum products.

In this chapter, quantitative, sensitive and precise methods for hydrocarbon-type determination of alkanes, naphthenes and total aromatics in gas oil have been presented, all of which are based on both conventional and high-efficiency TLC systems coupled to ultraviolet and fluorescence scanning densitometry. These methods are based on the use of berberine-impregnated silicagel plates for detection of saturated hydrocarbons. The nature and mechanisms of this fluorescent emission have been elucidated in detail.

Further research into the applications of TLC to petrochemical analysis should include: separation of aromatics according to ring number, separation between aromatics and sulfur compounds, and development of rapid, internal calibration techniques, which take full advantage of the versatility of detection provided by TLC systems.

6.   ACKNOWLEDGMENTS

This work was supported by the Spanish Ministery for Science and Technology (McyT, Plan Nacional de I+D+i, project PPQ2001-2388). One of us (M.M.) gratefully acknowledges a grant from the McyT (program "Stages for foreign scientists and technologists in Spain"), and a leave from the University of Metz (France).

Dr. Jesús Lázaro and Dr. José Mazón (CEPSA), and Dr. Robert Bacaud (IRC-CNRS) are also acknowledged for providing petrochemical samples and for fruitful discussions. Authors wish to thank Dr. Jesús Vela and Dr. Rosa Garriga (University of Zaragoza) for their contributions throughout the different stages of this work.

An oral presentation of a part of this work was made before the Petroleum Division of the American Chemical Society (Washington, D.C.; August, 2000).

7.   REFERENCES

1. Barman, B. N.; Cebolla, V. L.; Membrado, L. *Critical Rev. in Anal. Chem.* **2000**, 30 (2&3), 75-120.

2. Cebolla, V. L.; Membrado, L.; Vela, J. "Applications of Liquid Chromatography to petroleum" in *Encyclopedia of Separation Science*; Wilson, I. D.; Adlard, E. R.; Cooke, M.; Poole, C. F. (Eds.), Academic Press: London, 2000; pp. 3683-3690.

3. Fried, B.; Sherma, J. *Thin-Layer Chromatography*, 4th ed. (Revised and Expanded), Chromatographic Science Series, Vol. 81, Marcel Dekker: New York, 1999.

4. Nurok, D. *Anal. Chem.* **2000**, 72, 635-641A.

5. Poole, C. F. *J. Chromatogr. A* **1999**, 856, 399-427

6. Poole, C. F.; Poole, S. K. *J. Chromatogr. A* **1995**, 703, 573-612.

7. *Handbook of Thin Layer Chromatography*, 2nd ed., Sherma, J.; Fried, B. (Eds.), Marcel-Dekker: New York, 1996.

8. Fried, B.; Sherma, J. *Thin-Layer Chromatography-Techniques and Applications*, 3rd ed.; Chromatographic Science Series, Vol. 66; Marcel-Dekker: New York, 1994.

9. Fried, B.; Sherma, J. *Thin-Layer Chromatography-Techniques and Applications*. Chromatographic Science Series, Vol. 35, Marcel-Dekker: New York, 1986.

10. *Handbook of Thin Layer Chromatography*, Sherma, J.; Fried, B. (Eds.), Marcel Dekker: New York, 1991.

11. Guiochon, G.; Gonnord, M. F.; Siouffi, A.; Zakaria, M. *J. Chromatogr.* **1982**, 250, 1-20.

12. Kalmal, T.; Matsunaga, A. *Anal. Chem.* **1978**, 50, 268-270.

13. Harvey, T. G.; Matheson, T. W.; Pratt, K. C. *Anal. Chem.* **1984**, 56, 1277-1281.

14. Wang, Y. Y.; Yen, T. F. *J. Planar Chromatogr.* **1990**, 3, 376-380.

15. Herod, A.; Kandiyoti, R. *J. Chromatogr. A* **1995**, 708, 143-160.

16. Herod, A. *J. Planar Chromatogr.* **1994**, 7, 180-196.

17. Ranny, M. *Thin-Layer Chromatography with Flame Ionization Detection*; Riedel, D. Publishing Co.: Dordrecht, 1987.

18. Barman, B. N. *J. Chromatogr. Sci.* **1996**, 34, 219-225.

19. Vela, J.; Membrado, L.; Cebolla, V. L.; Ferrando, A. C. *J. Chromatogr. Sci.* **1998**, 36, 487-494.

20. Cebolla, V. L.; Membrado, L.; Domingo, M. P.; Henrion, P.; Garriga, R.; González, P.; Cossío, F. P.; Arrieta, A.; Vela, J. *J. Chromatogr. Sci.* **1999**, 37, 219-226.

21. Kubelka, P.; Munk, F. *Z.Tech. Phys.* **1931**, 12, 593.

22. Membrado, L.; Vela, J.; Ferrando, A.C.; Cebolla, V. L. *Energy & Fuels* **1996**, 10, 1005-1011.

23. *Standard Methods for Analysis and Testing of Petroleum and Related Products* 1992, vol. 2. Institute of Petroleum: London, 1992.

24. Cossío, F. P.; Arrieta, A.; Cebolla, V. L.; Membrado, L.; Domingo, M. P.; Henrion, P.; Vela, J. *Anal. Chem.* **2000**, 72, 1759-1766.

25. Cossío, F. P.; Arrieta, A.; Cebolla, V. L.; Membrado, L.; Garriga, R.; Vela, J.; Domingo, M. P. *Org. Letters* **2000**, 2, 2311-2313.

26. Bacaud, R.; Cebolla, V. L.; Membrado, L.; Matt, M.; Pessayre, S.; Gálvez, E. M. *I&EC Res*, submitted for publication.

Chapter 6

CHROMATOGRAPHIC ANALYSES OF FUELS

Frank P. Di Sanzo
ExxonMobil Research & Engineering Co.
Annandale, New Jersey USA

1. ANALYSES OF NAPHTHAS/MOTOR GASOLINES BY GAS CHROMATOGRAPHY

1.1 Introduction

A very important analysis in the petroleum industry is the analyses and characterization of low boiling petroleum fractions with a carbon number of C_1-C_{12}. These fractions are either obtained by the direct distillation of petroleum crudes (for example, "virgin naphthas") or as a result of catalytic processes from several high boiling feedstock conversions. Some of these light products may be further processed to increase octane (for example, conversion of virgin naphthas into "reformate") or reduce the sulfur levels to meet regulatory specifications. The resulting final products are the main sources used as blending components in the production of commercial automotive gasolines sold at the pumps. The main types of feedstocks and products can be summarized in the following Table 1:

Table 1. Light Products Used in Commercial Gasoline Blending. Streams in Italics are Commonly Analyzed for Composition.

Virgin naphthas (high cycloparaffins or naphthenes and paraffins; aromatics). Pretreated to remove S, N and used as feedstock for 'reformer'. The naphthas are obtained from light distillate (low octane) cut from crude oil. Virgin naphthas are reformed into a higher octane blending component "*reformate*" (high aromatics, low naphthenes and paraffins; low sulfur)
Isobutane and butenes converted into high octane "*alkylate*" (mostly paraffins, low sulfur). C_4's are obtained as by-product from cracking heavier petroleum fractions; distillation of crude oil.
Fluid catalytic cracking (FCC) naphtha (high olefin and aromatics; paraffins; high sulfur). Naphtha may be further processed to reduce sulfur level. Naphtha obtained from catalytic cracking of heavy petroleum fraction followed by distillation of vacuum gas oil fraction from crude oil

"coker naphtha" (high olefins; aromatics and saturates; high sulfur). May be further processed to reduce the sulfur level. Naphtha obtained from thermally cracked heavy petroleum fraction.
Oxygen Containing compounds (Ethanol, MTBE etc.)

In addition to the above and other "naphtha" streams, compounds containing oxygen, such as methyl t-butyl ether (MTBE), *t*-amyl methyl ether (TAME), ethanol etc. are or have been important as blending component into gasolines at percent levels. In selecting an analytical characterization approach, it is important to know if such oxygen containing compounds are present in the sample.

The two main techniques that have been used extensively in the characterization of naphthas and/or gasolines in the petroleum industry are mass spectrometry (MS) and gas chromatography (GC). Mass spectrometric method was developed through American Society for Testing and Materials (ASTM).[1] The latter method generally provides very accurate molecular type data such as monocycloparaffins, dicycloparaffins, alkyl benzenes, indans/tetralins, naphthalenes and total paraffins. Many laboratories also developed modified MS methods which included the olefinic molecular types. These MS methods provided valuable data without relying on the separation of components as is necessary in gas chromatography. Over the years, gas chromatography has played a significant role in the analyses of naphthas because of its simplicity and also because for many processes it is important to obtain isomeric information for the many individual components not obtainable by mass spectrometry alone. The following sections summarize the current approaches used in gas chromatography.

1.2 Classification of GC Methods for Naphtha Analyses

The gas chromatographic methods or approaches for the analyses of naphthas or gasolines can be classified as shown in Table 2.

Table 2. Classification of GC Methods for Analysis of Naphthas and Motor Gasolines

Single capillary	Multidimensional
• **Individual compound(s) analysis** • Relatively 'simple' • 50-60, 100, 150 meter methylsiloxane capillary column/FID most popular • Polar phases (Carbowax etc.) may be used for analysis of aromatics to complement methylsilicone capillary column approach • Analysis of virgin naphthas generally limited to C_1-C_8.	1. **Micropacked/Packed (*PIONA* Column Systems** • carbon number distribution only; compound classes up to C_{11}. Few individual isomers . • 'best' GC approach for overall analyses of hydrocarbon classes for most types of naphthas and blended gasolines (at least 6 modes available-see Table 3) • turnkey commercial GC system available. • O-PONA mode available for oxygenated

Single capillary	Multidimensional
• Analysis of olefinic naphthas (FCC) generally limited to C_1-C_7	(ethanol etc.) containing samples.
	• very good precision and accuracy
	• can replace mass spectrometry in many sample analyses
	• transferable to refinery labs
	2. **Capillary Column Multidimensional GC Systems**
	3. **Capillary Comprehensive 2-dimensional-GC (2D-GC)**

1.3   Terminology

Table 3 below gives abbreviations that are used to designate various GC methods for analysis of naphthas and blended motor gasolines. The 'modes of analysis' indicate the type of data that is obtainable by using various analytical configurations. These designations are used in the following sections.

Table 3. Modes of GC Analyses for Naphthas and Blended Motor Gasolines

Micropacked/Packed (PIONA)	
Multidimensional GC	
PNA	Paraffins, Naphthenes, Aromatics
nPiPNA	n-Paraffins, isoParaffins, Naphthenes, Aromatics
PONA	Paraffins, Olefins, Naphthenes, Aromatics
PIONA	Paraffins, Isoparaffins, Olefins, Naphthenes, Aromatics
nPiPONA	n-Paraffins, isoParaffins, Olefins, Naphthenes, Aromatic
O-PONA	Oxygenates, Paraffins, Olefins, Naphthenes, Aromatics
Single Capillary GC	
DHA	Detailed Hydrocarbon Analysis
PIANO	Paraffins, Isoparaffins, Aromatics, Naphthenes, Olefins

1.4   Single Capillary Methods

Single capillary column methods are relatively simple to implement and are used to obtain the analyses of individual hydrocarbons. Several approaches have been developed using 50-60 meter, 100 meter and 150 meter capillary columns coated with dimethylsiloxane stationary phases. Several of these methods recently have been adopted by the American Society for Testing and Materials (ASTM) (Table 4).[2-5]

Table 4. ASTM Single Capillary Methods

ASTM method	Application	Length of Capillary	Comments
D5134-98 Standard Test Method for Detailed Analysis of Petroleum Naphthas through n-	Virgin naphtha	50	Generally limited to the detailed composition up to C_7.

ASTM method	Application	Length of Capillary	Comments
Nonane by Capillary Gas Chromatography.[2]			Significant paraffin/naphthene compound overlap above C_7 yields incorrect P, N content.
D6733-01 Standard Test Method for Determination of Individual Components in Spark Ignition Engine Fuels by 50-Meter Capillary High Resolution Gas Chromatography.[3]	Blended gasolines	50	
D6730-01 Standard Test Method for Determination of Individual Components in Spark Ignition Engine Fuels by 100-Metre Capillary (with Precolumn) High-Resolution Gas Chromatography.[4]	Blended gasolines	100	Uses a precolumn for improved resolution of key components such as benzene, toluene
D6729-01 Standard Test Method for Determination of Individual Components in Spark Ignition Engine Fuels by 100 Meter Capillary High Resolution Gas Chromatography.[5]	Blended gasolines	100	Uses start temp of 0°C

The complexity of number and types of hydrocarbons present in naphthas and gasolines precludes complete separation of all compounds by the single capillary column approach. Although a large number of the individual hydrocarbons present are determined, some co-elution of compounds is encountered. These single capillary column methods are utilized to *estimate* total bulk hydrocarbon group-type composition (PONA) for the total sample; however, caution should be used because some error will be encountered due to the co-elution and a lack of identification of all components present.

For samples containing a significant amount of **olefins**, interfering coelution with the olefins above C_7 is possible, particularly if blending components or their higher boiling cuts such as those derived from fluid catalytic cracking (FCC) are analyzed, and the total olefin content may not be accurate. Samples containing significant amounts of **naphthenic** (cyclic paraffins) (for example, virgin naphthas) hydrocarbons above *n*-octane may reflect significant errors in PONA type groupings.

Tables 5-8 show comparisons between procedures ASTM D6730-01, D6729-01 and other methods, such as D5580 (an accepted method which yields accurate, low interference analysis of benzene, toluene and total C_9+ aromatics), for several compound types in ***blended motor gasolines***. For comparison, multidimensional *micropacked/packed PIONA* method (discussed later) is included since it tends to give reasonable accuracy with little or no interference for carbon number compound type groupings for total olefins, total paraffins and total naphthenes. The differences for benzene and toluene among the indicated methods are well within the reproducibility of the methods. For total olefins the agreement among the methods is reasonable except at the higher olefin levels where overlap among components reduces the accuracy of the DHA methods. The discrepancy for the total olefins is less pronounced for the blended gasolines (Tables 5-8) than the FCC streams used to blend gasolines (see Table 10) because the blended gasolines generally contain a diluted amount of the FCC streams and in some cases the more difficult to separate heavy FCC stream may be added only in small amounts in the blended gasoline.

Table 5. Comparison of Methods for Benzene

		BENZENE (wt %)	
Sample	ASTM D5580	ASTM Procedure 100 meter with precolumn (D6730-01)	ASTM Procedure 100 meter (no pre-column) (D6729-01)
1	1.52	1.58	1.61
2	1.05	1.12	1.12
3	1.10	1.15	1.16
4	1.13	1.19	1.18
5	0.14	0.17	0.16
6	0.62	0.69	0.70
Average	0.93	0.98	0.99

Table 6. Comparison of Methods for Toluene

		TOLUENE (wt %)	
Sample	ASTM D5580	ASTM Procedure 100 meter with precolumn (D6730-01)	ASTM Procedure 100 meter (no precolumn) (D6729-01)
1	4.3	4.5	4.6
2	2.1	2.0	1.9
3	10.1	10.3	11.4
4	5.0	5.2	6.1
5	3.3	3.3	2.9
6	4.4	4.7	5.3
Average	4.9	5.0	5.4

Table 7. Comparison of Methods for Total Aromatics

		TOTAL AROMATICS (wt %)		
Sample	ASTM D5580	Micropacked/ Packed *PIONA*	ASTM Procedure 100 meter with precolumn (D6730-01)	ASTM Procedure 100 meter (no precolumn) (D6729-01)
2	30.3	28.2	30.2	32.6
6	18.9	18.7	18.3	20.0
8	49.1	49.0	47.6	51.0
10	23.9	24.5	23.1	25.4
13	19.7	19.8	19.3	22.4
14	23.8	24.6	24.2	27.5
Average	27.6	27.5	27.1	29.8

Table 8. Comparison of Methods for Total Olefins for Blended Gasolines

		TOTAL OLEFINS (wt %)	
Sample	Micropacked/Packed *PIONA*	ASTM Procedure 100 meter with precolumn (D6730-01)	ASTM Procedure 100 meter (no precolumn) (D6729-01)
1	7.1	4.5	4.4
2	9.8	8.7	9.4
3	6.6	6.1	6.2
4	15.1	12.9	13.7
5	11.1	10.6	11.1
6	24.6	19.5	22.2
Average	12.4	10.9	11.2

Tables 9 and 10 show a larger discrepancy between DHA and micropacked/packed PIONA methods, in the cases of pure virgin and FCC naphtha, respectively. Virgin naphthas containing a large concentration of naphthenes, particularly in the higher boiling point range of the naphtha are more difficult to analyze by the DHA method. The problem with the olefins is clearly seen in Table 10. As the boiling point of the FCC naphtha is increased, the accuracy of the DHA is reduced significantly due to the overlap of olefins, paraffins and aromatics above C_7. Generally, these type of pure stream samples are best analyzed by a well-optimized *micropacked/packed PIONA* system.

Table 9. Comparison of DHA and Micropacked/Packed PIONA for Full Range Virgin Naphtha

Hydrocarbon Types	DHA (100 meter single capillary column) (wt%)	Micropacked/Packed PIONA (wt %)
Paraffins	52.2	44.1
Olefins	0.0	0.0
Naphthenes	23.8	40.1
Aromatics	21.2	15.8
Oxygenates	2.8	0.0

Table 10. Comparison of DHA and Micropacked/Packed PIONA for Three Cuts (Light, Medium, Heavy Boiling) FCC Naphthas

Method	Light FCC Naphtha Cut (Total Olefins)	Medium FCC Naphtha Cut (Total Olefins)	Heavy FCC Naphtha Cut (Total Olefins)
DHA (100 meter single capillary column) (wt%)	42.9	20.9	10.3
Micropacked/Packed PIONA (wt %)	50.2	37.9	30.0

1.5 "Pressurized" Naphtha Samples

Single capillary methods at times are used to analyze "pressurized" hydrocarbons samples. Analysis of "pressurized" naphthas as well as liquefied C_1-C_5 gas streams, containing a significant amount of C_1-C_4 hydrocarbons are generally sampled at the process units in vessels that withstand pressures up to 1000-3000 psi. These samples may be sampled according to ASTM or equivalent procedures using either a piston sampler[6] or water displacement [7] technique. The piston sampler technique is more expensive but more elegant and easier to operate. The water displacement technique may be used for components which are not water soluble. A schematic of the water displacement technique used with single capillary column DHA is shown in Figure 1. The sample (cylinder A) is pressurized from the bottom of the cylinder by introducing water from a water containing cylinder B at approximately 100 psi above the actual process unit sampling pressure to minimize the vapor phase in the sample A cylinder. The pressurized sample is then introduced into a liquid sampling valve containing one or two restriction valves at the exit line. The technique has been used successfully for the partial hydrocarbon analyses of pressurized C_1-C_{12} naphthas such as feeds hydrotreated virgin naphtha (Figure 2), product reformate (Figure 3) etc.

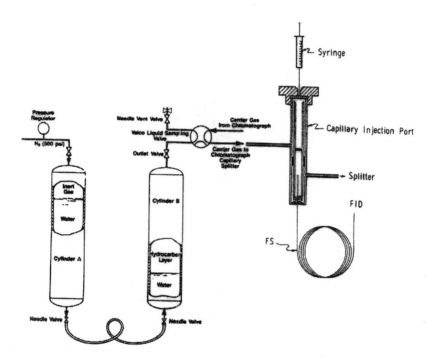

Figure 1. Instrumental Schematic for Analysis of Pressurized Samples by the Water Displacement Technique and Single Capillary Column GC

When a piston sampler is used, the water displacement is not used. Instead an inert gas, such as helium or nitrogen, under pressure at least equivalent to unit sample pressure on the piston is used to compress the sample and introduce it into the liquid sampling valve using a similar valve configuration shown in Figure 1.

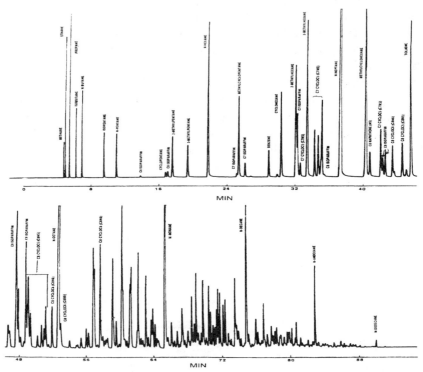

Figure 2. Gas Chromatogram of Full Range C_1-C_{12} Pressurized *Virgin Naphtha* Analyzed by the Set-up in Figure 1

1.6   Multidimensional Methods

Multidimensional GC is extensively used in the analysis of gasoline range petroleum fractions. The most popular approach is the use of a commercially available "micropacked/packed/capillary column (*PIONA* Column System" which provides hydrocarbon *carbon number* distributions for a variety of types of naphthas.[7,8] For example, when used in the PNA mode such a system provides carbon number lumps up to C_{12} for olefin-free virgin naphthas. Other modes (Table 3) available include PONA, PIONA, nPiPONA. Figure 4 shows a schematic of a commercial system.[8] A very polar (OV-275 or equivalent) phase is used to separate the aromatics from the paraffins/naphthas/olefins (PNO). Olefins if present are retained on silver impregnated trap while the PN are separated on a Molecular Sieve 13 X column. A Molecular Sieve 5Å trap is used also in the separation path when a carbon number distribution of the n-paraffins is desired. Olefins, if present, can be released from the olefin trap, hydrogenated and then separated as PN (to represent cyclic and non-cyclic olefins) by the MS 13X column. Similarly, non-branched olefins may be retained by the MS 5Å trap as hydrogenated n-paraffins and then separated by the MS 13X by carbon number. Carbon number distributions of the aromatic hydrocarbons are obtained on a nonpolar column, such as OV-101.

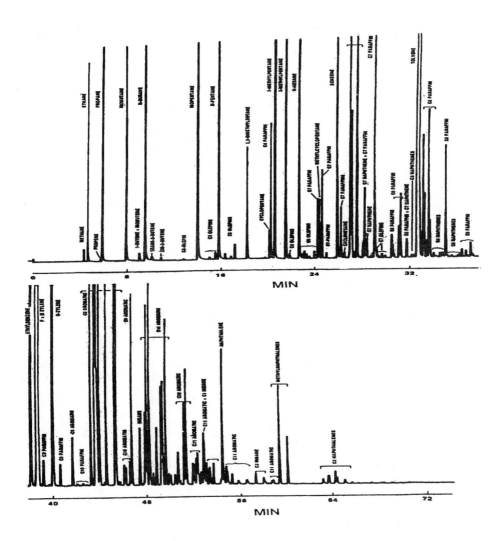

Figure 3. Gas Chromatogram of Full Range Pressurized *Reformate* (Reactor Product from Virgin Naphtha in Fig. 2 Analyzed by the Set-up in Figure 1

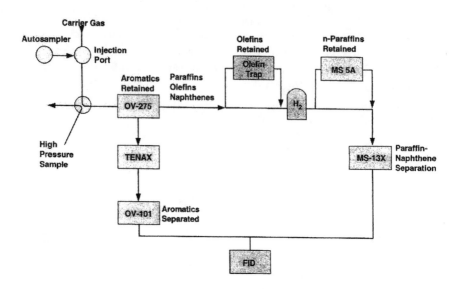

Figure 4. Schematic of a Commercial *"micropacked/packed PIONA"* System

Figures 5 and 6 show chromatograms of a virgin naphtha stream (feed to a "reformer" unit) containing a high concentration of naphthenes and a chromatogram of a reformate (the high octane product from reformer unit) containing a low concentration of naphthenes in the nPiPNA mode of analysis.

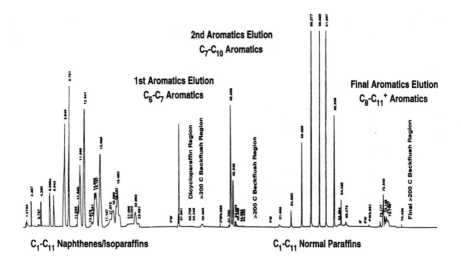

Figure 5. Gas Chromatogram of a Virgin Naphtha Analyzed by the nPiPNA Mode of *micropacked/packed PIONA* System

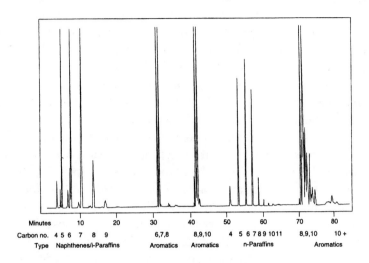

Figure 6. Gas Chromatogram of a Virgin Naphtha Analyzed by the nPiPNA Mode of
micropacked/packed PIONA System

Figures 7 (Part A) and 8 (Part B) show PIONA chromatograms of a FCC
naphtha sample containing a high concentration of olefins. For the olefinic
analyses the silver impregnated trap in Figure 4 is also used.

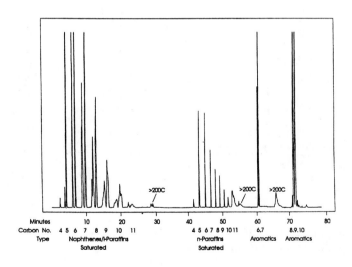

Figure 7. Gas Chromatogram of the First Half of a FCC Naphtha Analyzed by the nPiPNA Mode
of ***micropacked/packed PIONA*** System

Gasoline samples containing oxygenated blending components such as ethanol, MTBE etc. require a modification to the *micropacked/packed PIONA* instrumentation. A newer generation of commercially available PIONA analyzer ("O-PONA") is shown in Figure 9.[9] This type of instrument is similar to the previous generation except that an additional 'selective trap' to retain and release/analyze the oxygenated components is added. Figure 10 shows a chromatogram of a sample containing oxygenated component.

The micropacked/packed *PIONA* technique generally yields very accurate analyses up to 200°C. The results obtained have agreed very well with "batch inlet" mass spectrometric *total* compound class analysis. However, the *carbon number distributions* obtained by the micropacked/packed *PIONA* are more accurate than those estimated from MS type analysis. Tables 11 and 12 show a comparison between MS and micropacked/packed *PIONA* for bulk type analyses. Generally the agreement is excellent. In Table 11 note that the agreement for the single capillary DHA approach is excellent for the *reformate* sample. Generally, reformates are simpler streams containing very low concentration of olefins and the small amounts of naphthenes present are

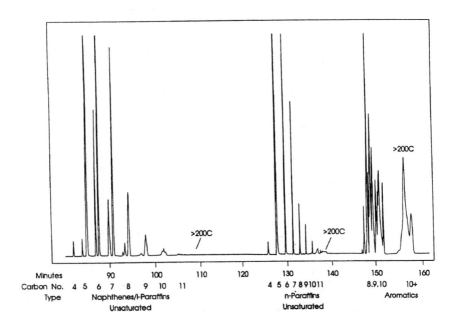

Figure 8. Gas Chromatogram of the Second Half of a FCC Naphtha Analyzed by the nPiPNA Mode of *micropacked/packed PIONA* System

O-PONA Analyzer

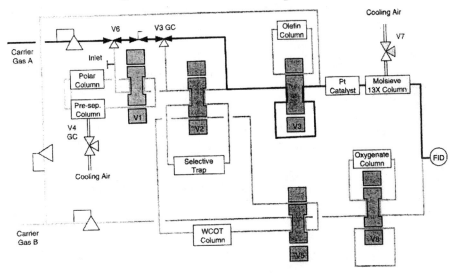

Figure 9. Schematic of a Commercial *"micropacked/packed PIONA"* System for Oxygenates Containing Gasolines

O-PONA (Gasoline)

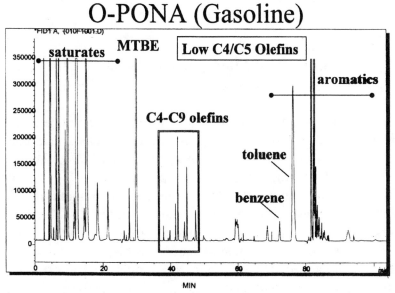

Figure 10. Gas Chromatogram of the MTBE containing Blended Gasoline by the O-PONA *micropacked/packed PIONA* System

generally below C_8 and can be resolved from other compounds. However, for virgin naphthas generally the agreement between MS and PIONA versus single capillary DHA is much poorer as the coelution of naphthenes, paraffins and aromatics above C_7 occurs with the DHA approach.

Table 11. Analysis of a Reformate

Parameter	Micropacked/ Packed *PIONA*	Single Capillary Column	Mass Spectrometry 'Type' Analysis
Paraffins	41.2	41.2	40.8
Olefins	----	0.3	----
Naphthenes	3.9	3.6	4.2
Aromatics	54.9	54.9	54.1

Table 12. Analysis of Virgin Naphtha

Parameter	Micropacked/Packed PIONA	Mass Spectrometry "Type" Analysis
Sample 1		
Paraffins	44.6	44.2
Naphthenes	45.4	45.0
Aromatics	10.0	10.8
Sample 2		
Paraffins	40.2	38.8
Naphthenes	39.3	40.5
Aromatics	20.4	20.4
Sample 3		
Paraffins	20.5	22.2
Naphthenes	43.3	43.5
Aromatics	36.2	34.3

1.7   Combination of Micropacked/Packed PIONA and Single Capillary Column Analyses

In the above sections the disadvantages and advantages of the single capillary column and the multidimensional PIONA approaches were demonstrated. Generally the micropacked/packed *PIONA* approach yields accurate total carbon lumps from approximately C_5-C_{12}. The single capillary column approach, depending on the amount of olefins and naphthenes present, show the most accuracy in the C_1-C_7 or C_1-C_8 carbon number range. As a result, to obtain very accurate analyses with as much hydrocarbon isomer composition as possible, the analyses from the micropacked/packed *PIONA* and the single DHA capillary column analyses may be combined into a single analysis. Generally, the micropacked/packed *PIONA* analyses may be considered the master (most accurate across the full range) analysis from C_5-C_{12}. The carbon number lumps quantitative values from the C_5-C_8 or C_5-C_7 micropacked/packed *PIONA* analysis may be further "broken down" by using the relative values for the corresponding carbon number from the single capillary column approach. This approach is shown schematically in Figure 11.

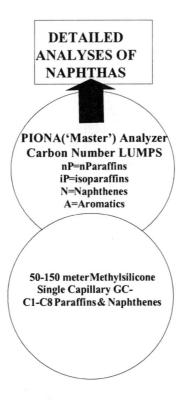

Figure 11. Scheme for Merging Micropacked/Packed PIONA and Single Capillary Column DHA Analyses to Maximize Detailed Hydrocarbon Composition and Accuracy for Naphthas

1.8   *Capillary Column* Multidimensional Systems

In the past, *capillary* column multidimensional GC systems have been used to obtain hydrocarbon type analyses as well as distribution of individual hydrocarbons. Generally two GC ovens are used, one oven containing a polar capillary column such as Carbowax phase and the second oven containing a dimethylsiloxane phase coated capillary. In some applications, it may be possible to place both columns in the same oven. Two configurations have been used by us.[10] Figure 12 shows a two oven system with a Carbowax and methylsilicone columns. Configuration A is simpler to use. The two capillary columns may be connected via a pressure switching device or by a valve. The chromatograms obtained by such configuration A are shown in Figures 13-14. Benzene and toluene are "heart cut" with the saturates to the second methylsilicone column to minimize their overlap with the higher boiling saturates on the first Carbowax column. Configuration B is used when olefins are present. A post column reactor, containing impregnated sulfuric acid is used on the saturates + olefins effluent to selectively remove the olefins from saturates. The olefins are quantitated by monitoring the signal difference between the unscrubbed (FID 2) and scrubbed (FID 3) effluents by using two flame ionization detectors. The aromatic fraction is monitored separately with a third flame ionization detector (FID 1). Figures 15-17 show the resulting three chromatogram from configuration B. Figure 18 shows an expanded chromatograms from the unscrubbed (FID 2) and scrubbed effluents (FID 3) obtained using the two flame ionization detectors to demonstrate the reaction and removal of the olefins. The capillary multidimensional GC system has been in used for several years and proved relatively rugged for a detailed isomer analysis of naphthas.

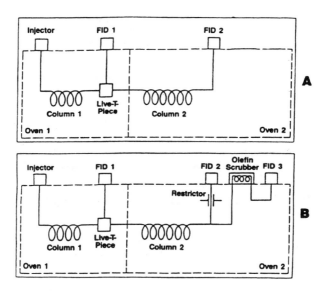

Figure 12. Capillary Column Multidimensional GC Configurations (Ref. 10): A) Column 1: Carbowax and Column 2: Methylsilicone; B) Similar to "A" but with Addition of Post Column Reactor for Removal of Olefins.

**REFORMATE AROMATIC FRACTION
ON DURAWAX**

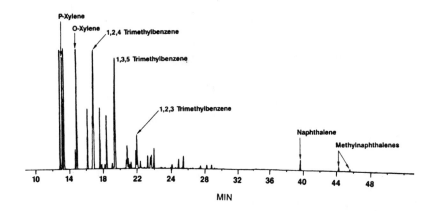

Figure 13. FID 1 Gas Chromatogram of Configuration "A": Carbowax Capillary Aromatic Hydrocarbons

**FID 2
SATURATES AND OLEFINS**

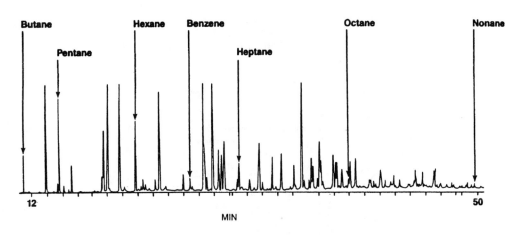

Figure 14. FID 2 Gas Chromatogram of Configuration 'A': Methylsilicone Capillary for Saturates + Olefins + Benzene + Toluene Hydrocarbons

FID 1
AROMATICS

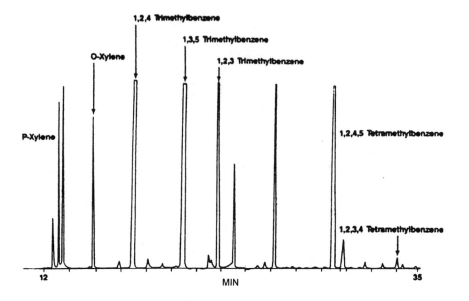

Figure 15. FID 1 Gas Chromatogram of Configuration "B": Aromatic Hydrocarbons

FID 2
SATURATES AND OLEFINS

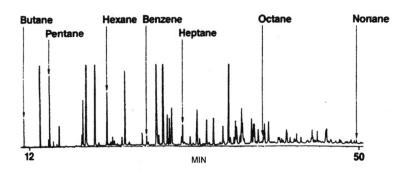

Figure 16. FID 2 Gas Chromatogram of Configuration "B": Saturates + Olefins Hydrocarbons

FID 3
SATURATES ONLY

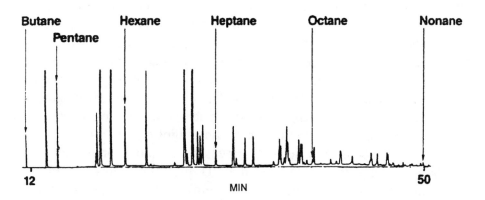

Figure 17. FID 3 Gas Chromatogram of Configuration "B": Olefins Hydrocarbons Only

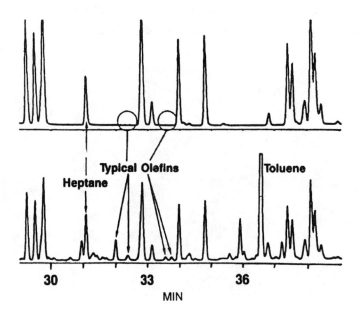

Figure 18. Expanded FID 2/3 Gas Chromatograms of Configuration 'B': Demonstration of Subtraction of the Olefins and Toluene

1.9   Comprehensive Two-dimensional GC (2D-GC)

Comprehensive two-dimensional gas chromatography (2D-GC or GCxGC) is a relatively new technique which offers promise for the analysis of naphthas.[11,12] Generally 2D-GC employs two columns of different selectivity (such as boiling point versus polarity) connected together with a refocusing/reinjection modulation at the point where the two columns are connected. 2D-GC differs from conventional "heartcutting" multidimensional approach described in Section 1.6 in that the "heart cutting" is simulated by modulating continuously over short periods of time such as 3-10 seconds. The resulting mini-chromatograms are plotted in a two dimensional mode as shown in Figure 19. The x-axis represents the retention time of the boiling point capillary column while the y-axis represents the second dimension of the shorter more polar capillary column. Each 'dot' on the 2D plot represents at least one compound. The technique may prove useful for the enhanced separation of compounds which coelute in a single dimension 50-150 meter polymethylsiloxane DHA capillary columns and will increase the number of resolved compounds. It is expected that 2D-GC will play an increasing role in the analysis of naphthas, jet and diesel fuels.

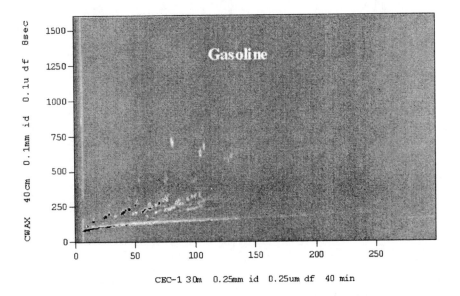

Figure 19. Comprehensive 2D-GC Image of a Gasoline Sample

1.10 Other GC Methods for Blended Gasoline Analysis

The chromatographic methods described above are generally used for a detailed compositional analysis of blended gasolines and naphthas. There are other methods for the analyses of gasolines that are used or have been used in regulation of gasolines parameters. The latter methods generally have more extensive calibration requirements and are used to quantitate specific components of the blended gasolines. Table 13 summarizes some of these methods:

Table 13. Other Useful Methods for Analyses of Selected Parameters in Naphthas

Method	Parameter determined	Technique	Comments
ASTM D5580-02	Benzene, toluene, total aromatics	Multidimensional GC	Some interference from $C_{12}+$ saturates if present in sample in the determination of total aromatics
ASTM D5769-98	Benzene, toluene and many other specified individual aromatics; total aromatics	GC/MS	The $C_{10}+$ aromatics are grouped by their carbon number and their concentration estimated from an average response factor which may yield slightly lower total aromatic results for blended gasolines than actually present.
ASTM D 5986-96	Oxygenates, Benzene, other specified individual aromatics, total aromatics	GC/FTIR	Response factors very stable over time
ASTM D3606-99	benzene	Packed Column/ Multidimensional GC	
ASTM D4815-99	Specific oxygenates in gasolines	Multidimensional GC	
ASTM D5599-00	Specific oxygenates in gasolines	Single capillary GC	Uses 'oxygen' specific flame ionization detection (O-FID)

2   ANALYSES OF NAPHTHAS, MOTOR GASOLINES, JET FUELS, DIESEL FUELS AND HIGHER PETROLEUM FRACTIONS BY SUPERCRITICAL FLUID CHROMATOGRAPHY (SFC) AND LIQUID CHROMATOGRAPHY (LC)

Several chromatographic methods have been developed for the analysis of gasolines, jet and diesel fuels using supercritical fluid chromatography (SFC) and high performance liquid chromatography (HPLC).

2.1   Supercritical Fluid Chromatography (gasolines, jet fuels and diesel fuels)

SFC is commonly used for the analysis of total saturates, and aromatic ring distribution.[13,14] The advantage of SFC with supercritical carbon dioxide as the mobile phase is that a flame ionization detector (FID) is used for the quantitation. *When the FID is properly optimized*, the response factor for most hydrocarbons is approximately unity, thus, simplifying quantitative analysis. ASTM test method D5186-99 provides total saturates (paraffins + naphthenes) and total aromatics in jet and diesel fuels using silica gel packed columns.[13] The aromatic rings are separated into one-ring (monoaromatics) and 2-ring (diaromatics + triaromatics) polynuclear aromatics. Figure 20 shows a chromatogram of the typical separation obtained for a diesel sample.

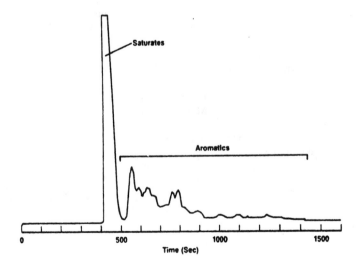

Figure 20. SFC Chromatogram of a Diesel Fuel

Table 14 shows the accuracy of the method as a function of total aromatics using actual diesel range petroleum derived saturated and aromatic hydrocarbons that were blended together at the indicated concentrations.[14] Generally, the accuracy of the method is limited by the resolution between the saturates and the monoaromatics. Table 14 shows that the accuracy deteriorates at concentrations below 10 wt % aromatics as the 'tailing' saturated hydrocarbons interfere with the smaller amounts of mono-aromatics.

Table 14. . Accuracy of SFC Analysis (ASTM D5186) of Diesel Fuel Blends

Diesel Blend sample	Actual (wt%)	SFC (wt%)	Relative difference
1	49.19	48.5	+1.4
2	39.86	39.3	+1.4
3	35.24	34.8	+1.2
4	20.40	20.5	+0.5
5	15.57	15.4	-1.1
6	13.71	13.7	0.0
7	12.75	12.5	-2.0
8	10.43	10.5	+0.7
9	9.81	10.3	+5.0
10	7.91	8.7	+10.0
11	7.27	7.8	+7.3
12	5.60	6.8	+21.4
13	3.56	4.8	+34.8

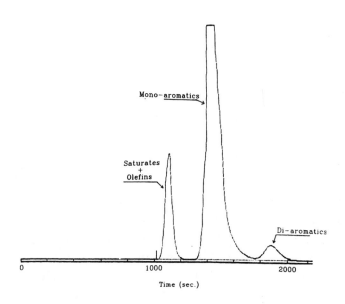

Figure 21. SFC Chromatogram of a *Reformate* Naphtha Using D5186 Silica Gel Column Configuration

SFC using the D5186-99 configuration may also be used for the analysis of total saturates and total aromatics in naphthas and gasolines. Generally for gasolines or naphthas the separation is improved significantly between the saturates and the aromatic hydrocarbons (Figure 21). In the analysis of wide range hydrocarbons by silica gel column SFC, generally the resolution between saturates and the aromatics decreases as the boiling point of the sample increases. Contrast the resolution of gasoline (Figure 21) with that of the diesel sample (Figure 20).

Table 15 summarizes the analysis of various naphthas by SFC and MS. The results are in good agreement with mass spectrometry batch analyses.

Table 15. SFC Analysis of Naphthas from Various Sources using D5186 SFC Configuration. Comparison with MS 'Type' Analysis

Naphtha Type	Saturates + Olefins	Mono-Aromatics	Di-Aromatics
Virgin Naphtha A			
SFC	75.8	24.2	0.0
MS	77.0	23.0	0.0
Virgin Naphtha B			
SFC	84.1	15.9	0.0
MS	84.6	15.4	0.0
Virgin Naphtha C			
SFC	96.5	3.5	0.0
MS	96.3	3.7	0.0
FCC Naphtha A			
SFC	60.9	34.0	5.1
MS	63.4	33.5	3.1
Gasoline 'A' from Methanol			
SFC	84.8	15.0	0.2
MS	84.4	15.2	0.4
Naphtha 'A' from Catalytic Dewaxing			
SFC	91.7	8.3	0.0
MS	92.0	8.0	0.0

The use of quantitative SFC for compositional or type analysis of high boiling petroleum streams has been limited to a few applications mainly due to poorer separations obtained between the bulk classes.[15]

SFC has also been used very successfully for the analysis of total olefins in gasolines. A relatively rugged method for this analysis is ASTM D6550-00.[16] D6550 is a multidimensional SFC method which uses a combination of a silica gel column and a silver impregnated ion exchange resin packed column to trap the olefins away from the saturates and aromatic hydrocarbons. The olefins are then released as a single group and quantitated by an external standard procedure. Figure 22 shows the valve configuration for D6550. Figures 23 and 24 show, respectively, a calibration curve and SFC chromatogram of a gasoline sample.

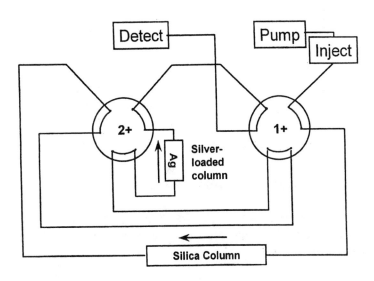

Figure 22. ASTM D6550 SFC Valve/Column Configuration for Analysis of Total Olefins in Gasolines and Naphthas

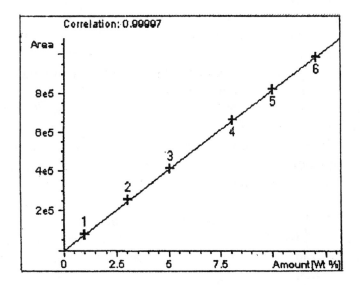

Figure 23. ASTM D6550 SFC Calibration for Analysis of Total Olefins in Gasolines and Naphthas

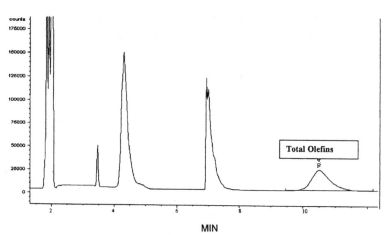

MIN

Figure 24. ASTM D6550 SFC Chromatogram of Blended California Gasoline

2.2 High Performance Liquid Chromatography (HPLC) for Higher Boiling Petroleum Fractions (Lube Feeds/Products, Vacuum Gas Oils)

Normal phase liquid chromatography (NP-LC) is used very extensively in the petroleum industry for the separation of a multitude of petroleum fractions into saturates, aromatics (including by aromatic rings in some cases) and polar components. In many cases non-instrumental manual preparative liquid chromatography is used whereby the isolated fractions are weighed gravimetrically and percent saturates, aromatics and 'polars' is reported.[17] Other methods have also provided aromatic ring distributions. *Instrumental* liquid chromatographic methods can be classified as ***preparative*** and ***analytical*** scale. The preparative scale methods may be automated with autosamplers, the mobile phases regulated precisely by pumping systems and in many cases detectors are used to establish and monitor cut points for the fractions separated. Inmost cases, the quantitation for the preparative scale is accomplished by weighing the fraction collected after solvent evaporation. Because of solvent evaporation, the petroleum fractions isolated have an initial boiling point of approximately 450-500F, mainly because the lighter boiling compounds are lost during the evaporation step. An instrumental preparative method is IP 368 which is used for the separation of total saturates and total aromatics.[18] The latter method uses precise pumping with a refractive index detector (RID) to determine the cut point between the saturates and the aromatics.

An improved separation of saturated hydrocarbons from the aromatic hydrocarbons can be accomplished with the use of silver impregnated columns.[19] The use of silver loaded columns allows excellent separation for difficult to separate hydrocarbon fractions such as very high boiling lube basestocks known as 'bright stocks' and other lubes. The effect of the use of silver loaded columns is shown in Figure 25 with LC analytical scale columns which demonstrates the dramatic improvement in the separation of saturates from the difficult to separate mono-aromatic hydrocarbons present in lube basestock. A preparative system using semi-preparative scale LC columns consisting of an amino column followed by a silver loaded strong ion exchange (SCX) column as described previously can be set-up with a dual UV and RI detector system to monitor the separation.[19] Figure 26 shows a chromatogram obtained from such a system. The separated fractions are semi-automatically evaporated and the solvent free fractions are weighed for quantitative analysis. Table 17 shows the precision obtained with a semi-automated (autosampler + fraction collector) silver impregnated preparative column system using 200 mg sample sizes of a high boiling lube 'bright stock' sample. Generally, preparative systems that use detectors to monitor the cut points are more precise that methods such as D2007 that use fixed volumes for cut determination.

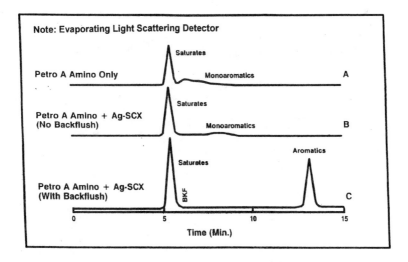

Figure 25. Enhanced Resolution of Aromatics on a Coupled Amino+ Silver-SCX LC Columns (Ref. 19)

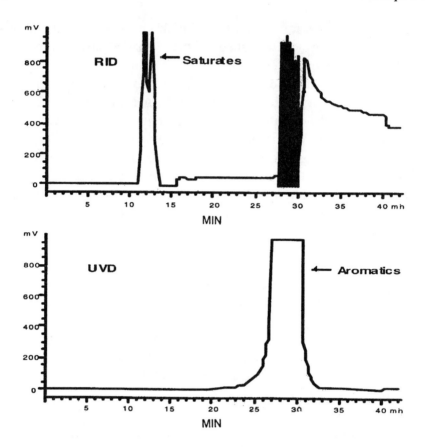

Figure 26. Semipreparative Separation (200 mg) of Saturates and Aromatics in High Boiling Lube Basestock Using Coupled Amino+ Silver-SCX LC Columns and Dual Detector

Table 17. Precision of Semipreparative Separation (200 mg) of Saturates and Aromatics in High Boiling Lube Basestock Using Coupled Amino+ Silver-SCX LC Columns and Dual Detector

Saturates %	Aromatics %	Recovery %
47.5	52.5	98.3
49.0	51.0	99.0
48.7	51.3	98.6
49.0	51.0	98.8
49.0	51.0	98.9
49.0	51.0	98.8
49.0	51.0	99.1
49.1	50.9	98.5
49.1	50.9	99.8
49.0	51.0	100
49.4	50.6	99.5
52.7	47.3	106
48.0	52.0	104

	48.4	51.6	100
	49.4	50.6	98.5
	49.4	50.6	100
Average	49.1	50.9	99.9
Std. Dev.	1.1	1.1	2.1
% RSD	2.2	2.1	2.1

Instrumental *analytical* scale techniques generally use a detector for the quantitation. Over the years this approach was limited because of a lack of mass detectors. Flame ionization detectors for liquid chromatography have been developed to partially resolve the quantitation problem.[19] However, these flame ionization detectors proved not sufficiently rugged and their applications were very limited.

The evaporative light scattering detector (ELSD) has proven more rugged and simpler to operate and has been used more extensively. One application describes the use of the ELSD for the determination of saturates, aromatic ring distributions and polars in high boiling petroleum streams.[20] Because of the nature of the ELSD which requires the removal of the solvent by a high flow of an inert gas, such as nitrogen, most of the light petroleum compounds with an initial boiling point below 300°C are not detected; therefore, such a detection system is limited to petroleum fractions with an initial boiling point greater than approximately 300°C. Samples appropriate for this analysis include petroleum fractions from lube processing (feeds and products) and feeds for catalytic process units such as fluid catalytic cracking (FCC) and resids.

2.3 High Performance Liquid Chromatography (HPLC) for Lower Boiling Petroleum Fractions (Jet Fuels, Diesels)

Analytical scale quantitative instrumental HPLC has been used for the analyses of total saturates, mono-aromatics, di-aromatics and tri-aromatics in jet and diesel fuels. The approach uses a refractive index detector (RID) for quantitation whereby the aromatics hydrocarbons are calibrated with several pure components and the saturates are obtained by difference. The quantitation is based on the fact that the refractive indices of the components within an aromatic ring number are not significantly different and a single component representing each of the aromatic rings may be used for calibration.

Two versions of the HPLC method have been accepted by ASTM which originally were developed by the European Institute Petroleum (IP) standards body. ASTM D6379-99 (21) covers the determination of mono-aromatic and di-aromatic hydrocarbon contents in aviation kerosenes and petroleum distillates boiling in the range from 50 to 300°C, such as Jet A or Jet A-1 fuels.. ASTM D6591-00 (22) is used for the determination of mono-aromatic, di-aromatic, and polyaromatic hydrocarbon contents in diesel fuels and petroleum distillates boiling in the range from 150 to 400°C. A backflush is used for the elution of the polyaromatic triaromatic + components. Figure 27 and 28, respectively, show a schematic of the HPLC system and a LC chromatogram of a diesel sample using an amino-cyano column for the separation. The total aromatic content for both methods is calculated from the sum of the individual aromatic hydrocarbon types. Note that the overall LC resolution between saturates and monoaromatics and among the aromatic rings is improved when compared to the SFC separation (Figure 20). Figure 29 shows a correlation comparison between the LC and SFC methods within one laboratory for total aromatics. Generally, the comparison between the two methods is reasonable.

2.4   Characterization of High Boiling Petroleum Fractions by Thin Layer Chromatography with FID Detection (TLC-FID)

An instrumental technique that has found a niche in the   analysis of petroleum fractions ranging from base stocks, distillates and *particularly 'resids'* is thin layer chromatography with FID detection (TLC-FID). This technique has also been called Iatroscan. It uses thin silica gel coated roads that are spotted and developed as in thin layer chromatography. Usually a multitude of rods may be worked up simultaneously on a holding rack. After a series of solvent evaporation steps, the rods loaded on a rack are automatically scanned through a flame ionization detector for quantitative analysis. The spotting and development of the rods require careful attention to details to ensure reproducibility and can be time consuming. The technique is particularly useful in the analysis of high boiling fractions that represent asphalt-like materials such as resids. The advantage of the technique is its ability to provide total saturates, aromatics, 'resin or polar' and asphaltene composition without the necessity and difficulty in eluting completely strongly bound polar components as would be necessary by LC. The main disadvantage of the technique is that the quantitation is closely dependent on the activity of the silica gel of the rods and requires frequent re-optimization of solvents to ensure that the separations are reproducible over long periods of time. Several applications of the technique are described in the literature.[23, 24]

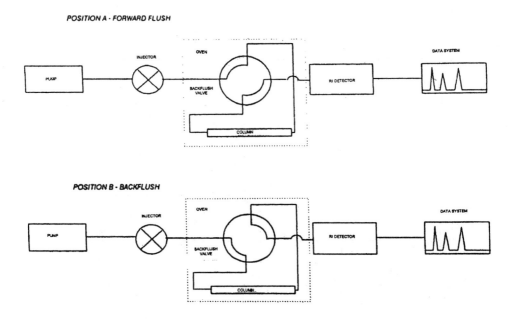

Figure 27. Schematic of HPLC Method for Analysis of Diesel Fuels Using a Column Backflush Configuration

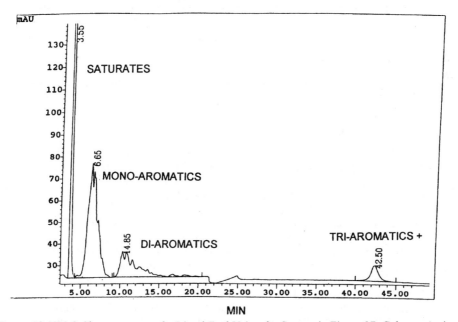

Figure 28. HPLC Chromatogram of a Diesel Fuel Using the System in Figure 27; Column: Amino-cyano

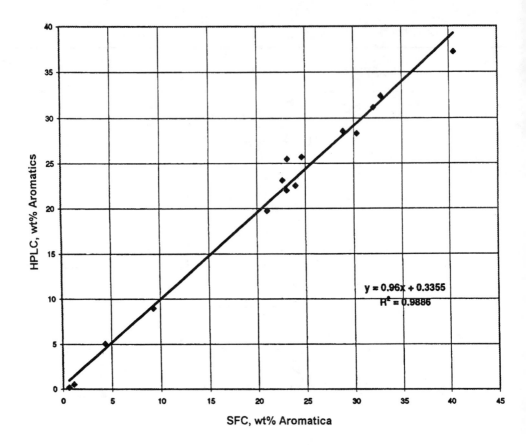

Figure 29. Correlation of Total Aromatic Hydrocarbons Determined by HPLC and SFC Methods.

3.   REFERENCES

1. ASTM D2789-95(2000) "Standard Test Method for Hydrocarbon Types in Low Olefinic Gasoline by Mass Spectrometry"
2. ASTM D5134-98 "Standard Test Method for Detailed Analysis of Petroleum Naphthas through n-Nonane by Capillary Gas Chromatography"
3. ASTM D6733-01 "Standard Test Method for Determination of Individual Components in Spark Ignition Engine Fuels by 50-Meter Capillary High Resolution Gas Chromatography"
4. ASTM D6730-01 "Standard Test Method for Determination of Individual Components in Spark Ignition Engine Fuels by 100-Metre Capillary (with Precolumn) High-Resolution Gas Chromatography"
5. ASTM D6729-01 "Standard Test Method for Determination of Individual Components in Spark Ignition Engine Fuels by 100 Meter Capillary High Resolution Gas Chromatography"
6. ASTM "D3700 Standard Practice for Containing Hydrocarbon Fluid Samples using a Floating Piston Cylinder"

7. Di Sanzo, F. P.; Giarrocco, V. J. "Analysis of Pressurized Gasoline Range Liquid Hydrocarbon Samples by Capillary Column and PIONA Analyzer Gas Chromatography", *J. Chromatogr. Sci.* **1988**, 28, 258.
8. ASTM D5443-93(1998) "Standard Test Method for Paraffin, Naphthene, and Aromatic Hydrocarbon Type Analysis in Petroleum Distillates Through 200°C by Multi-Dimensional Gas Chromatography"
9. ASTM D6293-98 "Standard Test Method for Oxygenates and Paraffin, Olefin, Naphthene, Aromatic (O-PONA) Hydrocarbon Types in Low-Olefin Spark Ignition Engine Fuels by Gas Chromatography"
10. Di Sanzo, F. P.; Lane, J. L.; Yoder, R. E. "Application of State-of-the-art Multidimensional High Resolution Gas Chromatography for Individual Component Analysis of Gasoline Range Hydrocarbons", *J. Chromatogr. Sci.* **1988**, 26, 206-209.
11. Philips, J. B.; Ledford, E. B. "Thermal Modulation: A Chemical InstrumentationPotential Value in Improving Portability" *Field Anal. Chem. Tech.* **1996**, 1, 20
12. Blomberg, J.; Schoenmakers, P. J.; Beens, J.; Tijissen, R. "Comprehensive Two-Dimensional Chromatography (GC x GC) and its Applicability to the Characterization of Complex Mixtures" *J. High Resol. Chrom.* 1997, 20, 539
13. ASTM D5186-99 "Standard Test Method for Determination of Aromatic Content and Polynuclear Aromatic Content of Diesel Fuels and Aviation Turbine Fuels by Supercritical Fluid Chromatography "
14. Di Sanzo, F. P.; Yoder, R. E. "Determination of Aromatics in Jet and Diesel Fuels by Supercritical Fluid Chromatography with Flame Ionization Detection (SFC-FID): A Quantitative Study", *J. Chromatogr. Sci.* **1991**, 29, 4-7.
15. Skaar, H.; Norli, H. R.; Lundanes, E.; Greibrokk, T. "Group Separation of Crude Oil by Supercritical Fluid Chromatography Using Packed Narrow Bore Columns, Column Switching and Backflushing", *J. Microcolumn Separations* 1990, 2(5), 222-228.
16. ASTM D6550-00 "Standard Test Method for Determination of Olefin Content of Gasolines by Supercritical-Fluid Chromatography"
17. ASTM D2007-01a "Standard Test Method for Characteristic Groups in Rubber Extender and Processing Oils and Other Petroleum-Derived Oils by the Clay—Gel Absorption Chromatographic Method"
18. IP 368/95 "Determination of hydrocarbon types in lubricating oil basestocks-Preparative high performance liquid chromatography method"
19. Di Sanzo, F. P., Herron F. P. , Chawla, B., Holloway, D. "Determination of Total Aromatic Hydrocarbons in Lube Base Stocks by Liquid Chromatography with Novel Thermospray Flame Ionization Detection", *Anal. Chem.* **1993**, 65, 3359
20. Robbins, W. K. "Quantitative Measurement of Mass and Aromaticity Distributions for Heavy Distillates 1. Capabilities of the HPLC-2 System", *J. Chromatogr. Sci.* **1998**, 36, 457-466
21. ASTM D6379-99 "Standard Test Method for Determination of Aromatic Hydrocarbon Types in Aviation Fuels and Petroleum Distillates-High Performance Liquid Chromatography Method with Refractive Index Detection"
22. ASTM D6591-00 "Standard Test Method for Determination of Aromatic Hydrocarbon Types in Middle Distillates-High Performance Liquid Chromatography Method with Refractive Index Detection"
23. Cebolla, V. L.; Vela, J,; Membrado, L.; Ferrando, A. C. "Suitability of Thin-Layer Chromatography-Flame Ionization Detection with Regard to Quantitative Characterization of Different Fossil Fuel Products. I. FID Performances and Response of Pure Compounds Related to Fossil Fuel Products" ", *J. Chromatographic Sci.* 1998, 36, 479
24. Vela, J.; Membrado, L.; Cebolla, V. L.; Ferrando, A. C.; "Suitability of Thin-Layer Chromatography-Flame Ionization Detection with Regard to Quantitative Characterization of

Different Fossil Fuel Products. II. Calibration Methods Concerning Quantitative Hydrocarbon-Group Type Analysis" *J. Chromatographic Sci.* **1998**, 36, 487.

Chapter 7

TEMPERATURE-PROGRAMMED RETENTION INDICES FOR GC AND GC-MS OF HYDROCARBON FUELS AND SIMULATED DISTILLATION GC OF HEAVY OILS

Chunshan Song, Wei-Chuan Lai, K. Madhusudan Reddy and Boli Wei
Clean Fuels and Catalysis Program
The Energy Institute, and Department of Energy & Geo-Environmental Engineering
The Pennsylvania State University
209 Academic Projects Building
University Park, PA 16802

1.   INTRODUCTION

It is important to clarify the exact chemical composition of various hydrocarbon streams for both fundamental studies of their composition-structure-property relationships and their practical applications for transportation fuels and organic chemical feedstocks. This chapter is a selective review of studies conducted in our laboratory on temperature-programmed retention indices for gas chromatography (GC) and GC coupled with mass spectrometry (GC-MS) analysis of distillate fuels, and high-temperature simulated distillation GC analysis of residual oils and their upgrading products. Liquid hydrocarbon fuels derived from petroleum and coal as well as natural gas through various processes contain a broad range of compounds that are different in boiling points, carbon number, molecular weight, compound class and chemical structure.

The first part of this chapter is concerned with retention index for temperature-programmed GC and GC-MS analysis of hydrocarbon fuels with emphasis on jet fuels and middle distillates. Modern GC-MS has contributed greatly to analysis of various mixtures of organic compounds with respect to their identification. GC and GC-MS are very useful tools for analysis of distillate fractions of refinery streams and liquid hydrocarbon fuels. However, many compounds in petroleum-derived and coal-derived fuels are difficult to identify by mass spectra alone. A common phenomenon in GC-MS of fuels is that two or more GC peaks have very similar or even identical mass spectra, although they have different retention times and different chemical nature. Examples of such compounds are the isomers of some cycloalkanes, the isomers of substituted aromatics such as trimethylbenzenes (MW 120), methylnaphthalenes (MW 142) and dimethylnaphthalenes (MW 156), different polyaromatics with same molecular weight but different ring structures, such as biphenyl and acenaphthene (MW 154), and phenanthrene and anthracene (MW 178), partially hydrogenated aromatics such as sym-octahydrophenanthrene and sym-octahydroanthracene (MW 186). Thus using GC-MS for compound identification often requires the assistance or confirmation of GC retention time. On the other hand, in many cases highly reliable identification can not be made by using retention times alone. This is because co-elution of two or more different compounds is possible under given conditions, and whether this occurs can not be determined by GC alone for unknown samples. For these reasons, the retention times and mass spectra are complementary to each other. The combined use of retention index and mass spectra can allow the identifications of individual compounds in complex mixtures to be made with high confidence.

The analytical GC and GC-MS work in our laboratory was conducted as a part of a research program at The Pennsylvania State University for investigating the compositional factors affecting thermal stability of jet fuels. The sources of the future hydrocarbon jet fuels may include petroleum, coal, and other fossil resources.[1-5] The future high-performance aircraft requires advanced thermally stable jet fuels.[6-10] Clarifying the hydrocarbon components in conventional and alternate jet fuels and identifying the thermally stable and unstable compounds are considered to be key steps in developing advanced jet fuels with high thermal stability and high density for future high-performance aircraft.[11-15] The need to characterize coal- and petroleum-derived jet fuels stems from the lack of knowledge on their molecular composition and the desire to establish the relationship between their composition/structure and thermal stability at high temperatures. Hayes and Pitzer[16,17] and Steward and Pitzer[18] have analyzed several petroleum- and shale-derived JP-4 jet fuels using capillary gas chromatography. These fuels are deceptively simple in appearance but are actually mixtures of hundreds of hydrocarbons. They indicated that, even for petroleum-derived jet fuels, the detailed hydrocarbon distribution is far too complicated to be unraveled by even the most efficient capillary column.[16] The coal-derived fuels are more complex than petroleum-derived fuels.[15] There is little published information on the detailed characterization of coal-derived jet fuels.

A common practice of qualitative GC analysis is the use of retention time. However, it is well known that retention time depends on several factors, e.g., temperature and flow rate. Thus retention time itself is not an ideal parameter for identification purposes, especially for complex samples such as jet fuels;

parameters that are less dependent on these factors are needed. A useful parameter is the relative retention time instead of the absolute retention time. Retention index [19,20] system is a measure of relative retention times referenced to ma homologous series of organic compounds, and is one of the successful parameters used for identification purposes [16-18, 21-25]. Hayes and Pitzer [16,17] have demonstrated the usefulness of determining retention indices for identifying compounds in complex mixtures. Though many GC and GC-MS analyses are carried out under isothermal conditions, temperature programming is known to improve separating complex mixtures such as liquid fuels whose components have widely varying vapor pressures.[26]

Kováts index relates (interpolates) the retention time of an unknown compound to that of reference standards eluting before and after it under isothermal conditions. Any homologous series of organic compounds (for example, n-alkanes) can be used as the retention index reference standards. Each of the standards is assigned an index I, for example, $I = 100n$ for a given alkane with a carbon number n. The retention index of any unknown compound is then calculated by logarithmic interpolation between the two relevant standards according to the following relationship under isothermal conditions

$$I_u = 100 \left[\frac{\log(t_u) - \log(t_n)}{\log(t_{n+1}) - \log(t_n)} + n \right] \tag{1}$$

where n and n+1 are the carbon numbers of the bracketing n-hydrocarbons, is the absolute retention time of the unknown under isothermal conditions, and are the absolute retention times of the n-hydrocarbons, eluted just prior to and just after the unknown respectively under isothermal conditions. However, in contrast to the logarithmetric relationship under isothermal conditions, a quasi-linear relationship exists for a non-isothermal GC analysis with a linear temperature program, as shown by Van den Dool and Kratz [27].

The work described in the first part of this chapter aimed at establishing retention indices for temperature-programmed GC and GC-MS analysis on two different capillary columns at three programmed heating rates for over 150 compounds that are components of coal and petroleum-derived liquid fuels. The effects of heating rate, temperature program, and polarity of the stationary phase on the retention indices as well as the relative elution order were studied. The temperature dependence and the column polarity dependence of retention indices have been examined for different compound classes. The data presented are temperature-programmed retention indices of 1-alkenes, n-alkanes, alkylcyclohexanes, alkylbenzenes, polycyclic aromatics, partially hydrogenated polycyclic aromatics, and N-, S-, and O-containing compounds. Also presented in this chapter are the retention indices on many compounds that were not used in previous studies but are generally found in coal-derived liquid fuels. The usefulness of using temperature-programmed retention indices for GC and GC-MS analysis was demonstrated in the detailed identification of components in two JP-8 jet fuels including a petroleum-derived JP-8P and a coal-derived JP-8C. The results reported here are believed to be useful to researchers in the field of chromatographic analysis of liquid fuels.

The second part of this chapter deals with simulated distillation gas chromatographic analysis of heavy oils and their upgrading products from catalytic hydroprocessing. Detailed analysis of heavy oils and residual oils is more difficult than that of distillate fuels in general. Residues of petroleum distillation, also called resids, are referred to as the bottom of the barrel and include two types of distillation tower bottoms. The term "atmospheric resid" describes the material at the bottom of the atmospheric distillation tower having a lower boiling limit of about 340°C; the term vacuum resid refers to the bottoms from vacuum distillation which has an atmospheric equivalent boiling point (AEBP) range above 540°C.[28]

The main purpose of petroleum resid upgrading is to reduce the boiling points as well as to remove sulfur, nitrogen and metals. Hydroprocessing is one of the important ways of resid upgrading. It is necessary to characterize resids and their products with various techniques that indicate their compositional features. Because petroleum resids are complex mixtures that contain hundreds of compounds with high molecular weights, characterizing resids is a tedious and complicated task. Laboratory-scale distillation tests are time-consuming but widely used for determining the boiling point range of crude oils and their products.[29] However, the boiling point range that can be determined by laboratory distillation operation is limited, particularly for resids.

Simulated distillation analysis by gas chromatography (SimDis GC) requires about one hour for each test. It achieved ASTM standard status in 1973 as test D2887. This test was revised in 1984, 1989 and 1993. ASTM D5307 was defined in 1992 for determining the crude oil boiling point range. However, these methods can only be applied to fractions with final boiling point up to 538°C (1000°F).[30,31] Consequently, various techniques have been tested to determine the boiling point distribution of heavy resids. Padlo and Kugler[32] studied SimDis by HPLC (high performance liquid chromatography) using an evaporative light scattering detector. Bacaud et al.[33] performed modeling based on SimDis GC to evaluate hydroprocessed resids and Klein and coworkers[34] used it as a part of their analytical data for resid structure modeling.

High-temperature simulated distillation using GC (HT-SimDis GC) has become possible recently with the development of thermally stable wide bore capillary columns. Some of the advantages of open tubular columns over packed columns are: better column stability and life, lower column bleed, faster analysis, elution of higher boiling petroleum fractions, compatibility with automated on-column injection, and improved reproducibility.[35] However, HT-SimDis GC has not been widely used, and no standard method is available for resids. We have established a method in our laboratory for HT-SimDis GC analysis of petroleum resids, with AEBP end points of up to 847°C (1557°F). This chapter presents our results on HT-SimDis GC analysis of two resids and their products from laboratory tests of catalytic hydroprocessing. The term "resid" in the remainder of this chapter refers to the feedstock, and the term "residue" refers to the uneluted portion of the resid in HT-SimDis GC.

2. EXPERIMENTAL

2.1 Reagents and Fuels

Reagent grade chemicals from Aldrich, Supelco, K&K Laboratories, TCI America, and Fisher Scientific were used for the determination of retention indices. Over 150 pure compounds were examined, covering aliphatic, olefinic, aromatic, and polycyclic aromatic hydrocarbons as well as O-, N-, and S-containing compounds that are related to petroleum- and coal-derived liquid hydrocarbon fuels.

The jet fuel samples used in this work were two JP-8 fuels supplied by the Air Force Wright Laboratory/Aero Propulsion and Power Directorate, unless otherwise mentioned. The coal-derived fuel, JP-8C, was produced by hydrotreating of tar liquids produced from the Great Plains Gasification plant.[14,36] The petroleum-derived JP-8P is a conventional military jet fuel, whose properties are usually selected to be identical to those of commercial jet fuel Jet A-1.[37] The JP-8C fuel is a blend of hydrotreated and hydrocracked stocks.[14] In addition, we have performed SimDis GC analysis of other military jet fuels (JP-8P, JP-8P2, JP-7, JP-TS), commercial jet fuels (Jet A, Jet A-1), middle distillates from Fischser-Tropsch synthesis (FT-MD), and middle distillates from direct catalytic coal liquefaction from Wilsonville plant (WI-MD) and from HRI (Hydrocarbon Research Inc.) plant (HRI-MD).[11,14,15] JP-8P2 is a different sample of petroleum-derived JP-8 jet fuel that is similar to JP-8P.

2.2 Retention Index

Standard mixtures of reagent grade chemicals were prepared for the determination of retention indices. The standard mixtures were analyzed on a Hewlett-Packard 5890 Series II GC coupled with HP 5971A Mass Selective Detector (MSD). The columns used were 30 m, 0.25 mm i.d., 0.25 μm film thickness, fused silica capillary columns (intermediately polar Restek Rtx-50 column coated with 50% diphenyl polysiloxane - 50% dimethyl polysiloxane; slightly polar J&W DB-5 column coated with 5% diphenyl - 95% dimethyl polysiloxane). In terms of stationary phase coating, the DB-5 column used in this work is similar to many other commercially available columns, such as Rtx-5, Rtx-5ms, HP-5, MXT-5, XTI-5, OV-5, DB-5.625, DB-5ht, AT-5, ATS-5, EC-5, 007-2, CP-Sil 8CB, PTE-5, PAS-5, RSL-200, SAC-5, SE-54, SPB-5, and ULTRA-2. The slightly polar DB-5 column contains 5% diphenylpolysiloxane (and 95% dimethyl polysiloxane) in its stationary phase compared to the non-polar DB-1 column which contains 100% dimethylpolysiloxane. The Rtx-50 column used in this work is similar (in terms of stationary phase chemical composition) to many other commercially available columns such as DB-17, HP-17, AT-50, 007-17, HP-50+, OV-17, SP-50, and SP-2250.

The column temperature was programmed linearly from 40°C to 310°C at three heating rates (2, 4, and 6°C/min), unless otherwise mentioned. Index values calculated from three different heating rates can be used as cross-references to known materials and aid in identifying unknown compounds. Other chromatographic operating conditions are as follows: detector temperature, 280°C; injector temperature, 280°C; average linear velocity of carrier gas (helium) through the column, 33 cm/sec; septum purge flow rate, 13 ml/min; volumetric column flow rate, 0.98 ml/min. The capillary column's sample capacity was also taken into consideration to avoid non-Gaussian peaks (tailing or fronting), which shift slightly in retention time. Symmetric (Gaussian) peaks were obtained by injecting appropriate amount (0.01 μL) standards under split mode (split ratio at 66); the amount of each standard compound entering the column was kept in the range of 1-10 ng/component.

The temperature-programmed retention indices of over 150 compounds were determined at three heating rates using the equation established by Van den Dool and Kratz [27]. They have shown that the retention indices for a linear temperature programmed GC can be calculated by the following quasi-linear relationship:

$$I_u = 100 \left[\frac{t_u - t_n}{t_{n+1} - t_n} + n \right] \qquad (2)$$

where the absolute retention times of the compounds are determined under conditions of linear temperature programming.

2.3   Chromatographic Separation of Distillate Fuels

The chromatographic fractions of JP-8P and JP-8C fuels were obtained using neutral alumina column and a series of elution solvents.[15] A 22 mm i.d. x 500 mm Pyrex glass column with removable Teflon stopcock was used. Small pieces of glass wool and 3 mm glass beads were used to retain the packing. The column was wet-packed using alumina and n-pentane. The neutral Al_2O_3 was added to the pentane-filled column slowly with continuous tapping to avoid trapping air and to insure a homogeneous sorbent bed. The jet fuel sample was placed directly on top of the Al_2O_3 column and was allowed to pre-adsorb on the Al_2O_3 gel at the top of the column. The solvents and their volume are as follows: n-pentane, 350 ml; 5% benzene-pentane, 200 ml; benzene, 250 ml; 1% ethanol-chloroform, 300 ml; 10 % ethanol-THF, 200 ml. The chromatographic fractions were recovered by rotary evaporation of the solvent.

2.4   Solvent Extraction of Petroleum Resids

Two petroleum resids, an atmospheric resid (AR) and a vacuum resid (VR), were used in this study which were obtained from Marathon Oil Company (as atmospheric resid #1, and vaccum resid #2, from Garyville Refinery). Their elemental compositions are as follows: for AR, C: 84.64; H: 11.14; S: 3.87; N: 0.23, O (by difference): 0.12 wt%; and for VR, C: 83.20; H: 9.83; S: 5.01; N: 0.54, O (by difference): 1.42 wt%. Asphaltenes, the hexane-insoluble but toluene-soluble fraction of the resid and were estimated for AR and VR. To do this estimation, 4 gram resid was dissolved in 400 mL hexane in an ultrasonic bath for 1 hour and then filtered through a 0.45 μm (pore size) membrane filter. The solid on filter paper was dried and weighed as hexane insoluble (HI) material. A similar procedure was used with toluene to determine the toluene insoluble (TI). The amounts of asphaltenes in AR and VR were determined from the difference of hexane and toluene insoluble.

2.5   High-temperature Simulated Distillation GC

HT-SimDis GC was performed using a Hewlett-Packard Model 5890 II Plus GC with a flame ionization detector and temperature programmable cool on-column injector. On-column injection avoids sample discrimination and decomposition that can occur in heated inlets that rely on flash vaporization.[8] The injector temperature was always maintained 3°C higher than the column temperature. Each sample was injected with an automatic sampler to maximize reproducibility. Raising the column temperature above 400°C can cause a considerable drop in the capillary column flow rate. This is caused by the increased carrier gas viscosity at higher temperatures. To avoid this effect, the GC on-column flow system is operated in a flow control mode with the carrier (He) flow constant (10.6 mL/min). The velocity of the carrier gas flow was 79.2 cm/second. A high-temperature aluminum-clad megabore capillary column (HT5 phase consisting of Carborane, 6 m length, 0.53 mm I.D., 0.78 mm O.D., and 0.1 μm film thickness; from SGE supplied by Supelco Inc.) was used for HT-SimDis GC analysis. A computer-based chromatography data system (Hewlett-Packard ChemStation software, B.02.04) was used for data acquisition. To obtain a sample area corrected for baseline drift and signal offset, the column compensation was performed and stored to do subtraction and give the right values. The SimDis Expert software (V 5.0) provided by Separations Systems Co. was used for post-run data analysis of HT-SimDis.

Prior to injection, the samples were diluted with CS_2 to about 2 wt%. High purity CS_2 was used as a solvent because of it s miscibility with resids, low boiling point, and low response factor in the FID. The injection volumes were 1.0 μL. The calibration standard containing n-paraffins (C_5-C_{100}) was prepared by dissolving liquid n-paraffins mixture (C_5-C_{20}) and polywax (Polywax 655) in CS_2 under an IR lamp in a sealed vial. Compounds used as true boiling point references were reagent-grade chemicals obtained from commercial suppliers. High boiling point lube oil standard (supplied by Separation Systems) was used as a reference standard to check the HT-SimDis GC calibration; it was used as an external standard for estimating the uneluted portion (undistillable residue) of the resids and their products.

Four different sets of GC conditions were tested in screening analysis. After the conditions had been set to meet the performance requirements, the column was conditioned; the blank run was then conducted and the data stored for column compensation. Using the same conditions as for the blank run, an appropriate aliquot of the calibration mixture was injected to obtain a retention time versus boiling point calibration. In general, the volume of the calibration mixture injected must be selected to avoid distortion of any component peak shapes caused by overloading the sample capacity of the column. Distorted peaks will result in erroneous retention times and hence in errors in the boiling point determination. The calibration curve (retention time of each peak versus the corresponding the boiling point of the component) should be essentially linear. Our calibration has been always linear in the range between C_{10} and C_{94} paraffin; the latter has a boiling point of 704°C. It was assumed that the calibration is linear beyond that point. Once a calibration is ready, it is then verified with reference standard. We have also measured boiling points of several types of pure model compounds (cycloalkanes, aromatics, heteroatom-containing compounds) and compared with their true boiling points.

2.6 Quantitative Calculations from SimDis GC Data

Both external standard and real samples were prepared after allowing them to reach room temperature (if they were stored in a refrigerator) to make them homogeneous. The exact amounts were weighed and dissolved in given amounts of CS_2 in order to obtain about a 2 wt% solution in CS_2.

The external standard and resid samples were analyzed separately using exactly the same conditions that were used in the blank and calibration runs. Exactly 1 μL of diluted samples and external standards were injected using the auto sample injector. With the total corrected cumulative area of the external standard and the samples, the residue (uneluted portion of the samples) was estimated using following equations:

$$W_E = M_E/(M_C + M_E);$$

where W_E = Weight fraction of external standard,
M_E = Wt. of external standard, and
M_C = Wt. of solvent CS_2.

$$W_S = M_S/(M_C + M_S);$$

where W_S = Weight fraction of resid sample,
M_S = Wt. of resid sample, and
M_C = Wt. of solvent CS_2.

$$ESAM = 100 \times (A_S/A_E) \times (W_E/W_S);$$

where $ESAM$ = Eluted portion (distillable) of the resid sample in wt%,
A_E = Cumulative GC peak area of ext. standard of 1 μL injection,
A_S = Cumulative GC peak area of resid sample of 1 μL injection.

$$ES = 100 - ESAM;$$

where RES = Uneluted portion (residue) of the resid sample in wt%.

Once the percentage of eluted portion of resid sample has been estimated, the boiling point distribution is calculated from the cumulative area at regular intervals of retention time. The IBP is defined as the temperature equivalent to the time when the first cumulative area is 0.5% of the total. The rest of the boiling point distribution is calculated from the cumulative area at a given time equivalent to the particular temperature and the tabulated percentage of sample boiling at the corresponding temperature.

2.7   Hydroprocessing of Resids

The hydroprocessing resids was performed in the absence and presence of a commercial Co-Mo/Al_2O_3 catalyst, Shell 344TL (Criterion 344TL). Prior to the catalytic tests, the Co-Mo/Al$_2$O$_3$ was ground up and presulfided. The hydroprocessing was carried out in a 25 mL horizontal tubing bomb reactor loaded with 3 g of resid and 0.3 g of the catalyst under an initial hydrogen pressure of 6.9 MPa for 60 min. The reaction temperature range between 350 and 450°C at intervals of 25°C. Details of catalyst sulfidation (using CS$_2$ in hydrocarbon solvent under H$_2$ pressure) and hydroprocessing procedures are given elsewhere.[38-41] After reaction, the reactor was quenched in a cold water bath and products were collected in CS$_2$. The solid and liquid products were filtered. The insoluble solid product is referred to as coke. The liquids were analyzed by HT-SimDis GC. The SimDis analysis data were processed and reported according to the boiling point ranges based on the classification proposed by Altgelt and Boduszinski [28], except that the FBP is 847°C in our work instead of 700°C. Resid conversion was calculated from the fractions of > 340°C and >540°C boiling point and the liquid yields.

3.   RESULTS AND DISCUSSION

3.1   GC and GC-MS of Distillate Fuels

3.1.1   Retention Index of Model Compounds

Alkanes, cycloalkanes, and aromatics are the three major classes of hydrocarbon components in liquid fuels as well as thermally stressed jet fuels.[11] Theoretically, any of the homologous series of n-alkanes, n-alkylbenzenes, and n-alkylcyclohexanes may be used as the reference standards for interpolation. Because of their common occurrence in most fuel samples, n-alkanes were used as bracketing hydrocarbons for calculating retention indices in this work. Figure 1 shows that the absolute retention times of n-alkanes are locally quasi-linear for large n-alkanes (> n-C$_{13}$) with linear temperature program. This implies that neighboring n-alkanes are needed to obtain more accurate retention indices in the lower alkanes region (< n-C$_{13}$) because of the poorer linearity. On the other hand, when large neighboring n-alkanes (> n-C$_{13}$) are not available, retention indices can be estimated by interpolation between large non-neighboring n-alkanes, and reasonably accurate results can be obtained, as shown in the following example. The retention indices of naphthalene calculated by interpolating between non-neighboring alkanes (n-tridecane and n-pentadecane) are 1388.2, 1403.3, and 1413.4, respectively, at 2, 4, and 6°C/min; these are comparable with those calculated from neighboring n-alkanes: 1386.8, 1401.2, and 1410.9.

Table 1 compares the retention time and retention indices of various hydrocarbons based on GC-MS using Rtx-50 column at two heating rates. Table

2 presents the retention indices determined using two columns at three heating rates (2, 4, and 6°C/min) for 154 compounds generally found in coal- and petroleum-derived liquid fuels. For calculation of indices shown in Table 1, neighboring n-alkanes were used as the reference standards (fixed points) for more accurate results. The compounds are arranged in increasing order of indices at the heating rate of 2°C/min on the Rtx-50 column. The values reported are the averages of 2-5 replicates. The retention times were reproducible to within about ±0.01 min and ±0.02 min for low- and high-boiling compounds, respectively. In general, the experimental error for the retention indices is of the order of 2 index units (i.u.). However, it is slightly larger for those low-boiling compounds eluted before n-octane; this can be rationalized from Equation 2 that due to the shorter retention times of these compounds a small difference in retention times can cause a larger error in the index calculation.

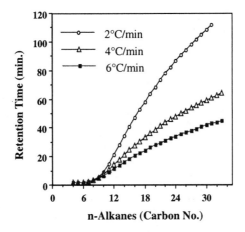

Figure 1. Retention times of n-alkanes at three different heating rates (Rtx-50 column, 40°C to 310°C)

Table 1. Comparison of retention time and retention indices of various hydrocarbons and standards from GC-MS on Rtx-50 at two heating rates

R.T. 4°C/min GC-MS	R.I. 4°C/min GC-MS	R.T. 2°C/min GC-MS	R.I. 2°C/min GC-MS	Compounds Identified GC-MS, GC	Mol.. Ion MS	Base Peak MS
1.59	**400.0**	**1.58**	**400.0**	**n-Butane (n-C4)**	**58**	**43**
1.63	466.7	1.62	466.7	Butane, 2-methyl-	72	43
1.65	**500.0**	**1.64**	**500.0**	**n-Pentane (n-C5)**	**72**	**43**
1.73	553.3	1.73	556.3	Pentane, 2-methyl-	86	43
1.80	**600.0**	**1.80**	**600.0**	**n-Hexane (n-C6)**	**86**	**57**
1.82	606.1	1.82	605.7	1-Hexene	84	56
1.92	636.4	1.93	637.1	Methylenechloride	84	49
2.13	**700.0**	**2.15**	**700.0**	**n-Heptane (n-C7)**	**100**	**43**

R.T. 4°C/min GC-MS	R.I. 4°C/min GC-MS	R.T. 2°C/min GC-MS	R.I. 2°C/min GC-MS	Compounds Identified GC-MS, GC	Mol.. Ion MS	Base Peak MS
2.15	702.7	2.18	703.6	Cyclohexane	84	56
2.19	708.1	2.21	707.2	1-Heptene	98	56
2.41	737.8	2.45	736.1	Cyclohexene	82	67
2.49	748.6	2.55	748.2	Cyclohexane, methyl-	98	83
2.62	766.2	2.68	763.9	Benzene	78	78
2.87	**800.0**	**2.98**	**800.0**	**n-Octane (n-C8)**	**114**	**43**
2.89	801.4	3.02	802.2	1,3-Dimethylcyclohexane, cis-	112	97
2.91	802.7	3.01	801.7	1,4-Dimethylcyclohexane, trans-	112	55,97
3.00	808.9	3.13	808.4	1-Octene	112	55
3.18	821.2	3.34	820.2	1,2-Dimethylcyclohexane, trans-	112	55,97
3.25	826.0	3.42	824.7	Cyclohexene, 1-methyl-	96	81
3.30	829.5	3.47	827.5	1,4-Dimethylcyclohexane, cis-	112	55,97
3.31	830.1	3.50	829.2	1,3-Dimethylcyclohexane, trans-	112	97
3.76	861.0	4.04	859.6	1,2-Dimethylcyclohexane, cis-	112	55,97
3.79	863.0	4.07	861.2	Cyclohexane, ethyl	112	83
3.89	869.9	4.21	869.1	Toluene	92	91
4.33	**900.0**	**4.76**	**900.0**	**n-Nonane (n-C9)**	**128**	**43**
4.49	906.9	4.97	906.5	Pyridine	79	79
4.53	908.7	5.02	908.0	1-Nonene	126	56
5.69	958.9	6.55	955.1	Cyclohexane, n-propyl-	126	83
5.92	968.8	6.89	965.5	Benzene, ethyl-	106	91
6.01	972.7	7.01	969.2	p-Xylene	106	91
6.05	974.5	7.08	971.4	m-Xylene	106	91
6.64	**1000.0**	**8.01**	**1000.0**	**n-Decane (n-C10)**	**142**	**43,57**
6.93	1009.6	8.38	1007.6	o-Xylene	106	91
6.96	1010.6	8.47	1009.4	1-Decene	140	56
7.03	1012.9	8.46	1009.2	1H-Indene, octahydro-, trans-	124	67
7.16	1017.2	8.67	1013.5	Cyclohexane, t-butyl-	140	56
7.54	1029.7	9.28	1026.0	Cumene (i-propylbenzene)	120	105
8.35	1056.4	10.43	1049.6	1H-Indene, octahydro-, cis-	124	67
8.46	1060.1	10.73	1055.7	Benzene, n-propyl-	120	91
8.53	1062.4	10.82	1057.6	Cyclohexane, n-butyl-	140	83
8.79	1071.0	11.28	1067.0	Benzene, 1-ethyl-3-methyl-	120	105
8.94	1075.9	11.47	1070.9	Benzene, 1,3,5-trimethyl-	120	105
9.66	1099.7	12.64	1094.9	Benzene, 1-ethyl-2-methyl-	120	105
9.67	**1100.0**	**12.89**	**1100.0**	**n-Undecane (n-C11)**	**156**	**57**
9.67	1100.0	12.69	1095.9	Benzene, t-butyl-	134	119
9.69	1100.6	12.66	1095.3	4,5,6,7-Tetrahydroindan	122	79
10.00	1109.7	13.15	1104.3	Benzene, 1,2,4-trimethyl-	120	105
10.06	1111.4	13.46	1109.5	1-Undecene	154	41
10.20	1115.5	13.57	1111.3	Benzene, sec-butyl-	134	105
10.31	1118.8	13.60	1111.8	trans-Decalin	138	138
11.15	1143.4	15.00	1135.2	1,1'-Bicyclopentyl-	138	68
11.41	1151.0	15.52	1143.8	Benzene, 1,2,3-trimethyl-	120	105
11.42	1151.3	15.72	1147.2	Phenylethylether	122	94
11.72	1160.1	16.18	1154.8	Benzene, methyl-propyl-	134	105
11.91	1165.7	16.52	1160.5	Benzene, n-butyl-	134	91
11.96	1167.2	16.58	1161.5	Cyclohexane, n-pentyl-	154	83
12.38	1179.5	17.10	1170.2	cis-Decalin	138	67
12.49	1182.7			Benzene, methyl-propyl-	134	105
12.52	1183.6	17.43	1175.7	Indan	118	117
13.08	**1200.0**	**18.89**	**1200.0**	**n-Dodecane (n-C12)**	**170**	**57**
13.54	1213.3	19.56	1210.3	1-Dodecene	168	55
13.62	1215.6	19.44	1208.5	Indan, 2-methyl-	132	117
13.98	1226.0	20.09	1218.5	Indan, 1-methyl-	132	117
14.40	1238.2	20.90	1230.9	Benzene, 1,2,4,5-tetramethyl-	134	119
15.60	1272.8	23.21	1266.5	Cyclohexane, n-hexyl-	168	83
15.94	1282.7	23.93	1277.5	Benzene, 1,4-diisopropyl-	162	147
16.06	1286.1	23.90	1277.1	Indan, 5-methyl-	132	117
16.20	1290.2	24.27	1282.8	Benzene, 1,2,3,4-tetramethyl-	134	119
16.54	**1300.0**	**25.39**	**1300.0**	**n-Tridecane (n-C13)**	**184**	**57**
16.56	1300.6	24.92	1292.8	Phenol, 2,6-dimethyl-	122	107
16.72	1305.3	25.11	1295.7	Indan, 4-methyl-	132	117

R.T. 4°C/min GC-MS	R.I. 4°C/min GC-MS	R.T. 2°C/min GC-MS	R.I. 2°C/min GC-MS	Compounds Identified GC-MS, GC	Mol.. Ion MS	Base Peak MS
16.99	1313.2	26.04	1310.0	1-Tridecene	182	55
17.56	1329.9	26.56	1318.1	Tetralin	132	104
17.96	1341.6	27.74	1336.3	Benzene, 1,3,5-triethyl-	162	147
18.29	1351.3	27.93	1339.3	Naphthalene, 1,2-dihydro-	130	130
19.06	1373.9	29.78	1367.9	Benzene, hexyl-	162	91,92
19.41	1384.2	30.08	1372.5	Bicyclohexyl	166	82
19.70	1392.7	30.53	1379.4	Naphthalene	128	128
19.95	**1400.0**	**31.86**	**1400.0**	**n-Tetradecane (n-C14)**	**198**	**57**
19.98	1400.9	31.36	1392.3	Phenol, 2,4,6-trimethyl-	136	121
20.36	1412.7	32.55	1411.0	1-Tetradecene	196	55
20.68	1422.6	32.37	1408.1	Benzothiophene	134	134
21.93	1461.3	34.84	1447.4	Indole, 2,3-dihydro-	119	118
22.61	1482.4	36.21	1469.2	Benzene, cyclohexyl-	160	104
22.61	1482.4	36.28	1470.3	1-Indanol	134	133
22.70	1485.1	36.72	1477.3	Cyclohexane, n-octyl-	196	83
23.18	**1500.0**	**38.15**	**1500.0**	**n-Pentadecane (n-C15)**	**212**	**57**
23.23	1501.6	37.27	1486.0	Quinoline	129	129
23.45	1508.8	37.69	1492.7	Naphthalene, 2-methyl-	142	142
23.60	1513.6	38.84	1511.4	1-Pentadecene	210	55
24.40	1539.6	39.45	1521.6	Naphthalene, 1-methyl-	142	142
25.27	1567.9	41.03	1547.8	Fluorene, dodecahydro-	178	97
25.37	1571.1	41.41	1554.1	Quinoline, 8-methyl-	143	143
25.53	1576.3	41.91	1562.4	1,2-Dicyclohexylethane	194	83,82
25.63	1579.5	42.00	1563.8	Quinoline, 5,6,7,8-tetrahydro-3-methyl-	147	146
25.79	1584.7	42.77	1576.6	Benzene, octyl-	190	92
26.26	**1600.0**	**44.18**	**1600.0**	**n-Hexadecane (n-C16)**	**226**	**57**
26.51	1608.5	43.61	1590.5	Quinoline, 1,2,3,4-tetrahydro-	133	132
26.53	1609.2	44.06	1598.0	2,6-Di-tert-butylphenol	206	191
26.63	1612.5	43.91	1595.5	Naphthalene, 2-ethyl-	156	141
26.70	1614.9	44.85	1611.7	1-Hexadecene	224	55
26.89	1621.4	44.42	1604.2	Naphthalene, 2,7-dimethyl-	156	156
26.92	1622.4	44.59	1607.1	1,1'-Biphenyl	154	154
26.93	1622.7	44.54	1606.3	Naphthalene, 2,6-dimethyl-	156	156
27.07	1627.5	44.97	1613.8	Biphenyl, 2-methyl-	168	168
27.13	1629.5	44.82	1611.1	Naphthalene, 1-ethyl-	156	141
27.17	1630.8	44.75	1609.9	Phenanthrene, tetradecahydro-	192	192
27.64	1646.8	45.86	1629.3	Naphthalene, 1,7-dimethyl-	156	156
27.80	1652.2	46.13	1634.0	Naphthalene, 1,3-dimethyl-	156	156
27.80	1652.2	46.13	1634.0	Naphthalene, 1,6-dimethyl-	156	156
28.01	1659.3	46.39	1638.5	Phenanthrene, tetradecahydro-	192	135
28.29	1668.8	47.19	1652.4	Naphthalene, 2-isopropyl-	170	155
28.44	1673.9	47.41	1656.3	Naphthalene, 2,3-dimethyl-	156	156
28.66	1681.4	47.78	1662.7	Naphthalene, 1,4-dimethyl-	156	141, 156
28.85	1687.8	48.10	1668.3	Naphthalene, 1,5-dimethyl-	156	156
28.98	1692.2	48.56	1676.3	Naphthalene, 1-isopropyl-	170	155
29.14	1697.6	49.24	1688.2	Cyclohexane, n-decyl-	224	83
29.21	**1700.0**	**49.92**	**1700.0**	**n-Heptadecane (n-C17)**	**240**	**57**
29.27	1702.1	48.90	1682.2	Naphthalene, 1,2-dimethyl-	156	141, 156
29.82	1721.5	49.81	1698.1	Phenanthrene, tetradecahydro-	192	96, 135
30.13	1732.4	50.87	1717.2	Biphenyl, 3-methyl-	168	168
30.15	1733.1	50.80	1715.9	Naphthalenol, 5,6,7,8-tetrahydro-	148	120
30.32	1739.1	50.90	1717.8	Naphthalene, 1,8-dimethyl-	156	156
30.40	1741.9	51.38	1726.4	Biphenyl, 4-methyl-	168	168
31.02	1763.7	52.17	1740.8	Acenaphthene	154	154
31.08	1765.8	52.65	1749.5	1,2-diphenylethane (bi-benzyl)	182	91
31.87	1793.7	53.83	1770.8	Dibenzofuran	168	168
31.90	1794.7	54.75	1787.5	Benzene, decyl-	218	92
32.00	**1800.0**	**55.44**	**1800.0**	**n-Octadecane (n-C18)**	**254**	**57**

Table 2. Retention indices of compounds determined for intermediately polar (Rtx-50) and slightly polar (DB-5) columns at three heating rates

Compounds	Mol. Ion	Base Peak	RI on Rtx-50			RI on DB-5		
			2°C/min	4°C/min	6°C/min	2°C/min	4°C/min	6°C/min
n-Pentane	72	43	500	500	500	500	500	500
n-Hexane	86	43	600	600	600	600	600	600
1-Hexene	84	56	606.4	607.0	607.7	588.2	587.9	587.9
n-Heptane	100	43	700	700	700	700	700	700
Cyclohexane	84	56	705.6	706.3	706.0	655.4	656.2	657.6
1-Heptene	98	56	708.3	708.4	709.5	688.0	689.0	689.4
Cyclohexene	82	67	736.1	740.0	739.3	677.1	676.7	678.8
Methylcyclohexane	98	83	748.1	749.5	752.4	718.3	720.0	720.3
Benzene	78	78	765.7	767.4	769.0	654.2	654.8	656.1
n-Octane	114	43	800	800	800	800	800	800
trans-1,4-Dimethylcyclohexane	112	97	801.8	802.3	803.5	775.6	776.7	777.3
cis-1,3-Dimethylcyclohexane	112	97	801.8	802.3	803.5	772.8	774.7	776.6
1-Octene	112	55	808.2	809.2	809.7	788.3	789.3	789.1
trans-1,2-Dimethylcyclohexane	112	97	820.9	822.4	825.0	796.7	798.0	798.4
1-Methylcyclohexene	96	81	824.1	827.0	829.2	763.9	765.3	766.4
cis-1,4-Dimethylcyclohexane	112	97	828.6	830.5	832.6	804.7	805.3	805.7
trans-1,3-Dimethylcyclohexane	112	97	830.9	832.2	835.4	804.1	805.3	806.2
cis-1,2-Dimethylcyclohexane	112	55	859.1	862.6	866.7	823.1	824.7	827.3
Ethylcyclohexane	112	83	862.3	865.5	867.4	826.9	829.1	830.9
Toluene	92	91	870.5	873.6	875.7	758.9	761.3	761.7
n-Nonane	128	43	900	900	900	900	900	900
Pyridine	79	79	908.9	910.3	913.4	735.6	736.7	739.1
1-Nonene	126	56	909.1	909.5	909.9	888.9	889.5	890.2
n-Propylcyclohexane	126	83	956.9	960.8	963.4	924.2	927.1	929.1
Ethylbenzene	106	91	968.1	971.5	975.2	853.2	856.3	857.2
1,4-Dimethylbenzene (p-Xylene)	106	91	972.3	975.7	979.2	861.7	864.4	866.5
1,3-Dimethylbenzene (m-Xylene)	106	91	974.2	977.9	981.2	861.7	864.4	866.5
n-Decane	142	43	1000	1000	1000	1000	1000	1000
1,2-Dimethylbenzene (o-Xylene)	106	91	1009.2	1012.5	1015.5	888.0	889.9	891.2
1-Decene	140	56	1010.3	1011.0	1011.3	989.2	989.5	990.7
trans-Octahydro-1H-indene	124	67	1011.6	1015.9	1018.8	947.7	952.0	955.3
Isopropylbenzene (cumene)	120	105	1028.2	1032.1	1034.7	919.4	922.2	923.6
cis-Octahydro-1H-indene	124	67	1053.0	1059.9	1063.2	981.0	985.8	989.5
n-Propylbenzene	120	91	1058.0	1063.0	1066.1	946.9	950.5	952.3
n-Butylcyclohexane	140	83	1059.5	1064.2	1066.5	1026.6	1030.2	1031.6
1-Ethyl-3-methylbenzene	120	105	1068.7	1073.1	1075.7	955.0	958.5	960.8
1-Ethyl-4-methylbenzene	120	105	1069.8	1074.0	1076.6	956.3	960.0	962.4
1,3,5-Trimethylbenzene (Mesitylene	120	105	1074.4	1079.2	1081.6	961.5	964.9	967.1
tert-Butylbenzene	134	119	1097.6	1102.3	1104.5	987.1	990.5	992.0
4,5,6,7-Tetrahydroindan	122	79	1097.8	1103.7	1106.9	1008.6	1011.4	1014.8
1-Ethyl-2-methylbenzene	120	105	1098.2	1103.1	1105.3	973.3	977.2	979.7
n-Undecane	156	57	1100	1100	1100	1100	1100	1100
1,2,4-Trimethylbenzene	120	105	1106.8	1111.9	1114.6	987.5	991.1	992.8
1-Undecene	154	41	1110.6	1111.0	1111.0	1090.9	1091.4	1091.4
sec-Butylbenzene	134	105	1112.9	1117.6	1120.3	1006.4	1008.9	1011.7
trans-Decalin	138	138	1114.7	1122.4	1127.2	1045.7	1051.8	1055.9
1,1'-Bicyclopentyl	138	68	1138.6	1145.9	1151.2	1071.8	1077.6	1080.9
1,2,3-Trimethylbenzene	120	105	1147.6	1155.0	1158.9	1016.0	1019.9	1022.7
Phenylethylether	122	94	1150.0	1154.1	1156.9	988.8	991.1	992.4
n-Butylbenzene	134	91	1163.2	1168.6	1171.1	1050.1	1053.7	1055.9
n-Pentylcyclohexane	154	83	1163.7	1169.4	1172.0	1130.2	1134.7	1136.0
cis-Decalin	138	67	1174.6	1184.4	1190.2	1089.5	1096.1	1100.4
Indan	118	117	1180.4	1188.7	1193.5	1027.4	1032.4	1035.5
n-Dodecane	170	57	1200	1200	1200	1200	1200	1200
1-Dodecene	168	55	1211.6	1212.4	1211.9	1191.3	1191.3	1191.3
2-Methylindan	132	117	1212.8	1220.8	1225.8	1074.3	1079.5	1082.0
1-Methylindan	132	117	1222.7	1231.2	1236.5	1079.3	1084.8	1087.1
1,2,4,5-Tetramethylbenzene	134	119	1235.1	1242.4	1247.5	1109.7	1114.1	1115.8

Compounds	Mol. Ion	Base Peak	RI on Rtx-50			RI on DB-5		
			2°C/min	4°C/min	6°C/min	2°C/min	4°C/min	6°C/min
n-Hexylcyclohexane	168	83	1269.0	1274.4	1276.6	1233.9	1237.7	1239.5
1,4-Diisopropylbenzene	162	147	1279.9	1285.1	1288.1	1170.1	1173.4	1174.7
5-Methylindan	132	117	1282.5	1292.7	1298.4	1131.1	1137.1	1139.9
1,2,3,4-Tetramethylbenzene	134	119	1286.4	1296.3	1302.6	1144.2	1149.9	1153.8
2,6-Dimethylphenol	122	107	1297.6	1305.3	1310.7	1102.3	1104.6	1105.5
n-Tridecane	184	57	1300	1300	1300	1300	1300	1300
4-Methylindan	132	117	1301.2	1312.0	1318.8	1141.6	1147.7	1151.4
1-Tridecene	182	55	1311.8	1312.0	1312.4	1291.5	1291.3	1292.2
Tetralin	132	104	1324.1	1336.0	1343.6	1151.7	1158.5	1162.5
1,3,5-Triethylbenzene	162	147	1338.5	1343.6	1346.2	1219.6	1222.3	1224.3
1,2-Dihydronaphthalene	130	130	1345.7	1358.8	1366.2			
n-Hexylbenzene	162	91	1370.1	1376.3	1379.9	1254.6	1258.4	1260.9
Bicyclohexyl	166	82	1377.5	1390.4	1397.4	1293.1	1301.5	1307.8
Naphthalene	128	128	1386.8	1401.2	1410.9	1171.5	1179.7	1183.4
2,4,6-Trimethylphenol	136	121	1397.4	1406.7	1412.2	1199.1	1202.5	1204.1
n-Tetradecane	198	57	1400	1400	1400	1400	1400	1400
1-Tetradecene	196	55	1412.5	1412.8	1412.7	1391.7	1391.4	1392.2
Benzothiophene	134	134	1415.7	1432.0	1442.5	1180.5	1188.3	1192.9
2,3-Dihydroindole	119	118	1455.5	1469.2	1479.2	1195.8	1202.2	1206.6
Cyclohexylbenzene	160	104	1475.5	1489.0	1497.7	1308.9	1317.7	1322.9
1-Indanol	134	133	1476.6	1489.0	1496.8	1224.7	1229.9	1232.9
n-Octylcyclohexane	196	83	1480.1	1485.7	1489.6	1442.4	1447.5	1449.1
Quinoline	129	129	1493.6	1510.6	1521.6	1224.7	1233.0	1237.4
n-Pentadecane	212	57	1500	1500	1500	1500	1500	1500
2-Methylnaphthalene	142	142	1501.0	1517.3	1528.6	1281.5	1290.5	1296.3
1-Pentadecene	210	55	1512.7	1513.1	1513.1	1492.4	1492.5	1492.3
1-Methylnaphthalene	142	142	1530.2	1548.7	1559.6	1297.1	1306.8	1313.4
Dodecahydrofluorene	178	97	1556.9	1576.9	1588.7			
8-Methylquinoline	143	143	1562.7	1580.1	1591.1	1304.5	1313.9	1319.5
1,2-Dicyclohexylethane	194	83	1568.2	1582.1	1589.2	1486.9	1496.3	1501.0
5,6,7,8-Tetrahydro-3-methylquinoline	147	146	1571.5	1586.5	1595.8	1330.2	1338.6	1343.7
n-Octylbenzene	190	92	1580.2	1587.2	1590.6	1461.6	1466.5	1468.2
n-Hexadecane	226	57	1600	1600	1600	1600	1600	1600
1,2,3,4-Tetrahydroquinoline	133	132	1600.0	1617.6	1627.8	1318.5	1326.8	1332.5
2,6-Di-tert-butylphenol	206	191	1603.1	1613.2	1619.7	1433.6	1440.1	1443.2
2-Ethylnaphthalene	156	141	1604.3	1621.6	1632.3	1380.6	1390.6	1396.1
1-Hexadecene	224	55	1613.1	1613.2	1613.1	1591.7	1591.6	1592.7
2,7-Dimethylnaphthalene	156	156	1613.3	1631.8	1641.9	1392.1	1402.2	1409.5
2,6-Dimethylnaphthalene	156	156	1615.5	1633.1	1643.9	1390.6	1400.9	1407.3
1,1'-Biphenyl (Diphenyl)	154	154	1615.7	1631.8	1641.9	1368.5	1377.3	1381.8
2-Methylbiphenyl	168	168	1620.2	1634.5	1644.4	1390.0	1397.6	1402.3
1-Ethylnaphthalene	156	141	1621.0	1639.2	1650.0	1383.6	1393.8	1400.0
1,3-Dimethylnaphthalene	156	156	1642.2	1661.8	1673.2	1405.3	1416.5	1422.3
1,6-Dimethylnaphthalene	156	156	1643.3	1662.5	1675.3	1408.4	1419.6	1427.3
2-Isopropylnaphthalene	170	155	1660.5	1677.4	1687.9	1443.9	1454.0	1459.1
1,4-Dimethylnaphthalene	156	156	1672.6	1692.9	1707.4	1424.2	1436.3	1443.2
1,5-Dimethylnaphthalene	156	156	1678.1	1699.0	1714.2	1428.3	1439.8	1446.4
1-Isopropylnaphthalene	170	155	1684.5	1701.8	1713.7	1451.1	1460.9	1465.9
n-Decylcyclohexane	224	83	1692.2	1698.0	1702.1	1650.7	1656.2	1658.4
1,2-Dimethylnaphthalene	156	156	1692.4	1713.6	1728.6	1439.6	1452.2	1459.1
n-Heptadecane	240	57	1700	1700	1700	1700	1700	1700
5,6,7,8-Tetrahydro-1-naphthol	148	120	1725.3	1743.8	1756.3	1440.8	1447.2	1450.9
3-Methylbiphenyl	168	168	1725.5	1742.3	1753.7	1474.5	1483.5	1487.7
1,8-Dimethylnaphthalene	156	156	1730.1	1752.7	1765.3	1459.7	1472.7	1480.0
4-Methylbiphenyl	168	168	1735.7	1753.4	1766.3	1482.8	1492.2	1497.3
Acenaphthene	154	154	1752.8	1777.6	1794.2	1468.0	1481.4	1488.6
1,2-Diphenylethane (Bibenzyl)	182	91	1758.7	1776.5	1787.9	1508.8	1519.2	1524.3
Dibenzofuran	168	168	1783.4	1808.2	1825.4	1499.4	1513.0	1520.9
n-Decylbenzene	218	92	1791.1	1799.6	1805.5	1669.5	1674.8	1677.7
n-Octadecane	254	57	1800	1800	1800	1800	1800	1800
1-Octadecene	252	55	1814.4	1814.5	1814.9	1792.6	1792.1	1793.5
3,3'-Dimethylbiphenyl	182	182	1837.3	1853.9	1865.7	1580.2	1589.0	1594.2
2-Naphthol	144	144	1849.4	1871.7	1885.6	1514.5	1520.5	1523.8
4,4'-Dimethylbiphenyl	182	182	1856.6	1875.1	1887.8	1596.8	1607.6	1614.2

Compounds	Mol. Ion	Base Peak	RI on Rtx-50			RI on DB-5		
			2°C/min	4°C/min	6°C/min	2°C/min	4°C/min	6°C/min
Fluorene	166	166	1869.3	1895.2	1913.9	1565.2	1579.5	1587.9
n-Nonadecane	268	57	1900	1900	1900	1900	1900	1900
2,6-Diisopropylnaphthalene	212	197	1925.5	1943.7	1956.1	1716.8	1728.0	1735.5
Dibenzylether	198	92	1943.4	1961.4	1974.0	1641.4	1650.3	1656.9
9,10-Dihydroanthracene	180	179	1983.3	2011.1	2028.7	1662.1	1676.6	1685.3
n-Eicosane	282	57	2000	2000	2000	2000	2000	2000
n-Dodecylbenzene	246	92	2005.0	2014.4	2019.5			
9,10-Dihydrophenanthrene	180	180	2010.2	2040.3	2059.1	1673.5	1689.8	1699.5
1,2,3,4,5,6,7,8-Octahydroanthracene	186	186				1680.2	1694.1	1703.2
1,2,3,4,5,6,7,8-Octahydrophenanthreı	186	186				1705.5	1721.1	1730.6
1,2,3,4,5,6,7,8-Octahydroacridine	187	186	2037.8	2067.9	2086.0	1712.2	1726.5	1736.6
1,2,3,4-Tetrahydroanthracene	182	182				1731.3	1748.7	1760.2
1,2,3,4-Tetrahydrophenanthrene	182	182				1737.6	1755.2	1767.7
n-Henicosane	296	57	2100	2100	2100	2100	2100	2100
Phenanthrene	178	178	2136.9	2174.0	2196.8	1756.9	1776.3	1789.2
Anthracene	178	178	2146.5	2183.0	2207.3	1767.0	1786.4	1800.0
1,2,3,4-Tetrahydrocarbazole	171	143	2190.6	2229.6	2250.0	1786.6	1800.4	1811.8
n-Docosane	310	57	2200	2200	2200	2200	2200	2200
1-Phenylnaphthalene	204	204	2224.1	2256.1	2276.0	1841.7	1858.2	1868.0
n-Tricosane	324	57	2300	2300	2300	2300	2300	2300
1-Methylanthracene	192	192				1934.8	1959.3	1976.3
n-Tetracosane	338	57	2400	2400	2400	2400	2400	2400
Fluoranthene	202	202	2494.8	2544.9	2578.2	2032.6	2060.1	2076.5
n-Pentacosane	352	57	2500	2500	2500	2500	2500	2500
Pyrene	202	202	2580.4	2636.3	2673.1	2082.6	2113.4	2132.5
9,10-Dimethylanthracene	206	206	2582.5	2632.1	2664.6	2107.6	2135.5	2152.6
n-Hexacosane	366	57	2600	2600	2600	2600	2600	2600
p-Terphenyl	230	230	2627.1	2665.3	2690.0	2171.1	2190.9	2204.0
n-Triacontane	422	57	3000	3000	3000	3000	3000	3000
Chrysene	228	228	3036.9	3107.4	3153.2	2434.2	2472.3	2494.9
n-Hentriacontane	436	57	3100	3100	3100	3100	3100	3100

3.1.2   Temperature Dependence of Retention Index

Several features can be seen from Table 2 regarding the dependence of retention indices on the column temperature. First, the retention indices of the compounds that are analogous to the reference standards, such as branched alkanes and 1-alkenes, show small temperature dependence. Table 3 gives the experimental results of retention indices and final elution temperatures for 1-alkenes (ranging from 1-hexene to 1-octadecene) at the three heating rates on Rtx-50 column. The difference in index values at different elution temperatures is less than ±1 index unit (i.u.), which is within the experimental error (of the order of 2 index units). These data demonstrate the nearly temperature-independent nature of the indices for these compounds even for a 30°C range (e.g., 157.9-187.8°C for 1-octadecene, Table 2).

Table 3. Final elution temperatures and retention indices of 1-alkenes on Rtx-50 column at three heating rates

Compounds	Heating rates					
	2°C/min		4°C/min		6°C/min	
	T (°C)	R.I.	T (°C)	R.I.	T (°C)	R.I.
1-Hexene	43.7	606.4	47.4	607.0	51.1	607.7
1-Heptene	44.8	708.3	49.3	708.4	53.7	709.5

			Heating rates			
1-Octene	47.1	808.2	53.4	809.2	59.1	809.7
1-Nonene	51.8	909.1	60.8	909.5	68.1	909.9
1-Decene	59.9	1010.3	71.7	1011.0	80.7	1011.3
1-Undecene	71.0	1110.6	84.9	1111.0	95.0	1111.0
1-Dodecene	83.9	1211.6	99.2	1212.4	109.9	1211.9
1-Tridecene	97.3	1311.8	113.4	1312.0	124.5	1312.4
1-Tetradecene	110.5	1412.5	127.1	1412.8	138.5	1412.7
1-Pentadecene	123.2	1512.7	140.2	1513.1	151.8	1513.1
1-Hexadecene	135.3	1613.1	152.6	1613.2	164.4	1613.1
1-Octadecene	1579	1814.4	175.6	1814.5	188	1814.9

Second, the other compounds do show measurable temperature dependence, as shown in Figure 2, which presents the retention index of several representative compounds as a function of elution temperature. The numerical results of the linear temperature dependence for cycloalkanes and benzenes are given in Table 4 for both Rtx-50 and DB-5 columns. The second column in Table 4 denotes the elution temperature range, which indicates the lowest (corresponding to at 2°C/min) and the highest (corresponding to at 6°C/min) elution temperatures for intermediately polar Rtx-50 column. The third column gives the corresponding indices at the lowest and the highest elution temperatures. The ΔR.I./ΔT ratios in Table 4 indicates the index change per 1°C, i.e., the average temperature coefficient of retention indices. Columns 5-7 in Table 3 show the data on the slightly polar DB-5 column.

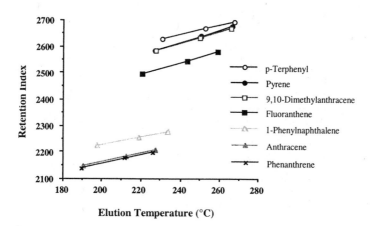

Figure 2. Retention index versus final elution temperature (Rtx-50 column, 40°C to 310°C).

Table 4. Temperature dependence of retention indices for cycloalkanes and benzenes

Compound	Rtx-50			DB-5		
	Temp. Range (°C)	R.I. Range	ΔR.I./ΔT	Temp. Range (°C)	R.I. Range	ΔR.I./ΔT
Ethylbenzene	56.4-76.1	968.1-975.2	0.36	52.5-69.1	853.2-857.2	0.24
1,4-Dimethylbenzene	56.7-76.5	972.3-979.2	0.35	53.1-70.2	861.7-866.5	0.28
1,3-Dimethylbenzene	56.8-76.8	974.2-981.2	0.35	53.1-70.2	861.7-866.5	0.28
1,2-Dimethylbenzene	59.8-81.3	1009.2-1015.5	0.29	54.9-73.1	888.0-891.2	0.18
trans-Octahydro-1H-Indene	60.1-81.8	1011.6-1018.8	0.33	60.7-81.9	947.7-955.3	0.36
tert-Butylcyclohexane	60.4-82.3	1015.1-1022.6	0.34	63.9-86.2	978.1-985.2	0.32
Isopropylbenzene	61.9-84.0	1028.2-1034.7	0.30	57.8-77.4	919.4-923.6	0.21
cis-Octahydro-1H-Indene	64.6-88.1	1053.0-1063.2	0.43	64.2-86.8	981.0-989.5	0.38
n-Propylbenzene	65.1-88.5	1058.0-1066.1	0.35	60.6-81.5	946.9-952.3	0.26
n-Butylcyclohexane	65.3-88.6	1059.5-1066.5	0.30	69.5-93.2	1026.6-1031.6	0.21
1-Ethyl-3-methylbenzene	66.2-89.9	1068.7-1075.7	0.29	61.5-82.7	955.0-960.8	0.27
1-Ethyl-4-methylbenzene	66.4-90.0	1069.8-1076.6	0.29	61.6-83.0	956.3-962.4	0.29
1,3,5-Trimethylbenzene	66.9-90.8	1074.4-1081.6	0.30	62.1-83.6	961.5-967.1	0.26
tert-Butylbenzene	69.4-94.1	1097.6-1104.5	0.28	64.8-87.2	987.1-992.0	0.21
4,5,6,7-Tetrahydroindan	69.4-94.4	1097.8-1106.9	0.36	67.2-90.6	1008.6-1014.8	0.27
1-Ethyl-2-methylbenzene	69.5-94.2	1098.2-1105.3	0.29	63.4-85.4	973.3-979.7	0.29
1,2,4-Trimethylbenzene	70.5-95.6	1106.8-1114.6	0.31	64.8-87.3	987.5-992.8	0.24
sec-Butylbenzene	71.3-96.4	1112.9-1120.3	0.29	67.0-90.1	1006.4-1011.7	0.23
trans-Decalin	71.5-97.4	1114.7-1127.2	0.49	72.0-96.9	1045.7-1055.9	0.41
1,1'-Bicyclopentyl	74.6-101.0	1138.6-1151.2	0.48	75.3-100.7	1071.8-1080.9	0.35
1,2,3-Trimethylbenzene	75.7-102.1	1147.6-1158.9	0.43	68.2-91.8	1016.0-1022.7	0.28
n-Butylbenzene	77.7-103.9	1163.2-1171.1	0.30	72.5-96.9	1050.1-1055.9	0.24
n-Pentylcyclohexane	77.7-104.0	1163.7-1172.0	0.31	83.1-109.1	1130.2-1136.0	0.22
cis-Decalin	79.1-106.7	1174.6-1190.2	0.57	77.6-103.7	1089.5-1100.4	0.42
Indan	79.9-107.2	1180.4-1193.5	0.48	69.6-93.8	1027.4-1035.5	0.34
2-Methylindan	84.1-111.9	1212.8-1225.8	0.47	75.6-100.9	1074.3-1082.0	0.30
1-Methylindan	85.4-113.5	1222.7-1236.5	0.49	76.3-101.7	1079.3-1087.1	0.31
1,2,4,5-Tetramethylbenzene	87.0-115.1	1235.1-1247.5	0.44	80.3-106.1	1109.7-1115.8	0.24
n-Hexylcyclohexane	91.6-119.4	1269.0-1276.6	0.28	97.3-124.6	1233.9-1239.5	0.21
1,4-Diisopropylbenzene	93.0-121.1	1279.9-1288.1	0.29	88.6-115.0	1170.1-1174.7	0.18
5-Methylindan	93.4-122.6	1282.5-1298.4	0.54	83.2-109.7	1131.1-1139.9	0.33
1,2,3,4-Tetramethylbenzene	93.9-123.2	1286.4-1302.6	0.55	85.0-111.8	1144.2-1153.8	0.36
4-Methylindan	95.9-125.4	1301.2-1318.8	0.60	84.6-111.5	1141.6-1151.4	0.37
1,3,5-Triethylbenzene	100.8-129.3	1338.5-1346.2	0.27	95.4-122.4	1219.6-1224.3	0.17
n-Hexylbenzene	105.0-134.0	1370.1-1379.9	0.34	100.2-127.7	1254.6-1260.9	0.23
Bicyclohexyl	106.0-136.5	1377.5-1397.4	0.65	105.4-134.5	1293.1-1307.8	0.50
n-Octylcyclohexane	119.1-148.7	1480.1-1489.6	0.32	124.9-153.8	1442.4-1449.1	0.23
1,2-Dicyclohexylethane	129.9-161.5	1568.2-1589.2	0.67	130.5-160.6	1486.9-1501.0	0.47
n-Octylbenzene	131.4-161.7	1580.2-1590.6	0.34	127.3-156.3	1461.6-1468.2	0.23
n-Decylcyclohexane	144.5-175.0	1692.2-1702.1	0.32	149.9-179.7	1650.7-1658.4	0.26
n-Decylbenzene	155.4-186.8	1791.1-1805.5	0.46	152.1-182.0	1669.5-1677.7	0.27
n-Dodecylbenzene	177.4-209.3	2005.0-2019.5	0.45			

In the same fashion, Table 5 presents the numerical results of the linear temperature dependence for aromatics and NSO (N-, S-, and O-containing) compounds. Take the data on Rtx-50 as an example, it can be seen that in general, alkylated cyclohexanes and benzenes displayed a ΔR.I./ΔT of about 0.33±0.05 index units/°C while benzenes with multi-substitution (such as tetramethyl-) and long side-chains (decyl and dodecyl) exhibit slightly higher values (ca. 0.46). Compared with alkylated cyclohexanes and benzenes, the indices for compounds with two-ring structure exhibit higher temperature dependence.

Table 5. Temperature dependence of retention indices for aromatics, and N-, S-, and O-containing compounds

Compounds	Rtx-50 Temp. Range (°C)	R.I. Range	ΔR.I./ΔT	DB-5 Temp. Range (°C)	R.I. Range	ΔR.I./ΔT
Tetralin	98.9-128.9	1324.1-1343.6	0.65	86.0-113.1	1151.7-1162.5	0.40
1,2-Dihydronaphthalene	101.8-132.1	1345.7-1366.2	0.68			
Naphthalene	107.2-138.3	1386.8-1410.9	0.77	88.8-116.3	1171.5-1183.4	0.43
2-Methylnaphthalene	121.8-153.8	1501.0-1528.6	0.87	103.8-132.9	1281.5-1296.3	0.51
1-Methylnaphthalene	125.3-157.7	1530.2-1559.6	0.91	106.0-135.3	1297.1-1313.4	0.56
Dodecahydrofluorene	128.6-161.4	1556.9-1588.7	0.97			
2-Ethylnaphthalene	134.3-166.7	1604.3-1632.3	0.86	117.0-146.7	1380.6-1396.1	0.52
2,7-Dimethylnaphthalene	135.3-167.9	1613.3-1641.9	0.88	118.5-148.5	1392.1-1409.5	0.58
2,6-Dimethylnaphthalene	135.6-168.1	1615.5-1643.9	0.87	118.3-148.2	1390.6-1407.3	0.56
1,1'-Biphenyl	135.6-167.9	1615.7-1641.9	0.81	115.4-144.8	1368.5-1381.8	0.45
2-Methylbiphenyl	136.1-168.2	1620.2-1644.4	0.76	118.3-147.6	1390.0-1402.3	0.42
1-Ethylnaphthalene	136.2-168.8	1621.0-1650.0	0.89	117.4-147.3	1383.6-1400.0	0.55
1,7-Dimethylnaphthalene	138.3-171.2	1638.6-1670.2	0.96			
1,3-Dimethylnaphthalene	138.7-171.6	1642.2-1673.2	0.94	120.2-150.2	1405.3-1422.3	0.57
1,6-Dimethylnaphthalene	138.8-171.8	1643.3-1675.3	0.97	120.6-150.9	1408.4-1427.3	0.62
2-Isopropylnaphthalene	140.8-173.3	1660.5-1687.9	0.84	125.1-155.1	1443.9-1459.1	0.51
2,3-Dimethylnaphthalene	141.4-174.6	1665.9-1698.5	0.98			
1,4-Dimethylnaphthalene	142.2-175.6	1672.6-1707.4	1.04	122.6-153.0	1424.2-1443.2	0.63
1,5-Dimethylnaphthalene	142.9-176.4	1678.1-1714.2	1.08	123.1-153.4	1428.3-1446.4	0.60
1-Isopropylnaphthalene	143.6-176.3	1684.5-1713.7	0.89	126.0-156.0	1451.1-1465.9	0.49
1,2-Dimethylnaphthalene	144.5-178.1	1692.4-1728.6	1.08	124.6-155.1	1439.6-1459.1	0.64
3-Methylbiphenyl	148.2-180.9	1725.5-1753.7	0.86	128.9-158.9	1474.5-1487.7	0.44
1,8-Dimethylnaphthalene	148.7-182.2	1730.1-1765.3	1.05	127.1-157.8	1459.7-1480.0	0.66
4-Methylbiphenyl	149.3-182.3	1735.7-1766.3	0.93	130.0-160.1	1482.8-1497.3	0.48
Acenaphthene	151.2-185.5	1752.8-1794.2	1.21	128.1-159.0	1468.0-1488.6	0.67
1,2-Diphenylethane (Bibenzyl)	151.8-184.8	1758.7-1787.9	0.89	133.2-163.5	1508.8-1524.3	0.51
3,3'-Dimethylbiphenyl	160.3-193.3	1837.3-1865.7	0.86	141.8-172.1	1580.2-1594.2	0.46
4,4'-Dimethylbiphenyl	162.4-195.7	1856.6-1887.8	0.94	143.8-174.5	1596.8-1614.2	0.57
Fluorene	163.7-198.5	1869.3-1913.9	1.28	140.0-171.3	1565.2-1587.9	0.72
2,6-Diisopropylnaphthalene	169.5-202.8	1925.5-1956.1	0.92	157.4-188.6	1716.8-1735.5	0.60
9,10-Dihydroanthracene	175.2-210.2	1983.3-2028.7	1.30	151.2-182.9	1662.1-1685.3	0.73
9,10-Dihydrophenanthrene	177.9-213.2	2010.2-2059.1	1.39	152.5-184.6	1673.5-1699.5	0.81
1,2,3,4,5,6,7,8-Octahydroanthracene				153.3-185.0	1680.2-1703.2	0.73
1,2,3,4,5,6,7,8-Octahydrophenanthrene				156.1-188.1	1705.5-1730.6	0.79
1,2,3,4-Tetrahydroanthracene				158.9-191.4	1731.3-1760.2	0.89
1,2,3,4-Tetrahydrophenanthrene				159.6-192.2	1737.6-1767.7	0.93
Phenanthrene	189.9-226.4	2136.9-2196.8	1.64	161.7-194.6	1756.9-1789.2	0.98
Anthracene	190.8-227.4	2146.5-2207.3	1.66	162.8-195.8	1767.0-1800.0	1.00
1-Phenylnaphthalene	197.8-233.6	2224.1-2276.0	1.45	170.7-203.1	1841.7-1868.0	0.81
1-Methylanthracene				180.2-214.2	1934.8-1976.3	1.22
Fluoranthene	220.6-259.0	2494:8-2578.2	2.17	189.7-224.1	2032.6-2076.5	1.28
Pyrene	227.3-266.4	2580.4-2673.1	2.37	194.4-229.4	2082.6-2132.5	1.43
9,10-Dimethylanthracene	227.5-265.8	2582.5-2664.6	2.14	196.7-231.2	2107.6-2152.6	1.30
p-Terphenyl	230.9-267.8	2627.1-2690.0	1.71	202.4-236.0	2171.1-2204.0	0.98
Chrysene	259.9-300.2	3036.9-3153.2	2.88	224.6-260.9	2434.2-2494.9	1.67
Phenylethylether	76.0-101.8	1150.0-1156.9	0.27	65.0-87.2	988.8-992.4	0.16
2,6-Dimethylphenol	95.4-124.3	1297.6-1310.7	0.45	79.2-104.5	1102.3-1105.5	0.13
2,4,6-Trimethylphenol	108.6-138.5	1397.4-1412.2	0.50	92.6-119.4	1199.1-1204.1	0.19
Benzothiophene	110.9-142.5	1415.7-1442.5	0.85	90.0-117.8	1180.5-1192.9	0.45
2,3-Dihydroindole	116.0-147.3	1455.5-1479.2	0.76	92.1-119.8	1195.8-1206.6	0.39
1-Indanol	118.7-149.7	1476.6-1496.8	0.65	96.1-123.6	1224.7-1232.9	0.30
Quinoline	120.9-152.9	1493.6-1521.6	0.88	96.1-124.3	1224.7-1237.4	0.45

Compounds	Rtx-50			DB-5		
	Temp. Range (°C)	R.I. Range	ΔR.I./ ΔT	Temp. Range (°C)	R.I. Range	ΔR.I./ ΔT
8-Methylquinoline	129.3-161.7	1562.7-1591.1	0.87	107.0-136.1	1304.5-1319.5	0.51
5,6,7,8-Tetrahydro-3-methylquinoline	130.3-162.3	1571.5-1595.8	0.76	110.4-139.5	1330.2-1343.7	0.47
1,2,3,4-Tetrahydroquinoline	133.8-166.2	1600.0-1627.8	0.86	108.8-137.9	1318.5-1332.5	0.48
2,6-Di-tert-butylphenol	134.2-165.2	1603.1-1619.7	0.53	123.8-153.0	1433.6-1443.2	0.33
5,6,7,8-Tetrahydro-1-naphthol	148.2-181.2	1725.3-1756.3	0.94	124.7-154.0	1440.8-1450.9	0.35
Dibenzofuran	154.6-188.9	1783.4-1825.4	1.22	132.1-163.1	1499.4-1520.9	0.69
2-Naphthol	161.6-195.5	1849.4-1885.6	1.07	133.9-163.4	1514.5-1523.8	0.31
Dibenzylether	171.3-204.7	1943.4-1974.0	0.92	148.9-179.6	1641.4-1656.9	0.50
1,2,3,4,5,6,7,8-Octahydroacridine	180.5-215.9	2037.8-2086.0	1.36	156.9-188.7	1712.2-1736.6	0.77
1,2,3,4-Tetrahydrocarbazole	194.8-231.2	2190.6-2250.0	1.63	164.9-197.1	1786.6-1811.8	0.79

For example, indan, 1- and 2-methylindan, bicyclopentyl, and trans-decalin all have values about 0.49; 4- and 5-methylindan, cis-decalin, and bicyclohexyls show even higher index temperature dependence (ca. 0.60). Multi-ring aromatics exhibit the largest temperature dependence of the retention index among the studied compounds. The 16 alkylated (C_1~C_3) naphthalenes and 6 biphenyls in Table 5 show an average ΔR.I./ΔT of 0.94 and 0.86, respectively, with a standard deviation of about 0.08. The values for three-ring aromatics in general range from 1.20 to 1.65. Four-ring aromatics all display significant temperature dependence judging from the large values of ΔR.I./ΔT (larger than 2.0); chrysene shows a value as large as 2.88. The ΔR.I./ΔT ratios in Table 4 are quite useful in that they may be easily used to estimate the retention indices of these compounds over the studied temperature range by interpolation.

On the other hand, the large temperature dependence of indices for multi-ring aromatics implies that using the ΔR.I./ΔT ratios in Table 5 to estimate their retention indices over the studied temperature range by interpolation will be less accurate. As to the NSO compounds, they in general show larger retention indices and exhibit slightly greater temperature dependence (ΔR.I./ΔT) than their corresponding hydrocarbons.

Third, an important consequence of the observed difference in the different temperature dependence (ΔR.I./ΔT) among the compounds is that compounds which co-eluted at one temperature may be separated at some other temperature(s). This means that co-elution problems occurred when using a given heating rate may be solved by using a different heating rate. For example, 4-methylindan and 1-tridecene co-elute on Rtx-50 at the heating rate of 4°C/min, but they can be resolved by decreasing or increasing the heating rate (to 2 or 6°C/min). Similarly, 1,2-dimethylnaphthalene and 2-isopropylnaphthalene co-elute on DB-5 at the heating rate of 6°C/min, but they can be resolved by decreasing the heating rate into 2°C/min. 1-Alkenes co-elute (or almost co-elute) with some compounds at the heating rates of 2°C/min but they can be easily separated from their respective co-eluting compounds at 6°C/min, e.g., 1-decene and 1,2-dimethylbenzene; 1-undecene and sec-butylbenzene; 1-dodecene and 2-

methylindan;   1-hexadecene and 2,7-dimethylnaphthalene.   Some other examples, which can be found in Table 1, include 2-methylbiphenyl and 1-ethylnaphthalene;   1,2-dimethylnaphthalene   and   n-decylcyclohexane; acenaphthene   and   1,2-diphenylethane;   9,10-dihydroanthracene   and dodecylbenzene; pyrene and 9,10-dimethylanthracene.

Although the retention indices in Table 2 were measured under linear temperature programming without isothermal initial holding period, the results may also be applied to GC analysis with a short initial holding time. Table 6 presents the retention times and retention indices measured with 5 minutes of initial isothermal holding time at 40 °C as well as those without initial holding time for 60 representative (out of 154 in Table 1) compounds. The data were collected using the heating rate of 4 °C/min on the DB-5 column. The retention indices with 5-minute holding time were also approximated by Equation 2. It was found that adding a 5-minute holding time at low temperature (40 °C) has only small effects on retention indices although the retention times may differ by as many as 5 minutes. It was apparent from Table 6 that adding initial holding time only serves to delay the elution of heavier compounds which have retention indices greater than 1400 by as much time as added, for example, 5 minutes in this study. For the 154 compounds studied, the retention index was lowered, on average, by about 1 index unit when 5 minutes of holding time was added at the beginning temperature; the retention index decrement for heavier compounds is in general negligible.

Therefore, the temperature-programmed retention indices in Table 2 may also be applied with minor adjustment to the analysis which incorporates a short initial holding time.

Table 6. Retention times and retention indices on DB-5 column measured at the heating rate of 4 °C/min with and without isothermal initial holding

Compounds	Retention Time (min)			Retention Index		
	IH^a=0	IH=5	Difference	IH=0	IH=5	difference
Benzene	2.21	2.34	0.13	654.8	653.8	-0.9
Cyclohexane	2.22	2.35	0.13	656.2	654.9	-1.2
1-Octene	3.88	4.82	0.94	789.3	788.4	-0.9
cis-1,2-Dimethylcyclohexane	4.65	6.09	1.44	824.7	824.4	-0.2
Ethylcyclohexane	4.76	6.25	1.49	829.1	828.4	-0.8
Ethylbenzene	5.43	7.36	1.93	856.3	855.5	-0.8
n-Propylcyclohexane	7.39	10.36	2.97	927.1	926.4	-0.7
1-Ethyl-3-methylbenzene	8.41	11.80	3.39	958.5	958.6	0.2
4,5,6,7-Tetrahydroindan	10.17	14.07	3.90	1011.4	1009.9	-1.5
n-Butylcyclohexane	10.85	14.88	4.03	1030.2	1028.9	-1.3
trans-Decalin	11.63	15.77	4.14	1051.8	1049.8	-2.0
n-Butylbenzene	11.70	15.92	4.22	1053.7	1053.3	-0.5
5-Methylindan	14.74	19.30	4.56	1137.1	1135.1	-2.0
1,2,3,4-Tetramethylbenzene	15.21	19.82	4.61	1149.9	1148.2	-1.6
2,3-Dihydroindole	17.14	21.89	4.75	1202.2	1200.5	-1.7
2,4,6-Trimethylphenol	17.15	21.93	4.78	1202.5	1201.6	-0.9
1,3,5-Triethylbenzene	17.86	22.68	4.82	1222.3	1222.0	-0.4
1-Indanol	18.13	22.92	4.79	1229.9	1228.5	-1.4
n-Hexylcyclohexane	18.41	23.24	4.83	1237.7	1237.1	-0.6
1-Tridecene	20.33	25.26	4.93	1291.3	1291.9	0.5

Compounds	Retention Time (min)			Retention Index		
	IH[a]=0	IH=5	Difference	IH=0	IH=5	difference
Bicyclohexyl	20.69	25.60	4.91	1301.5	1301.2	-0.3
1-Methylnaphthalene	20.87	25.75	4.88	1306.8	1305.5	-1.3
1,2,3,4-Tetrahydroquinoline	21.55	26.43	4.88	1326.8	1325.1	-1.7
1,1'-Biphenyl	23.26	28.17	4.91	1377.3	1375.4	-1.9
2-Ethylnaphthalene	23.71	28.66	4.95	1390.6	1389.6	-1.0
1-Ethylnaphthalene	23.82	28.78	4.96	1393.8	1393.1	-0.7
2,6-Dimethylnaphthalene	24.06	29.02	4.96	1400.9	1400.0	-0.9
5,6,7,8-Tetrahydronaphthalenol	25.55	30.51	4.96	1447.2	1446.1	-1.1
n-Octylcyclohexane	25.56	30.54	4.98	1447.5	1447.1	-0.5
1,2-Dimethylnaphthalene	25.71	30.67	4.96	1452.2	1451.1	-1.1
2-Isopropylnaphthalene	25.77	30.75	4.98	1454.0	1453.6	-0.5
n-Octylbenzene	26.17	31.16	4.99	1466.5	1466.3	-0.2
1,8-Dimethylnaphthalene	26.37	31.35	4.98	1472.7	1472.1	-0.5
Acenaphthene	26.65	31.62	4.97	1481.4	1480.5	-0.9
4-Methylbiphenyl	27.00	32.00	5.00	1492.2	1492.3	0.0
1,2-Dicyclohexylethane	27.13	32.09	4.96	1496.3	1495.0	-1.2
Dibenzofuran	27.65	32.62	4.97	1513.0	1512.0	-1.0
2-Naphthol	27.88	32.87	4.99	1520.5	1520.1	-0.3
Fluorene	29.70	34.68	4.98	1579.5	1578.9	-0.6
3,3'-Dimethylbiphenyl	29.99	34.98	4.99	1589.0	1588.6	-0.3
1-Hexadecene	30.07	35.08	5.01	1591.6	1591.9	0.3
Dibenzylether	31.79	36.77	4.98	1650.3	1649.7	-0.7
n-Decylcyclohexane	31.96	36.95	4.99	1656.2	1655.9	-0.3
n-Decylbenzene	32.50	37.49	4.99	1674.8	1674.5	-0.3
9,10-Dihydroanthracene	32.55	37.54	4.99	1676.6	1676.2	-0.3
9,10-Dihydrophenanthrene	32.92	37.90	4.98	1689.3	1688.6	-0.7
1,2,3,4,5,6,7,8-Octahydroanthracene	33.06	38.06	5.00	1694.1	1694.1	0.0
1,2,3,4,5,6,7,8-Octahydroacridine	33.97	38.96	4.99	1726.5	1726.2	-0.4
2,6-Diisopropylnaphthalene	34.01	39.01	5.00	1728.0	1728.0	0.0
1,2,3,4-Tetrahydroanthracene	34.59	39.59	5.00	1748.7	1748.7	0.0
Phenanthrene	35.36	40.36	5.00	1776.3	1776.3	0.0
1-Octadecene	35.80	40.80	5.00	1792.1	1792.1	0.0
1,2,3,4-Tetrahydrocarbazole	36.03	41.02	4.99	1800.4	1800.0	-0.4
1-Phenylnaphthalene	37.54	42.54	5.00	1858.2	1858.2	0.0
Fluoranthene	42.59	47.59	5.00	2060.1	2060.1	0.0
Pyrene	43.85	48.85	5.00	2113.4	2113.4	0.0
9,10-Dimethylanthracene	44.36	49.35	4.99	2135.5	2135.1	-0.4
Dibenzylsulfid	44.62	49.60	4.98	2146.8	2145.9	-0.9
p-Terphenyl	45.64	50.64	5.00	2190.9	2190.9	0.0
Chrysene	51.64	56.63	4.99	2472.3	2471.8	-0.5

3.1.3   **Dependence of Retention Index on Polarity of GC Column**

Retention indices exhibit significant dependence on column polarity as shown in Table 2. There are several characteristics to be pointed out. First, Table 2 shows that the retention indices of all the compounds studied decrease as the column polarity decreases (from Rtx-50 to DB-5). Similar to temperature dependence, the decrement of the retention indices depends on the structure of the compounds. For example, Figure 3 displays the retention index difference between Rtx-50 column and DB-5 column (Δ_{pol}R.I.) for 20 representative compounds. The retention index difference ranges from as small as 20 to over 600. Compared with other compounds, 1-alkenes, which are analogous to the reference standards (n-alkanes in this work), display the smallest polarity dependence. Δ_{pol}R.I. is about 20±2 for 1-alkenes ranging from 1-hexene to 1-

octadecene. Alkylcyclohexanes (from methylcyclohexane to n-decylcyclohexane) also show small polarity dependence; Δ_{pol}R.I. is about 36±6. On the other hand, the retention indices of alkylbenzenes, aromatics, and N-, S-, and O-containing compounds are highly dependent on the column polarity. Multi-ring aromatics exhibit the largest polarity dependence of the retention index among the studied compounds. It is worth noting that the more polar column (Rtx-50) retains more polar compounds longer than less polar isomer compounds and results in larger Δ_{pol}R.I. This can be demonstrated by 2,6-dimethylnaphthalene and 1,8-dimethylnaphthalene; the latter is relatively more polar and display larger Δ_{pol}R.I.than the former. Overall, the behavior of the polarity dependence of retention index can be summarized by the facts that the retention of polar compounds decreases, but that of non-polar compounds such as n-alkanes increases, when the polarity of stationary phase decreases. Figure 4 shows the differences in retention index between RTx-50 column and DB-5 column as a function of the Z number (Δ_{pol}R.I. versus Z number) [Z as in $C_nH_{2n-z}X$] for the 20 representative compounds plotted in Figure 3. Z number is representative of double bond equivalence. It appears that there is a good relationship in general between (Δ_{pol}R.I. and Z number of the compounds.

Second, the change or shift of compound elution order on different columns can be observed owing to the fact that different compounds display different degrees of retention index changes with column polarity. For example, the effects of column polarity on the elution order of compounds are very evident by comparing 1-alkenes and n-alkanes. The 1-alkene, which eluted after the n-alkane with the same carbon number on Rtx-50 column, eluted before n-alkane on DB-5 column. More examples of the shift in elution order can be found in Tables 1 and 2. Based on the knowledge on the shift in compound elution order, compound identification can be improved by using columns of different polarity. Another application of the knowledge on the retention indices is that they can facilitate the choice of column phase to reduce or even eliminate co-elution within complex samples. For example, n-propylbenzene and n-butylcyclohexane co-elute on Rtx-50 column but resolve on DB-5 column at any of the three different heating rates (2,4,6°C/min). On the other hand, 2,4,6-trimethylphenol and 2,3-dihydroindole co-elute on DB-5 but resolve on Rtx-50 column.

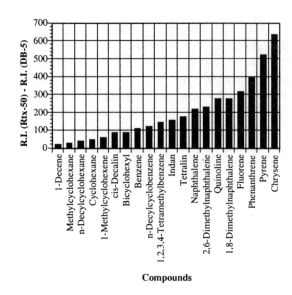

Compounds

Figure 3. Retention index difference between Rtx-50 column and DB-5 column (ΔpolR.I.) for 20 representative compounds

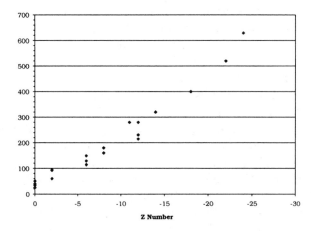

Z Number

Figure 4. Retention index difference between Rtx-50 column and DB-5 column (Δ_{pol}R.I.) versus Z number [Z as in CnH2n-zX] for 20 representative compounds.

Third, for the two capillary columns studied, in general the values of ΔR.I./ΔT are larger for Rtx-50 than for DB-5; i.e., almost all the compounds studied display higher temperature sensitivity on the more polar column (Rtx-50) than on the less polar column (DB-5), as can be seen from Tables 4 and 5.

3.1.4   Characterization of JP-8 Jet Fuels Using RI

Retention indices determined in this work can be directly used in identifying compounds in complex mixtures such as jet fuels. In many cases, the unknown compounds may be identified by calculating their retention indices and comparing with standards shown in Table 2, if the match is within ± 2 index units for a given heating rate. The choice of 2 index units is based on and close to the experimental error for the retention indices. In fact, coal-derived JP-8C fuel was analyzed by GC under the same flow conditions as those for the standard compounds, and two heating rates (4 and 6 °C/min) were used for the purpose of cross references. The capillary column's sample capacity was again taken into consideration and symmetric GC peaks were obtained. Figure 5 presents the GC-MS total ion chromatogram (TIC) of coal-derived JP-8C fuel on Rtx-50 column, together with the structures of major components. Besides the n-alkane components, more than 50 representative compounds present in JP-8C were identified by matching retention indices, as shown in Table 7. The first column in Table 6 denotes the representative compounds identified. The second and third columns display the retention indices (at 4°C/min) of known standard compounds (denoted Std) and those of the JP-8C, respectively. Similarly, the data under 6°C/min are given in the last two columns. The reported data are the mean indices of three replicates. The deviations between R.I.$_{Std}$ and R.I.$_{JP-8C}$ in Table 7 are less than 1 index unit, which is within the experimental error for the retention indices (of the order of 2 index units). We also analyzed the petroleum-derived JP-8P fuel in a similar fashion. Figure 6 shows the GC-MS TIC of petroleum-derived JP-8P fuel on Rtx-50 column, together with the structures of major components.

The GC-MS analysis of the whole jet fuel also reveals that many compounds are present in trace amounts or co-eluted with other compounds, making them difficult to be identified from analyzing the whole fuels. Another approach used in this work is to separate the jet fuels into chemically similar compound classes by liquid column chromatographic separation, which also serves to concentrate the compounds present in whole fuels in very low concentrations and to eliminate (or reduce) the co-elution of different compounds from capillary GC column, followed by the GC-MS analysis of the chromatographic fractions. The fuels were separated into several fractions using a neutral alumina gel column and a series of elution solvents, as described elsewhere.[15] The identifications of compounds in JP-8P fraction 1 (n-pentane eluate), fraction 2 (5% benzene-pentane eluate), and fraction 3 (benzene eluate) are described as follows.

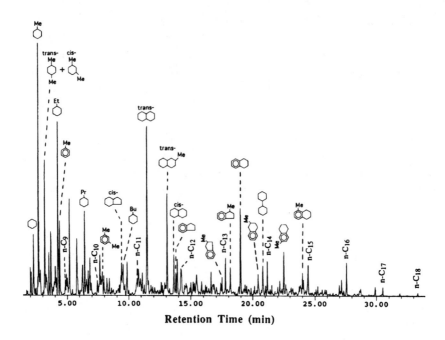

Figure 5. GC-MS total ion chromatogram of unfractionated JP-8C jet fuel (Rtx-50 column, 40°C to 310°C at 4°C/min, split)

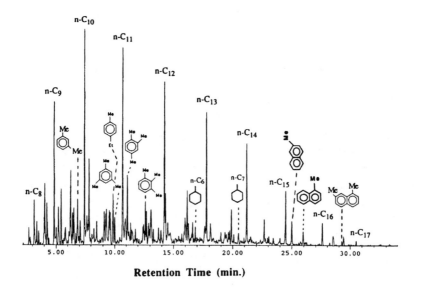

Figure 6. GC-MS total ion chromatogram of unfractionated JP-8P jet fuel (Rtx-50 column, 40°C to 310°C at 4°C/min, split)

Table 7. Hydrocarbons in coal-derived JP-8C jet fuel identified by matching retention index on Rtx-50 column at two heating rates

Compounds[a]	R.I. at 4°C/min		R.I. at 6°C/min	
	Std	JP-8C	Std	JP-8C
Cyclohexane	706.3	705.4	706.0	706.0
Methylcyclohexane	749.5	750.0	752.4	751.8
Benzene	767.4	767.4	769.0	768.7
Trans-1,4-Dimethylcyclohexane	802.3	802.4[b]	803.5	803.5[b]
cis-1,3-Dimethylcyclohexane	802.3	802.4[b]	803.5	803.5[b]
Trans-1,2-Dimethylcyclohexane	822.4	822.9	825.0	824.6
cis-1,4-Dimethylcyclohexane	830.5	831.2	832.6	832.4
Trans-1,3-Dimethylcyclohexane	832.2	832.9	835.4	835.2
Ethylcyclohexane	865.5	865.3	867.4	866.9
Toluene	873.6	873.5	875.7	875.4
n-Propylcyclohexane	960.8	960.9	963.4	963.5
Ethylbenzene	971.5	971.6	975.2	975.5
1,4-Dimethylbenzene	975.7	975.5	979.2	978.5
1,3-Dimethylbenzene	977.9	977.4	981.2	980.5
1,2-Dimethylbenzene	1012.5	1012.3	1015.5	1015.6
Trans-Octahydro-1H-indene	1015.9	1016.0	1018.8	1019.4
Isopropylbenzene	1032.1	1032.4	1034.7	1035.0
cis-Octahydro-1H-indene	1059.9	1060.2	1063.2	1063.7
n-Propylbenzene	1063.0	1063.3	1066.1	1066.2
n-Butylcyclohexane	1064.2	1064.5	1066.5	1067.1
1-Ethyl-4-methylbenzene	1074.0	1074.4	1076.6	1076.8
1,3,5-Trimethylbenzene	1079.2	1079.3	1081.6	1082.3
1-Ethyl-2-methylbenzene	1103.1	1103.1	1105.3	1105.7
1,2,4-Trimethylbenzene	1111.9	1112.2	1114.6	1115.1
Trans-Decalin	1122.4	1122.4	1127.2	1127.8
1,2,3-Trimethylbenzene	1155.0	1155.0	1158.9	1159.2
n-Butylbenzene	1168.6	1168.3	1171.1	1171.8
n-Pentylcyclohexane	1169.4	1168.8	1172.0	1172.2
cis-Decalin	1184.4	1183.9	1190.2	1190.6
Indan	1188.7	1188.4	1193.5	1193.9
2-Methylindan	1220.8	1219.8	1225.8	1225.5
1-Methylindan	1231.2	1230.9	1236.5	1236.2
1,2,4,5-Tetramethylbenzene	1242.4	1242.5	1247.5	1247.3
n-Hexylcyclohexane	1274.4	1274.2	1276.6	1277.4
5-Methylindan	1292.7	1292.4	1298.4	1298.4
1,2,3,4-Tetramethylbenzene	1296.3	1296.3	1302.6	1302.6
4-Methylindan	1312.0	1312.0	1318.8	1318.9
Tetralin	1336.0	1335.9	1343.6	1343.3
n-Hexylbenzene	1376.3	1376.1	1379.9	1380.3
Bicyclohexyl	1390.4	1389.5	1397.4	1397.9
Naphthalene	1401.2	1401.2	1410.9	1410.8
n-Octylcyclohexane	1485.7	1485.9	1489.6	1489.6
Cyclohexylbenzene	1489.0	1489.0	1497.7	1497.3
2-Methylnaphthalene	1517.3	1517.4	1528.6	1528.0
1-Methylnaphthalene	1548.7	1548.6	1559.6	1559.7
1,2-Dicyclohexylethane	1582.1	1582.0	1589.2	1589.6
n-Octylbenzene	1587.2	1587.5	1590.6	1591.0
2,7-Dimethylnaphthalene	1631.8	1631.2[b]	1641.9	1642.2[b]
1,1'-Biphenyl	1631.8	1631.2[b]	1641.9	1642.2[b]
2,6-Dimethylnaphthalene	1633.1	1632.5	1643.9	1644.2
1,3-Dimethylnaphthalene	1661.8	1662.0	1673.2	1673.4
1,6-Dimethylnaphthalene	1662.5	1662.7	1675.3	1674.9
n-Decylcyclohexane	1698.0	1698.0	1702.1	1702.1
3-Methylbiphenyl	1742.3	1742.9	1753.7	1753.7

a) Excluding the n-alkanes that are also present in JP-8C jet fuel.
b) Identified with the assistance of mass spectra.

Figure 7 shows the TIC as well as the detailed identification results of the JP-8P fraction 1 (saturates). All the identifications in Figure 7 were made by using retention indices and further confirmed by using mass spectrometry. Combined use of retention indices and mass spectra allows compound identification to be performed with higher confidence. The JP-8P saturate fraction consists primarily of open chain compounds. In their mass spectra, straight-chain alkanes show weak molecular ions but typical and relatively strong $C_nH_{2n+1}^{+}$ fragment ions, and specific compounds can be accurately identified using GC-MS. Many branched alkanes exhibit very weak molecular ions, and in some cases such ions disappear from their mass spectra. It should be noted that Hayes and Pitzer [16,17] have published the retention indices of some branchedalkanes and alkylated benzenes. Their data were also used to help identify several more peaks. It can be seen that the dominant constituents in fraction 1 of JP-8P are the long-chain alkanes with carbon number ranging from C_7 to C_{17} with most falling between C_9 and C_{14}. Straight-chain (normal) alkanes are predominant; many branched alkanes are also present, and the positions of the side chain are also indicated for them in Figure 7.

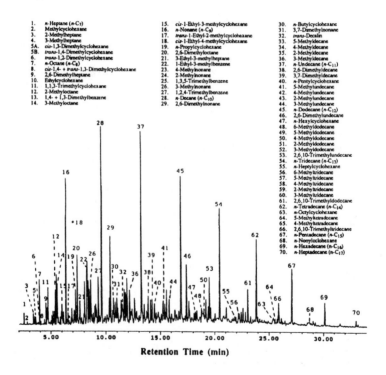

1.	*n*-Heptane (*n*-C$_7$)
2.	Methylcyclohexane
3.	2-Methylheptane
4.	3-Methylheptane
5A.	*cis*-1,3-Dimethylcyclohexane
5B.	*trans*-1,4-Dimethylcyclohexane
6.	*trans*-1,2-Dimethylcyclohexane
7.	*n*-Octane (*n*-C$_8$)
8.	*cis*-1,4- + *trans*-1,3-Dimethylcyclohexane
9.	2,6-Dimethylheptane
10.	Ethylcyclohexane
11.	1,1,3-Trimethylcyclohexane
12.	2-Methyloctane
13.	1,4- + 1,3-Dimethylbenzene
14.	3-Methyloctane

15.	*cis*-1-Ethyl-3-methylcyclohexane
16.	*n*-Nonane (*n*-C$_9$)
17.	*trans*-1-Ethyl-2-methylcyclohexane
18.	*cis*-1-Ethyl-4-methylcyclohexane
19.	*n*-Propylcyclohexane
20.	2,6-Dimethyloctane
21.	3-Ethyl-3-methylheptane
22.	1-Ethyl-3-methylbenzene
23.	4-Methylnonane
24.	2-Methylnonane
25.	1,3,5-Trimethylbenzene
26.	3-Methylnonane
27.	1,2,4-Trimethylbenzene
28.	*n*-Decane (*n*-C$_{10}$)
29.	2,6-Dimethylnonane

30.	*n*-Butylcyclohexane
31.	3,7-Dimethylnonane
32.	*trans*-Decalin
33.	5-Methyldecane
34.	4-Methyldecane
35.	2-Methyldecane
36.	3-Methyldecane
37.	*n*-Undecane (*n*-C$_{11}$)
38.	2,6-Dimethyldecane
39.	3,7-Dimethyldecane
40.	5-Methylundecane
41.	5-Methylundecane
42.	4-Methylundecane
43.	2-Methylundecane
44.	3-Methylundecane
45.	*n*-Dodecane (*n*-C$_{12}$)
46.	2,6-Dimethylundecane
47.	*n*-Hexylcyclohexane
48.	6-Methyldodecane
49.	5-Methyldodecane
50.	4-Methyldodecane
51.	2-Methyldodecane
52.	3-Methyldodecane
53.	2,6,10-Trimethylundecane
54.	*n*-Tridecane (*n*-C$_{13}$)
55.	*n*-Heptylcyclohexane
56.	6-Methyltridecane
57.	5-Methyltridecane
58.	4-Methyltridecane
59.	2-Methyltridecane
60.	3-Methyltridecane
61.	2,6,10-Trimethyldodecane
62.	*n*-Tetradecane (*n*-C$_{14}$)
63.	*n*-Octylcyclohexane
64.	5-Methyltetradecane
65.	4-Methyltetradecane
66.	2,6,10-Trimethyltridecane
67.	*n*-Pentadecane (*n*-C$_{15}$)
68.	*n*-Nonylcyclohexane
69.	*n*-Hexadecane (*n*-C$_{16}$)
70.	*n*-Heptadecane (*n*-C$_{17}$)

Figure 7. GC-MS total ion chromatogram of fraction 1 (saturate fraction, n-pentane eluate) of JP-8P jet fuel (DB-5 column, 40°C to 310°C at 4°C/min, split)

Figures 8 and 9 show the detailed GC-MS results for JP-8P fractions 2 and 3, respectively. Fraction 2 is composed of monoaromatics, and is free of saturates. Alkylbenzenes are the major components in this fraction, and minor components include tetralins and indans, whose concentrations are much lower than those found in coal-derived jet fuel. The analysis shows that multi-substituted alkylbenzenes (instead of long-chain n-alkylbenzenes) are the major compounds. Fraction 3 from JP-8P represents a concentrated diaromatics fraction, and consists of mainly naphthalene and alkylnaphthalenes. The identifications of ethylnaphthalenes and dimethylnaphthalenes were based on our own retention indices and GC-MS results; however, trimethylnaphthalenes were identified by comparing calculated retention indices or relative retention time in this study with those reported in literature for similar capillary columns.[42,43]

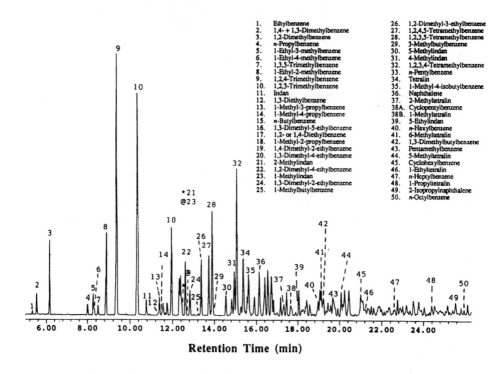

1.	Ethylbenzene	26.	1,2-Dimethyl-3-ethylbenzene
2.	1,4- + 1,3-Dimethylbenzene	27.	1,2,4,5-Tetramethylbenzene
3.	1,2-Dimethylbenzene	28.	1,2,3,5-Tetramethylbenzene
4.	n-Propylbenzene	29.	3-Methylbutylbenzene
5.	1-Ethyl-3-methylbenzene	30.	5-Methylindan
6.	1-Ethyl-4-methylbenzene	31.	4-Methylindan
7.	1,3,5-Trimethylbenzene	32.	1,2,3,4-Tetramethylbenzene
8.	1-Ethyl-2-methylbenzene	33.	n-Pentylbenzene
9.	1,2,4-Trimethylbenzene	34.	Tetralin
10.	1,2,3-Trimethylbenzene	35.	1-Methyl-4-isobutylbenzene
11.	Indan	36.	Naphthalene
12.	1,3-Diethylbenzene	37.	2-Methyltetralin
13.	1-Methyl-3-propylbenzene	38A.	Cyclopentylbenzene
14.	1-Methyl-4-propylbenzene	38B.	1-Methyltetralin
15.	n-Butylbenzene	39.	5-Ethylindan
16.	1,3-Dimethyl-5-ethylbenzene	40.	n-Hexylbenzene
17.	1,2- or 1,4-Diethylbenzene	41.	6-Methyltetralin
18.	1-Methyl-2-propylbenzene	42.	1,3-Dimethylbutylbenzene
19.	1,4-Dimethyl-2-ethylbenzene	43.	Pentamethylbenzene
20.	1,3-Dimethyl-4-ethylbenzene	44.	5-Methyltetralin
21.	2-Methylindan	45.	Cyclohexylbenzene
22.	1,2-Dimethyl-4-ethylbenzene	46.	1-Ethyltetralin
23.	1-Methylindan	47.	n-Heptylbenzene
24.	1,3-Dimethyl-2-ethylbenzene	48.	1-Propyltetralin
25.	1-Methylbutylbenzene	49.	2-Isopropylnaphthalene
		50.	n-Octylbenzene

Figure 8. GC-MS total ion chromatogram of fraction 2 (monoaromatic fraction, 5% benzene-pentane eluate) of JP-8P jet fuel (DB-5 column, 40°C to 310°C at 4°C/min, split)

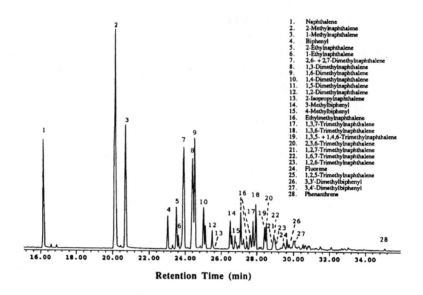

Figure 9. GC-MS total ion chromatogram of fraction 3 (aromatic fraction, benzene elute) of JP-8P jet fuel (DB-5 column, 40°C to 310°C at 4°C/min, split)

The above results indicate that the coal-derived JP-8C fuel is significantly different from petroleum-based JP-8P in composition. The saturate fraction constitutes 78% of JP-8C and 85% of JP-8P, and the aromatic fractions make up 20% of JP-8C and 12% of JP-8P.[15] The major compounds in JP-8P saturates are straight-chain and branched alkanes ranging from C_7 to C_{17}. However, the JP-8C saturates consist mainly of monocyclic, bicyclic and tricyclic alkanes, together with some long-chain alkanes. The aromatics in JP-8P are dominated by alkylbenzenes and alkylnaphthalenes, whereas the JP-8C aromatics are rich in hydroaromatic compounds such as tetralin, alkyltetralins, indans, cyclohexylbenzene, and some partially hydrogenated three-ring compounds, together with some alkylbenzenes and alkylnaphthalenes. These results show that jet fuels derived from different sources may have distinctly different molecular composition.

3.1.5   Potential Applications of Temperature-Programmed RI

The temperature-programmed RI method and general results from the present work, as summarized in Tables 1-7, are expected to be useful to a wide variety of analytical studies concerning hydrocarbon fuels and related samples. This is because detailed GC-MS analysis of distillate fuel samples and various hydrocarbons continues to be an important area of active study with respect to the following application categories:

(1) Fuel analysis or distillate fraction analysis and identifications [44-51], structure-composition-reactivity relationships [52,53] or composition-property correlations [54,55], and quality control in refinery processing or physico-chemical treatments of gasoline, jet fuel and diesel fuels [56-61], as well as structural analysis of solid organic materials coupling flash pyrolysis with GC-MS [62,63];

(2) Identification and quantification of pre-scribed components such as priority-pollutant sulfur compounds (e.g., thiophene, dibenzothiophene, and dimethyldibenzothiophenes) [64,65], nitrogen compounds [66,67], and polyaromatics in gasoline, diesel, jet fuels and fuel oils [68,69] or additives such as phenolics [70-73] or other additives [74] or blending components (such as MTBE) in fuels [75];

(3) Environmental monitoring such as analysis of organics in fuel- or oil-contaminated soils [76-80], fuel or oil spilled in waters [81-83] and volatiles in air or PAHs emitted from vehicles or plants [84-88] as well as airborne particulates containing soluble hydrocarbons [89-91];

(4) Medical analysis such as determination of volatile fuel compounds in human bloods [92-94], and toxicological studies such as mutagenicity analysis [95];

(5) Forensic autopsies related to organics such as volatile aromatic and aliphatic hydrocarbons. [96-98]

There are comprehensive references and reviews that cover two or more aspects of analytical work mentioned above. [99, 100]

It should be noted that pristane (2,6,10,14-tetramethylpentadecane; $C_{19}H_{40}$; MW 268; bp 296 °C) and phytane ($C_{20}H_{42}$; MW 282) are commonly used biomarkers in GC and GC-MS analysis of crude oils or their fractions and in pyrolysis-GC-MS analysis of kerogen and coal. These two isoprenoid hydrocarbon molecules are known to originate from breakdown of dihydrophytol which is a side chain of the chlorophyll molecule. Although these two molecules were not found in the jet fuel samples in any significant concentrations, it is worthy to discuss them briefly. On DB-5 type column (with stationary phase), pristane elutes immediately after n-C_{17}, and phytane elutes immediately after n-C_{18}. The same order of elution applies also to DB-1 type column. However, the elution order with RTx-50 column is different. On Rtx-50 type column (with stationary phase), pristane elutes before n-C_{17}, and phytane elutes before n-C_{18}. Based on preliminary analysis, the RI for pristane with Rtx-50 type column is estimated to be around 1670-1680, which is smaller than that of n-C_{17} (1700). On DB-5 type column, the RI for pristane is estimated to be around 1708-1720, which is slightly larger than that of n-C_{17} (1700).

3.2 SimDis GC and GC-MS of Middle Distillate Fuels

There are established standard methods for SimDis GC analysis of distillate fuels. Table 8 shows the SimDis GC results from our laboratory for petroleum-derived jet fuels, coal-derived jet fuels, and middle distillates from Fischer-Tropsch synthesis (FT-MD), and from direct catalytic coal liquefaction from Wilsonville plant (WI-MD) and from HRI plant (HRI-MD).[11,14] Figure 10 shows the simulated distillation profiles for coal-derived JP-8 jet fuels (JP-8C, JP-8C/E, JP-8C/D), petroleum-derived jet fuels (JP-8P, JP-8P2, JP-7, JP-TS) and commercial jet fuels (Jet A, Jet A-1).

Table 8. Simulated distillation (°C) of jet fuels and middle distillates by using ASTM D2887 method with HP-5 Column, 40-280 °C @3°C/min, IH=5.

Rec. wt%	JP-8P	JP-8P2	Jet A	Jet A-1	JP-7	JPTS	JP-8C	JP-8C/1	JP-8C/	FT-MD	HRI-M	WI-MD
IBP	113.8	118.2	113.6	154.4	170.7	117.2	76.0	76.3	78.0	151.0	8.4	196.3
5	140.6	153.7	150.7	166.7	185.1	151.1	103.0	103.0	103.1	174.1	82.6	238.6
10	151.1	167.0	163.0	173.4	190.5	159.7	116.4	115.9	121.1	188.1	102.1	254.8
15	158.6	174.3	171.1	176.2	195.6	165.3	126.8	126.7	132.4	200.7	132.0	264.7
20	165.6	182.7	176.4	179.7	196.2	169.7	136.5	134.3	146.4	210.3	166.1	273.9
25	171.3	189.6	183.4	183.4	199.2	174.1	146.8	146.4	156.8	222.7	190.2	281.5
30	176.5	196.0	189.6	186.6	203.3	178.2	157.8	156.7	169.3	234.5	209.6	287.4
35	181.8	198.8	196.0	189.5	207.3	180.5	169.1	168.7	180.4	244.6	226.6	295.8
40	188.4	204.6	198.7	192.4	210.1	183.1	179.7	177.8	183.9	253.8	241.0	301.6
45	196.0	209.2	205.8	196.1	213.3	185.9	184.0	183.8	193.1	262.9	254.0	308.3
50	199.6	214.5	211.9	196.5	216.2	188.5	192.7	191.7	199.2	271.6	265.4	314.7
55	207.4	218.2	216.4	199.0	216.5	191.6	199.4	199.0	206.4	280.3	274.4	321.0
60	214.5	223.2	222.4	202.6	219.0	196.0	206.5	206.3	214.0	288.3	283.7	327.0
65	219.0	228.7	229.7	206.0	222.5	199.0	214.6	213.9	223.3	296.5	291.9	333.3
70	227.4	233.6	235.4	209.1	226.7	205.1	224.7	223.9	231.7	303.8	300.9	339.8
75	234.3	238.1	241.4	213.3	230.3	212.0	233.1	232.7	238.3	311.8	309.1	344.9
80	241.0	246.2	249.6	216.4	235.4	219.0	241.5	241.4	247.2	319.1	320.0	350.0
85	251.7	253.6	254.9	220.2	240.9	229.7	251.9	251.9	256.4	327.3	328.8	354.5
90	260.6	260.8	264.7	228.4	250.1	240.1	262.6	262.8	266.6	335.6	342.3	362.5
95	273.3	271.1	273.9	236.7	261.0	253.7	278.0	278.2	284.0	345.8	358.6	370.1
FBP	295.0	290.0	290.0	256.0	277.0	271.1	304.0	304.0	306.0	359.0	402.2	392.5

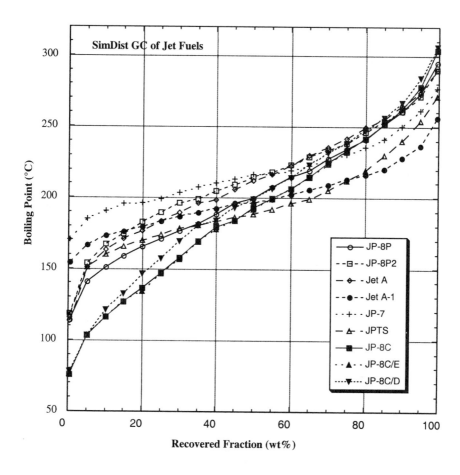

Figure 10. Simulated distillation profiles for coal-derived JP-8 jet fuels (JP-8C, JP-8C/E, JP-8C/D), petroleum-derived jet fuels (JP-8P, JP-8P2, JP-7, JP-TS) and commercial jet fuels (jet A, Jet A-1).

Table 9 shows the results from quantitative GC and GC-MS analysis of JP-8P2 fuel including the contents of individual components in the original fuel and in the thermally stressed fuels (450 °C under 100 psig UHP N_2). This sample is representative of many petroleum-derived JP-8 jet fuels that we have analyzed and tested. Table 10 shows the results from quantitative GC and GC-MS analysis of JP-8C fuel including the contents of individual components in the original fuel and in the thermally stressed fuels (450 °C under 100 psig UHP N2). This sample is representative of several coal-derived JP-8 jet fuels that we have analyzed in our laboratory. The contents of compounds in the whole fuels were determined by using response factors, as listed in Tables 9 and 10. The results in Tables 9 and 10 also show that JP-8C is thermally more stable than JP-8P2, in terms of less decomposition, more liquid remained and less solid deposits after thermal stressing at 450 °C.

Table 9. Identified Compounds and Their Contents in Neat JP-8P2 and Thermally Stressed JP-8P2 at 450°C for 1.0-8.0 hours in 100 psig N2

GC R.T.	GC-MS R.T.	Mol Ion	Base Peak	Compound Identified	Response Factor	100 Orig. JP-8P2 Neat	95.8 TS22 450°C 1 hr	82.0 TS48 450°C 2.5 hr	70.6 TS9 450°C 4 hr
2.84	2.64	114	57	Branched alkane (C8)	1.02	0.17	0.56	0.37	0.45
2.88	2.68	98	83	Cyclohexane, methyl-	1.04	0.04	0.46	0.65	1.30
2.99	2.79	128	57	Branched alkane (C9)	1.00	0.14	0.21	0.25	0.15
3.38	3.16	114	43	n-Octane	1.02	0.46	1.21	1.42	1.34
3.42	3.20	112	97	Cyclohexane, c-1,3- & t-1,4-dimethyl-	1.02	0.16	0.28	0.50	0.65
3.67	3.47	128	57	**2,6-Dimethylheptane**	1.00	0.31	0.29	0.15	0.04
3.86	3.67	112	97	Cyclohexane, trans-1,2-dimethyl-	1.02	0.25	0.28	0.25	0.34
4.52	4.31	126	111	Cyclohexane, trimethyl-	1.02	0.41	0.63	0.62	0.65
4.63	4.41	128	43	Branched alkane (C9)	1.00	0.68	0.67	0.42	0.29
4.82	4.63	128	57	Branched alkane (C9)	1.00	0.59	0.67	0.47	0.26
4.86	4.68	112	83	Cyclohexane, ethyl-	1.03	0.19	0.31	0.35	0.26
5.72	5.60	128	57	n-Nonane	1.00	1.75	1.50	1.24	0.60
5.90	5.78	126	55, 69	Cyclohexane, C3-	1.02	0.23	0.19	0.17	0.14
6.17	6.07	126	97	Cyclohexane, 1-ethyl-?-methy	1.02	0.51	0.49	0.46	0.38
6.35	6.27	142	57	Nonane, 4-methyl-	0.99	0.20	0.12		
6.60	6.57	142	57	**2,6-Dimethyloctane**	0.99	0.73	0.70	0.21	0.13
7.06	7.03	126	97	Cyclohexane, 1-ethyl-?-methy	1.02	0.24	0.37	0.29	0.33
7.12	7.08	142	57	Branched alkane (C10)	0.99	0.34	0.25	0.12	
7.40	7.36	140	69	Cyclohexane, C4-	1.00	0.04			
7.56	7.55	140	69	Cyclohexane, C4-	1.00	0.10	0.06	0.04	
7.61	7.61	142	57	Branched alkane (C10)	0.99	0.18	0.12	0.03	
7.79	7.80	142	57	Nonane, 2-methyl-	0.99	1.49	1.35	0.54	0.20
7.95	8.00	126	83	Cyclohexane, n-propyl-	1.02	0.27	0.41	0.25	0.16
8.00	8.03	140	69	Cyclohexane, C4-	1.00	0.15	0.13	0.14	0.10
8.19	8.22	142	57	Nonane, 3-methyl-	0.99	0.69	0.59	0.26	0.14
8.50	8.56	106	91	Ethyl-benzene and p- and m-Xylene	1.10	0.48	1.08	3.00	3.40
8.54	8.60	140	69	Cyclohexane, 1-ethyl-dimethy	1.00	0.17	0.15	0.12	0.08
8.70	8.77	140	55, 69	Cyclohexane, C4-	1.00	0.37	0.32	0.19	0.12
8.94	9.02	140	55, 69	Cyclohexane, C4-	1.00	0.08	0.10	0.07	0.03
9.36	9.48	142	57	n-Decane	0.99	3.83	3.74	1.79	0.93
9.53	9.65	140	97	Cyclohexane, 1-methyl-?-prop	1.00	0.87	0.70	0.42	0.27
9.66	9.81	140	56	1-Decene	0.99	0.10	0.19	0.13	0.05
9.66	9.81	170	57	Branched alkane (C12)	0.95	0.30	0.20	0.13	0.06
9.80	10.00	106	91	0-Xylene	1.10	0.32	0.72	1.25	1.22
9.91	10.11	156	57	**2,6-Dimethylnonane**	0.97	1.91	0.98	0.17	0.08
10.11	10.28	140	69	Cyclohexane, C4-	1.00	0.07	0.04	0.01	
10.11	10.28	156	57	Branched alkane (C11)	0.97	0.10	0.08	0.02	
10.25	10.43	140	69, 97	Cyclohexane, C4-	1.00	0.19	0.19	0.12	0.07
10.31	10.49	156	57	Branched alkane (C11)	0.97	0.26	0.15	0.04	
10.47	10.65	156	57	Branched alkane (C11)	0.97	0.20	0.11	0.03	
10.68	10.98	156	57	Branched alkane (C11)	0.97	0.38	0.23	0.05	0.02
10.73	11.03	120	105	Benzene, i-propyl- (Cumeme)	1.10	0.12	0.16	0.23	0.25
10.95	11.24	154	69	Branched olefin (C11)	0.97	0.20	0.09	0.03	0.02
11.20	11.50	170	57	Branched alkane (C12)	0.95	0.15	0.14	0.05	
11.20	11.50	154	55, 69	Branched olefin (C11)	0.97	0.30	0.25	0.15	0.09
11.44	11.74	156	57	Decane, 5-methyl-	0.97	0.72	0.49	0.11	0.05
11.50	11.80	154	55, 69	Cyclohexane, C5-	0.98	0.17	0.12	0.05	0.03
11.60	11.90	156	71	Branched alkane (C11)	0.97	0.78	0.64	0.21	
11.70	12.00	156	57	Decane, 4-methyl-	0.97	1.00	0.79	0.25	0.15
11.90	12.20	140	83	Cyclohexane, n-butyl-	0.98	0.72	0.60	0.20	0.10
11.92	12.23	120	91	Benzene, n-propyl-	1.10	0.35	0.48	0.46	0.37
12.07	12.38	156	57	Decane, 3-methyl-	0.97	0.83	0.64	0.16	0.06
12.10	12.40	154	69	Cyclohexane, C5-	0.98	0.20	0.16	0.07	0.03
12.38	12.68	120	105	3-Ethyltoluene & 1,3,5-trimethylbenzene	1.10	1.24	1.67	2.63	2.95

GC R.T.	GC-MS R.T.	Mol Ion	Base Peak	Compound Identified	Response Factor	Weight % of feed			
						100 Orig. JP-8P2 Neat	95.8 TS22 450°C 1 hr	82.0 TS48 450°C 2.5 hr	70.6 TS9 450°C 4 hr
12.72	13.02	168	69	Cyclohexane, C6-	0.94	0.19	0.12	0.07	0.03
13.00	13.27	156	57	Branched alkane (C11)	0.97	0.06	0.02		
13.00	13.27	154	69	Branched olefin (C11)	0.97	0.09	0.06	0.02	
13.20	13.48	154	69	Cyclohexane, C5-	0.98	0.17	0.10	0.04	
13.20	13.48	156	57	Branched alkane (C11)	0.97	0.17	0.08	0.02	
13.39	13.69	156	57	n-Undecane	0.97	5.59	4.66	1.74	0.72
13.46	13.80	120	105	Toluene, 2-ethyl-	1.10	0.35	0.61	0.58	0.67
13.60	13.90	154	55, 97	Cyclohexane, C5-	0.98	0.69	0.39	0.17	0.06
13.68	13.98	154	55	1-Undecene	0.97	0.40	0.29	0.13	
13.70	14.01	170	57	**2,6-Dimethyldecane**	0.95	1.02	0.50	0.09	
13.77	14.08	120	105	Benzene, 1,2,4-trimethyl-	1.10	1.76	2.10	2.15	2.23
13.81	14.12	134	91	Benzene, iso-butyl-	1.05	0.37	0.27	0.08	
13.95	14.24	154	69	Cyclohexane, C5-	0.98	0.18	0.10	0.04	
14.03	14.30	170	57	Branched alkane (C12)	0.95	0.18	0.10	0.04	
14.20	14.47	138	138	trans-Decalin	1.00	0.23	0.19	0.14	0.10
14.24	14.51	134	105	Benzene, C4-	1.05	0.16	0.18	0.19	0.15
14.30	14.57	170	57	**3,7-Dimethyldecane**	0.95	1.02	0.50	0.05	
14.34	14.67	154	83	Cyclohexane, C5-	0.98	0.57	0.19	0.03	
14.44	14.81	170	57	Branched alkane (C12)	0.95	0.26	0.15	0.02	
14.53	14.95	154	69	Cyclohexane, C5-	0.98	0.30	0.15	0.03	
14.69	15.12	134	119	Benzene, C4-	1.05	0.28	0.48	0.56	0.73
14.76	15.18	140	69	Cyclohexane, C4-	1.00	0.12	0.08	0.03	
14.76	15.18	170	57	Branched alkane (C12)	0.95	0.12	0.07	0.02	
15.02	15.43	140	69	Cyclohexane, C4-	1.00	0.04	0.05	0.02	
15.19	15.60	170	57	Branched alkane (C12)	0.95	0.12	010	0.04	
15.31	15.74	168	57	3-Undecene, 6-methyl-	0.95	0.71	0.56	0.13	
15.35	15.78	170	57	Undecane, 5-methyl-	0.95	0.61	0.32	0.09	0.03
15.48	15.86	120	105	Benzene, 1,2,3-trimethyl-	1.10	0.44	0.69	0.69	0.77
15.60	15.96	170	71	Undecane, 4-methyl-	0.95	0.72	0.52	0.12	0.02
15.70	16.07	170	57	Undecane, 2-methyl-	0.95	1.12	0.99	0.29	0.04
15.82	16.20	152	81, 152	Decalin, ?-methyl-	1.00	0.57	0.52	0.47	0.10
15.91	16.31	134	105	Toluene, ?-propyl-	1.05	0.32	0.53	0.46	0.63
16.00	16.38	134	105	Toluene, ?-propyl-	1.05	0.14	0.36	0.24	0.31
16.05	16.44	170	57	Undecane, 3-methyl-	0.95	1.12	0.65	0.12	0.03
16.07	16.46	154	83	Cyclohexane, n-pentyl-	0.98	0.75	0.53	0.14	0.04
16.11	16.50	134	91, 105	Benzene, n-butyl- + C4-	1.05	0.35	0.43	0.30	0.52
16.31	16.69	134	119	Benzene, C4-	1.05	0.37	0.68	0.80	0.93
16.50	16.88	168	55, 69	Branched olefin (C12)	0.95	0.16	0.15		
16.50	16.88	170	57	Branched alkane (C12)	0.95	0.23	0.17		
16.75	17.15	152	95	Decalin, ?-methyl-	1.00	0.58	0.48	0.19	0.06
16.82	17.20	134	105	Toluene, ?-propyl-	1.05	0.49	0.54	0.47	0.42
17.33	17.73	134	105	Benzene, ethyl-dimethyl-	1.05	0.37	0.64	1.13	1.09
17.33	17.73	170	57	n-Dodecane	0.95	4.86	3.74	0.65	0.20
17.52	17.90	184	57	**2,6-Dimethylundecane**	0.92	2.31	1.54	0.28	0.06
17.60	17.98	168	97	Cyclohexane, C6-	0.94	0.41	0.16	0.06	
17.60	17.98	168	55	1-Dodecene	0.95	0.46	0.20	0.11	
17.67	18.09	134	119	Benzene, C4-	1.05	0.74	0.82	0.77	0.77
17.76	18.18	148	105	Benzene, C5-	1.00	0.14	0.11	0.07	0.06
17.76	18.18	184	57	Branched alkane (C13)	0.92	0.35	0.21	0.02	
17.76	18.18	168	55, 69	Cyclohexane, C6-	0.94	0.35	0.13	0.04	
17.88	18.34	168	55, 69	Cyclohexane, C6-	0.94	0.11	0.16		
17.88	18.34	148	119	Benzene, C5-	1.00	0.22	0.19	0.19	0.10
17.88	18.34	184	57	Branched alkane (C13)	0.92	0.46	0.21		
18.00	18.47	148	119	Benzene, C5-	1.00	0.11	0.22	0.08	0.05
18.23	18.68	168	55, 97	Cyclohexane, C6-	0.94	0.14	0.12	0.04	
18.24	18.69	134	119	Benzene, C4-	1.05	0.05	0.08	0.10	0.10
18.29	18.73	166	81	Decalin, C2-	1.00	0.29	0.18	0.09	0.06
18.45	18.91	148	133	Benzene, C5-	1.00	0.15	0.17	0.11	0.14
18.55	19.01	132	117	Indan, 1-methyl-	1.00	0.19	0.27	0.30	0.34
18.80	19.20	168	83	Cyclohexane, C6-˙	0.94	0.19	0.12	0.01	
18.85	19.24	134	119	Benzene, C4-	1.05	0.14	0.18	0.20	0.14
18.98	19.38	134	119	Benzene, C4-	1.05	0.17	0.21	0.20	0.21
19.10	19.51	182	69	Cyclohexane, butyltrimethyl-	0.94	0.29	0.15	0.02	
19.14	19.55	184	57	Dodecane, 6-methyl-	0.92	0.53	0.41	0.02	
19.20	19.62	134	119	Benzene, C4-	1.05	0.37	0.41	0.40	0.38
19.26	19.68	184	57	Dodecane, 5-methyl-	0.92	0.67	0.36	0.03	
19.36	19.79	148	133	Benzene, C5-	1.00	0.15	0.24	0.27	0.29
19.36	19.79	184	71	Dodecane, 4-methyl-	0.92	0.37	0.31		
19.48	19.89	198	57	**2,6,10Trimethylundecane**	0.90	2.48	1.00	0.16	
19.56	20.00	148	119	Benzene, C5-	1.00	0.17	0.20	0.18	0.21
19.75	20.20	182	55, 69	Branched olefin (C13)	0.92	0.16	0.06		

GC R.T.	GC-MS R.T.	Mol Ion	Base Peak	Compound Identified	Response Factor	100 Orig. JP-8P2 Neat	95.8 TS22 450°C 1 hr	82.0 TS48 450°C 2.5 hr	70.6 TS9 450°C 4 hr
						<---------- Weight % of feed ---------->			
19.75	20.20	184	57	Branched alkane (C13)	0.92	0.21	0.06		
19.75	20.20	148	119	Benzene, C5-	1.00	0.24	0.10	0.25	0.36
19.83	20.28	184	57	Dodecane, 3-methyl-	0.92	0.63	0.39		
19.89	20.34	148	105, 133	Benzene, C5-	1.00	0.17	0.27	0.47	0.37
20.04	20.46	148	105	Benzene, C5-	1.00	0.13	0.08	0.13	0.15
20.10	20.52	148	91	Benzene, n-pentyl-	1.00	0.05	0.06	0.08	0.04
20.10	20.52	168	83	Cyclohexane, n-hexyl-	0.94	0.55	0.30	0.11	0.05
20.33	20.73	148	119	Benzene, C5-	1.00	0.16	0.15	0.14	0.12
20.60	21.00	148	119	Benzene, C5-	1.00	0.19	0.20	0.17	0.14
20.68	21.11	148	105, 133	Benzene, C5-	1.00	0.24	0.17	0.19	0.20
20.79	21.22	132	117	Indan, 5- or 6-methyl-	1.00	0.10	0.07	0.12	0.18
20.79	21.22	168	69	Cyclohexane, C6-	0.94	0.14	0.04		
20.90	21.30	134	119	Benzene, C4-	1.05	0.18	0.36	0.47	0.43
20.94	21.34	196	55, 69	Branched olefin (C14)	0.90	0.32	0.21		
21.06	21.46	184	57	n-Tridecane	0.92	3.53	2.90	0.55	0.18
21.10	21.50	148	119	Benzene, C5-	1.00	0.24	0.38	0.40	0.43
21.20	21.60	198	57	Branched alkane (C14)	0.90	0.22	0.13		
21.50	21.88	132	117	Indan, 5- or 6-methyl-	1.00	0.16	0.29	0.38	0.39
21.50	21.88	182	55	1-Tridecene	0.92	0.26	0.10		
21.50	21.88	198	57	Branched alkane (C14)	0.90	0.86	0.53		
21.60	22.00	162	119, 147	Benzene, C6-	1.00	0.06	0.06	0.08	0.07
21.64	22.04	182	97	Cyclohexane, C7-	0.94	0.05	0.03		
22.19	22.61	182	69	Cyclohexane, C7-	0.94	0.10			
22.19	22.61	148	133	Benzene, C5-	1.00	0.15	0.10	0.08	0.07
22.32	22.75	146	131	Indan, C2-	1.00	0.29	0.19	0.25	0.37
22.40	22.84	132	104	Tetralin	1.00	0.29	0.19	0.10	0.02
22.56	23.00	162	119	Benzene, C6-	1.00	0.07	0.05	0.03	0.02
22.68	23.12	198	57	Tridecane, 6-methyl-	0.90	0.48	0.21	0.05	
22.80	23.22	198	43	Tridecane, 5-methyl-	0.90	0.32	0.11		
22.85	23.26	148	133	Benzene, C5-	1.00	0.19	0.19	0.13	0.18
22.95	23.36	146	131	Indan, C2-	1.00	0.24	0.19	0.25	0.26
23.00	23.39	160	145	Indan, C3-	1.00	0.04	0.08	0.08	0.04
23.00	23.39	198	57,71	Tridecane, 4-methyl-	0.90	0.32	0.16	0.03	
23.10	23.50	198	57	Tridecane, 2-methyl-	0.90	0.54	0.33	0.06	
23.32	23.73	212	57	**2,6,10-Trimethyldodecane**	0.86	1.42	0.66	0.05	
23.50	23.88	198	57	Tridecane, 3-methyl-	0.90	0.37	0.19	0.04	
23.63	24.00	162	105, 133	Benzene, C6-	1.00	0.10	0.07	0.07	0.06
23.81	24.21	146	104	Tetralin, methyl-	1.00	0.22	0.10	0.01	
23.88	24.28	182	83	Cyclohexane, n-Heptyl-	0.94	0.57	0.20	0.03	
24.14	24.53	148	133	Benzene, C5-	1.00	0.12	0.06	0.03	0.04
24.22	24.61	162	119	Benzene, C6-	1.00	0.12	0.06	0.02	0.03
24.35	24.73	162	105	Benzene, C6-	1.00	0.10	0.05	0.01	0.01
24.45	24.82	212	57	Tetradecane, 6-methyl-	0.88	0.17	0.06		
24.60	24.98	198	57	n-Tetradecane	0.90	3.12	2.11	0.26	
24.64	25.02	128	128	Naphthalene	1.00	0.29	0.44	1.06	1.29
24.77	25.13	146	117	Indan, C2-	1.00	0.06	0.01	0.02	0.02
24.85	25.21	162	119	Benzene, C6-	1.00	0.13	0.02	0.02	0.01
24.98	25.34	146	131	Indan, C2-	1.00	0.06	0.03	0.03	0.04
25.12	25.48	162	133	Benzene, C6-	1.00	0.14	0.10	0.05	0.03
25.16	25.52	196	55	1-Tetradecene	0.90	0.15	0.02		
25.22	25.58	162	133	Benzene, C6-	1.00	012	0.02	0.02	0.01
25.62	25.98	146	131	Tetralin, C1	1.00	0.29	0.14	0.05	0.04
25.72	26.08	160	145	Indan, C3-	1.00	0.12	0.07	0.03	0.02
25.93	26.29	160	145	Indan, C3-	1.00	0.03	0.02	0.02	0.01
25.93	26.29	148	133	Benzene, C5-	1.00	0.04	0.03	0.02	0.01
26.06	26.42	146	131	Tetralin, methyl-	1.00	0.42	0.29	0.03	
26.09	26.45	212	57	Tetradecane, 6-methyl-	0.88	0.52	0.14		
26.16	26.53	212	43	Tetradecane, 5-methyl-	0.88	0.77	0.54		
26.28	26.65	160	145	Indan, C3-	1.00	0.19	0.28	0.17	0.15
26.40	26.77	212	71	Tetradecane, 4-methyl-	0.88	0.37	0.13		
26.50	26.87	212	57	Tetradecane, 2-methyl-	0.88	0.55	0.24	0.01	
26.77	27.15	160	145	Indan, C3-	1.00	0.25	0.14	0.09	0.09
26.83	27.21	212	57	Tetradecane, 3-methyl-	0.88	0.29	0.15	0.01	
27.25	27.62	160	118	Tetralin, dimethyl-	1.00	0.09	0.06		
27.34	27.71	160	118	Tetralin, dimethyl-	1.00	0.18	0.10		
27.50	27.84	194	83	Cyclohexane, 1,1'-(1,2-ethanediyl)bis-	0.94	0.13	0.10		
27.56	27.90	146	131	Tetralin, methyl-	1.00	0.50	0.29	0.04	0.01
27.67	28.02	176	119	Benzene, C7-	1.00	0.10	0.06	0.03	0.01
27.78	28.15	176	105	Benzene, C7-	1.00	0.15	0.10		
27.88	28.25	160	145	Indan, C3-	1.00	0.12	0.10	0.08	0.06

GC R.T.	GC-MS R.T.	Mol Ion	Base Peak	Compound Identified	Response Factor	100 Orig. JP-8P2 Neat	95.8 TS22 450°C 1 hr	82.0 TS48 450°C 2.5 hr	70.6 TS9 450°C 4 hr
						<------- Weight % of feed ------->			
27.91	28.28	212	57	n-Pentadecane	0.88	1.65	1.06	0.24	
28.33	28.70	142	142	Naphthalene, 2-methyl-	1.00	0.53	0.69	1.72	2.05
28.63	29.00	160	145	Indan, C3-	1.00	0.22	0.06	0.07	0.05
28.99	29.35	158	129	C2-Dihydro-naphthalene	1.00	0.04	0.04	0.08	0.04
28.99	29.35	174	159	Indan, C4-	1.00	0.10	0.02	0.03	0.03
29.43	29.75	142	142	Naphthalene, 1-methyl-	1.00	0.47	0.61	1.20	1.11
29.60	29.94	174	159	Indan, C4-	1.00	0.15			0.01
30.76	31.02	160	145	Indan, C3-	1.00	0.16	0.11	0.02	
31.18	31.37	226	57	n-Hexadecane	0.86	0.38	0.34	0.02	
31.77	32.02	156	141	Naphthalene, 2-ethyl-	1.00	0.12	0.13	0.29	0.35
31.86	32.09	156	156	Naphthalene, 2,7-dimethyl-	1.00	0.16	0.27	0.26	0.30
31.96	32.19	156	156	Naphthalene, 2,6-dimethyl-	1.00	0.14	0.21	0.23	0.28
32.07	32.31	154	154	Biphenyl	1.00	0.08	0.10	0.15	0.26
32.32	32.60	156	141	Naphthalene, 1-ethyl-	1.00	0.09	0.17	0.23	0.26
32.38	32.66	174	145	Indan, C4-	1.00	0.07			
32.74	32.98	156	156	Naphthalene, 1,7-dimethyl-	1.00	0.07	0.24	0.26	0.34
32.87	33.09	156	156	Naphthalene, 1,3- and 1,6-dimethyl-	1.00	0.38	0.56	0.89	1.00
33.47	33.70	156	156	Naphthalene, 2,3-dimethyl-	1.00	0.07	0.10	0.13	0.15
34.02	34.22	156	156	Naphthalene, 1,5-dimethyl-	1.00	0.02	0.10	0.15	0.16
34.97	35.16	170	155	Naphthalene, C3-	1.00	0.03	0.03	0.07	0.08
35.24	35.42	168	168	Biphenyl, 3-methyl-	1.00	0.05	0.05	0.12	0.15
35.95	36.13	170	155, 170	Naphthalene, C3-	1.00	0.10	0.19	0.29	0.34

Table 10. Identified Compounds and Their Contents in Coal-derived JP-8C (Neat) and Thermally Stressed JP-8C at 450°C for 1.0-8.0 hours in 100 psig N_2

GC R.T.	GC-MS R.T.	Mol Ion	Base Peak	Compound Identified	Response Factor	100 Orig. JP-8C Neat	96.3 TS21 450°C 1 hr	94.9 TS47 450°C 2.5 hr	86.5 TS8 450°C 4 hr
						<------- Weight % of feed ------->			
1.81	1.71	86	43	Pentane, 2-methyl-	1.06	0.09	0.25	0.19	0.26
1.90	1.80	86	57	n-Hexane (n-C6)	1.06	0.09	0.63	0.57	0.93
2.11	2.06	84	56	Cyclopentane, methyl-	1.06	0.29	0.44	0.52	1.58
2.11	2.06	100	43	Hexane, 2-methyl-	1.04	0.59	0.35	0.29	0.34
2.17	2.11	100	43	Hexane, 3-methyl-	1.04	0.78	0.40	0.39	0.21
2.28	2.27	100	43	n-Heptane (n-C7)	1.04	0.49	0.40	0.29	0.36
2.36	2.35	84	56	Cyclohexane	1.04	2.92	3.25	3.11	3.85
2.57	2.58	112	55	Cyclopentane, trimethyl-	1.02	0.40	0.20	0.15	0.21
2.68	2.71	112	70	Cyclopentane, trimethyl-	1.02	0.60	0.41	0.30	0.36
2.79	2.82	98	83	Cyclohexane, methyl-	1.04	6.95	6.36	6.80	5.45
2.86	2.94	98	69	Cyclopentane, ethyl-	1.04	0.49	0.50	0.29	0.49
2.89	2.97	114	43	Branched alkane (C8)	1.02	0.50	0.41	0.30	0.18
2.95	3.02	78	78	Benzene	1.10	0.18	0.24	0.18	0.57
3.25	3.31	112	70	Cyclopentane, C3-	1.02	0.20	0.30	0.25	0.18
3.30	3.36	114	43	n-Octane (n-C8)	1.02	0.15	0.30	0.30	0.18
3.36	3.42	112	97	Cyclohexane, c-1,3- or t-1,4-dimethy	1.02	3.58	3.06	3.81	3.68
3.51	3.62	112	55,83	Cyclopentane, ethyl-methyl-	1.02	1.11	0.85	0.64	0.89
3.78	3.91	112	97	Cyclohexane, trans-1,2-dimethyl-	1.02	1.04	0.78	0.57	0.57
3.98	4.20	112	97	Cyclohexane, trans-1,3-dimethyl-	1.02	2.16	1.49	2.36	1.30
4.36	4.63	126	69, 111	Cyclohexane, trimethyl-	1.02	0.60	0.50	0.40	0.42
4.47	4.74	128	43	Branched alkane (C9)	1.00	0.69	0.52	0.32	0.14
4.65	4.95	126	69	Cyclohexane, C3-	1.02	0.70	0.46	0.40	0.27
4.76	5.05	126	83	Cyclohexane, ethyl-	1.03	5.13	4.28	3.39	2.56
4.95	5.24	92	91	Toluene	1.10	1.47	2.20	2.83	2.83
5.66	5.97	128	57	n-Nonane (n-C9)	1.00	0.51	0.52	0.40	0.28
5.86	6.17	126	55,69	Cyclohexane, C3-	1.02	1.00	1.02	0.64	0.27
6.11	6.47	126	97	Cyclohexane, 1-ethyl-?-methyl-	1.02	2.69	2.32	2.24	1.46
6.27	6.64	142	57	Branched alkane (C10)	0.99	0.26	0.13	0.10	0.04
6.51	7.00	142	57	2,6-Dimethyloctane	0.99	0.24	0.14	0.10	0.04
6.96	7.48	126	97	Cyclohexane, 1-ethyl-?-methyl-	1.02	2.11	1.68	1.54	0.91
7.48	8.02	140	69, 111	Cyclohexane, 1-ethyl-dimethyl-	1.00	0.25	0.31	0.20	0.19
7.61	8.15	126	83,55	Cyclohexane, 1-ethyl-methyl-	1.02	0.30	0.41	0.24	0.18
7.67	8.21	142	57	Nonane, 2-methyl-	0.99	0.36	0.31	0.18	0.04
7.88	8.44	126	83	Cyclohexane, n-propyl-	1.02	1.57	1.69	1.58	0.78

GC R.T.	GC-MS R.T.	Mol Ion	Base Peak	Compound Identified	Response Factor	100 Orig. JP-8C Neat	96.3 TS21 450°C 1 hr	94.9 TS47 450°C 2.5 hr	86.5 TS8 450°C 4 hr
						<----- Weight % of feed ----->			
8.12	8.68	124	81,67	Pentalene, octahydro-C!-	1.00	0.12	0.25	0.15	0.19
8.12	8.68	142	57	Nonane, 3-methyl-	0.99	0.15	0.26	0.14	0.09
8.12	8.68	126	55,97	Cyclohexane, C3-	1.02	0.35	0.36	0.14	0.09
8.45	9.00	106	91	Ethylbenzene + p- and m-Xylene	1.10	1.29	2.00	2.48	3.82
8.54	9.10	140	69	Cyclohexane, C4-	1.00	0.30	0.21	0.20	0.09
8.66	9.22	140	69	Cyclohexane, C4-	1.00	0.10	0.10	0.08	0.06
9.25	9.85	140	55,97	Cyclohexane, 1-methyl-?-propyl-	1.00	0.20	0.21	0.14	0.09
9.35	9.94	142	57	n-Decane (n-C10)	0.99	0.31	0.51	0.36	0.21
9.54	10.10	140	97	Cyclohexane, 1-methyl-?-propyl-	1.00	2.09	1.42	1.01	0.57
9.74	10.30	140	55,69	Cyclohexane, C4-	1.00	0.20	0.16	0.15	0.09
9.79	10.45	106	91	o-Xylene	1.10	0.74	0.85	0.84	0.88
9.89	10.57	124	67	Indene, octahydro-, trans-	1.00	0.60	0.78	0.45	0.46
10.25	10.90	140	69,97	Cyclohexane, C4-	1.00	0.68	0.51	0.36	0.23
10.50	11.21	140	97	Cyclohexane, C4-	1.00	0.61	0.45	0.27	0.12
10.69	11.44	120	105	Benzene, i-propyl- (Cumeme)	1.10	0.06	0.14	0.11	0.13
10.69	11.44	140	97	Cyclohexane, C4-	1.00	0.30	0.26	0.15	0.09
10.82	11.62	140	69	Cyclohexane, C4-	1.00	0.15	0.16	0.10	0.06
10.98	11.80	154	69	Cyclohexane, C5-	0.98	0.10	0.11	0.04	0.02
11.22	12.04	154	55,69	Cyclohexane, C5-	0.98	0.36	0.16	0.10	0.05
11.60	12.39	138	67	Indene, methyloctahydro-	1.00	0.81	0.73	0.61	0.37
11.71	12.50	154	69	Cyclohexane, C5-	0.98	0.31	0.11		
11.78	12.57	124	81	Indene, octahydro-, cis-	1.00	0.91	0.73	0.66	0.33
11.90	12.70	140	83	Cyclohexane, n-butyl-	1.00	0.56	0.78	0.50	0.25
11.92	12.72	120	91	Benzene, n-propyl-	1.10	0.32	0.40	0.46	0.47
12.07	12.88	156	57	Decane, 3-methyl-	0.97	0.16	0.09	0.05	0.03
12.07	12.88	154	69	Cyclohexane, C5-	0.98	0.31	0.13	0.08	0.06
12.33	13.16	120	105	3-Ethyltoluene + 1,3,5-trimethylbenzene	1.10	0.86	1.23	1.56	2.24
12.75	13.52	154	69	Cyclohexane, C5-	0.98	0.10	0.10	0.06	0.04
13.05	13.77	154	69	Cyclohexane, C5-	0.98	0.05	0.04	0.03	0.02
13.34	14.09	138	81	Cyclohexene, 1-butyl-	0.99	0.40	0.43	0.44	0.23
13.46	14.22	156	57	n-Undecane (n-C11)	0.97	0.37	0.27	0.26	0.14
13.50	14.25	138	81	Cyclohexene, 1-butyl-	0.99	0.51	0.52	0.56	0.38
13.55	14.31	120	105	Toluene, 2-ethyl-	1.10	0.18	0.38	0.46	0.42
13.65	14.44	154	97,55	Cyclohexane, C5-	0.98	0.62	0.32	0.26	0.16
13.81	14.59	120	105	Benzene, 1,2,4-trimethyl-	1.10	0.55	0.80	1.02	0.90
14.06	14.79	154	69	Cyclohexane, C5-	0.98	0.34	0.24	0.15	0.10
14.25	15.06	138	138	trans-Decalin	1.00	5.63	4.70	4.96	3.08
14.35	15.17	154	97	Cyclohexane, C5-	0.98	0.10	0.05	0.04	
14.41	15.23	154	83	Cyclohexane, C5-	0.98	0.21	0.11	0.06	
14.63	15.45	154	69	Cyclohexane, C5-	0.98	0.26	0.16	0.21	0.05
14.78	15.60	134	119	Benzene, C4-	1.05	0.10	0.25	0.38	0.44
14.83	15.65	154	97	Cyclohexane, C5-	0.98	0.10	0.11	0.02	
14.96	15.78	152	81,95	Decalin, methyl-	1.00	0.30	0.31	0.24	0.06
15.14	15.96	140	69	Cyclohexane, C4-	1.00	0.05	0.05	0.06	
15.42	16.24	152	95	Decalin, methyl-	1.00	0.10	0.21	0.16	0.05
15.53	16.37	120	105	Benzene, 1,2,3-trimethyl-	1.10	0.12	0.29	0.41	0.38
15.88	16.75	152	152	Decalin, methyl-	1.00	2.54	2.94	2.21	1.58
15.98	16.81	134	105	Toluene, ?-propyl-	1.05	0.29	0.37	0.53	0.53
16.10	16.92	134	105	Toluene, ?-propyl-	1.05	0.10	0.15	0.21	0.27
16.16	17.00	154	83	Cyclohexane, n-pentyl-	0.98	0.21	0.16	0.10	0.08
16.20	17.02	134	91, 105	Benzene, n-butyl- + C4-	1.05	0.34	0.39	0.48	0.40
16.33	17.22	134	119	Benzene, C4-	1.05	0.18	0.22	0.27	0.40
16.73	17.63	138	67	cis-Decalin	1.00	1.25	0.90	0.85	0.37
16.87	17.77	152	95	Decalin, methyl-	1.00	1.22	0.78	0.71	0.46
16.96	17.85	118	117	Indan	0.90	0.45	0.58	0.84	0.93
17.33	18.23	170	57	n-Dodecane (n-C12)	0.95	0.53	0.39	0.30	0.15
17.36	18.26	134	119	Benzene, C4-	1.05	0.19	0.26	0.39	0.53
17.36	18.26	152	152	Decalin, methyl-	1.00	0.41	0.50	0.38	0.28
17.52	18.42	184	57	2,6-Dimethylundecane	0.92	0.28	0.10	0.07	
17.52	18.42	166	81	Decalin, C2-	1.00	0.30	0.21	0.25	0.19
17.64	18.54	168	97,55	Cyclohexane, C6-	0.94	0.27	0.09	0.04	
17.71	18.61	134	119	Benzene, C4-	1.05	0.24	0.25	0.33	0.35
17.95	18.85	148	119	Benzene, C5-	1.00	0.10	0.10	0.10	0.05
18.18	19.06	152	95	Decalin, methyl-	1.00	0.41	0.31	0.29	0.16
18.20	19.08	132	117	Indan, 2-methyl-	1.00	0.22	0.25	0.29	0.31
18.26	19.14	148	133	Benzene, C5-	1.00	0.05		0.04	0.05
18.34	19.22	152	81	Decalin, methyl-	1.00	0.30	0.32	0.26	0.13
18.45	19.32	166	81,95	Decalin, C2-	1.00	0.81	0.54	0.60	0.37
18.66	19.54	132	117	Indan, 1-methyl-	1.00	0.81	0.92	1.21	1.30
18.88	19.77	168	83,55	Cyclohexane, C6- .	0.94	0.11	0.11	0.05	

GC R.T.	GC-MS R.T.	Mol Ion	Base Peak	Compound Identified	Response Factor	<-------- Weight % of feed -------->			
						100 Orig. JP-8C Neat	96.3 TS21 450°C 1 hr	94.9 TS47 450°C 2.5 hr	86.5 TS8 450°C 4 hr
18.90	19.80	134	119	Benzene, C4-	1.05	0.10	0.08	0.06	0.05
18.96	19.87	166	81,95	Decalin, C2-	1.00	0.20	0.31	0.25	0.11
19.06	19.96	134	119	Benzene, C4-	1.05	0.05	0.10	0.14	0.18
19.23	20.14	134	119	Benzene, C4-	1.05	0.10	0.12	0.10	0.07
19.36	20.27	166	81,95	Decalin, C2-	1.00	0.13	0.21	0.14	0.06
19.43	20.33	148	119	Benzene, C5-	1.00	0.13	0.10	0.16	0.19
19.51	20.41	184	57	Dodecane, 2-methyl-	0.92	0.20	0.11	0.07	
19.51	20.41	166	81,95	Decalin, C2-	1.00	0.20	0.13	0.10	0.06
19.63	20.53	148	119	Benzene, C5-	1.00	0.05	0.12	0.10	0.07
19.73	20.64	166	81,95	Decalin, C2-	1.00	0.25	0.21	0.20	0.05
19.82	20.73	148	119	Benzene, C5-	1.00	0.10	0.05	0.12	0.11
19.92	20.85	148	105	Benzene, C5-	1.00	0.30	0.	0.22	0.17
20.04	20.97	148	105	Benzene, C5-	1.00	0.15	0.12	0.18	0.10
20.08	21.00	146	131	Indan, C2-	1.00	0.06	0.06	0.06	0.07
20.14	21.06	148	91	Benzene, n-pentyl-	1.00	0.30	0.23	0.20	0.09
20.14	21.06	168	83	Cyclohexane, n-hexyl-	0.94	0.32	0.28	0.27	0.08
20.34	21.28	148	119	Benzene, C5-	1.00	0.05	0.04	0.06	0.07
20.48	21.42	166	137	Decalin, C2-	1.00	0.25	0.30	0.25	0.19
20.59	21.53	148	119	Benzene, C5-	1.00	0.05	0.04	0.04	0.07
20.59	21.53	166	95	Decalin, C2-	1.00	0.05	0.04	0.03	0.02
20.72	21.64	148	105, 133	Benzene, C5-	1.00	0.06	0.05	0.06	0.07
20.81	21.75	132	117	Indan, 5- or 6-methyl-	1.00	0.30	0.52	0.75	0.67
20.90	21.84	134	119	Benzene, C4-	1.05	0.04	0.04	0.08	0.04
21.05	21.98	184	57	n-Tridecane (n-C13)	0.92	0.72	0.45	0.37	0.10
21.10	22.02	148	119	Benzene, C5-	1.00	0.20	0.12	0.10	0.12
21.20	22.12	166	81	Decalin, C2-	1.00	0.15	0.15	0.12	0.07
21.25	22.17	180	81	Decalin, C3-	1.00	0.10	0.06	0.06	0.05
21.52	22.44	132	117	Indan, 5- or 6-methyl-	1.00	0.77	0.75	0.74	0.96
21.95	22.86	146	131	Indan, C2-	1.00	0.10	0.04	0.10	0.12
22.34	23.24	146	131	Indan, C2-	1.00	0.12	0.17	0.28	0.46
22.43	23.32	146	131	Indan, C2-	1.00	0.32	0.36	0.44	0.56
22.54	23.44	132	104	Tetralin	1.00	3.61	2.33	2.20	0.56
22.75	23.64	146	117	Indan, ?-ethyl-	1.00	0.15	0.06	0.10	0.11
22.98	23.90	146	131	Indan, C2-	1.00	0.30	0.52	0.48	0.58
23.05	23.97	196	111	Cyclohexane, C8-	0.94	0.16	0.06	0.06	0.07
23.12	24.04	198	57	Tridecane, 2-methyl-	0.90	0.11	0.06	0.04	
23.16	24.08	162	119	Benzene, C6-	1.00	0.05	0.02	0.04	0.04
23.63	24.53	162	105	Benzene, C6-	1.00	0.27	0.23	0.18	0.05
23.68	24.58	160	145	Indan, C3-	1.00	0.05	0.01	0.02	0.02
23.81	24.72	162	105	Benzene, C6-	1.00	0.36	0.21	0.23	0.06
23.88	24.79	146	104	Tetralin, ?-methyl-	1.00	0.51	0.47	0.40	0.07
23.95	24.86	182	83	Cyclohexane, C7-	0.94	0.11	0.06	0.03	
24.02	24.93	180	81	Decalin, C3-	1.00	0.10	0.06	0.06	0.05
24.27	25.17	162	119, 133	Benzene, C6-	1.00	0.05	0.03	0.02	0.03
24.27	25.17	166	82	Bicyclohexyl	1.00	0.81	0.54	0.54	0.15
24.41	25.35	162	105	Benzene, C6-	1.00	0.08	0.05	0.04	0.04
24.62	25.50	146	131, 117	Indan, C2-	1.00	0.25	0.21	0.30	0.22
24.62	25.50	198	57	n-Tetradecane (n-C14)	0.90	0.90	0.46	0.28	0.06
24.69	25.57	128	128	Naphthalene	1.00	0.20	0.62	0.81	2.23
24.80	25.67	146	117	Indan, C2-	1.00	0.10	0.15	0.16	0.17
24.90	25.77	162	119	Benzene, C6-	1.00	0.10	0.02	0.04	0.06
25.04	25.90	146	131	Indan, C2-	1.00	0.21	0.21	0.30	0.27
25.64	26.49	176	106	Benzene, methyl-hexyl-	1.00	0.30	0.15	0.14	0.02
25.69	25.54	146	131	Indan, C2-	1.00	0.36	0.22	0.32	0.30
25.82	26.67	160	145	Indan, C3-	1.00	0.10	0.08	0.10	0.09
26.02	26.87	160	145	Indan, C3-	1.00	0.10	0.06	0.08	0.09
26.15	27.02	146	131	Tetralin, ?-methyl-	1.00	1.57	1.09	1.01	0.17
26.27	27.10	212	43	Tetradecane, 5-methyl-	0.88	0.12	0.05	0.02	
26.29	27.15	160	131	Indan, C3-	1.00	0.05	0.08	0.11	0.07
26.36	27.20	160	145	Indan, C3-	1.00	0.15	0.21	0.27	0.30
26.61	27.45	212	57	Tetradecane, 2-methyl-	0.88	0.12	0.05	0.02	
26.70	27.52	146	131	Indan, C2-	1.00	0.30	0.26	0.28	0.27
26.84	27.68	160	145	Indan, C3-	1.00	0.10	0.11	0.09	0.11
27.16	28.00	160	145	Indan, C3-	1.00	0.03	0.04	0.05	0.07
27.31	28.16	160	118	Tetralin, dimethyl-	1.00	0.14	0.20	0.14	0.02
27.42	28.27	160	118, 145	Tetralin, dimethyl-	1.00	0.26	0.20	0.18	0.07
27.61	28.45	146	131	Tetralin, ?-methyl-	1.00	0.41	0.48	0.36	0.13
27.70	28.53	160	104	Tetralin, C2-	1.00	0.41	0.48	0.42	0.13
27.76	28.60	194	81,95	Decalin, C4-	1.00	0.10	0.10	0.12	0.09
27.81	28.65	160	131	Indan, C3-	1.00	0.15	0.05	0.08	0.09
27.96	28.80	160	145	Indan, C3-	1.00	0.15	0.31	0.35	0.28

| GC R.T. | GC-MS R.T. | Mol Ion | Base Peak | Compound Identified | Response Factor | <----- Weight % of feed -----> | | | |
						100 Orig. JP-8C Neat	96.3 TS21 450°C 1 hr	94.9 TS47 450°C 2.5 hr	86.5 TS8 450°C 4 hr
27.96	28.80	212	57	n-Pentadecane (n-C15)	0.88	0.58	0.47	0.47	0.08
28.12	28.95	160	145	Tetralin, C2-	1.00	0.20	0.16	0.18	0.04
28.50	29.30	142	142	Naphthalene, 2-methyl-	1.00	0.09	0.57	0.71	2.45
28.78	29.60	160	145	Tetralin, C2-	1.00	0.25	0.21	0.20	0.07
28.98	29.80	174	159, 104	Tetralin, C3-	1.00	0.05	0.05	0.05	0.05
29.08	29.90	174	159	Tetralin, C3-	1.00	0.05	0.05	0.04	0.03
29.50	30.31	190	119	Benzene, C8-	1.00	0.05			
29.50	30.31	142	142	Naphthalene, 1-methyl-	1.00	0.05	0.41	0.50	1.28
29.62	30.43	160	131, 145	Tetralin, C2-	1.00	0.30	0.24	0.24	0.05
30.30	31.12	192	192	Phenanthrene, tetradecahydro-	1.00	0.15	0.05	0.10	0.09
30.45	31.26	160	131	Tetralin, C2-	1.00	0.10	0.05	0.04	0.01
30.65	31.45	174	159, 145	Tetralin, C3-	1.00	0.10	0.05	0.04	
30.80	31.62	160	145	Tetralin, dimethyl-	1.00	0.36	0.27	0.24	0.04
30.88	31.68	190	119	Benzene, C8-	1.00	0.16	0.10	0.10	
31.00	31.80	192	135	Phenanthrene, tetradecahydro-	1.00	0.51	0.31	0.25	0.14
31.03	31.82	174	131	Tetralin, C3-	1.00	0.20	0.08	0.05	0.05
31.12	31.88	226	57	n-Hexadecane (n-C16)	0.86	0.59	0.22	0.18	
31.22	31.98	174	131	Tetralin, C3-	1.00	0.04	0.04	0.05	0.02
31.44	32.20	174	159, 145	Tetralin, C3-	1.00	0.05	0.05	0.07	
31.81	32.55	174	159	Tetralin, C3-	1.00	0.10	0.04		
31.84	32.58	156	141	Naphthalene, 2-ethyl-	1.00	0.01	0.10	0.17	0.46
31.91	32.65	156	156	Naphthalene, 2,7-dimethyl-	1.00	0.02	0.10	0.15	0.40
32.01	32.75	156	156	Naphthalene, 2,6-dimethyl-	1.00	0.02	0.10	0.13	0.37
32.26	33.00	192	135	Phenanthrene, tetradecahydro-	1.00	0.08	0.16	0.08	0.04
32.30	33.04	188	118, 145	Tetralin, C4-	1.00	0.05	0.03	0.02	
32.46	33.20	174	145, 159	Tetralin, C3-	1.00	0.18	0.10	0.08	0.03
32.53	33.27	174	145	Tetralin, C3-	1.00	0.10	0.16	0.08	0.04
32.53	33.27	172	129	Dihydronaphthalene, C3-	1.00	0.15	0.10	0.06	0.03
32.85	33.57	156	156	Naphthalene, 1,7-dimethyl-	1.00	0.03	0.15	0.16	0.33
32.95	33.67	156	156	Naphthalene, 1,3- and 1,6-dimethyl-	1.00	0.06	0.31	0.40	0.60
33.57	34.31	254	57	Branched alkane (C18)	0.82	0.25	0.13	0.12	
33.88	34.62	156	156	Naphthalene, 1,4-dimethyl-	1.00	0.03	0.03	0.03	0.05
34.08	34.82	240	57	n-Heptadecane (n-C17)	0.84	0.15	0.12	0.19	
35.74	36.48	188	145	Tetralin, C4-	1.00	0.07	0.02	0.04	
37.16	37.65	254	57	n-Octadecane (n-C18)	0.82	0.06	0.03	0.02	
37.78	38.43	172	144	Dihydronaphthalene, C3-	1.00	0.04	0.03	0.02	
39.71	40.41	268	57	n-Nonadecane (n-C19)	0.80	0.05			
41.68	42.38	186	186	Anthracene, 1,2,3,4,5,6,7,8-octahydr	1.00	0.03	0.01	0.01	
42.83	43.42	186	186	Phenanthrene, 1,2,3,4,5,6,7,8-octahydro-	1.00	0.03			

Table 11 shows the identified compounds and their contents in neat HRI-MD from direct catalytic coal liquefaction and in thermally stressed HRI-MD at 450°C for 1.0-4.0 hours in 100 psi N_2.[11] The abundance of cyclic structures and aromatic as well as hydroaromatic compounds are characteristic of liquids derived from coal. Thermal stressing of such liquids caused less decomposition and most of the liquids remain as liquid after severe stressing at 450 °C. Table 12 shows the identified compounds and their contents in neat FT-MD from Fischer-Tropsch synthesis using the synthesis gas (CO + H_2) and in thermally stressed FT-MD at 450°C for 1.0-4.0 hours in 100 psi N_2.[11] The FT middle distillate is rich in paraffinic components, which are relatively more sensitive to thermal stressing that causes decomposition and loss of liquid.

Table 11. Identified Compounds and Their Contents in Neat HRI-MD from Direct Coal Liquefaction and in Thermally Stressed HRI-MD at 450°C for 1.0-4.0 hours in 100 psi N2

GC R.T.	GC-MS R.T.	Mol Ion	Base Peak	Compound Identified	Response Factor	100 Orig. HRI-MD Neat	96.7 TS38 450°C 1 hr	92.9 TS55 450°C 2.5 hr	89.6 TS63 450°C 4 hr
1.68	1.48	58	43	n-Butane	1.06	1.18	1.09	1.98	2.57
1.72	1.52	72	43	Butane, 2-methyl-	1.06	0.19	0.72	0.76	0.86
1.74	1.54	72	43	n-Pentane	1.06	1.48	2.59	3.55	3.27
1.84	1.63	86	43	Pentane, 2-methyl-	1.06	0.39	0.27	0.66	0.69
1.89	1.68	86	57	Pentane, 3-methyl-	1.06	0.29	0.09	0.35	0.28
1.94	1.72	86	57	n-Hexane	1.06	1.77	1.83	2.04	2.50
2.18	1.96	84	56	Cyclopentane, methyl-	1.06	1.55	1.68	2.69	3.14
2.36	2.13	100	43	n-Heptane	1.04	0.47	1.20	0.85	1.16
2.44	2.22	84	56	Cyclohexane	1.04	4.29	4.73	5.91	5.92
2.87	2.68	98	83	Cyclohexane, methyl-	1.04	4.20	5.28	5.33	4.88
2.95	2.77	98	69	Cyclopentane, ethyl-	1.04	0.40	0.57	0.54	0.87
3.06	2.85	78	78	Benzene	1.10	0.33	0.44	0.42	0.59
3.36	3.16	114	43	n-Octane	1.02	0.30	1.06	1.12	0.85
3.39	3.20	112	97	Cyclohexane, c-1,3- & t-1,4-dimethyl-	1.02	0.64	1.47	0.99	0.53
3.85	3.67	112	97	Cyclohexane, trans-1,2-dimethyl-	1.02	0.15	0.29	0.24	0.17
4.09	3.92	112	97	Cyclohexane, trans-1,3-dimethyl-	1.02	0.44	0.51	0.38	0.35
4.85	4.71	112	83	Cyclohexane, ethyl-	1.03	1.65	2.00	1.33	1.19
5.05	4.91	92	91	Toluene	1.10	0.80	1.51	1.76	2.29
5.69	5.60	128	57	n-Nonane	1.00	0.36	0.61	0.56	0.48
6.14	6.07	126	97	Cyclohexane, 1-ethyl-?-methyl-	1.02	0.33	0.39	0.35	0.31
7.05	7.03	126	97	Cyclohexane, 1-ethyl-?-methyl-	1.02	0.35	0.34	0.29	0.23
7.93	8.00	126	83	Cyclohexane, n-propyl-	1.02	0.75	0.93	0.53	0.38
8.14	8.23	126	55, 97	Cyclohexane, C3-	1.02	0.12	0.13	0.11	
8.34	8.50	106	91	Benzene, ethyl-	1.10	0.27	0.51	0.37	0.51
8.40	8.56	106	91	p- and m-Xylene	1.10	0.52	0.95	1.56	1.84
9.32	9.48	142	57	n-Decane	0.99	0.37	0.64	0.55	0.41
9.51	9.65	140	97	Cyclohexane, 1-methyl-?-propyl-	1.00	0.30	0.41	0.29	0.22
9.81	10.00	106	91	o-Xylene	1.10	0.25	0.45	0.45	0.51
9.93	10.10	124	67	Indene, octahydro-, trans-	1.00	0.28	0.49	0.34	0.21
10.25	10.43	140	69, 97	Cyclohexane, C4-	1.00	0.08	0.13	0.08	0.03
10.54	10.79	140	97	Cyclohexane, C4-	1.00	0.06	0.06	0.03	0.01
10.70	11.03	120	105	Benzene, i-propyl- (Cumeme)	1.10	0.07	0.09	0.09	0.10
11.58	11.91	138	67	Methyloctahydroindene	1.00	0.07	0.06	0.04	0.04
11.75	12.06	124	81	Indene, octahydro-, cis-	1.00	0.85	0.71	0.45	0.26
11.89	12.20	140	83	Cyclohexane, n-butyl-	0.98	0.26	0.36	0.17	0.11
11.95	12.25	120	91	Benzene, n-propyl-	1.10	0.23	0.33	0.32	0.30
12.30	12.68	120	105	3-Ethyltoluene &1,3,5-trimethylbenzene	1.10	0.36	0.80	1.03	1.35
13.18	13.56	138	81	Cyclohexene, 1-butyl-	1.00	0.13		0.06	
13.38	13.69	156	57	n-Undecane	0.97	0.80	1.14	0.52	0.38
13.56	13.83	120	105	Toluene, 2-ethyl-	1.10	0.09	0.19	0.26	0.17
13.77	14.08	120	105	Benzene, 1,2,4-trimethyl-	1.10	0.15	0.46	0.41	0.45
13.93	14.24	154	69	Cyclohexane, C5-	0.98	0.08	0.04		
14.17	14.47	138	138	trans-Decalin	1.00	0.37	0.41	0.29	0.24
14.70	15.12	134	119	Benzene, C4-	1.05	0.19	0.15	0.18	0.21
15.44	15.90	120	105	Benzene, 1,2,3-trimethyl-	1.10	0.20	0.10	0.10	0.11
15.61	16.10	94	94	Phenol	1.00	0.46	0.48	0.45	0.45
15.71	16.20	152	152	Decalin, ?-methyl-	1.00	0.11	0.05	0.09	0.09
15.84	16.31	134	105	Toluene, ?-propyl-	1.05	0.13	0.08	0.22	0.18
15.96	16.46	154	83	Cyclohexane, n-pentyl-	0.98	0.28	0.10		
16.00	16.50	134	91, 105	Benzene, C4-	1.05	0.25	0.09	0.20	0.28
16.24	16.71	134	119	Benzene, C4-	1.05	0.19	0.15	0.23	0.28
16.59	17.06	138	67	cis-Decalin	1.00	0.13	0.06	0.03	
16.85	17.32	118	117	Indan	0.90	0.68	1.01	0.98	1.14
17.26	17.71	170	57	n-Dodecane	0.95	0.62	0.47	0.28	0.15
17.63	18.12	134	119	Benzene, C4-	1.05	0.11	0.26	0.16	0.16
17.84	18.34	148	119	Benzene, C5-	1.00	0.05	0.06	0.03	
18.06	18.53	132	117	Indan, 2-methyl-	1.00	0.22	0.18	0.14	0.13
18.43	18.91	148	133	Benzene, C5-	1.00	0.08	0.04	0.01	
18.52	19.01	132	117	Indan, 1-methyl-	1.00	0.53	0.79	0.84	0.82
18.72	19.22	108	108	Phenol, ?-methyl-	1.00	0.31	0.29	0.23	0.18
19.12	19.62	134	119	Benzene, C4-	1.05	0.06	0.05	0.03	0.03
19.32	19.81	148	119	Benzene, C5-	1.00	0.07	0.10	0.08	0.08
19.40	19.99	108	107	Phenol, ?-methyl-	1.00	0.62	0.58	0.42	0.45
19.53	20.12	146	131	Indan, C2-	1.00	0.04	0.11	0.05	0.04
19.70	20.25	148	119	Benzene, C5-	1.00	0.11	0.13	0.15	0.15
19.85	20.34	148	105	Benzene, C5-	1.00	0.11	0.26	0.13	0.09
19.99	20.46	148	105	Benzene, C5-	1.00	0.09	0.13	0.05	0.03
20.06	20.54	168	83	Cyclohexane, n-hexyl-	0.94	0.34	0.20	0.04	0.02

GC R.T.	GC-MS R.T.	Mol Ion	Base Peak	Compound Identified	Response Factor	100 Orig. HRI-MD Neat	96.7 TS38 450°C 1 hr	92.9 TS55 450°C 2.5 hr	89.6 TS63 450°C 4 hr
						<----------- Weight % of feed ---------->			
20.39	20.87	166	137	Decalin, C2-	1.00	0.08	0.05	0.01	
20.39	20.87	146	131	Indan, C2-	1.00	0.09	0.05	0.07	0.07
20.66	21.11	148	105	Benzene, C5-	1.00	0.08	0.07	0.04	
20.77	21.22	132	117	Indan, 5- or 6-methyl-	1.00	0.26	0.29	0.43	0.50
21.14	21.48	184	57	n-Tridecane	0.92	0.47	0.31	0.09	0.04
21.45	21.91	132	117	Indan, 5- or 6-methyl-	1.00	0.46	0.57	0.55	0.61
21.90	22.35	146	131	Indan, C2-	1.00	0.04	0.07	0.05	0.04
22.09	22.54	122	107	Phenol, ?-ethyl-	1.00	0.09	0.09	0.06	
22.11	22.56	162	105	Benzene, C6-	1.00	0.03			
22.29	22.76	146	131	Indan, C2-	1.00	0.30	0.54	0.65	0.50
22.43	22.89	132	104	Tetralin	1.00	1.49	1.19	0.31	0.04
22.66	23.11	146	117	Indan, ?-ethyl-	1.00	0.04	0.06	0.06	0.04
22.92	23.36	146	131	Indan, C2-	1.00	0.16	0.41	0.30	0.29
23.28	23.72	122	107	Phenol, ?-ethyl-	1.00	0.26	0.22	0.26	0.18
23.60	24.00	162	105	Benzene, C6-	1.00	0.03	0.05	0.04	
23.65	24.05	160	104	Tetralin, C2-	1.00	0.06			
23.84	24.25	146	104	Tetralin, ?-methyl-	1.00	1.02	0.32	0.06	
24.17	24.61	166	82	Bicyclohexyl	0.94	0.33	0.18	0.06	
24.52	24.96	146	131	Indan, C2-	1.00	0.30	0.10		
24.54	24.98	198	57	n-Tetradecane	0.90	0.52	0.21	0.10	0.04
24.63	25.07	128	128	Naphthalene	1.00	0.56	1.53	2.27	2.90
24.90	25.33	146	131	Indan, C2-	1.00	0.18	0.10	0.12	0.14
25.64	26.08	160	145	Indan, C3-	1.00	0.05	0.16	0.15	0.17
25.90	26.29	136	121	Phenol, C3-	1.00	0.03	0.06	0.07	0.11
26.07	26.46	146	131	Tetralin, ?-methyl-	1.00	1.47	1.08	0.15	0.03
26.27	26.65	160	145	Indan, C3-	1.00	0.03	0.15	0.17	0.12
26.61	27.00	146	131	Indan, C2-	1.00	0.05	0.15	0.10	0.10
27.30	27.64	160	118	Tetralin, dimethyl-	1.00	0.38	0.23	0.07	
27.39	27.75	160	118	Tetralin, dimethyl-	1.00	0.45	0.30		
27.57	27.93	146	131	Tetralin, ?-methyl-	1.00	0.45	0.43	0.15	0.04
27.66	28.01	160	104	Benzene, cyclohexyl-	1.00	0.47	0.45	0.19	0.06
27.76	28.11	160	131	Indan, C3-	1.00	0.18	0.04	0.04	0.09
27.87	28.25	160	145	Indan, C3-	1.00	0.31	0.28	0.19	0.13
27.94	28.30	212	57	n-Pentadecane	0.88	0.70	0.61	0.16	
28.04	28.40	160	145	Tetralin, C2-	1.00	0.29	0.28		
28.40	28.74	142	142	Naphthalene, 2-methyl-	1.00	1.03	1.51	3.58	4.42
28.70	29.04	160	145	C3-indan + C2-Tetralin	1.00	0.29	0.05	0.08	0.09
28.92	29.26	174	104	Tetralin, C3-	1.00	0.29	0.18	0.05	
29.01	29.35	174	104	Tetralin, C3-	1.00	0.11	0.05		
29.43	29.75	142	142	Naphthalene, 1-methyl-	1.00	0.29	0.31	0.92	0.88
29.50	29.87	160	131	Tetralin, ?-ethyl-	1.00	0.50	0.34	0.03	
30.11	30.47	174	104	Tetralin, C3-	1.00	0.09	0.06		
30.37	30.73	160	131	Tetralin, C2-	1.00	0.19	0.06		
30.58	30.90	174	145	Tetralin, C3-	1.00	0.22	0.05		
30.73	31.03	160	145	Tetralin, dimethyl	1.00	0.19	0.16		
30.73	31.03	174	145	Tetralin, C3-	1.00	0.19	0.16		
30.87	31.19	192	135	Phenanthrene, tetradecahydro-	1.00	0.21	0.06	0.03	
30.93	31.27	174	131	Benzene, 1-cyclohexyl-?-methyl-	1.00	0.84	0.48	0.13	0.06
31.08	31.41	226	57	n-Hexadecane	0.86	0.58	0.22	0.08	0.04
31.12	31.45	174	131	Tetralin, C3-	1.00	0.50	0.27	0.07	
31.22	31.56	174	131	Tetralin, C3-	1.00	0.16	0.08	0.19	
31.32	31.68	174	159	Tetralin, C3-	1.00	0.11	0.28		
31.74	32.02	156	141	Naphthalene, 2-ethyl-	1.00	0.21	0.56	0.79	1.41
31.80	32.09	156	156	Naphthalene, 2,7-dimethyl-	1.00	0.21	0.52	0.56	0.84
31.88	32.19	156	156	Naphthalene, 2,6-dimethyl	1.00	0.21	0.38	0.56	0.77
32.01	32.31	154	154	Biphenyl	1.00	1.00	0.87	0.56	0.74
32.15	32.44	188	118, 145	Tetralin, C4-	1.00	0.52	0.60	0.08	
32.28	32.56	168	168	Biphenyl, 2-methyl	1.00	0.56	0.35	0.07	0.13
32.32	32.60	156	141	Naphthalene, 1-ethyl	1.00	0.06	0.12	0.13	0.29
32.41	32.73	174	145	Indan, C4-	1.00	0.52	0.65		
32.71	32.98	156	156	Naphthalene, 1,7-dimethyl	1.00	0.11	0.15	0.18	0.22
32.84	33.09	156	156	Naphthalene, 1,3- and 1,6-dimethyl	1.00	0.15	0.53	0.71	0.74
33.16	33.52	188	145	Tetralin, C4-	1.00	0.53	0.17	0.06	
33.33	33.70	156	156	Naphthalene, 2,3-dimethyl	1.00	0.10	0.27	0.31	0.30
33.37	33.73	170	155	Naphthalene, 2-isopropyl	1.00	0.06	0.27	0.31	0.30
33.37	33.73	188	145	Tetralin, C4-	1.00	0.10			
33.48	33.84	174	145	Tetralin, C3-	1.00	0.11	0.03		
33.71	33.96	188	117	Tetralin, C4-	1.00	0.11	0.06		
33.90	34.14	174	145	Tetralin, C3-	1.00	0.22	0.04		
33.99	34.22	188	159	Tetralin, C4-	1.00	0.30	0.11	0.02	
34.14	34.36	240	57	n-Heptadecane	0.84	0.67	0.33	0.03	

GC R.T.	GC-MS R.T.	Mol Ion	Base Peak	Compound Identified	Response Factor	100 Orig. HRI-MD Neat	96.7 TS38 450°C 1 hr	92.9 TS55 450°C 2.5 hr	89.6 TS63 450°C 4 hr
34.37	34.59	188	131	Tetralin, C4-	1.00	0.12	0.05		
34.42	34.64	188	104	Tetralin, C4-	1.00	0.28	0.05		
34.61	34.84	170	141	Naphthalene, C3-	1.00	0.26	0.11	0.28	0.30
34.97	35.16	170	155	Naphthalene, C3-	1.00	0.06	0.22	0.30	0.29
35.09	35.27	170	155	Naphthalene, C3-	1.00	0.23	0.35	0.41	0.37
35.19	35.48	168	168	Biphenyl, 3-methyl-	1.00	1.84	1.90	1.68	1.77
35.52	35.78	168	168	Biphenyl, 4-methyl-	1.00	0.77	0.83	0.72	0.73
35.66	35.92	188	145	Tetralin, C4-	1.00	0.79	0.41		
36.26	36.45	172	144	Dihydronaphthalene, C3-	1.00	0.27			
36.28	36.48	154	154	Acenaphthene	1.00	0.14	0.21	0.22	0.25
36.55	36.71	202	159	Tetralin, C5-	1.00	0.39	0.19		
36.66	36.80	170	170	Naphthalene, C3-	1.00	0.14	0.13	0.21	0.09
36.73	36.87	202	159	Tetralin, C5-	1.00	0.18	0.05		
37.08	37.17	254	57	n-Octadecane	0.82	0.24	0.18		
37.31	37.41	202	145	Tetralin, C5-	1.00	0.15	0.06		
37.70	37.85	172	144	Dihydronaphthalene, C3-	1.00	0.15	0.06		
37.78	37.94	184	141	Naphthalene, ?-butyl-	1.00	0.72	0.44	0.11	0.05
37.88	38.13	182	167	Biphenyl, C2-	1.00	0.24	0.38	0.33	0.26
38.09	38.32	182	182	Biphenyl, 3,4-dimethyl-	1.00	0.65	0.51	0.39	0.36
38.23	38.45	182	182	Biphenyl, 3,3'-dimethyl-	1.00	1.29	1.10	0.81	0.75
38.58	38.77	182	182	Biphenyl, 3,4'-dimethyl-	1.00	0.71	1.08	0.86	0.74
38.88	39.05	182	182	Biphenyl, 4,4'-dimethyl-	1.00	0.11	0.10	0.11	0.07
39.06	39.23	200	185	Dihydronaphthalene, C5-	1.00	0.04			
39.26	39.46	166	166	Fluorene	1.00	0.09	0.25	0.33	0.37
39.63	39.81	268	57	n-Nonadecane	0.80	0.32	0.11		
39.77	39.93	182	167	Biphenyl, C2-	1.00	0.26	0.43	0.45	0.52
39.90	40.06	182	168	Biphenyl, C2-	1.00	0.31	0.10	0.14	0.18
40.00	40.17	168	168	Naphthalene, 1-(2-propenyl)-	1.00	0.54	0.58	0.49	0.39
40.21	40.38	182	167	Biphenyl, C2-	1.00	0.01	0.15	0.22	0.31
40.51	40.70	196	181	Biphenyl, C3-	1.00	0.15	0.13	0.09	0.09
40.54	40.73	198	155	Naphthalene, C5-	1.00	0.30	0.12	0.03	0.02
40.58	40.78	182	167	Biphenyl, C2-	1.00	0.20	0.24	0.23	0.23
40.64	40.85	198	155	Naphthalene, C5-	1.00	0.44	0.15	0.08	0.02
40.70	40.89	182	167	Biphenyl, C2-	1.00	0.10	0.19	0.14	0.18
40.73	40.93	196	181	Biphenyl, C3-	1.00	0.57	0.17	0.14	0.14
40.95	41.14	196	196	Biphenyl, C3-	1.00	0.66	0.43	0.32	0.44
41.33	41.50	196	181, 196	Biphenyl, C3-	1.00	0.30	0.25	0.21	0.25
41.60	41.77	186	186	Anthracene, 1,2,3,4,5,6,7,8-octahydro-	1.00	0.28	0.07		
42.24	42.30	180	165	Fluorene, ?-methyl-	1.00	0.26	0.19	0.23	0.20
42.29	42.35	282	57	n-Eicosane	0.80	0.26	0.04		
42.46	42.51	180	165	Fluorene, ?-methyl-	1.00	0.10	0.10	0.17	0.23
42.75	42.80	186	186	Phenanthrene, 1,2,3,4,5,6,7,8-octahydro-	1.00	0.41	0.05		
42.82	42.86	180	165	Fluorene, ?-methyl-	1.00	0.41	0.38	0.51	0.45
42.95	42.99	182	167	Biphenyl, dimethyl-	1.00	0.31	0.29	0.19	0.09
42.99	43.03	196	196	Biphenyl, C3-	1.00	0.10	0.05	0.09	0.14
43.09	43.13	196	181	Biphenyl, C3-	1.00	0.21	0.10	0.07	0.05
43.16	43.20	180	180	Phenanthrene, 9,10-dihydro-	1.00	0.10			
43.29	43.33	200	200	Octahydroanthracene, methyl-	1.00	0.57	0.05		
43.34	43.39	210	181	Biphenyl, C4-	1.00	0.21	0.10	0.11	0.09
43.53	43.57	200	200	Octahydroanthracene, methyl-	1.00	0.54	0.10		
43.83	43.87	182	167	Biphenyl, C2-	1.00	0.15	0.05		
44.26	44.29	182	182	Biphenyl, C2-	1.00	0.32	0.23	0.03	0.02
44.75	44.74	182	154	Phenanthrene, 1,2,3,4-tetrahydro-	1.00	0.67	0.63		
44.79	44.79	296	57	n-Henicosane (n-C21)	0.80	0.26	0.07		
45.03	45.03	196	196	Biphenyl, C3-	1.00	0.11	0.06	0.05	
45.31	45.30	196	196	Biphenyl, C3-	1.00	0.15	0.10		
45.42	45.36	194	179	Fluorene, C2-	1.00	0.10	0.14	0.24	0.08
45.61	45.58	194	179	Fluorene, C2-	1.00	0.08	0.13	0.26	0.04
45.67	45.67	194	194	Fluorene, C2-	1.00	0.35	0.13	0.19	0.07
45.99	45.95	194	179	Fluorene, dimethyl-	1.00	0.20	0.19	0.23	0.06
46.06	46.07	194	194	Fluorene, C2-	1.00	0.29	0.10		
46.28	46.25	178	178	Phenanthrene	1.00	1.33	1.65	2.76	2.73
46.67	46.64	194	179	Fluorene, C2-	1.00	0.09			
46.98	46.95	196	196	Biphenyl, C3-	1.00	0.29	0.15		
47.11	47.08	310	57	n-Docosane	0.80	0.45	0.05		
47.22	47.19	196	196	Biphenyl, C3-	1.00	0.46	0.24		
47.45	47.43	196	196	Biphenyl, C3-	1.00	0.53	0.22	0.03	
47.70	47.66	210	210	Biphenyl, C4-	1.00	0.21	0.10	0.04	
47.85	47.81	210	210	Biphenyl, C4-	1.00	0.10	0.02	0.05	
47.96	47.91	210	210	Biphenyl, C4-	1.00	0.11	0.05		
47.96	47.91	208	179	Fluorene, C3-	1.00	0.11	0.05		

GC R.T.	GC-MS R.T.	Mol Ion	Base Peak	Compound Identified	Response Factor	100 Orig. HRI-MD Neat	96.7 TS38 450°C 1 hr	92.9 TS55 450°C 2.5 hr	89.6 TS63 450°C 4 hr
48.02	47.97	210	210	Biphenyl, C4-	1.00	0.11	0.03		
48.02	47.97	208	208	Fluorene, C3-	1.00	0.12	0.03		0.04
48.13	48.07	210	210	Biphenyl, C4-	1.00	0.21	0.04		
48.17	48.11	208	193	Fluorene, C3-	1.00	0.15	0.05		
48.41	48.34	208	165	Fluorene, C3-	1.00	0.23			
48.46	48.39	208	193	Fluorene, C3-	1.00	0.48	0.20	0.12	0.10
48.86	48.77	192	192	Anthracene or Phenanthrene, methyl-	1.00	1.50	1.40	1.56	1.43
49.13	49.00	192	192	Anthracene or Phenanthrene, methyl-	1.00	1.52	1.91	2.24	2.12
49.44	49.31	324	57	n-Tricosane	0.80	0.40	0.02		
49.59	49.46	208	193	Fluorene, C3-	1.00	0.08	0.05		
49.74	49.60	222	180	Fluorene, C4-	1.00	0.10			
49.74	49.60	208	208	Fluorene, C3-	1.00	0.11	0.05	0.02	
50.13	49.97	192	192	Anthracene or Phenanthrene, methyl-	1.00	0.33	0.19	0.29	0.37
50.16	50.00	212	212		1.00	0.15	0.10		
50.22	50.06	212	212		1.00	0.28	0.05		
50.61	50.46	210	181	Biphenyl, C4-	1.00	0.08			
50.84	50.69	222	222	Fluorene, C4-	1.00	0.21			
50.95	50.82	206	191	Anthracene or Phenanthrene, ethyl-	1.00	0.33	0.12	0.14	0.14
51.27	51.13	206	206	Anthracene or Phenanthrene, dimethyl-	1.00	0.57	0.25	0.23	0.22
51.60	51.42	206	191, 206	Anthracene or Phenanthrene, C2-	1.00	0.84	1.04	0.91	0.84
51.82	51.61	206	206	Anthracene or Phenanthrene, dimethyl-	1.00	0.60	0.48	0.41	0.44
52.23	52.04	206	206	Anthracene or Phenanthrene, dimethyl-	1.00	0.06	0.04	0.04	
52.23	52.04	222	222	Fluorene, C4-	1.00	0.08	0.02		
52.34	52.15	206	206	Anthracene or Phenanthrene, dimethyl-	1.00	0.36	0.27	0.20	0.17
52.54	52.32	206	206	Anthracene or Phenanthrene, dimethyl-	1.00	0.16	0.23	0.16	0.11
52.59	52.37	226	226		1.00	0.19	0.12	0.08	0.07
52.88	52.66	206	206	Anthracene or Phenanthrene, dimethyl-	1.00	0.46	0.51	0.41	0.31
53.16	53.01	220	205	Anthracene or Phenanthrene, C3-	1.00	0.32	0.11	0.07	
53.82	53.61	220	205, 220	Anthracene or Phenanthrene, C3-	1.00	0.08	0.12	0.06	0.07
54.10	53.84	220	205	Anthracene or Phenanthrene, C3-	1.00	0.16	0.13	0.09	0.05
54.32	54.02	204	204	Pyrene, dihydro-	1.00	0.32			
54.52	54.22	220	220	Anthracene or Phenanthrene, C3-	1.00	0.10	0.02	0.02	
54.99	54.72	220	220	Anthracene or Phenanthrene, C3-	1.00	0.03	0.03	0.03	
55.14	54.89	220	220	Anthracene or Phenanthrene, C3-	1.00	0.27	0.07	0.07	
55.42	55.15	220	220	Anthracene or Phenanthrene, C3-	1.00	0.12	0.12	0.12	0.09
56.10	55.77	202	202	Pyrene	1.00	1.17	1.18	1.03	0.85
56.73	56.39	218	218	Dihydropyrene, methyl-	1.00	0.18			
57.79	57.42	216	216	Pyrene or Fluoranthene, methyl-	1.00	0.04	0.04	0.07	0.09
58.52	58.11	216	216	Pyrene or Fluoranthene, methyl-	1.00	0.66	0.65	0.55	0.41
58.83	58.49	218	218	Dihydropyrene, methyl-	1.00	0.25	0.07	0.08	
59.56	59.06	216	216	Pyrene or Fluoranthene, methyl-	1.00	0.07	0.05	0.08	0.09
60.44	60.01	230	230	Pyrene or Fluoranthene, C2-	1.00	0.07	0.06	0.07	0.05
60.83	60.31	230	230	Pyrene or Fluoranthene, dimethyl-	1.00	0.09	0.12	0.12	0.05

Table 12. Identified Compounds and Their Contents in Neat FT-MD from Fischer-Tropscl Synthesis and in Thermally Stressed FT-MD at 450°C for 1.0-4.0 hours in 100 psi N2

GC R.T.	GC-MS R.T.	Mol Ion	Base Peak	Compound Identified	Response Factor	100 Orig. FT-MD Neat	83.3 TS25 450°C 1 hr	71.3 TS51 450°C 2.5 hr	64.7 TS16 450°C 4 hr
4.59	4.64	128	43	Branched alkane (C9)	1.00	0.24	1.45	0.89	1.21
4.85	4.90	128	57	Branched alkane (C9)	1.00	0.14	0.84	0.55	0.67
5.73	5.92	128	57	n-Nonane	1.00	0.33	1.94	1.30	1.55
6.39	6.63	142	57	Nonane, 4-methyl-	0.99	0.27	0.34	0.07	0.03
6.72	6.98	142	57	2,6-Dimethyloctane	0.99	0.24	0.17	0.04	0.03
7.09	7.39	142	57	Branched alkane (C10)	0.99	0.13	0.07	0.04	
7.63	7.98	142	57	Branched alkane (C10)	0.99	0.30	0.17	0.09	0.07
7.83	8.20	142	57	Nonane, 2-methyl-	0.99	2.36	1.56	0.71	0.79
8.17	8.63	142	57	Nonane, 3-methyl-	0.99	1.44	1.01	0.40	0.37
9.35	9.91	142	57	n-Decane	0.99	2.85	2.02	1.04	1.14
9.58	10.10	156	57	Branched alkane (C11)	0.97	0.61	0.39	0.18	0.07
9.97	10.50	156	57	2,6-Dimethylnonane	0.97	0.70	0.34	0.15	0.03
10.15	10.68	156	57	Branched alkane (C11)	0.97	0.37	0.09	0.03	0.01
10.36	10.90	156	57	Branched alkane (C11)	0.97	0.60	0.21	0.05	0.02
10.55	11.10	156	57	Branched alkane (C11)	0.97	0.52	0.22	0.08	0.05
10.82	11.38	156	57	Branched alkane (C11)	0.97	0.89	0.10	0.05	
11.29	11.87	170	57	Branched alkane (C12)	0.95	0.21	0.12	0.02	

GC R.T.	GC-MS R.T.	Mol Ion	Base Peak	Compound Identified	Response Factor	100 Orig. FT-MD Neat	83.3 TS25 450°C 1 hr	71.3 TS51 450°C 2.5 hr	64.7 TS16 450°C 4 hr
						<----------- Weight % of feed ----------->			
11.39	11.97	184	43	Undecane, dimethyl-	0.92	0.43	0.19	0.04	
11.50	12.10	156	57	Branched alkane (C11)	0.97	0.25	0.14	0.02	
11.60	12.20	156	57	Decane, 5-methyl-	0.97	1.10	0.52	0.24	0.17
11.80	12.40	156	57	Decane, 4-methyl-	0.97	2.96	1.01	0.28	0.25
12.20	12.83	156	57	Decane, 3-methyl-	0.97	1.93	0.77	0.24	0.33
12.75	13.40	170	57	Branched alkane (C12)	0.95	0.12			
13.02	13.68	170	57	Branched alkane (C12)	0.95	0.20	0.06		
13.36	14.15	156	57	n-Undecane	0.97	1.87	1.52	0.59	0.70
13.68	14.46	170	57	2,6-Dimethyldecane	0.95	1.49	0.31	0.07	0.02
13.72	14.50	170	43	Branched alkane (C12)	0.95	0.31	0.18		
13.95	14.73	170	43	Branched alkane (C12)	0.95	0.33	0.12	0.02	
14.05	14.83	170	57	Branched alkane (C12)	0.95	0.62	0.27	0.03	
14.26	15.02	170	57	3,7-Dimethyldecane	0.95	0.29	0.15	0.02	
14.38	15.14	170	57	Branched alkane (C12)	0.95	0.20	0.04	0.02	
14.47	15.23	170	57	Branched alkane (C12)	0.95	0.57	0.18	0.02	0.01
14.61	15.36	212	43	Branched alkane (C15)	0.88	0.16	0.06		
14.84	15.61	184	57	Branched alkane (C13)	0.92	0.30	0.14	0.02	
15.06	15.83	170	57	Branched alkane (C12)	0.95	0.30	0.13	0.02	0.03
15.26	16.03	170	57	Branched alkane (C12)	0.95	0.32	0.12	0.02	
15.42	16.19	168	57	3-Undecene, 6-methyl-	0.95	0.41	0.18	0.05	0.04
15.50	16.27	170	43	Undecane, 5-methyl-	0.95	1.11	0.38	0.12	0.07
15.63	16.43	170	71	Undecane, 4-methyl-	0.95	0.73		0.12	0.04
15.73	16.53	170	43, 57	Undecane, 2-methyl-	0.95	1.66	0.47	0.15	0.16
15.99	16.80	170	43	Branched alkane (C12)	0.95	0.24	0.11		
16.12	16.93	170	57	Undecane, 3-methyl-	0.95	1.41	0.44	0.17	0.12
16.82	17.62	170	57	Branched alkane (C12)	0.95	0.07			
17.00	17.79	184	57	Branched alkane (C13)	0.92	0.54	0.09		
17.04	17.83	170	57	Branched alkane (C12)	0.95	0.12			
17.12	17.91	170	57	Branched alkane (C12)	0.95	0.26	0.09		
17.37	18.17	170	57	n-Dodecane	0.95	3.09	1.14	0.45	0.53
17.57	18.36	184	57	2,6-Dimethylundecane	0.92	0.87	0.25	0.06	0.01
17.70	18.51	184	43	Branched alkane (C13)	0.92	0.49	0.18	0.02	
17.85	18.67	184	57	Branched alkane (C13)	0.92	0.30	0.16	0.02	
18.22	19.06	212	57	Branched alkane (C15)	0.88	0.33	0.16	0.02	
18.72	19.55	212	57	Branched alkane (C15)	0.88	0.29	0.09	0.02	
18.80	19.63	184	57	Branched alkane (C13)	0.92	0.31	0.06	0.02	
19.06	19.87	198	43	Branched alkane (C14)	0.90	0.28	0.09	0.01	
19.17	20.01	184	57	Dodecane, 6-methyl-	0.92	0.69	0.22	0.05	
19.29	20.13	184	43	Dodecane, 5-methyl-	0.92	1.04	0.29	0.05	0.02
19.40	20.24	184	43	Dodecane, 4-methyl-	0.92	0.72	0.25	0.02	0.03
19.53	20.38	184	57	2,6,10-Trimethylundecane	0.92	1.35	0.34	0.10	0.06
19.82	20.66	184	57	Branched alkane (C13)	0.92	0.38	0.07	0.02	
19.93	20.77	184	57	Dodecane, 3-methyl-	0.92	1.20	0.42	0.09	0.05
20.49	21.33	198	57	Branched alkane (C14)	0.90	0.07			
20.60	21.44	198	57	Branched alkane (C14)	0.90	0.30	0.04		
20.82	21.67	184	57	Branched alkane (C13)	0.92	0.27	0.05		
21.06	21.93	184	57	n-Tridecane	0.92	2.16	0.89	0.23	0.28
21.22	22.09	198	57	Branched alkane (C14)	0.90	0.48	0.10	0.01	
21.31	22.18	198	57	Branched alkane (C14)	0.90	0.48	0.16	0.01	
21.44	22.30	226	57	Branched alkane (C16)	0.86	0.20	0.06		
21.52	22.37	198	57	Branched alkane (C14)	0.90	0.98	0.19	0.02	
21.70	22.55	226	57	Branched alkane (C16)	0.86	0.44	0.11		
21.89	22.73	226	57	Branched alkane (C16)	0.86	0.33	0.10		
22.00	22.84	198	57	Branched alkane (C14)	0.90	0.28	0.13		
22.25	23.07	198	57	Branched alkane (C14)	0.90	0.29	0.13		
22.40	23.22	212	57	Branched alkane (C15)	0.88	0.55	0.16		
22.66	23.48	240	43	Branched alkane (C17)	0.84	0.24	0.04		
22.76	23.58	198	57	Tridecane, 6-methyl-	0.90	0.92	0.23	0.03	0.01
22.90	23.72	198	43	Tridecane, 5-methyl-	0.90	0.94	0.24	0.03	0.02
23.07	23.89	198	43	Tridecane, 4-methyl-	0.90	0.71	0.17	0.02	
23.18	24.00	198	57	Tridecane, 2-methyl-	0.90	1.09	0.27	0.06	0.01
23.42	24.24	212	57	2,6,10-Trimethyldodecane	0.86	0.37	0.06	0.02	
23.56	24.38	198	57	Tridecane, 3-methyl-	0.90	0.86	0.22	0.04	0.03
23.98	24.80	212	57	Branched alkane (C15)	0.88	0.17			
24.10	24.92	212	57	Branched alkane (C15)	0.88	0.29			
24.32	25.14	226	57	Branched alkane (C16)	0.86	0.43			
24.48	25.30	212	57	Branched alkane (C15)	0.88	0.15	0.03		
24.63	25.45	198	57	n-Tetradecane	0.90	2.73	0.56	0.11	0.07
24.78	25.60	212	57	Branched alkane (C15)	0.88	0.72	0.04	0.01	
24.93	25.74	212	57	Branched alkane (C15)	0.88	0.29	0.03		
25.32	26.13	212	57	Branched alkane (C15)	0.88	0.20	0.04		

| GC R.T. | GC-MS R.T. | Mol Ion | Base Peak | Compound Identified | Response Factor | <-------------- Weight % of feed --------------> | | | |
						100 Orig. FT-MD Neat	83.3 TS25 450°C 1 hr	71.3 TS51 450°C 2.5 hr	64.7 TS16 450°C 4 hr
25.44	26.25	212	57	Branched alkane (C15)	0.88	0.22	0.03		
25.60	26.40	212	57	Branched alkane (C15)	0.88	0.17	0.04		
25.74	26.54	212	57	Branched alkane (C15)	0.88	0.20	0.04		
25.85	26.65	212	57	Branched alkane (C15)	0.88	0.20	0.04		
26.10	26.90	212	57	Tetradecane, 6-methyl-	0.88	1.30	0.27	0.05	
26.26	27.07	212	57	Tetradecane, 5-methyl-	0.88	0.80	0.15	0.03	
26.42	27.24	212	43	Tetradecane, 4-methyl-	0.88	0.59	0.14	0.02	0.01
26.53	27.35	212	57	Tetradecane, 2-methyl-	0.88	0.64	0.18	0.02	0.01
26.73	27.55	240	57	Branched alkane (C17)	0.84	0.31	0.01		
26.91	27.73	212	57	Tetradecane, 3-methyl-	0.88	0.69	0.13	0.03	0.01
27.28	28.10	212	57	Branched alkane (C15)	0.88	0.19	0.04		
27.50	28.32	240	57	Branched alkane (C17)	0.84	0.28	0.03		
27.93	28.75	212	57	n-Pentadecane	0.88	2.25	0.49	0.12	0.12
28.13	28.95	226	57	Branched alkane (C16)	0.86	0.83	0.12	0.02	
28.25	29.07	226	57	Branched alkane (C16)	0.86	0.26	0.16		
28.38	29.20	282	57	Branched alkane (C20)	0.80	0.23	0.06		
28.73	29.55	226	57	Branched alkane (C16)	0.86	0.45	0.09	0.01	
28.93	29.75	226	57	Branched alkane (C16)	0.86	0.24	0.09		
29.30	30.10	226	57	Branched alkane (C16)	0.86	1.22	0.46	0.05	
29.43	30.23	268	43	Branched alkane (C19)	0.80	0.59	0.18	0.02	
29.63	30.43	226	43	Pentadecane, 4-methyl-	0.86	0.48	0.17	0.02	
29.74	30.52	226	57	Pentadecane, 2-methyl-	0.86	0.61	0.15		
29.94	30.72	226	57	Branched alkane (C16)	0.86	0.09	0.07		
30.06	30.84	226	57	Branched alkane (C16)	0.86	0.47	0.11		
30.63	31.42	282	57	Branched alkane (C20)	0.80	0.23	0.03		
30.67	31.46	296	57	Branched alkane (C21)	0.80	0.25	0.05		
30.87	31.66	226	57	Branched alkane (C16)	0.86	0.89	0.09		
31.09	31.88	226	57	n-Hexadecane	0.86	1.33	0.29	0.04	0.02
31.15	31.94	240	57	Branched alkane (C16)	0.84	0.37	0.02		
31.67	32.46	240	57	Branched alkane (C16)	0.84	0.24	0.06		
32.00	32.78	254	57	Branched alkane (C18)	0.82	0.29	0.07		
32.24	33.02	240	57	Branched alkane (C17)	0.84	0.58	0.10	0.01	
32.33	33.09	254	57	Branched alkane (C18)	0.82	0.54	0.07	0.01	
32.48	33.24	240	43	Hexadecane, 5-methyl-	0.84	0.41	0.07		
32.66	33.42	240	57	Hexadecane, 4-methyl-	0.84	0.41	0.08		
32.75	33.50	240	57	Hexadecane, 2-methyl-	0.84	0.50	0.12		
33.12	33.87	240	57	Hexadecane, 3-methyl-	0.84	0.61	0.07		
33.54	34.30	254	57	Branched alkane (C18)	0.82	0.15	0.03		
33.65	34.41	254	57	Branched alkane (C18)	0.82	0.19	0.03		
33.86	34.62	254	57	Branched alkane (C18)	0.82	0.50	0.07		
34.03	34.80	240	57	n-Heptadecane	0.84	1.53	0.22	0.02	
34.59	35.32	254	57	Branched alkane (C18)	0.82	0.19			
34.72	35.46	254	57	Branched alkane (C18)	0.82	0.30			
35.12	35.87	254	57	Branched alkane (C18)	0.82	0.77	0.09		
35.22	35.97	254	57	Branched alkane (C18)	0.82	0.50	0.05		
35.33	36.08	254	43	Heptadecane, 5-methyl-	0.82	0.35	0.04		
35.56	36.29	254	43	Heptadecane, 4-methyl-	0.82	0.40	0.03		
35.63	36.36	254	57	Heptadecane, 2-methyl-	0.82	0.49	0.04		
35.99	36.70	254	57	Heptadecane, 3-methyl-	0.82	0.75	0.05		
36.53	37.23	268	57	Branched alkane (C19)	0.80	0.18			
36.66	37.36	268	57	Branched alkane (C19)	0.80	0.43	0.02		
36.91	37.61	254	57	n-Octadecane	0.82	1.45	0.11	0.02	
37.32	38.02	282	57	Branched alkane (C20)	0.80	0.26			
37.54	38.24	282	57	Branched alkane (C20)	0.80	0.31			
37.89	38.60	268	57	Branched alkane (C19)	0.80	0.72	0.03		
37.99	38.70	268	57	Branched alkane (C19)	0.80	0.38	0.01		
38.14	38.85	282	57	Branched alkane (C20)	0.80	0.28	0.01		
38.28	38.99	268	57	Branched alkane (C19)	0.80	0.20	0.01		
38.39	39.09	268	57	Branched alkane (C19)	0.80	0.34	0.03		
38.74	39.45	268	57	Branched alkane (C19)	0.80	0.63	0.04		
39.35	40.02	282	57	Branched alkane (C20)	0.80	0.26			
39.60	40.27	268	57	n-Nonadecane	0.80	1.17	0.07		
39.91	40.60	282	57	Branched alkane (C20)	0.80	0.09			
40.13	40.82	282	57	Branched alkane (C20)	0.80	0.07			
40.46	41.16	282	57	Branched alkane (C20)	0.80	0.52	0.02		
40.59	41.29	282	57	Branched alkane (C20)	0.80	0.29			
40.73	41.43	282	57	Branched alkane (C20)	0.80	0.14			
40.88	41.58	282	57	Branched alkane (C20)	0.80	0.15			
40.97	41.67	282	57	Branched alkane (C20)	0.80	0.27			
41.30	42.00	282	57	Branched alkane (C20)	0.80	0.43			
42.12	42.79	282	57	n-Eicosane	0.80	0.59	0.06		

GC R.T.	GC-MS R.T.	Mol Ion	Base Peak	Compound Identified	Response Factor	<-------------- Weight % of feed -------------->			
						100 Orig. FT-MD Neat	83.3 TS25 450°C 1 hr	71.3 TS51 450°C 2.5 hr	64.7 TS16 450°C 4 hr
44.50	45.17	296	57	n-Heneicosane	0.80	0.28	0.02		
46.81	47.46	310	57	n-Docosane	0.80	0.06			

3.3   High-Temperature SimDis GC for Petroleum Resids

3.3.1   High-Temperature SimDis GC Method

There were no known standard methods for SimDis GC of resids at the onset of this work. To establish conditions for HT-SimDis GC analysis of resids and their products, analysis was performed at different GC conditions. Table 13 lists the four GC methods used. In the first two (TP1 and TP2), the lower detector temperature (350°C) was used. In TP2, the final oven temperature was higher but the heating rate lower than that TP1. The heating rate in TP3 was lower than that in TP4. Figure 9 and Figure 10 show the high-temperature gas chromatograms of resids AR and VR. Comparison of Figure 9B (and Figure 10A) with Figure 9A indicates that at a low detector temperature and lower final column temperature (TP1) the GC is not sensitive enough to detect the heavy portion of resids. Increasing the final temperature and total time (in method TP2) improved the elution of heavy materials, but the lower detector temperature presented problems. In third method TP3, we increased the detector temperature from 350 to 435°C. The gas chromatograms using method TP3 (Figure 9C and Figure 10B) show that the high boiling portion of resid produces a significant second peak. To test for further improvement in the elution of resid, we increased the heating rate from 10 to 15°C/min and the final holding time from 15.5 to 28.8 min in TP4. Comparing Figure 9D, with Figure 9C or comparing Figure 10C with Figure 10B show that the increased heating rate reduced the time needed for resid elution. However, the higher heating rate gave poorer reproducibility during calibration. Table 2 indicates that method TP3 performed best in detector sensitivity, elution of high molecular weight compounds, reproducibility and the correct calibration. Hence this method was used for subsequent GC analysis.

Table 13. Different GC Methods for HT-SimDis analysis

Method	Detector Temp. (°C)	Column Temperature Program				
		Initial Temp. (°C)	Initial Time (min)	Rate (°C/min)	Final Temp. (°C)	Final Time (min)	Total Time (min)
TP1	350	40	1	15	355	18.0	40
TP2	350	40	1	10	425	15.5	55
TP3	435	40	1	10	425	15.5	55
TP4	435	40	1	15	425	28.8	60

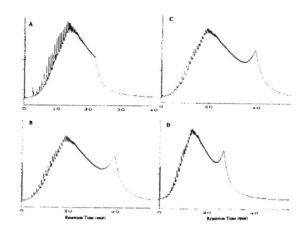

Figure 11. HT-SimDis gas chromatograms of AR obtained using GC methods A) TP1, B) TP2, C) TP3 and D) TP4.

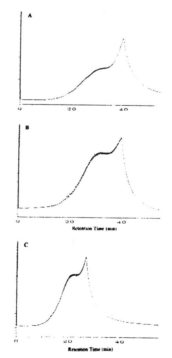

Figure 12. HT-SimDis gas chromatograms of VR obtained using GC methods A) TP2, B) TP3, and C) TP4.

To further confirm the general applicability of the method TP3, it was checked with a reference standard (high-temperature lube oil) and various model compounds. Figure 11 shows the high-temperature gas chromatograms of the reference standard, AR and VR, respectively, along with calibration curves. It is clear from Figure 11 that calibration curve is linear except lower paraffins (C5-C9). For the analysis of resids, it is necessary to calibrate to as high boiling point as possible. In this work it was possible to calibrate through C92 paraffin which has a boiling point of 704°C. The reference standard eluted within the range covered by the calibration curve (Figure 11A). The atmospheric equivalent boiling point (AEBP, Table 14) distribution obtained for the reference standard using our calibrated HT-SimDis is within ± 3% of the values supplied by the vendor for that standard. Though a portion of the chromatograms for AR and VR were beyond the range of the calibration standard (Figures 11B and 11C), the calibration was assumed to be linear through the range of resid elution.

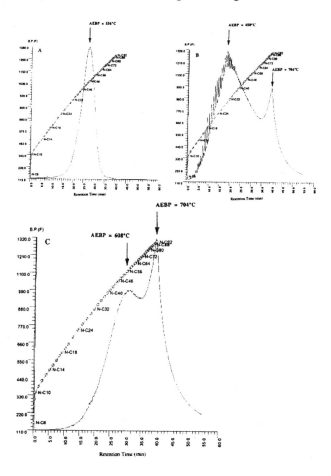

Figure 13. HT-SimDis gas chromatograms of A) high temperature lube oil (reference standard), B) atmospheric resid (AR) and C) vacuum resid (VR).

Table 14. Comparison of HT-SimDis derived boiling points with the true values of various compounds

Compounds	True Boiling Point °C	True Boiling Point °F	SimDis-BP °C	SimDis-BP °F	Deviation °C	Deviation % in °C
Cycloalkanes:						
Methylcyclohexane	101	214	104	220	3	3.3
trans-1,3-Dimethylcyclohexane	124	255	133	271	9	7.0
n-Hexylcyclohexane	221	430	223	434	2	1.8
cis-Decalin	193	379	197	387	4	2.3
trans-Decalin	185	365	194	381	9	4.8
Ethylcyclohexane	130	266	128	262	-2	-1.8
n-Propylcyclohexene	155	311	156	313	1	0.6
n-Butylcyclohexane	180	356	179	354	-1	-0.6
Aromatics:						
Toluene	110	230	118	244	8	6.8
p-xylene	139	282	140	283	1	0.4
Ethylbenzene	136	277	143	289	7	4.9
Naphthalene	218	424	218	425	0	0.1
Cyclohexylbenzene	239	462	240	464	1	0.4
2-Methylnaphthalene	241	466	237	459	-4	-1.6
1-Methylnaphthalene	240	464	227	441	-13	-5.2
2,6-Dimethylnaphthalene	262	504	275	527	13	5.0
1,5-Dimethylnaphthalene	266	511	278	532	12	4.4
Fluorene	298	568	291	556	-7	-2.2
Biphenyl	255	491	258	497	3	1.2
p-Terphenyl	389	732	383	721	-6	-1.6
Phenanthrene	336	637	320	607	-16	-4.8
Anthracene	342	648	340	643	-2	-0.7
Pyrene	395	743	382	719	-13	-3.4
Chrysene	447	837	418	784	-29	-6.6
Heteroatom-containing compounds:						
Phenol	182	360	172	341	-10	-5.7
m-Cresol	203	397	197	386	-6	-3.2
2-Naphthol	285	545	284	543	-1	-0.3
Dibenzofuran	285	545	298	569	13	4.6
Quinoline	237	459	227	441	-10	-4.2
1.2.3.4-Tetrahydroquinoline	249	480	247	477	-2	-0.7
Acridine	346	655	349	659	3	0.7
Benzothiophene	221	430	219	426	-2	-0.8
Dibenzothiophene	332	630	317	603	-15	-4.5

In order to check the HT-SimDis applicability to various compounds, we also analyzed different types of pure compounds and estimated their boiling points using HT-SimDis method. For a pure compound the initial boiling point (where 0.5 wt% sample is distilled) from HT-SimDis was taken as the estimated boiling point of that compound. Table 14 lists the estimated and the true boiling points of the compounds. The boiling points derived from SimDis GC are consistent with the true boiling points within ± 5% for most compounds, and ± 7% for some compounds. It seems that the SimDis GC-derived boiling points tend to be slightly lower for many polycyclic aromatic compounds and polar compounds containing heteroatoms. However, the accuracy is considered to be acceptable for analyzing such complicated mixtures as resids.

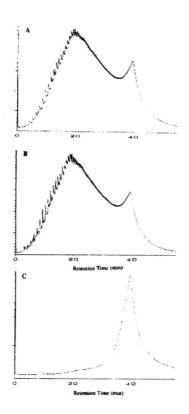

Figure 14. HT-SimDis gas chromatograms of A) high temperature lube oil (reference standard), B) atmospheric resid (AR) and C) vacuum resid (VR).

3.3.2   HT-SimDis GC Analysis of Resids

HT-SimDis gas chromatograms of AR and VR resids contain two major peaks (Figure 11B and 11C). With VR, the second peak at high temperature is larger than that of indicating that the VR has a larger high boiling fraction. The first peak and the second peak in Figure 11B (AR) correspond to AEBP's of about 450°C and 704°C, respectively; the two peaks in Figure 11C (VR) correspond to AEBP's of about 608°C and 704°C, respectively.

Table 15shows the AEBP distribution of AR and VR obtained using HT-SimDis, according to the boiling ranges shown in Table 16. The fractions with boiling point of >340°C (cut-off point for atmospheric distillation) in AR and VR are 93.9 and 99.1 wt%, respectively. The fractions with AEBP's of > 540°C (cut-off point for vacuum distillation) in AR and VR are 57.0 and 88.6 wt%, respectively. Table 15 also shows the amount of residue (uneluted portion of the sample) estimated by HT-SimDis. The amounts of residue (with AEBP of >847°C) in AR and VR feedstocks are 16.5 and 21.6 wt%, respectively. The amount of asphaltenes for AR and VR feeds were estimated from solvent extraction method and the values are 9.22 and 14.02 wt%, respectively. It is interesting to note that these residue values are considerably higher than the asphaltene contents. AR is completely soluble in toluene, and VR only contains 0.12 wt% toluene-insoluble material.

Table 15. HT-SimDis GC analysis of AR and VR resid feedstocks

Feed	Atmospheric Equivalent Boiling Point Distribution (%)						
	IBP-220°C	220-340°C	340-450°C	450-540°C	540-700°C	700°C-FBP[a]	Residue[b]
AR	0.5	5.6	17.7	19.2	27.4	13.1	16.5
VR	0.3	0.6	2.0	8.5	43.1	23.9	21.6

a) Final boiling point, FBP: = 847°C; b) Residue: uneluted portion of the resid with AEBP of >847°C.

Table 16. Boiling point range of different fractions

Fraction	Atmospheric Equivalent Boiling Point	
	°C	°F
Light Naphtha	IBP-130	IBP-266
Heavy Naphtha	130-220	266-428
Atmospheric Gas Oil	220-340	428-644
Light Gas Oil	340-450	644-842
Heavy Vacuum Gas Oil	450-540	842-1004
Super Heavy Gas Oil	540-847	1004-1557
Nondistillable Residue	> 847	> 1557

Figure 12 and Figure 13 illustrate the high-temperature gas chromatograms of feedstock, hexane soluble (HS) and hexane insoluble (HI) fractions of AR and VR, respectively. HT-SimDis gas chromatograms of feedstocks and HS fractions of AR (Figure 12) and VR (Figure 13) show two peaks, but hexane insoluble (HI) fractions of both AR (Figure 12C) and VR (Figure 13C) resids show a single peak at high temperature. Essentially all HI is asphaltenes in these resids. We have found no literature report on HT-SimDis GC of asphaltene. The peak AEBP for asphaltene in AR (Figure 12) is 704°C, which corresponds to a GC oven temperature of 425°C. The majority of the asphaltene peak eluted between 32 and 48 min at 425°C, which corresponds to an AEBP range of 620-840°C. A question that arises is, is this material asphaltene or is it the products of asphaltene decomposition. It is known that decomposition of asphaltene can occur in this temperature range (41). It is difficult to clarify this issue for asphaltene, because no standard reference is available. However, what we have established is that this peak is characteristic of the presence of very heavy materials, and this peak can disappear after upgrading. The eluted materials in

Figure 12C only represent 50.1 wt% of the asphaltene fraction in AR. Similarly, the eluted materials in Figure 13C represent only 39.7 wt% of the HI (mostly asphaltene) in VR.

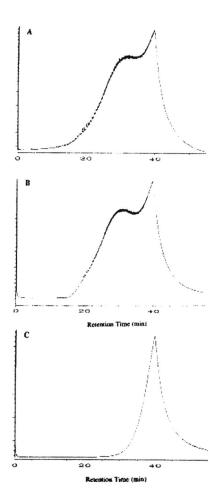

Figure 15. HT-SimDis gas chromatograms of A) AR, B) hexane soluble AR and C) hexane insoluble AR.

Another interesting feature is that the HS fraction of AR, the asphaltene-free materials, also displays the GC peak with AEBP of about 704°C (Figure 12B), which is almost identical to that for asphaltene (Figure 12C). This trend can also be observed for VR in Figure 5. Apparently, asphaltene fractions contribute only partially to the second peak of the whole resids. These results show that asphaltene-free resids contain a significant amount of materials which have the same AEBP range as asphaltene. Such elutable heavy materials largely disappear in HT-SimDis GC of the upgrading products, as described below. HT-SimDis gas chromatograms of HI fractions (Figure 12C, Figure 13C) and Table 16 clearly show that VR feed contains more high boiling fraction (second peak) and more asphaltenes compared to AR feed.

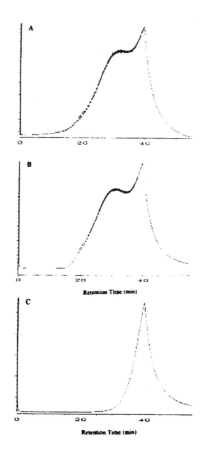

Figure 16. HT-SimDis gas chromatograms of A) VR, B) hexane soluble VR and C) hexane insoluble VR.

3.3.3   Analysis of Upgraded Products

Figure 14 and Figure 15 compare the high-temperature gas chromatograms of AR and VR feeds and their products obtained over Co-Mo/Al$_2$O$_3$ catalyst (Table 17) at 425°C, respectively. Compared to the AR feedstock, the chromatogram of products has more lighter components, and the second GC peak (due to heavy fraction of the feedstocks) eluting at higher temperature (Figure 6) almost disappeared. This trend is more remarkable for VR (Figure 7). Complete AEBP distribution of the liquid products, liquid yield, coke yield, and conversions of >340°C and >540°C fractions at different conditions are shown in Tables 18 and 19 for AR and VR, respectively. We have done several repeated analyses for some samples to examine the reproducibility of the HT-SimDis analysis of upgraded products. Table 20 shows the good reproducibility of the AEBP distribution of AR, VR and their products from repeated analyses. AEBP distribution from Tables 18 and 19 reveals that both catalytic and noncatalytic upgrading of AR and VR produced lighter fraction but the extent of lighter fraction production varied with the type of feedstock and the temperature. Increasing the reaction temperature increased the conversion of >340°C and >540°C fractions both in the presence and absence of catalyst.

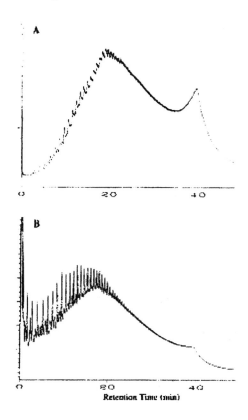

Figure 17. HT-SimDis gas chromatograms of A) AR feed and B) product obtained over Co-Mo/Al₂O₃ (Shell 344TL) at 425°C.

Table 17. Properties of the Shell 344TL Co-Mo/Al$_2$O$_3$ Catalyst

Material Type	Co-Mo/Al$_2$O$_3$
CoO, wt%	2.9
MoO$_3$, wt%	13.5
Surface area, m^2/g	208
Pore volume, cc/g	0.53
Median pore diam., Å	110

Table 18. HT-SimDis GC of Hydroprocessed AR Resid With and Without Co-Mo/Al2O3

Expt ID	Cat.	RxnTemp. (°C)	IBP-220°C	220-340°C	340-450°C	450-540°C	540-700°C	700°C-FBP[a]	Residue[b]	Liquid Yield (wt%)	Coke Yield (wt%)	>40°C Con3v. (%)
1	No	375	0.9	6.5	18.6	19.5	28.1	13.4	13.0	97.2	0.1	4.1
5	No	400	1.0	9.6	20.8	19.3	24.6	10.6	14.1	95.5	0.2	9.1
7	No	425	12.3	19.1	21.1	13.2	13.7	6.1	14.5	97.7	0.4	28.6
27	Yes	375	0.7	7.0	19.9	20.3	28.2	12.9	11.0	97.2	0.1	4.4
20	Yes	400	1.1	8.1	19.8	19.3	24.9	10.8	16.0	97.6	0.1	5.6
21	Yes	425	5.4	13.7	22.7	18.5	19.9	7.1	12.7	94.8	0.2	18.3

a) Final boiling point, FBP: = 847°C; b) Residue: uneluted portion of the resid with AEBP of >847°C.

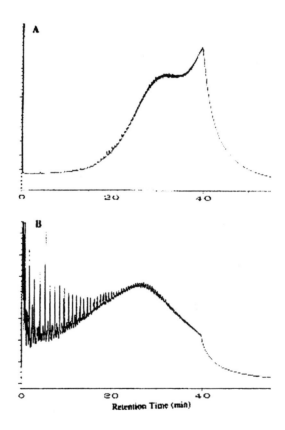

Figure 18. HT-SimDis gas chromatograms of A) VR feed and B) product obtained over Co-Mo/Al$_2$O$_3$ (Shell 344TL) at 425°C.

Table 19. HT-SimDis GC of Hydroprocessed VR Resid With and Without Co-Mo/Al$_2$O$_3$

Expt ID	Cat.	Rxn.Temp (°C)	Atmospheric Equivalent BP Distn (%)							Liquid Yield (wt%)	Coke Yield (wt%)	>340°C Conv. (%)	>540°C Conv.(%)
			IBP-220°C	220-340°C	340-450°C	450-540°C	540-700°C	700°C-FBP[a]	Residue[b]				
40	No	400	2.4	4.4	7.1	12.8	39.7	18.9	14.7	97.3	0.2	8.5	16.0
31	No	425	6.2	13.0	14.2	13.4	21.3	7.7	24.2	95.2	1.9	22.4	40.3
37	No	450	8.6	27.2	21.5	12.5	13.3	6.0	10.9	77.6	13.6	49.8	72.4
47	Yes	350	0.9	3.5	4.4	10.0	42.4	21.4	17.4	97.6	0.1	5.8	6.6
46	Yes	375	1.2	3.6	4.4	10.2	40.0	20.4	20.2	97.2	0.1	6.6	7.7
42	Yes	400	2.6	5.0	7.0	12.9	39.8	18.1	14.6	97.0	0.2	9.6	17.2
33	Yes	425	5.6	10.7	14.3	17.2	29.1	9.5	13.6	94.6	0.4	20.1	41.8
38	Yes	450	13.9	22.9	20.6	14.0	15.3	6.1	7.2	85.4	1.5	45.6	71.2

a) Final boiling point, FBP: = 847°C; b) Residue: uneluted portion of the resid with AEBP of >847°C.

Table 20. Reproducibility of HT-SimDis GC of AR and VR and Products

Expt. ID	Temp (°C)	Atmospheric Equivalent Boiling Point Distribution (%)						
		IBP-220°C	220-340°C	340-450°C	450-540°C	540-700°C	700°C-FBP[a]	Residue[b]
AR	Feed	0.5	5.6	17.7	19.2	27.4	13.1	16.5
AR	Feed	0.6	5.4	17.5	19.2	27.3	13.4	16.6
VR	Feed	0.3	0.6	2.0	8.5	43.1	23.9	21.6
VR	Feed	0.4	0.6	2.0	8.6	44.4	25.4	18.6
37/A	450	9.2	28.4	22.2	12.4	12.4	4.8	10.6
37/B	450	8.9	27.5	21.9	12.6	13.0	5.3	10.8
37/C	450	8.6	27.5	21.5	12.5	13.3	5.7	10.9
31/A	425	6.7	12.1	13.8	13.4	20.5	8.6	24.9
31/B	425	6.7	11.7	13.6	13.4	20.7	9.4	24.5
31/C	425	6.2	13.0	14.2	13.4	21.3	7.7	24.2
46/A	375	0.7	3.0	4.3	9.9	38.9	18.4	24.8
46/B	375	1.2	3.6	4.4	10.2	40.0	20.4	20.2

a) Final boiling point, FBP: = 847°C; b) Residue : uneluted portion of the resid with AEBP of >847°C.

However, the presence of catalyst has two positive effects: 1) more heavy materials (>847°C fraction) is converted, and 2) the coke formation is suppressed, particularly in the case of VR upgrading. Thermal runs produced higher amounts of coke with lower liquid yield at high temperature of 450°C, which indicate condensation of reactive fragments in the absence of catalyst. In the case of VR resid upgrading, more of the >540°C fraction is converted, both in catalytic and noncatalytic runs, compared to those of AR resid. Consequently the fractions between IBP-220°C and 220-340°C increased significantly during the hydroprocessing of VR resid at 425-450°C.

Figure 16 clearly illustrates that the reaction temperature has a significant influence on product distribution, particularly on >540°C fractions. At <400°C, there is no significant change in >540°C fractions. But an increase in temperature from 400°C to 425°C led to significant conversion of 540-700°C AEBP materials to lighter fractions (<450°C). This indicates that increasing the reaction temperature causes the production of large amounts of lighter fractions (<540°C) at the expense of heavier fractions (>540°C). When the temperature is further increased to 450°C, both the >540°C fraction and the 450-540°C fraction converted further. These results indicate that in hydroprocessing of VR, the temperature required to activate >540°C fraction is about 425°C, whereas 450-540°C fraction is activated at about 450°C.

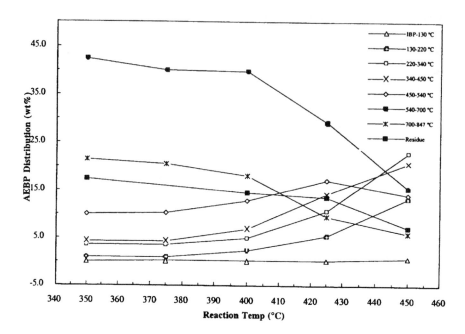

Figure 19. Effect of temperature on VR hydroprocessing over Co-Mo/Al₂O₃ catalyst. The residue fraction refers to the uneluted resid products with AEBP of >847°C.

Reduction in boiling points is also accompanied by desulfurization during resid upgrading. Figure 17 demonstrates the effect of temperature on sulfur removal in hydroprocessing of VR in the presence and absence of Co-Mo/Al₂O₃ catalyst. Reaction temperature has a significant effect on desulfurization, particularly in the presence of catalyst. Unlike the AEBP distribution, the extent of catalytic desulfurization increased almost monotonically with increase in temperature from 350°C to 450°C. This trend is different from those observed for hydrodesulfurization of distillate fuels, where reaction temperature is typically between 350 ~ 400°C.

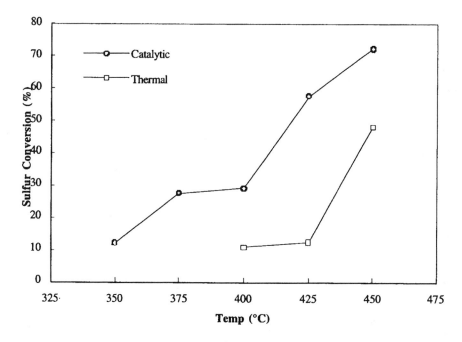

Figure 20. Effect of temperature on sulfur conversion by HDS during VR hydroprocessing with and without sulfided Co-Mo/Al₂O₃ catalyst.

It appears that catalytic hydrodesulfurization of resids also depends on thermally driven hydrocracking reactions. In other words, the C-C bond cleavage reactions are largely thermally driven in the catalytic upgrading. The catalyst has two roles. First, it promotes hydrogen transfer to produce stable molecules and suppress coke formation. Second, it enhances the desulfurization of relatively smaller molecules derived from the resids. In the noncatalytic reaction, coke formation is significant at 450°C (Table 19).

4. CONCLUSIONS

A broad range of hydrocarbon fuel streams from petroleum and coal as well as natural gas can be characterized by gas chromatographic techniques, using temperature-programmed retention indices for GC and GC-MS of light and middle distillate fuels, and high-temperature simulated distillation with temperature programming for heavy oils and their upgrading products.

The retention characteristics of various hydrocarbons and N-, S-, and O-containing compounds have been established on different columns with different phase polarity at three different heating rates. Aliphatic compounds give nearly constant indices at different heating rates, but the retention indices of polycyclic aromatic compounds exhibit relatively large temperature dependence. The use of a short isothermal holding (5 min) prior to the programmed heat-up does not cause any significant difference in the retention indices. Their dependences on the heating rates and temperature program, and on column phase polarity have been demonstrated.

Retention indices as well as their sensitivity to temperature program decrease with decreasing column polarity. The column polarity can affect the retention behavior significantly, depending on the compound type. There also exist relationships between the temperature dependence (or column polarity dependence) of retention indices and the compound type. Almost all the compounds studied display higher temperature sensitivity on the more polar column (Rtx-50) than on the less polar column (DB-5), i.e., the values of ΔR.I./ΔT are larger for Rtx-50 than for DB-5.

The knowledge on the effects of temperature and column polarity on retention behavior of various compounds can be applied to improving compound identification by using temperature programming rates or using columns of different polarity.

The temperature-programmed retention indices are very useful for GC and GC-MS analysis of petroleum-derived and coal-derived liquid fuels, as have been demonstrated in identifying components in two JP-8 jet fuels and their liquid chromatographic fractions. Over 100 fuel components were identified in detail. The results revealed that the coal-derived JP-8C fuel is significantly different from petroleum-based JP-8P in composition.

The present results can also be used for selecting appropriate column coating phase and temperature program for more efficient analysis of given samples and for eliminating or reducing co-elution of certain compounds. In addition, the present results of retention indices can be used to build a computer-assisted library of retention indices, which in combination with mass spectral library can lead to faster and more reliable automatic peak identification through computer library search.

For simulated distillation of heavy petroleum fractions using GC, it is important to use the right GC conditions and a high-temperature column. After screening tests at various GC conditions in our work, a HT-SimDis GC method has been established, which can be used to determine AEBP distributions up to 847°C. It has been successfully applied for analyzing the AEBP distribution of an atmospheric resid (AR) and a vacuum resid (VR) as well as their products from catalytic hydroprocessing.

HT-SimDis GC analysis of AR and VR feedstocks clearly indicated the difference in AEBP distribution between the two feedstocks, which is partially responsible for different levels of upgrading. Such analysis also revealed that both thermal and catalytic hydpropocessing of AR and VR caused changes in

AEBP distribution of hihger boiling fractions (>450°C AEBP). For the runs over a sulfided Co-Mo/Al$_2$O$_3$ catalyst, the yields of products with AEBP range of 540-700°C remained almost constant between reaction temperatures of 350-400°C, but decreased monotonically upon increasing reaction temperature to 450°C. At the high reaction temperature (450°C), the 450-540°C fraction is also converted significantly. The extent of catalytic hydrodesulfurization over Co-Mo/Al$_2$O$_3$ catalyst increased monotonically with an increase in reaction temperature from 350 to 450°C.

5.   ACKNOWLEDGMENTS

We are most grateful to Prof. H. H. Schobert for his encouragement, support and helpful discussions. This work was supported by the Air Force Wright Laboratory/Aero Propulsion and Power Directorate, Wright-Patterson AFB, and the U.S. Department of Energy, Pittsburgh Energy Technology Center. We thank Mr. W.E. Harrison III and Dr. D.M. Storch of WL and Dr. S. Rogers of PETC for providing technical support and jet fuel samples, and Dr. P. G. Hatcher for his instrumental support to the maintenance of analytical equipments. We also thank Dr. Mark Plummer of Marathon Oil Company for providing the petroleum resid samples. One of the authors, C.S. wishes to thank Dr. J. Shiea for helpful discussions on GC-MS analysis of branched alkanes. B. Wei wishes to acknowledge the Committee of Higher Education of The People's Republic of China for financial support.

6.   REFERENCES

1.   Nowack, C. J.; Solash, J.; Delfosse, R. J. *Coal Processing Technol.* **1977**, 3, 122.
2.   Erwin, J.; Sefer, N. R. *Prepr. Pap.-Am. Chem. Soc., Div. Pet. Chem.* **1989**, 34 (4), 900.
3.   Zhou, P.-Z.; Marano, J. J.; Winschel, R. A. *Prepr. Pap.-Am. Chem. Soc., Div. Fuel Chem.* **1992**, 37 (4), 1847.
4.   TeVelde, J.; Spadaccini, L. J.; Szetela, E. J.; Glickstein, M. R. *Thermal Stability of Alternative Aircraft Fuels*, AIAA-83-1143, American Institute of Aeronautics and Astronautics: New York, 1983.
5.   Sullivan, R. J; Frumkin, H. A. *Prepr. Pap.-Am. Chem. Soc., Div. Fuel Chem.* **1986**, 31 (2), 325
6.   Roquermore, W. M.; Pearce, J. A.; Harrison III; W. E., Krazinski, J. L.; Vanka, S. P. *Prepr. Pap.-Am. Chem. Soc., Div. Pet. Chem.* **1989**, 34 (4), 841
7.   Lee, C. M.; Niedzwiecki, R. W. *Prepr. Pap.-Am. Chem. Soc., Div. Pet. Chem.* **1989**, 34 (4), 911
8.   Moler, J. L.; Steward, E. M. *Prepr. Pap.-Am. Chem. Soc., Div. Pet. Chem.* **1989**, 34 (4), 837.
9.   Watkins, J. J.; Krukonis, V. J. *Supercritical Fluid Fractionation of JP-8*, Final Report, U.S. Air Force Aero Propulsion and Power Laboratory, WL-TR-91-2083, 1992, 159 pp (Available from NTIS).
10.   Hazlett, R. N. *Thermal Oxidation Stability of Aviation Turbine Fuels*, ASTM Monograph 1, American Society for Testing and Materials, Philadelphia, PA, 1991, 163 pp.

11. Lai, W.-C.; Song, C.; Schobert, H. H.; Arumugam, R. *Prepr. Pap.-Am. Chem. Soc., Div. Fuel Chem.* **1992**, 37 (4), 1671.
12. Song, C.; Lai, W.-C.; Schobert, H. H. *Ind. Eng. Chem. Res.* **1994**, 33, 534.
13. Song, C.; Lai, W.-C.; Schobert, H. H. *Ind. Eng. Chem. Res.* **1994**, 33, 548.
14. Song, C.; Eser, S.; Schobert, H. H.; Hatcher, P. G. *Energy & Fuels* **1993**, 7, 234.
15. Song, C.; Hatcher, P. G. *Prepr. Pap.-Am. Chem. Soc., Div. Petrol. Chem.* **1992**, 37 (2), 529.
16. Hayes, P. C., Jr.; Pitzer, E. W. *J. Chromatogr.* **1982**, 253, 179.
17. Hayes, P. C., Jr.; Pitzer, E. W. *J. High Resol. Chromatogr. & Chromatogr. Comm.* **1985**, 8, 230.
18. Steward, E. M.; Pitzer, E. W. *J. Chromatogr. Sci.* **1988**, 26, 218.
19. Kováts, E. sz. *Helv. Chim. Acta.* **1958**, 41, 1915.
20. Kováts, E. sz. *Advances in Chromatography Vol. 1*, Giddings, J. C.; Keller, R. A. (Eds), Marcel Dekker, Inc.: New York, 1965; pp. 229-247.
21. Budahegyi, M. V.; Lombosi, E. R.; Lombosi, T. S. Mészáros, S. Y.; Nyiredy, Sz.; Tarján, G.; Timár, I.; Takács, J. M. *J. Chromatogr.* **1983**, 271, 213.
22. Guiochon, G. *Anal. Chem.* **1964**, 36, 661
23. Habgood, H. W.; Harris, W. E. *Anal. Chem.* **1964**, 36, 663.
24. Lee, M. L.; Vassilaros, D. L.; White, C. M.; Novotny, M. *Anal. Chem.* **1979**, 51, 768.
25. Whalen-Pedersen, E. K.; Jurs, P. C. *Anal. Chem.* **1981**, 53, 2184
26. Harris, W. E.; Habgood, H. W. *Programmed Temperature Gas Chromatography*, Wiley: New York, 1966.
27. Ven Den Dool, H.; Kratz, P. D. *J. Chromatogr.* **1963**, 11, 463.
28. Altgelt, K. H.; Boduszynski, M. M. *Composition and Analysis of Heavy Petroleum Fractions*, Marcel Dekker, Inc., 1994; p. 37, 77.
29. ASTM Designation: D 2892-84 Drews A., *Manual on Hydrocarbon Analysis*, 4th Ed., ASTM, 1989, 559.
30. ASTM Designation: D 2887-93 Annual Book of ASTM Standards, 05. 02, ASTM, 1994, 194.
31. ASTM Designation: D 5307-92 Annual Book of ASTM Standards, 05. 03, ASTM, 1994, 564.
32. Padlo, D. M.; Kugler, E. L. *Energy & Fuels* **1996**, 10, 1031.
33. Bacaud, R.; Rouleau, L.; Bacaud, B. *Energy & Fuels* **1996**, 10, 915.
34. Trauth, D.M.; Stark, S. M.; Petti, T. F.; Neurock, M.; Klein, M. *Energy & Fuels* **1994**, 8, 76.
35. Hewlett-Packard Gas Chromatography Application Note 228-60, Hewlett-Packard, January 1988.
36. Furlong, M.; Fox, J.; Masin, J. *Production of Jet Fuels from Coal-Derived Liquids*, Vol. IX, Interim Report, AFWAL-TR-87-2042, 1989, 52 pp (available from NTIS).
37. Martel, C. R. *Military Jet Fuels 1944-1987*, Summary Report for period Oct. 85-Oct. 87, U.S. Air Force Aero Propulsion and Power Laboratory, AFWAL-TR-87-2062, 1987, 62 pp.
38. Song, C.; Reddy, K. M. *Am. Chem. Soc. Div. Petrol. Chem. Prepr.* **1996**, 41 (3), 567.
39. Song, C.; Reddy, K. M. *Appl. Catal.* **1999**, 176 (1), 1.
40. Reddy, K. M.; Boli W.; Song, C. *Am. Chem. Soc. Div. Petrol. Chem. Prepr.* **1997**, 42 (2), 336.
41. Song, C.; Nihonmatsu T.; Nomura, M. *Ind. Eng. Chem. Res.* **1991**, 30, 1726.
42. Rowland, S. J.; Alexander, R.; Kagi, R. I. *J. Chromatogr.* **1984**, 294, 407.
43. Borrett, V.; Charlesworth, J. M.; Moritz, A. G. *Ind. Eng. Chem. Res.* **1991**, 30, 1971.
44. Lai, W. C.; Song, C. S. *Fuel* **1995**, 74 (10), 1436-1451.
45. Elizalde G. M. P.; Hutfliess, M.; Hedden, K. *HRC-J. High Res. Chrom.* **1996**, 19 (6), 345-352.
46. Akiyama K. *Bunseki Kagaku* **1996**, 45 (5), 441-446.
47. Robbins, W. K.; Hsu, C. S. "Petroleum: Composition" in *Kirk-Othmer Encyclopedia of Chemical Technology*, 4th ed., John Wiley & Sons: New York, 1996; pp. 352-370.
48. Hsu, C. S.; Drinkwater, D. "GC/MS in the Petroleum Industry" in *Current Practice in GC/MS* (Chromatogr. Sci. Series, Vol. 86), W. W. A. Niessen, ed. New York: Dekker Marcel, 2001; pp. 55-94.

49. Hsu, C. S.; Green, M. *Rapid Comm. Mass Spectrom.* **2001**, 15, 236-239.
50. Veriotti, T.; Sacks, R. *Anal. Chem.* **2000**, 72 (14), 3063-3069.
51. Aczel, T.; Hsu, C. S. *Int. J. Mass Spectrom. Ion Proc.* **1989**, 92, 1-7.
52. Song et al., Energy Fuels, 1993, 7 (2), 234-243.
53. Andresen, J. M.; Strohm, J. J.; Sun, L.; Song, C. *Energy Fuel* **2001**, 15 (3), 714-723.
54. Wynne, J. H,; Stalick, W. M.; Mushrush, G. W. *Petrol. Sci. Technol.* **2000**, 18 (1-2), 221-229.
55. Hsu, C. S.; Genowitz, M. W.; Dechert, G. J.; Abbott, D. J.; Barbour, R. "Molecular Characterization of Diesel Fuels by Modern Analytical Techniques" in *Chemistry of Diesel Fuels*, Song, C.; Hsu, C. S.; Mochida, I. (Eds.), Taylor & Francis: New York, 2000; Chap.2, pp. 61-76.
56. Pal, R.; Juhasz, M.; Stumpf, A. *J. Chromatogr. A* **1998**, 819 (1-2), 249-257.
57. Hsu, C. S. "Diesel Fuel Analysis" in *Encyclopedia of Analytical Chemistry*, John Wiley: New York, 2000; pp. 6613-6622.
58. Malhatra, R.; Coggiola, M. J.; Young, S. E.; Hsu, C. S.; Dechert, G. J.; Rahimi, P. M.; Briker. Y. "Rapid Detailed Analysis of Diesel Fuels by GC-FIMS" in *Chemistry of Diesel Fuels*, Song, C.; Hsu, C. S.; Mochida, I. (Eds.), Taylor & Francis: New York, 2000; Chap. 3, pp. 77-92.
59. Pillon, L. Z. *Petrol. Sci. Technol.* **2001b**, 19 (9-10), 1109-1118.
60. Yang, H.; Ring; Z.; Briker, Y.; McLean, N.; Friesen, W.; Fairbridge, C. *Fuel* **2002**, 81 (1), 65-74.
61. Qian, K.; Hsu, C. S. *Anal. Chem.* **1992**, 64 (20), 2327-2333.
62. Song, C.; Hou, L.; Saini, A. K.; Hatcher, P. G.; Schobert, H. H. *Fuel Proc. Technol.* **1993**, 34 (3), 249-276.
63. Song, C.; Saini, A. K.; Schobert, H. H. *Energy Fuels* **1994**, 8 (2), 301-312.
64. Miki, Y.; Suggimoto, Y.; Yamadaya S. *Nippon Kagaku Kaishi* **1993**, (1), 79-85.
65. Onaka, T.; Kobayashi, M.; Ishii, Y.; Okumura, K.; Suzuki, M. *J. Chromatogr. A*, **2000**, 903 (1-2), 193-202.
66. Walls, C. L.; Beal, E. J.; Mushrush, G. W. *J. Environ. Sci. Health A*, **1999**, 34 (1), 31-51.
67. Dinh, H. T.; Mushrush, G. W.; Beal, E. *J. Petrol. Sci. Technol.* **1999**, 17 (3-4), 383-427.
68. Lewis, A. C.; Askey, S. A.; Holden, K. M.; Bartle, K. D.; Pilling, M. J. *HRC-J. High Resol. Chromatogr.* **1997**, 20 (2), 109-114.
69. Zoccolillo, L.; Babi, D,; Felli, M. *Chromatographia* **2000**, 52 (5-6), 373-376.
70. Yoshida, T.; Chantal, P. D.; Sawatzky, H. *Energy Fuel* **1991**, 5 (2), 299-303.
71. Green, J. B.; Yu, S. K. T.; Vrana, R. P. *HRC-J. High Res. Chromatogr.* **1994**, 17 (6), 439-451.
72. Laespada, M. E. F.; Pavon, J. L. P.; Cordero, B. M. *J. Chromatogr. A*, **1999**, 852 (2), 395-406.
73. Bernabei, M.; Bocchinfuso, G.; Carrozzo, P.; De Angelis, C. J. *Chromatogr. A* **2000**, 871 (1-2), 235-241.
74. Bernabei, M.; Spila, E.; Sechi, G. *Anal. Lett.* **1997**, 30 (11), 2085-2097.
75. Cummins, T. M.; Robbins, G. A.; Henebry, B. J.; Goad, C. R.; Gilbert, E. J.; Miller, M. E.; Stuart, J. D. *Environ. Sci. Technol.* **2001**, 35 (6), 1202-1208.
76. Wang, Z.D.; Fingas, M.; Sigouin, L. *LC-GC* **2000**, 18 (10), 1058.
77. Loren, A.; Hallbeck, L.; Pedersen, K.; Abrahamsson, K. *Environ. Sci. Technol.* **2001**, 35 (2), 374-378.
78. Mielke, H. W.; Wang, G.; Gonzales, C. R.; Le, B.; Quach, V. N.; Mielke, P. W. *Sci. Tatal Environ.* **2001**, 281 (1-3), 217-227.
79. Abbondanzi, F.; Antonellini, R.; Campisi, T.; Gagni, S.; Malaspina, F.; Iacondini, A. *Annali Di Chimica* **2001**, 91 (7-8), 391-400.
80. Siddiqui, S.; Adams, W. A.; Schollion, J. J. *Plant Nutr. Soil Sci.* **2001**, 164 (6): 631-635 DEC 2001
81. Wang, Z. D,; Fingas, M.; Sigouin, L. *J. Chromatogr. A* **2001**, 909 (2), 155-169.
82. Ali, L. N.; Mantoura, R. F. C.; Rowland, S. J. *Marine Environ. Res.* **1995**, 40 (1), 1-17.

83. Pillon, L. Z. *Petrol. Sci. Technol.* **2001a**, 19 (7-8), 875-884.
84. Lee, M. R.; Chen, C. M.; Wang, D. W.; Hwang, B. H.; Yang, T. C. *Anal. Lett.* **2000**, 33 (14), 3077-3092.
85. Gratz, L. D.; Bagley, S. T.; Leddy, D. G.; Johnson, J. H.; Chiu, C.; Stommel, P. *J. Hazard Matter* **2000**, 74 (1-2), 37-46.
86. Muir, B.; Hursthouse, A.; Smith, F. *J Environ. Monitor* **2001**, 3 (6), 646-653.
87. Shojania, S.; McComb, M. E.; Oleschuk, R. D.; Perreault, H.; Gesser, H. D.; Chow, A. *Can. J. Chem.-Revue Canadiene de Chimie* **1999**, 77 (11), 1716-1727.
88. Li, C. T.; Mi, H. H.; Lee, W. J.; You, W. C.; Wang, Y. F. *J. Hazardous Materials* **1999**, 69 (1), 1-11.
89. Koziel, J. A.; Odziemkowski, M.; Pawliszyn, J. *Anal. Chem.* **2001**, 73 (1), 47-54.
90. Nicol, S.; Dugay, J.; Hennion, M. C. *Chromatographia* **2001**, 53, S464-S469.
91. Sauvain, J. J.; Due, T. V.; Huynh, C. K. *Fresen J. Anal. Chem.* **2001**, 371 (7), 966-974.
92. Liu, J. T.; Hara, K.; Kashimura, S.; Kashiwagi, M.; Hamanaka, T.; Miyoshi, A.; Kageura. M. *J. Chromatogr. B* **2000**, 748 (2), 401-406.
93. Liu, S. M.; Pleil, J. D. *J Chromatogr. B* **2001**, 752 (1), 159-171.
94. Melikian, A. A.; Malpure, S.; John, A.; Meng, M.; Schoket, B.; Mayer, G.; Vincze, I.; Kolozsi-Ringelhann, A.; Hecht, S. S. *Polycyclic Aromatic Compounds* **1999**, 17 (1-4), 125-134.
95. Eide, I.; Neverdal, G.; Thorvaldsen, B.; Shen, H. L.; Grung, B.; Kvalheim, O. *Environ. Sci. Technol.* **2001**, 35 (11), 2314-2318.
96. Kavouras, I. G.; Koutrakis, P.; Tsapakis, M.; Lagoudaki, E.; Stephanou, E. G.; Von Baer, D.; Oyola, P. *Environ. Sci. Technol.* **2001**, 35 (11), 2288-2294.
97. Rodgers, R. P.; Blumer, E. N.; Freitas, M. A.; Marshall, A. G. *J. Forensic Sci.* **2001**, 46 (2), 268-279.
98. Takayasu, T.; Ohshima, T.; Kondo, T.; Sato, Y. *J. Forensic Sci.* **2001**, 46 (1), 98-104.
99. Hsu, C. S. "Hydrocarbons" in *Encyclopedia of Analytical Science*, premiere edition, Academic Press: London, United Kingdom, 1995; pp. 2028-2034.
100. *Chemistry of Diesel Fuels*, Song, C.; Hsu, C. S.; Mochida, I. (Eds.), Taylor & Francis: New York, 2000.

Chapter 8

MASS SPECTROMETRIC ANALYSES FOR ELEMENTAL SULFUR AND SULFUR COMPOUNDS IN PETROLEUM PRODUCTS AND CRUDE OILS

Vincent P. Nero
Texaco Inc, 3901 Briarpark, Houston, Texas 77042

1.   INTRODUCTION

The environmental restrictions that limit sulfur in petroleum products are becoming increasing more stringent and are continuously driving the analytical technology towards better sulfur analysis procedures.   The comprehensive analysis of the various individual types of sulfur compounds in petroleum products, especially crude oil, is challenging.   The known sulfur components found in natural and refined petroleum sources have very different physical and chemical properties.   The concentration of these components is also extremely variable and dependent upon the refining processes and crude oil source. Furthermore, the exact structures of many higher boiling sulfur components remain undetermined.   Because of this complexity, variability and uncertainty, most traditional analytical methods generally lack precision and specificity. Many recent mass spectrometric methods have been developed for identifying and measuring various sulfur components.

This chapter will focus on the specialized analyses for sulfur species in petroleum: 1) Rapid and accurate analysis for elemental sulfur, S_8, using direct insertion probe mass spectrometry - mass spectrometry in conjunction with an isotopic sulfur internal standard.   Elemental sulfur can account for ten percent of the sulfur content in a crude oil, but in many cases it is overlooked.   2) The rapid analysis for selective thioaromatics components using direct insertion probe mass spectrometry - mass spectrometry.   Thiophenic components account for most of the thio-organic content of a crude oil and refined petroleum products.

3) Monitoring changes in thio-organic composition that occur during various refining processes using gas chromatography - mass spectrometry. Many refining processes, such as hydrotreating exist primarily to reduce the thio-organic content of petroleum products. Modern gas chromatography methods, especially when combined with the identification capabilities of mass spectrometry, will provide quick solutions to complex refinery problems. 4) Monitoring reactions of elemental sulfur with hydrocarbons using gas chromatography mass spectrometry in conjunction with an isotopic sulfur internal standard. The presence of elemental sulfur in a petroleum stream complicates refining operations because sulfur can slowly react with other petroleum components even at relatively low temperatures.

These methods will be compared to and evaluated against more traditional mass spectrometric methods and alternative non-mass spectrometric methods. In general, the best analytical results are achieved when combinations of several analytical techniques are applied.

2. ANALYSIS FOR ELEMENTAL SULFUR BY MASS SPECTROMETRY - MASS SPECTROMETRY

Elemental sulfur, S_8, can be a major sulfur source in crude oil, but is often overlooked and is difficult to measure. The reason sulfur is a major analytical problem is that it slowly reacts with hydrocarbons under relative mild thermal conditions. A novel method was recently developed that can rapidly analyze elemental sulfur in crude oil at room temperatures.[1] The whole crude oil is weighed and is homogeneously spiked with a measured amount of isotopic elemental sulfur-34. The sample is then directly inserted into the mass spectrometer using a thermal probe, which is maintained isothermally at 20°C. The bulk of the oil sample will not evaporate at this temperature even at the very low pressure in the mass spectrometer's source. However, the natural sulfur, the isotopic sulfur and some of the lighter components in the crude oil quickly evaporate from the probe. This volatile fraction is then ionized under electron impact ionization (EI) conditions using ionization energy of 70 eV. The ion source temperature is held at 150°C. The sample was analyzed using direct insertion probe mass spectrometry - mass spectrometry (DIP MS-MS).

The alternative DIP MS analysis in the full scan mode will fail since the light crude oil components interfere with the major natural ions for the $^{32}S_8$ at 160 and 256 m/z and for corresponding major isotopic ions for $^{34}S_8$ at 170 and 272 m/z. However, tandem mass spectrometry can be used to selectively separate the sulfur ions from the interfering hydrocarbon ions in the crude oil. The S_5 fragmentation ions, resulting from the loss of three sulfur atoms from the parent S_8 molecular ions are dominant in the MS-MS mode. The other ionization conditions remain the same. Argon at 1.2 millitorr was used as the collision gas.

The collision-offset energy was held at 10 eV. The natural sulfur daughter ions at 160 m/z, $^{32}S_5$, are measured against the isotopic daughter ions at 170 m/z, $^{34}S_5$, from the enriched sulfur. Each daughter ion was scanned sequentially with a cycle time of 1.0 seconds per scan.

A response factor between natural sulfur and the isotopically enriched sulfur must be determined since the abundances of sulfur-32 and sulfur-34 will not be 100% in either case. The response factors will depend upon the extent that the isotopic sulfur was enriched with ^{34}S. Traditional full-scan mass spectrometry was used to illustrate isotopic distribution for the major ions of natural sulfur and S-34 enriched sulfur. These distributions in the molecular ion region are represented in Figure 1. The response factor is generated using sulfur-free oil that was spiked with natural sulfur and sulfur-34 at several concentrations.

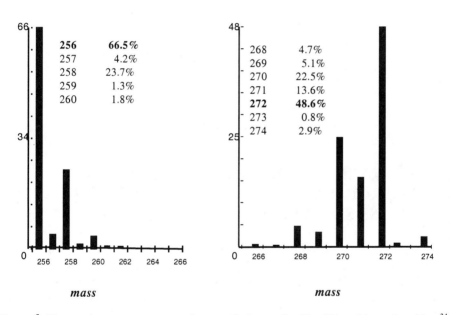

Figure 1. Electron impact mass spectra for natural elemental sulfur (S_8) and isotopic sulfur ($^{34}S_8$) showing mass distributions in the molecular ion region.

The relative amount of elemental sulfur in a sour crude oil (a crude oil with a high total sulfur content) can vary greatly. Some very sour crude oils contain no elemental sulfur. The daughter ion mass spectra for a sour crude oil that contained a relatively high free sulfur content are shown in Figure 2. The S_5 daughter ions are the major ions in the spectra. The crude oil was determined to contain 0.4% sulfur. The analysis is relatively quick. Sample preparation, calibration, and acquisition takes about one hour for the first sample and then about 10 minutes per sample. The results are accurate and precise to within .03% total sulfur.

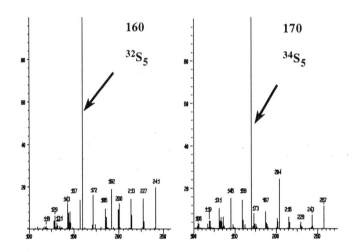

Figure 2. Daughter ions for most prominent molecular ion in natural elemental sulfur (256) and elemental sulfur-34 (272)

3.   ANALYSIS OF THIOPHENIC COMPOUNDS IN PETROLEUM STREAMS BY MASS SPECTROMETRY - MASS SPECTROMETRY

Thioaromatics components, especially benzothiophenes, and dibenzothiophenes are the major sulfur components in a numerous variety of petroleum streams. The thiophene content of any petroleum liquid can be very effectively and quickly analyzed using direct insertion probe mass spectrometry - mass spectrometry. First the petroleum liquid is separated into saturate, aromatic, resin and asphaltene fractions using multiple columns, high performance liquid chromatography (HPLC). The aromatic fraction was then placed in a quartz direct insertion tube, which is inserted into the mass spectrometer, operating in full scan mode for comparison. The sample is heated from 50°C to 450°C at 50°C per minute and then held for several minutes. The mass spectrometer is operated under EI conditions using ionization energy at 70 eV. The ion source temperature is held at 200°C. For illustration, Figure 3 shows the resulting conventional total ion mass spectrum of the aromatic fraction. The mass distribution is clearly bimodal. The first distinguishable region is composed of primarily odd ions and represents ion fragments of molecules, most of which do not contain any sulfur atoms. The second dominant region is composed of primarily even ions and represents molecular ions, again of mostly non-sulfurous molecules. The significant thioaromatic information is obscured in the matrix of other aromatic hydrocarbons typically present in crude oil.

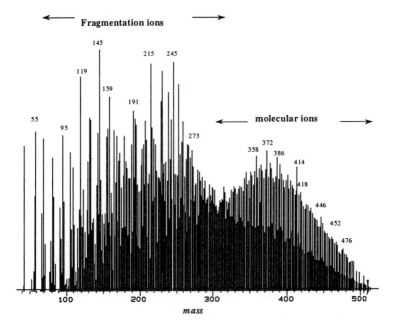

Figure 3. Direct insertion probe electron impact total ion mass spectrum of an aromatic crude oil faction.

The analysis is repeated using the same mass spectrometric conditions but in the MS-MS parent ion mode. Alkyl distributions for various thiophenes, naphthothiophenes benzothiophenes, naphthobenzo-thiophenes dibenzothiophenes, and naphthodibenzothiophenes were determined by sequentially measuring the parent ions of the major common benzylic fragments. Various proportions of these thiophene homologies represent most of the thio-organic content of typical crude oils and refinery streams. Figure 4 shows the parent ion mass spectrum of 197. This mass spectrum partially represents the dibenzothiophene series. The total sum of the parent ion series of 197, 211, 225, 239...etc. more completely represent that series.

The relative amounts of the various thioalkyl distributions can be estimated from the total ion current for the individual series and are similar to the traditional type distributions generated by ASTM D-3239.[2] These MS-MS data additionally provide molecular weight distributions within a given thioalkyl series. These parent ion molecular profiles, when used in conjunction with the traditional ASTM D-3239 class percentages, provide a quite accurate description of the total sulfur in the aromatic cut.

Direct Insertion Probe MS / MS

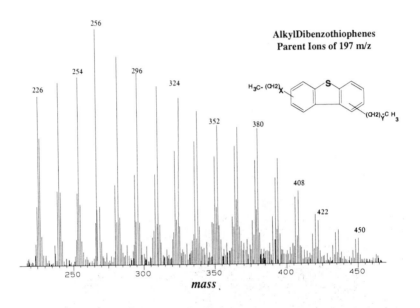

Figu

re 4. Parents ion electron impact mass spectrum, of 197 m/z, representing a dibenzothiophenes series

The total sulfur calculation estimated from the ASTM D-3239 analysis corresponds well with the total sulfur directly measured by the dispersive X-ray fluorescence technique, ASTM Method D-2622.[3] However, these various thiophene types, which eluted in the aromatic HPLC fraction, accounted for only 49% of the sulfur containing components in this particular crude oil as measured by ASTM D-2622. Elemental sulfur accounted for an additional 11% of the sulfur content for the same crude oil. Some of the missing sulfur content is found in aliphatic fractions. However, most of the missing 40% of the sulfur content in this crude oil is in the heavier and generally more complex heterocyclic sulfur aromatic types, many of which contain more than one sulfur atom. For example, phenanthrothiophene and thianthrene series are not measured by ASTM D-2622 but have been detected by DIP MS-MS.

Table 1 shows the original lubricant base oil and the same lubricant oil in three successive stages of hydrotreating. As the hydrotreating process progressed, the various sulfur components decreased as expected. However, the more highly alkylated thiophenes became increasingly more resistant to hydrogenation. As a result, the thiophenic components, which were measured by ASTM D-3239, represented an increasing proportion of the total remaining sulfur content of the oil. Generally the highly alkylated benzothiophenes and dibenzothiophenes are more resistant to hydrotreating and even persist at low levels in commercial petroleum products.

Table 1. Hydrotreated aromatic fractions showing mass spectrometric types obtained using ASTM D3239 and comparing actual and calculated total sulfur content

Thiophenic Types by ASTM D-3239	Initial	#1	#2	#3
Benzothiophenes	3.0	1.9	1.6	1.2
Dibenzothiophenes	0.8	0.8	0.7	0.6
Naphthobenzothiophenes	0.2	0.1	0.1	0.0
Total	4.1	2.8	2.4	1.8
SULFUR calculated from Thiophene Types	0.3	0.2	0.2	0.1
SULFUR by X-ray Fluorescence	1.1	0.6	0.4	0.2
Percentage Represented by Thiophenic Types	30.6	39.0	44.5	68.6

4. MONITORING THIOAROMATICS IN REFINERY PROCESSES

Both individual sulfur components and various classes of sulfur compounds can be monitored throughout refinery processing using gas chromatography - mass spectrometry (GC-MS). Figure 5 is a mass chromatogram of the whole crude oil that was discussed in the previous section. The selected ion mass chromatograms for dimethyl and trimethyl dibenzothiophenes, traces at 212 m/z and 226 m/z have also been superimposed on the total ion mass chromatogram. Methyl, tetramethyl, and pentamethyl dibenzothiophenes also produced clearly identifiable selective ion chromatograms. Each selected ion trace is distinctive, being composed of various alkyl isomers, all eluting or co-eluting at various retention times. Different crude oils will each have their own unique thiophenic patterns. However, as the hydrotreating process progresses, the selected ion traces will change their characteristic pattern. Some isomers are more sterically hindered and resist hydrogenation but all of the thiophenic peaks will decrease in intensity to some extent. The total ion mass chromatogram, which represents mainly aliphatic molecules, will show very little change.

In a complex mixture like crude oil, each individual GC peak is composed of multiple molecular components with the similar physical and sometimes similar structural properties. Many components will have the same GC retention characteristics. Complete chromatographic resolution is rarely achieved, even when high-resolution GC conditions are employed. The mass spectrum at any given GC retention time will be a weighed average of these various molecules eluting at that retention time. However, alkyl dibenzothiophenes have usually dominant molecular and benzylic ions that can be clearly identified in the selected ion chromatographic peaks even without complete component resolution.

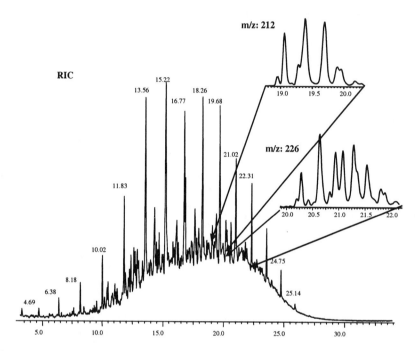

Figure 5. Total ion mass chromatogram of an oil using GC-MS with selected ion chromatograms of dimethyl and trimethyl dibenzothiophenes superimposed

Figure 6 shows individual mass spectra of dimethyl (or ethyl) and trimethyl (or methyl ethyl) dibenzothiophenes that were obtained from the selected ion traces for their molecular ions at 212 and 226 m/z. Strong benzylic ions, representing the loss of a methyl group from the aromatic molecules can also be easily detected at 197 and 211 m/z. These ions provide additional confirmation for alkyl dibenzothiophenes.

The identifying capability of GC-MS can be very effectively complimented by simultaneous performing a parallel analysis using gas chromatography with sulfur chemiluminescence detection, GC/SCD. Chemiluminescence spectroscopy measures sulfur in thio-organics with excellent accuracy and minimal calibration. Hydrocarbons are not detected and do not interfere with the analysis. Chemiluminescence is an excellent method for monitoring quantitative changes in the sulfur content of individual sulfur types. Figure 7 compares the sulfur content of a crude oil before and after hydrodesulfurization. The chromatographic elution order is the same as for GC-MS. The prominent peaks, shown eluting with retention range between 12 and 15 minutes on the GC/SCD chromatograph, were clearly identified by GC-MS as methyl, dimethyl, and trimethyl dibenzothiophenes. The earlier peaks are primarily alkyl benzothiophenes. The later less intensive peaks are heavier more complex polycyclic aromatic thiophenes.

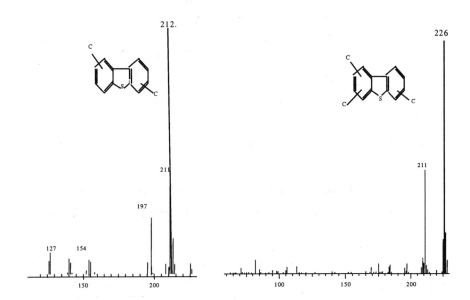

Figure 6. Mass Spectra a dimethyl dibenzothiophene at 19.3 minutes and a trimethyl-dibenzothiophene at 23.3 minutes obtained from the previous mass chromatogram in Figure 5.

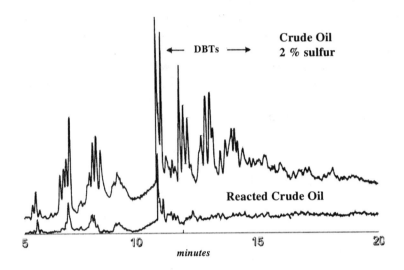

Figure 7. Monitoring reactions using simulated distillation gas chromatography with a sulfur chemiluminescence detector.

5. MONITORING REACTION PRODUCTS OF ELEMENTAL SULFUR WITH HYDROCARBONS

Major upstream environmental problems in the oil fields, transportation problems on tankers and at terminals and downstream processing problems at the refineries can occur when a crude oil contains dissolved hydrogen sulfide. Hydrogen sulfide is corrosive and very toxic. The hydrogen sulfide content in oil is generally measured at the oil field and, if its concentration is found to be too high, is purged from the oil prior to being transported. However, hydrogen sulfide can unexpectedly evolve from a crude oil during transportation or refinery operations when elemental sulfur is present. Under fairly mild thermal conditions, elemental sulfur will react with hydrocarbons in the crude oil to form hydrogen sulfide and sometimes-volatile thio-organic compounds. Unfortunately it is necessary to heat crude many times during producing and shipping. Of course, most refinery operations will involve heating.

It is important to understand the mechanism for these sulfur reactions to control or eliminate the unexpected evolution of hydrogen sulfide. The reaction was studied by adding sulfur-34 to various crude oils and model systems. The reactors were purged with argon, sealed and reacted. Various naphthenic acids, mineral acids, organic amines, salts and water content were added at various levels. The headspace above the reactor, containing the volatile reaction products of isotopically enriched sulfur and natural sulfur, were then analyzed using GC-MS. In all cases hydrogen sulfide was generated slowly but typically in low yield. Figure 8 shows a GC-MS chromatogram of the light volatile end of a crude oil, which contained both natural sulfur and spiked sulfur-34.

Potential sulfur components were singled out using selected ion traces. Hydrogen sulfide, sulfur dioxide, carbon disulfide, and carbonyl sulfide are detected and quantified down to the low parts-per-million (ppm) levels. Mass spectrometry is necessary to distinguish between the natural and isotopically enriched components, which will elute with the identical GC retention times. The major product was hydrogen sulfide in all case. Both natural and enriched sulfur reacted to form these volatile sulfur compounds. The product distributions of natural and enriched sulfur compounds were proportional to relative amounts of each sulfur present. No natural or isotopically enriched sulfur trioxide, light mercaptans, or alkyl sulfides were detected. It is unclear what the oxygen source was for the observed products. Table II lists the relative distributions of the natural and enriched sulfur components detected in the mass chromatogram shown in Figure 8.

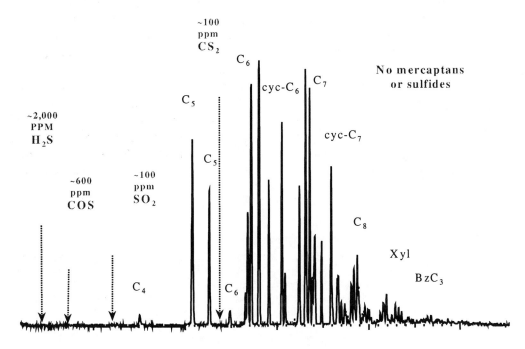

Figure 8. Mass Chromatogram showing reaction products of sulfur

Table 2. Distribution of natural and isotopic components, which were present in the crude oil, represented in Figure 8.

Compound	ppm
$H_2{}^{32}S$	700
$H_2{}^{34}S$	1300
$CO^{32}S$	220
$CO^{34}S$	370
$^{32}SO_2$	40
$^{34}SO_2$	60
$C^{32}S_2$	10
$C^{32}S^{34}S$	40
$C^{34}S_2$	50

6. SUMMARY

The recent environmental requirements and the reliance on heavier sour crude oil present the petroleum industry with some difficult challenges and interesting opportunities. This chapter focuses on a few specialized analyses for sulfur species in petroleum: 1) The rapid and accurate analysis for elemental sulfur, S_8, using direct insertion probe mass spectrometry - mass spectrometry in conjunction with an isotopic sulfur internal standard. 2) The rapid analysis for selective thioaromatics components using direct insertion probe mass spectrometry - mass spectrometry. 3) Monitoring changes in thio-organic composition that occur during various refining processes using gas chromatography - mass spectrometry. 4) Monitoring reactions of elemental sulfur with hydrocarbons using gas chromatography mass spectrometry in conjunction with an isotopic sulfur internal standard. In general, the best analytical results are achieved when mass spectrometry is applied with a variety of analytical techniques.

7. REFERENCES

1. Nero, V. P.; Drinkwater, D. E. "The Rapid Analysis of Elemental Sulfur in Complex Petroleum Matrices Using Tandem Mass Spectrometry" in *Proceedings of the 46th ASMS Conference on Mass Spectrometry and Allied Topics*, Orlando, FL, 1998; pp. 1474.
2. ASTM, "Hydrocarbon Types in Gas-Oil Aromatic Fractions by High Ionization Voltage Mass Spectrometry", Method D3239-91 American Society for Testing and Materials (ASTM): Washington, D.C., 1994.
3. ASTM, "Sulfur in Petroleum Products by X-Ray Spectrometry", Method D-2622-92 American Society for Testing and Materials: Washington, D.C., 1994.

Chapter 9

BIOMARKER ANALYSIS IN PETROLEUM EXPLORATION

C. S. Hsu[1], C.C. Walters[2], G. H. Isaksen[3] , M. E. Schaps[3], and K.E. Peters[3]
1. *ExxonMobil Research and Engineering Co.*
   Baton Rouge, LA 07821
2. *ExxonMobil Research and Engineering Co.*
   Annandale, NJ 08801
3. *ExxonMobil Upstream Research Co.*
   Houston, TX 77252

1. INTRODUCTION

The accumulation of economic volumes of petroleum (oil and/or gas) in the subsurface requires that several essential geological elements and processes be present in time and space.[i] *Source rocks* generate and expel petroleum when sufficient thermal energy is imparted to the sedimentary organic matter (kerogen) to break chemical bonds. This heating is induced usually by burial by *overburden rock*. Once expelled, petroleum migrates either along faults and/or highly permeable strata. Accumulations form only when high porosity strata (*reservoir rocks*) are charged with migrating petroleum and the petroleum is prevented from further migration. These petroleum *traps* are formed only when geologic movements result in subsurface topographies (structural and stratigraphic) that block migration and when the reservoir rocks are covered by low permeability strata (*seal rocks*). The mere presence of these geologic elements is insufficient to form petroleum reserves. Traps must be available at the time of oil expulsion and once charged, their integrity must be preserved until exploited. These elements and processes constitute the *Petroleum System* (Figure 1).

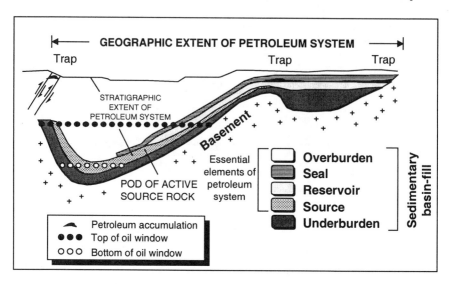

Figure 1. The *Petroleum System.*

Petroleum geochemistry is an essential science in understanding a petroleum system. Source rocks may be directly sampled and analyzed to reveal their source potential and thermal maturity. Extracted bitumens from naturally and artificially matured source rocks are then compared with oils, providing a direct correlation. All too often, rock samples are unavailable for study. This is particularly in frontier and offshore basins, where most wells target specific reservoir structures and fail to penetrate the deeper source formations. Even when samples of potential source formations are obtainable, they may not be representative of the true source facies. The type and generative quality of individual samples within a source unit may vary considerably, reflecting regional and temporal differences in depositional setting (e.g., the influx of organic and inorganic material and the conditions favorable for preservation).

As the characteristics of a source are reflected in the molecular and isotopic composition of the petroleum that they generate, analyses of petroleum can be used to determine what source rocks are present and their level of exposure to thermal stress. Oils are advantageous as they represent an average composition of the source(s) and may be more obtainable in frontier basins, as surface seeps, stains, or well tests. Armed with the knowledge of the type and maturity of a source that is active in the petroleum system, its location within a basin can be inferred through geologic models and seismic reconstruction.

Once a basin is proved to have a viable petroleum system, successful exploration requires defining the migration pathways, the course petroleum followed as it moved from source to reservoir and from reservoir to reservoir. Such pathways may be defined by analyzing petroleum from multiple reservoirs and locations and determining their relationship to each other. Such oil-oil correlations are performed on basin to field scales. Basin-wide studies can lead to new play and prospects along inferred migration pathways; single field studies provide understanding of reservoir compartmentalization needed for proper reservoir development and management.

Economics ultimately determine whether a petroleum accumulation is developed and brought to market. One of the key factors in financial assessment is the quality of the petroleum. Biogenic input and depositional environment determines initial petroleum quality. Marine and evaporate source rocks tend to generate high sulfur, asphaltic crudes while lacustrine (lake) and terrestrially influenced (coals, coastal plains, and deltas) tend to generate low sulfur, paraffinic crudes. Petroleum can be altered greatly by physical, chemical, and biological processes. Increased thermal stress, either during generation in the source or while accumulated in the reservoir, leads to hydrocarbon cracking forming lighter hydrocarbons and solid char. As petroleum migrates, changes in pressure and temperature may result in a single-phase fluid separating into distinct liquid and gas phases that may then migrated independently. Once in the reservoir, petroleum can be altered by microbial activity (biodegradation), chemical reactions (e.g., thermochemical sulfate reduction), water washing, and the addition of light hydrocarbons causing asphaltene precipitation. Each of these alterations leaves a distinct chemical signature and analyses of known petroleum samples are used to predict trends in oil quality in unexplored regions.

The questions imposed by the demands of exploration and production have profound influence on the manner in which petroleum geochemists analyze oil. The interplay of source, thermal maturity, and secondary alteration processes are often revealed by subtle variations in the isomer distributions of trace molecules and the isotopic ratios of individual compounds. Consequently, petroleum geochemists tend to analyze small classes of hydrocarbon compounds in extraordinary detail, while ignoring others classes that are less diagnostic. In contrast, refinery chemists are interested mostly how oil behaves as a feedstock to be processed into marketable products. Their analyses attempt to characterize all major components and tend to group compounds according to their behavior during refinery processes and its refined products.

2. BIOLOGICAL MARKERS IN OILS

Nearly all of the Earth's natural hydrocarbons and all of its economic accumulations of petroleum (oil and/or natural gas) and coals are derived from the thermal alteration of sedimentary organic matter.[ii,iii] On the surface and near sub-surface, autotropic organisms utilize either solar or chemical energy to convert inorganic carbon into organic matter, while other organism live off this organic matter by converting the biomass back into inorganic species. The short-term carbon cycle between the atmosphere, oceans, biota, and soils is not totally efficient and a small portion of this organic matter (~4 x 10^9 g/year) become sequestered as peat and in sedimentary rocks as insoluble kerogen (Figure 2). Over the millenium, the pool of sedimentary organic matter has accumulated to more than 15 x 10^6 Gt of carbon (1Gt = 1 x 10^{15}g). A very small portion of this sedimentary carbon has been converted to coal (~3.5 x 10^3 Gt) and petroleum (4.7 x 10^2 Gt)[iv]. As petroleum is derived from the organic carbon of once living organisms, these resources are correctly termed *fossil fuels*.

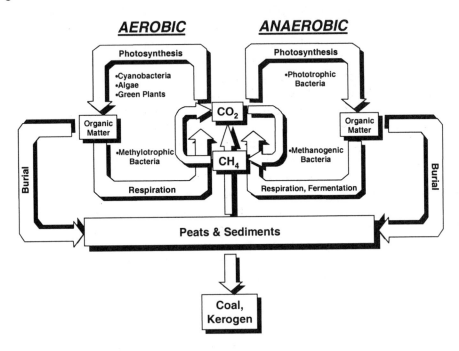

Figure 2. The biogeochemical carbon cycle.

Few biological molecules escape substantial alteration and rearrangement during the conversion from biomass to kerogen to oil. Proteins, sugars, nucleic acids, and most biomolecules are degraded rapidly by microbes and chemical processes. Lipids, waxes, and polymeric material in cell walls and membranes tend to be more resistive and may survive relatively intact through diagenesis. These compounds become incorporated into the kerogen matrix and lose functional groups. When thermally cleaved and expelled in oil, some of these compounds retain enough of their original carbon skeleton to identified their biological precursors. These molecules are termed *biomarker compounds*, or *biomarkers* for short.

The product-precursor relationship is the foundation of biomarker applications in geochemistry.   Figure 3 illustrates biosynthesis of several biological compounds from a common precursor: squalene, a C_{30} isoprenoid. Using reactions that do not require the presence of oxygen, bacterial cyclize squalene into diploptene. This pentacyclic triterpanoid can be further modified to produce wide array of biochemicals. For example, with the addition of D-pentose, diploptene is converted to C_{35} bacteriohopanetetral.   Eukaryotic organisms use oxygen and synthesize tetracyclic steroids instead of triterpanoids. Lanosterol is a precursor to many common sterols, but numerous other modifications to the basic structure occur in nature. Both bacterial hopanols and eukaryotic sterols are used primarily to modify the properties of cellular membranes. Higher land plants are particular adept in the synthesis of unique biochemicals from a squalene precursor. These are used for a wide array of functions ranging from growth hormones to defenses against insects and fungi. All functional groups are lost when these compounds are buried and undergo diagenesis; however the carbon skeleton is preserved and the relationship between the biological precursor and the geological product is evident.

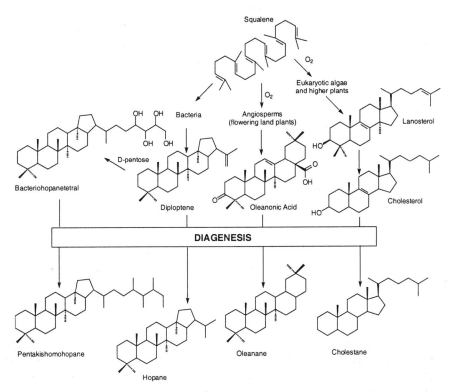

Figure 3. Biosynthesis of numerous biochemicals from squalene and their equivalents found in the geosphere after diagenesis.

Many compounds in petroleum and other fossil fuels are unambiguously biomarkers.[v,vi,vii] The most common and widely studied are acyclic isoprenoids and selected branched alkanes, and cyclic terpanoids and steroids. In immature bitumens and very low maturity oils, unsaturated and functionalized forms may still be present; under thermal conditions most associated with oil generation, defunctionalized saturated and aromatic hydrocarbon forms are present. Biomarkers also may contain heteroatoms. Sulfur may become incorporated into the above mentioned biomarkers during early diagenesis under anoxic, marine conditions, acids are a byproduct of microbial degradation, and porphyrins containing nitrogen and metals are common in low maturity oils.

Other petroliferous compounds may be considered to be biomarkers only under specific conditions. Their relationship to specific biological precursor(s) may be ambiguous and supplemental information, such as its isotopic composition or relative distribution, may be required. Normal alkanes are usually not considered to be biomarkers because they have so many potential precursor species that it is not possible to define a specific biological source. However, *n*-alkanes can be biomarkers if their $\delta^{13}C$ values or distribution that corresponds to specific biota. For example, oil with a normal alkane distribution seen in Figure 4 is derived in part from *Gloeocapsomorpha prisca*, an extinct green algae that flourished during the Ordovician.

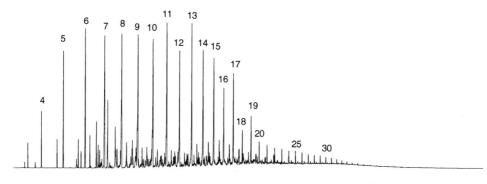

Figure 4. Gas chromatogram (FID) of an Ordovician oil from the Michigan Basin showing an enrichment of odd carbon numbered *n*-alkanes over the range of *n*-C_{11} to *n*-C_{19}. Such distributions are characteristic of G. prisca, an extinct organism believed to be related to *Botyroccoccus*, a modern green microalgae.

Only a few petroleum biomarkers are unique to a specific organism. One such group of compounds includes saturated equivalents of botyrococene and associated polymethysqualenes that are produced by *Botryococcus braunii*, a freshwater, green microalgae[viii]. More prevalent are biomarkers that are characteristic of group of organisms that share common biochemical pathways.

For example, highly branched isprenoids are diagnostic of diatoms, C_{40} biphytanes are only produced by methnogenic archaea, bicadinanes indicate input from Dipterocarpaceae (a family of resinous higher plants), and oleananes are produced by Angiosperms (flowering plants that flourished since the Late Cretaceous).

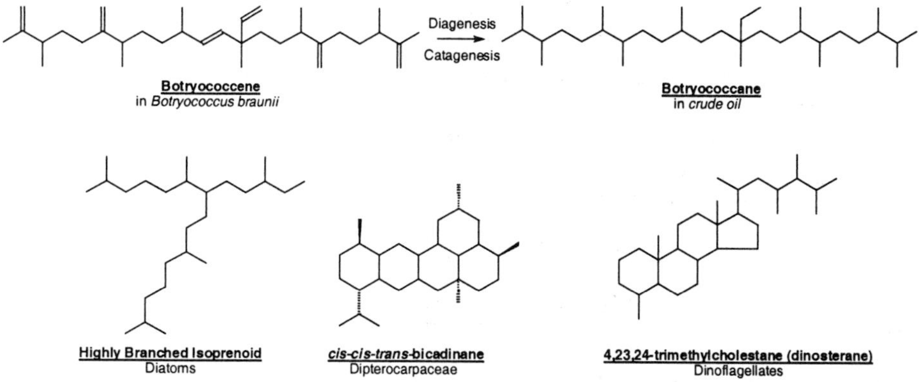

Figure 5. Botryococcane, a species specific biomarker, and several biomarkers indicative of specific families of organisms

Steranes and triterpanes are ubiquitous in most oils and rock extracts. Some of these are diagnostic for specific biota. Dinosteranes are produced by dinoflagellates, *n*-propylcholestane indicates input from chrysophyte marine algae, and gammacerane is thought to be derived from tetrahymanol, a triterpanoid that is produced by anaerobic green-sulfur bacteria and found in ciliates that consume such organisms. Much of the information gleaned from steranes and triterpanes arise not from the occurrence but from their relative distribution. For example, the ratio of hopanes-to-steranes can serve as an indicator of source facies.[ix] Low ratios were found in marine sources, while high ratios were observed for lacustrine and terrigenous sources, shown in Figure 6.

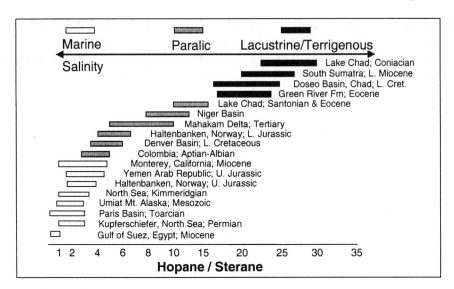

Figure 6. Correlation of source facies with the total hopane/sterane ratio.

Most petroleum biomarkers are produced from biochemicals that follow ancient synthetic pathways. Steranes and hopanes have been identified in Precambrian rocks ~2.7 billion years old.[x] Some biochemical evolved with new biota the arose in the Phanaerozoic and can be used as age-diagnostic biomarkers. For examples, norcholestanes are derived from diatoms and their relative abundance in marine oils increases slowly in the Cretaceous and rapidly in the Tertiary.[xi] Similarly, oleananes are derived from Angiosperm land plants. Although these flowering plants evolved in the early Mesozoic, they become the dominant land plants only in the Late Cretaceous. Oils with high oleanane concentrations are identified as relatively young and derived from source rocks that received a high proportion of land plant debris. Such environments are found in Tertiary deltaic shales (e.g., Niger Delta) and paralic coast plains (e.g., Southeast Asia).

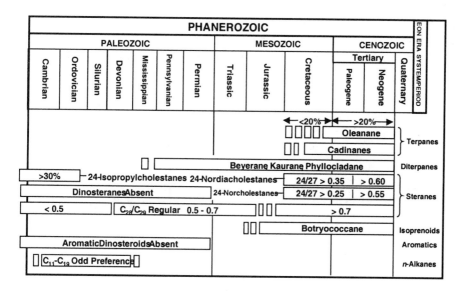

Figure 7. Age-diagnostic biomarkers

Biological compounds in general have particular stereochemical configurations (epimers). Defunctionation and maturation of the biomolecules lead to formation of thermodynamically more stable epimers. Hence, both epimers with biological configurations and new configurations from geotransformation are present in oils and rock extracts. Although the mechanism is unclear, configurational isomerization or stereoisomerization in saturated biomarkers occurs only at asymmetric carbon atoms where one of the four substituents is a hydrogen. Depending on the kinetics of the reaction and the stability of the products, the isomerization produces two configurations: R versus S when the asymmetric center is not part of a ring or α versus ß when part of a ring. Configurational isomerization can occurs only cleavage and renewed formation of the bonds results in an inverted configuration compared to the starting asymmetric center.

For example, bacteriohopanetetral has multiple chiral carbons. Three of these optical centers occur at C-14 and C-21 in the terminal ring and at C-22 in the alkyl-sidechain. Upon diagenesis, the functional groups are lost, but the resulting geolipid retains the chirality of the initial biomolecule (14ß,17ß-22R). This configuration is relatively unstable and isomerizes into 14α,17ß, 14ß,17α and 22S isomers (Figure 8) Epimers with 22S and 22R configurations reach an equilibrium ratio of ~3:2 in mature oils and rock bitumens . Similarly, sterols have optical ring-carbons at C-5, 14, and 17 and a chiral carbon at C-20 in the alkyl sidechain. Most sterols have a double bond at C-5 and hydrogenation results in the formation of 5ß and 5α-steranes. The latter are more stable and are prevalent in geological epimers. Biological sterols have a initial configuration of 14α,17α-20R. These are converted during early diagenesis into 14α,17α-20S, 14β,17β-20R and 14β,17β-20S epimers (Figure 9). Equilibrium values for S and R isomers is approximately 1:1.

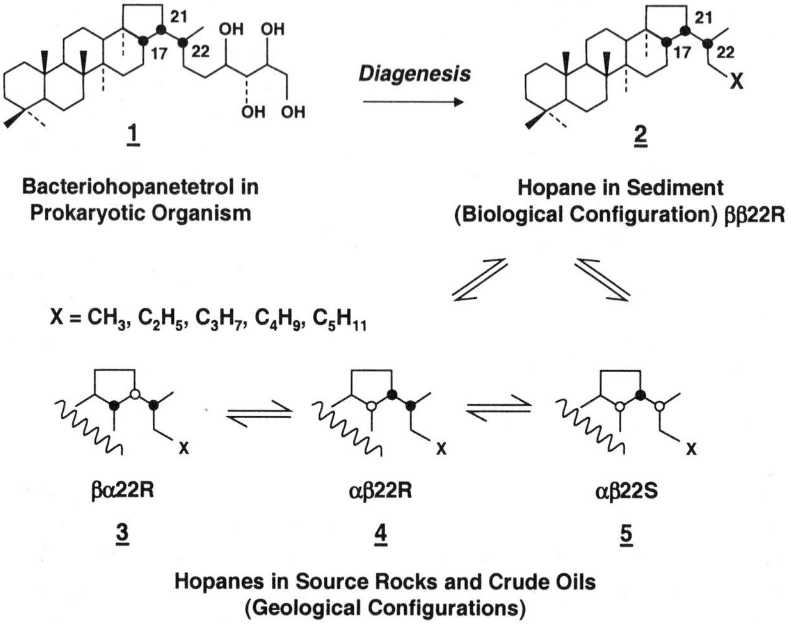

Bacteriohopanetetrol in Prokaryotic Organism   **Hopane in Sediment (Biological Configuration) ββ22R**

X = CH₃, C₂H₅, C₃H₇, C₄H₉, C₅H₁₁

βα22R   αβ22R   αβ22S

3   **4**   **5**

Hopanes in Source Rocks and Crude Oils
(Geological Configurations)

Figure 8. Origin of hopanes in petroleum from bacteriohopanetetrol (1) found in the lipid membranes of prokaryotic organisms. Stereochemistry is indicated by open (α) and solid (ß) dots where the hydrogen is directed into and out of the page, respectively. The biological configuration (17ß,21ß,22R) imposed on bacteriohopanetetrol and its immediate saturated product (2) by enzymes in the living organism is unstable during catagenesis and undergoes isomerization to geological configurations (e.g., 3,4,5). The 17ß,21α-hopanes (e.g., 3) are called moretanes, while all others are hopanes (e.g., 2, 4, 5).

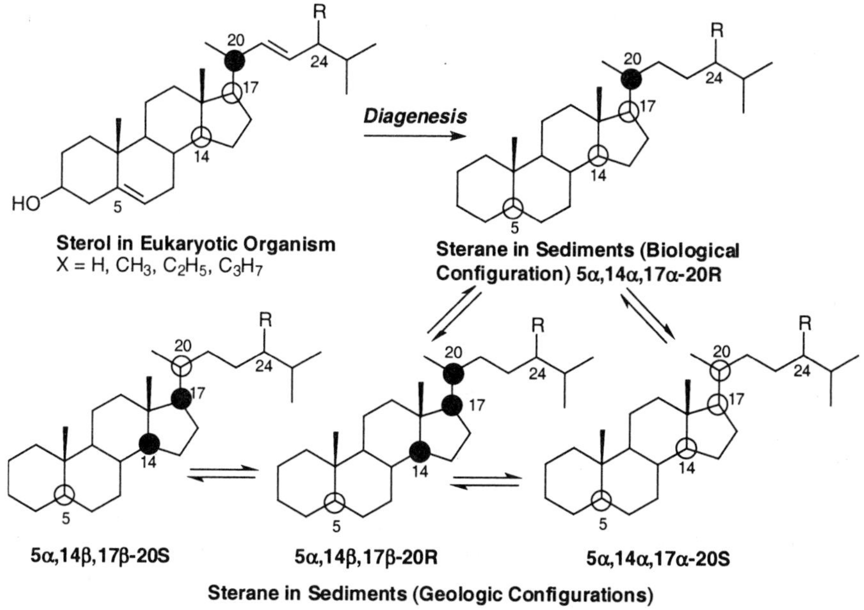

Figure 9. Diagenesis and isomerization of sterols into geological epimers. Stereochemistry is indicated by open (○) and solid (●) dots (hydrogen directed into and out of the page, respectively).

Confirmational isomerization result from rearrangements of the biological carbon skeleton. For example, diasteranes are formed during early diagenesis in depositional environments rich in detrital clays. The acid catalyzed rearrangement of methyl groups from the C-10 and C-13 positions to the C-5 and C-14 positions yielding 13α(H),17β(H) and 13β(H),17α(H) isomers (Figure 12).

The epimerization reactions involving the transformation from biochemical to geochemical configurations can be used as thermal maturity indicators. Other such biomarker indicators involve the selective preservation of more stable over less stable compounds or the conversion of low maturity biomarkers into other forms. As these reactions are considered under kinetic control, their extent of completion can be used to evaluate the maturity of oil or rock bitumen and to model their geothermal history.

3. BIOMARKER ANALYSIS BY GC AND GC-MS

The majority of major components in petroleum and fossil fuels have been extensively converted from their biological origins; thus, it would be very difficult to use them for inference of their biological precursors. Gas chromatography (GC) has been used for biomarker analysis, particularly for *n*-alkanes, branched alkanes and acyclic isoprenoid hydrocarbons, as they are usually abundant in crude oils.[xii] The napthenic biomarkers are present in much lower quantities and are poorly resolved from interfering non-biomarker compounds (Figure 10)

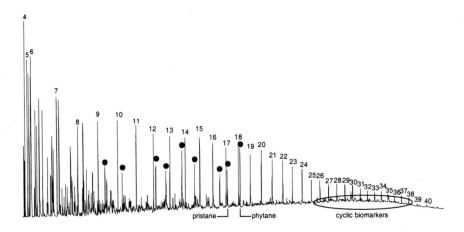

Figure 10. Gas chromatogram of a whole oil from onshore California (Monterey Fm.). Normal alkanes are indicated by their carbon number. Except for the isoprenoid hydrocarbons (●),biomarkers are largely undiscernable.

Despite their low abundance compared to fractions such as the aromatics, individual biomarkers are some of the most abundant structurally defined compounds in petroleum. Individual biomarkers occur in crude oils with typical concentrations ranging from ~1 to 500 ppm. Figure 11 shows the relative abundance of various compound fractions and biomarkers in a biodegraded oil (17°API, 3.2 wt.% sulfur) generated from the Permian Phosphoria Formation and produced from Hamilton Dome, Wyoming. Biomarkers used as examples in the figure (far right) include: C_{29} 5α,14α,17α(H) 24S- and 24R-ethylcholestane, C_{30} 17α,21ß-hopane, C_{29} 5α 20R, 24R- and 24S-monoaromatic steroid, and C_{28} 20S, 24R- and 24S-triaromatic steroid.

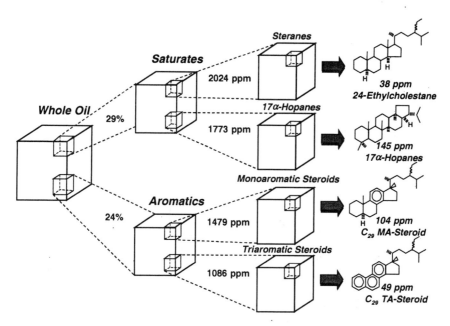

Figure 11. Biomarkers occur in parts per million (ppm) in most petroleums. The example above shows the concentration for selected oil fractions in four individual biomarkers in a biodegraded oil from Hamilton Dome, Wyoming.

Saturated and aromatic biomarkers are analyzed routinely by GC-MS, either on whole oils, or more commonly, on hydrocarbon fractions. The separation of individual isomers, including epimers, using modern capillary columns with non-polar stationary phases is sufficient to resolve many of the biomarkers of interest. A mass spectrometer is used for detection, providing a molecular "fingerprints" that identify the eluting compounds. Electron impact (typically 70 ev) is the most commonly used ioniziation techinique. Under these conditions, most biomarkers yield ion fragments that are characteristic of their class (Figure 12). All hopanes (and most other triterpanes and diterpanes) yield a characteristic ion with a m/z (mass-to-charge) of 191. Steranes yield characteristic ions with m/z of 217 and 218. The former is routinely used as the general diagnostic ion as steranes with the αββ-configuration yield enhanced 218 ions. To enhance the sensitivity, i.e., signal to noise, selected ion monitoring of diagnostic ions are used routinely rather than full scans over the entire mass range of interest.

All types of mass spectrometers have been used for biomarker analysis. By far the most common are benchtop single quadrupole instruments, but magnetic sector, ion trap, and time-of-flight instruments have individual advantages. Regardless of the detector design, the ability to reconstruct selected ion response as a function of time demands computerized data processing. In fact, the rapid development of petroleum geochemistry and its applications to oil potential assessment in petroleum exploration since the 1960's has largely been due to the availability of GC-MS, especially after its computerization since the late 1970's.[12]

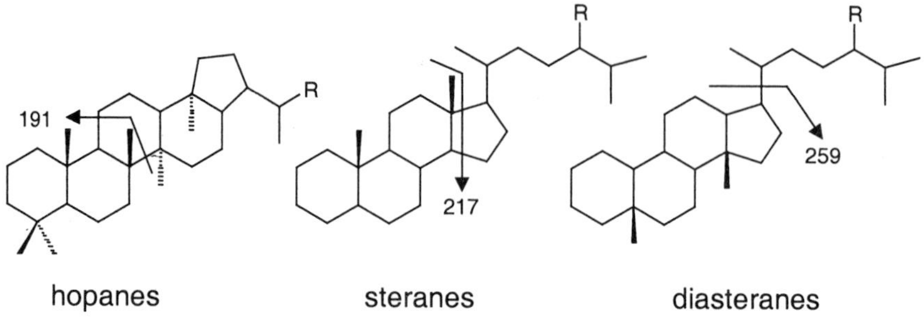

Figure 12. Formation of characteristic fragment ions of m/z 191 for hopanes and m/z 217 for steranes.

Figure 13 shows typical reconstructed ion chromatograms (m/z 191 and 217) for a saturated fraction of an oil from onshore California (a high sulfur crude derived from the Monterey Formation). The distribution of pentacyclic hopanes is typical of the Monterey source facies. The oil is enriched with C_{28} bisnorhopane and C_{35} pentakishomohopane, characteristic of marine deposition under highly reducing conditions. The ratio of two C_{27} hopanes (Ts & Tm) is used an indicator of thermal maturity. The C_{31}-C_{35} homohopanes are doublets, representing the 22S and 22R epimers. These are in equilibrium proportions, as is typically observed in oils.

The tricyclic terpanes and the steranes are derived from algal marine phytoplanton. The steranes are dominated by four major isomers of C_{27}, C_{28}, and C_{29} steranes. The middle two peaks, the 5α,14β,17β-isomers, are more abundant than the flanking 5α,14α,17α-isomers; consistent with deposition under anoxic, saline conditions. When summed by carbon number, the steranes are approximately equal in proportion, typical of a Tertiary marine source with mixed phytoplanktonic input that includes diatoms. The ratios of ααα-20S to ααα-20R isomers is a maturity indicator. In this oil, the ratio is slightly below equilibrium indicating generation early in the oil window. Diasteranes (rearranged) in this oil are low compared to the normal (non-rearranged) steranes. This observation is consistent with deposition in a clay-poor environment (the Monterey Fm. is mostly siliceous chert derived from diatoms) and with low thermal maturity.

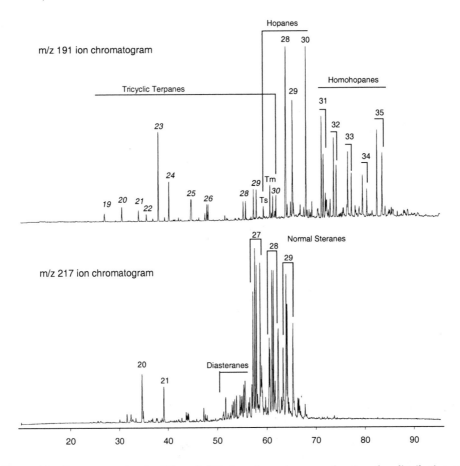

Figure 13. Reconstructed m/z 191 and 217 ion chromatograms showing the distribution of saturated biomarkers in an oil from the Monterey Formation, onshore Califorina. Tricyclic terpanes are indicated by their carbon number in *italic*. The distribution of pentacyclic hopanes (indicated by their carbon number) is typical of an anoxic source facies. Four major isomers of C_{27}, C_{28}, and C_{29} regular steranes are observed in the m/z 217 ion chromatogram.

Conventional GC-MS resolves and identifies most of the major saturated and aromatic biomarkers. The steranes, however, are particularly problematic as there are numerous compounds that co-elute and yield that same diagnostic fragment (Figure 14). The C_{26} steranes and diasteranes are very useful indicators of geological age. These compounds are typically an order of magnitude less abundant than C_{27} diasteranes that co-elute. C_{30} steranes occur as *n*-propylcholestanes, isopropylcholestanes, and numerous methylated variations. The latter biomarkers are characterized by the m/z 231 ion, but also yield an interferring m/z 217 ion fragment The C_{30} steroidal hydrocarbons occur in highly variable amounts, ranging from total absence to concentrations greater than the regular C_{28}-C_{29} steranes which may co-elute. In oils from lacustrine and paralic environments, the steranes may be substantially less abundant then hopanes, bicadinanes, and other hydrocarbons. In such samples the m/z 217 ion trace may be dominated by the response from these non-steroidal hydrocarbons.

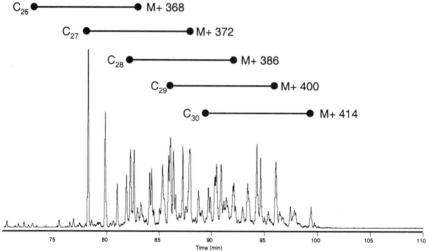

Figure 14. Reconstructed (m/z 217) ion chromatogram showing the elution range of diasteranes and steranes by carbon number.

4.   GC-MS-MS ANALYSIS OF STERANES

While the co-elution of specific sterane isomers can be resolved by optimizing gas chromatographic conditions, current column technology is inadequate for improving most problematic separations. The co-elution problems involving steranes can be largely resolved by using GC coupled with tandem MS (GC-MS-MS) or a magnetic sector MS operated in linked scan modes.

Mass spectrometry-mass spectrometry (MS-MS), coupling two mass analyzers in tandem, can perform daughter scans (product ion scans), parent scans (precursor ion scans) and neutral loss scans for selected ions. When MS-MS is coupled with GC, the additional mass selection provides even higher selectivity than GC-MS. Commonly used MS-MS's include double focusing sector MS [xiii], tandem quadrupole MS [xiv], and ion trap MS[xv].

The use of GC-MS-MS provides isomer specificity and quantitative distributions of biomarkers, which is illustrated in Figure 15. A parent scanning mode is used for resolving overlapping biomarker classes and isomers. In this mode, the second stage MS is used as a filter to allow selected ions characteristic to biomarkers, e.g. m/z (mass-to-charge ratio) 217 ions for steranes, to fly through while the first stage MS is scanned to monitor the molecular ions of biomarkers that produce the selected characteristic ions. By linking the molecular ions and characteristic fragment ions together, only biomarker compounds that yield *both* ions are detected while other co-eluting non-biomarker compounds are filtered out. Thus, GC-MS-MS provides interference-free measurement of biomarker compounds, resulting in baseline resolution of the chromatographic peaks.

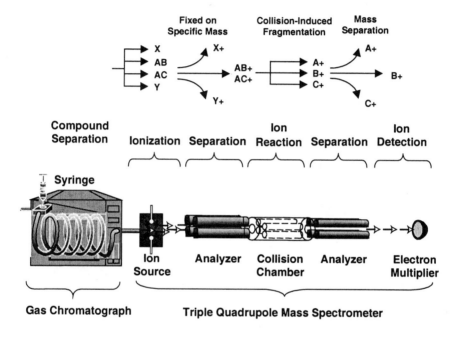

Figure 15. Isomer specificity provided by GC-MS-MS. Co-eluting compounds can be resolved by a first stage mass filter that passes only molecular ions of a specific mass (AB and AC). These are fragmented (by collosion with a gas in second quadrupole). Ions diagnostic for a biomarker are passed through to the detector.

The advantage of GC-MS-MS lies in the ability to distinguish between homologous series of steranes by linking their molecular ions with their characteristic fragment ion at m/z 217. Figure 16 shows GC-MS-MS ion chromatograms for the C_{26}-C_{30} steranes for a mature, clastic-source, oil. The linked parent-daughter ($M^+ \rightarrow 217^+$) traces are scaled to show their relative abundance. Each parent ion is resolved into eight distinct epimers. These are labeled in the C_{27} (372 → 217) trace. Peaks A through D are diasteranes (13β,17α-20S, 13β,17α-20R, 13α,17β-20S, and 13α,17β-20S, respectively). Peaks E through F are regular steranes (14α,17α-20S, 14β,17β-20S, 14β,17β-20R, and 14α,17α-20R, respectively).

The GC-MS m/z 217 ion trace for this same sample is shown in Figure 14. A comparison of the two clearly shows the advantage of using MS-MS over MS dectection. Most, but not all, of the major C_{27}-C_{29} peaks are comparable, but lack baseline resolution. Quantifying the C_{26} and C_{30} steranes is nearly impossible in the GC-MS analysis, but easily accomplished using GC-MS-MS. Another characteristic more apparent in the GC-MS-MS is the partial separation of several of the C_{28} (386 → 217) peaks into doublets. These are 24S and 24R epimers that are not chromatographically resolved in the C_{29} and C_{30} steranes.

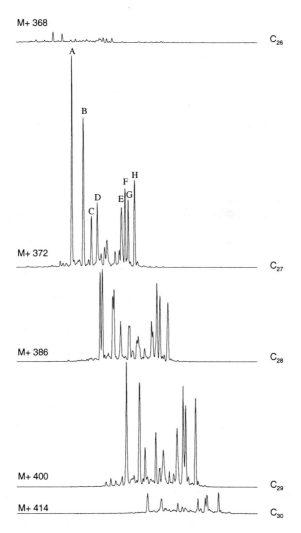

Figure 16. GC-MS-MS ion chromatograms showing the distribution of C_{26}-C_{30} steranes. The GC-MS m/z 217 trace for this same oil is shown in Figure 14.

In the presence of relatively larger amounts of hopanes, the GC-MS measurement for sterane distribution based on the m/z 217 chromatogram can be interfered by hopanes, shown in top trace of Figure 17; yielding misleading thermal maturity assessment of oil and/or source rock extracts. The ratio of C_{29} $5\alpha(H),14\alpha(H),17\alpha(H)$ 20S epimer to the sum of the C_{29} $5\alpha(H),14\alpha(H),17\alpha(H)$ 20S and 20R epimers, i.e., Peak 1/(Peak 1 + Peak 2), is determined by GC-MS as 0.5, with corresponding level of maturity (LOM) of 9. The oil is considered early mature. However, the ratio determined by the GC-MS-MS chromatogram, with no hopane interference, shown as the bottom trace of Figure 16, is 0.3. The corresponding LOM of the oil is 7, indicating the oil is actually immature.

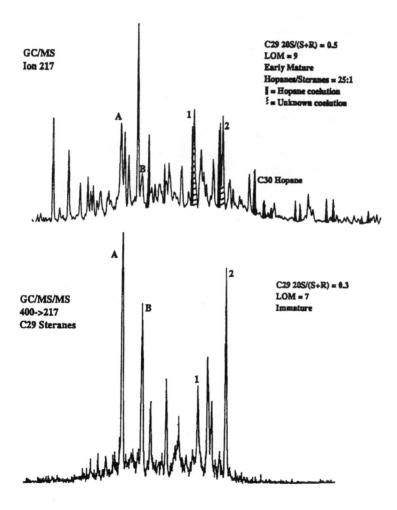

Figure 17. GC-MS-MS resolves interfering biomarkers for thermal maturity assessment.

5.   PRINCIPAL COMPONENT ANALYSIS OF GC-MS AND GC-MS-MS DATA

When analyzing a large volume of data, chemometrics can provide un-biased assessment of the data. Chemometrics utilizes mathematical and statistical methods for handling, interpreting and predicting chemical data and associated physical/chemical properties.[xvi] These statistical methods are now being routinely used in interpreting biomarkers[xvii] and geologic data.[xviii]

To illustrate the technique, we have applied principal component analysis (PCA; or multivariate analysis) to the sterane dataset from GC-MS-MS analysis of 39 Western Canadian Basin source rock extracts. Our goals from this chemometric analysis are to identify key biomarkers that correlate with age, maturity and depositional environment, and to correlate oil/oil and oil/source rock with multiple molecular parameters for migration and accumulation studies. The data matrix is split into score and loading matrices: loadings represent linear combinations of the components in mutually orthogonal vector space (axes), while scores represent the projections of samples on the axes. Figures 18 and 19 show the score and loading plots of first two principal components (PC's) of the data set. PC1 and PC2 represent 56% and 34% of variance, respectively. Careful examination of the component and sample distributions, PC1 appears to correlate with facies changes and PC2 with maturity.

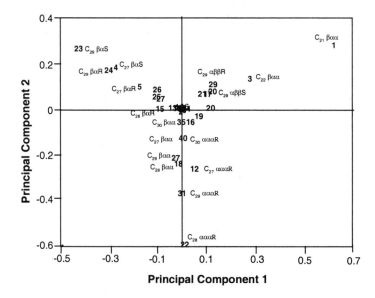

Figure 18. Loading plot of Western Canadian Basin source rock extracts for the first two principal components.

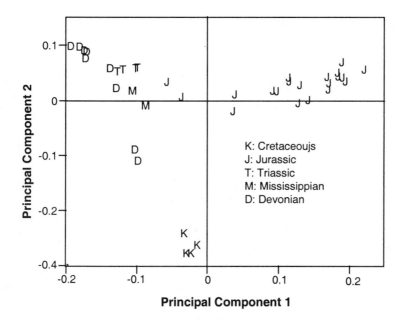

Figure 19. Score plot of Western Canadian Basin source rock extracts for the first two principal components. Rock sources - K: Cretaceous Manville Formation, J: Jurassic Nordegg Formation, T: Triassic Doig Formation, M: Mississippian Exshaw/Bakken Formation, and D: Devonian Duvernay Formation.

6. FUTURE PROSPECTIVES

Biomarker science is a rich field to organic matter input, depositional environments, thermal history, and reservoir alteration of source rocks and oils. With the advancement of analytical technology, many biomarkers have been found and studied. However, there are still many biomarker classes remain to be discovered and studied for the understanding of paleo-transformation of ancient living organisms to current geological products. Hence, abundant opportunities exist for exploring new classes of biomarkers and their geochemical significance.

The development of GC-MS is responsible for the explosive growth of biomarker science. GC-MS-MS further refines biomarker distribution by eliminating interfering components. In MS electron-impact ionization (EI) is the most widely used ionization technique for the characterization of biomarkers. However, under EI conditions some biomarkers, such as aromatized steranes, do not yield molecular ions. Attempts have been made to use chemical ionization for obtaining protonated molecular ions that may enhance the abilities of GC-MS-MS for biomarkers.[xix] The use of other soft ionization techniques (e.g., field ionization, laser desorption) have not yet been investigated.

Improvements can be made in the chromatographic separation of biomarkers. Comprehensive two-dimensional GC can separate co-eluting compounds using two different stationary phases. Preliminary results prove that the technique is possible;[xx] but, advances in 2-dimensional gas chromatography coupled with time-of-flight mass spectrometry (GCxGC/TOFMS) instrumentation and computerized data processes must be made before this technique becomes practical.

Another potential area is the use of coupled liquid chromatography with mass spectrometry or tandem mass spectrometry (LC-MS or LC-MS-MS) for analyzing polar compounds that contain oxygenated and unsaturated functionality. These analytical techniques have been used to determine the functionalized structures of microbial lipids[xxi] and would provide opportunities for discovering intermediates between biological origins and geological products for the understanding of molecular transformation in depositional environment under various geological settings.[xxii]

Perhaps the greatest breakthroughs are emerging in compound-specific isotope analysis. [xxiii,xxiv] This hyphenated-analytical technique couples a gas chromatograph with an isotope ratio mass spectrometer. In extracts of immature source rocks, biomarkers are sufficiently abundant and resolved to obtain its carbon isotope ratio. Isotopic compositions are preserved through diagenesis and reflect both the biological origin and depositional conditions. The $\delta^{13}C$ of individual biomarkers can vary greatly within sediments. For example, biomarkers in the Eocene Messel shale show a broad range in carbon isotopic compositions from -20.9 to $-73.4‰$. These biologically controlled isotopic compositions can be used to identify specific sources for some compounds and help reconstruct the carbon cycling within the lake and sediments that formed the Messel Shale. Systems are now commercially available for continuous flow measurements of carbon, nitrogen, and hydrogen isotopes.

Exploitation of compound specific isotopic analysis (CSIA) techniques have paralleled those made earlier using GC-MS and GC-MS-MS in that the development of a new analytical method rapidly opened new insights and applications (see Chapter 10). CSIA, however, requires fairly high signal-to-noise and baseline resolution for accurate measurements. Improved methods of separation and enrichment are needed to extent CSIA to the full range of biomarkers seen in most mature crude oils and rock extracts.

7. REFERENCES

1. Magoon, L.B.; Beaumont, E.A. "Petroleum Systems". In *Handbook of Petroleum Geology: Exploring for Oil and Gas Traps*, Beumont, E.A.; Foster, H.N. (eds). AAPG 1999, 3-1 to 3-34.Hunt, J. M. *Petroleum Geochemistry and Geology*, W. H. Freeman and Co.: San Francisco, 1979
2. Tissot, B. P.; Welte, D. H. *Petroleum Formation and Occurance*, 2nd ed. Spinger-Verlag: Berlin, 1984
3. Hunt, J. M. *Petroleum Geochemistry and Geology*, 2nd ed. W. H. Freeman and Co.: San Francisco, 1996
4. Falkowski, P.; Scholes, R.J.; Boyle, E.; et al. *Science*, **2000**, 290(5490), 251-259
5. Philp, R. P. *Fossil Fuel Biomarkers: Applications and Spectra*, Elsevier: Amsterdam, 1985.
6. Johns, R.B. (ed.) *Biological markers in the sedimentary record.*, Methods in Geochemistry and Geophysics 24 Elsevier, Amsterdam, Netherlands, 1986.
7. Peters, K.E.; Moldowan J.M. *The Biomarker Guide; Interpreting Molecular Fossils in Petroleum and Ancient Sediments* Prentice Hall, Englewood Cliffs, NJ, United States, 1993
8. Summons, R.E.; Metzger, P.; Largeau, C.; Murray, A.P.;Hope, J.M. *Organic Geochemistry* **2002** 33(2), 99-109.
9. Isaksen, G. H. *Oil & Gas J.* 3/18/91; pp. 130
10. Brocks J. J.; Logan, G.A.; Buick, R.; Summons, R. E. *Science,* **1999** 285, 1033-1036.
11. Holba A.G.; Dzou, L.I.P.; Masterson, W. D.; et al (1998). *Organic Geochemistry,* **1998**, 29, 1269-1283.
12. Hsu, C. S.; Drinkwater, D. "Gas Chromatography-Mass Spectrometry in the Petroleum Industry" In *Current Practice of Gas Chromatography-Mass Spectrometry*, W. M. A. Niessen (Ed.), Marcel Dekker: New York, 2001.
13. Warburten, G. A.; Zumberge, J. E. *Anal. Chem.* **1983**, 55, 123-126.
14. Watson, J. T. *Introduction to Mass Spectrometry*, 3rd ed. Lippincott-Raven: Philadelphia, 1997
15. Karr, D.E.; Walters C.C.Rapid Comm. Mass Spec. **1996**, 10, 1088-1092.
16. Kramer, R. *Chemometric Techniques for Quantitative Analysis*, Marcel Dekker: New York, 1998
17. Christie, O.H.J. *Chemometrics and Intelligent Laboratory System* **1992**,14, 319-329.
18. Peters, K.E., Sarg, J.F., Enrico, R.J., Snedden, J.W. & Sulaeman, A. *AAPG Bulletin* **2000**, **84,** 12-44.
19. Hsu, C. S.; Dechert, G. J.; Schaps, M. E.; Hieshima, G. B. *Proc. 43rd ASMS Conf. on Mass Spectrom. and Allied Top.*, Atlanta, GA, May 21-26, 1995; P. 474
20. Frysinger, G.S.;Gaines, R.B. *Journal of Separation Science* 2001, **24**, 87-96.
21. Talbot, H.M.; Watson, D.F.; Murrell, J.C.; Carter, J.F.; Farrimond, P. *Journal of Chromatography A* **2001**, 921, 175-185.
22. Watson, D.F.; Farrimond, P. *Organic Geochemistry* **2000**, 31, 1247-1252.
23. Hayes J. M.; Takigiku, R.; Ocampo, R.; Callot, H. J.; Albrecht, A. *Nature* **1987**, 329, 48-51.
24. Freeman K. H.; Hayes, J. M.; Trendel, J. M.; Albrecht, P. *Nature* **1990**, 343, 254-256.

Chapter 10

APPLICATIONS OF LIGHT HYDROCARBON MOLECULAR AND ISOTOPIC COMPOSITIONS IN OIL AND GAS EXPLORATION

Clifford C. Walters[1], Gary H. Isaksen[2] and Kenneth E. Peters[2]
1. *ExxonMobil Research & Engineering Company*
   Annandale, New Jersey, 08801-0998
2. *ExxonMobil Upstream Research Company*
   P.O. Box 2189, Houston, Texas, 77252

1.   INTRODUCTION

Light hydrocarbons (C_4-C_{11}) are a significant portion of most crude oils. It is therefore desirable to establish interpretation guidelines from molecular parameters within this boiling range that are diagnostic of source characteristics, thermal maturity, and reservoir alteration processes. This is obviously true for high-maturity or phase-separated liquids that have little or no high-molecular weight biomarkers (see Chapter 9). However, even when $C_{15}+$ biomarkers are present, evaluation of the light hydrocarbons is considered essential as reservoir rock may be charged by multiple sources. Consequently, the origins and geohistories of the light and heavy hydrocarbons may differ.

Although studied for decades, the mechanisms of light hydrocarbon generation are still unresolved. While thermal cracking of oil is widely accepted for forming light hydrocarbons under late-stage catagenesis and metagenesis, the reactions involved under the thermal conditions associated with oil generation are disputed. Recent sediments lack light hydrocarbons, except for trace amounts that may be biogenic or migrated. Systematic changes in the abundance and distribution of light hydrocarbons occur with increasing temperature and depth (Figure 1). These changes have been attributed to various chemical processes including diagenetic carbocationic rearrangements, and catagenic free radical cracking and metal-catalyzed steady-state reactions.[1-5] Regardless of chemical origin, the distribution of light hydrocarbons in petroleum is known to be a function of organic facies and thermal maturity of the source rock and post-generative alteration effects. Migration and reservoir processes included thermal sulfate reduction, gas-liquid and aqueous phase fractionation cracking, thermochemical, and biodegradation.

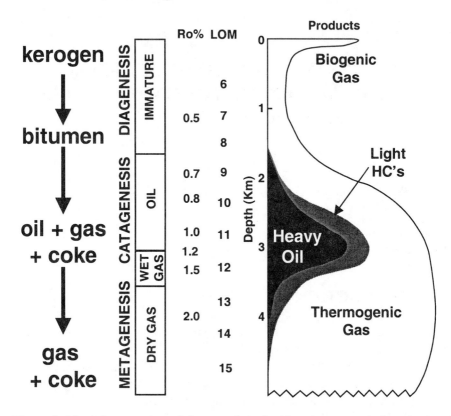

Figure 1. Thermal maturation of kerogen into liquid and gaseous hydrocarbons. Light hydrocarbons are generated at the top of the oil window and become progressive enriched with increasing thermal stress. Typical depth, vitrinite reflectance and level of maturity (LOM) at which these reactions occur are illustrated. Actual values are dependent on the specific geothermal history and the chemical composition of the kerogen.

Light hydrocarbon composition is determined initially by the type of organic matter in the source rock.[6,11] Kerogens derived from algal remains, such as those deposited in marine carbonates and deep-freshwater lake sediments, are enriched in normal alkanes. Kerogens derived from higher land plant debris, such as coals and delatic shales, are enriched in cycloparaffin and aromatic hydrocarbons.

In primary oils, the amounts of light hydrocarbons correlate with thermal maturity. Low-maturity, early-expelled oils may have less than 15 wt.%, typical mid-oil window marine oils have about 25 to 40 wt.%, and high-maturity or phase-separated condensates may be nearly 100 wt.% light hydrocarbons. These changes involve the thermal breakdown of high molecular weight compounds into small hydrocarbons. Thermal alteration, either in the source or reservoir rocks, may change the ratio of specific light hydrocarbon isomers.[4-8] Many of these ratios are empirically determined from observations of wellbore strata or of oils and condensates at varying levels of thermal stress. Others are based on theories of light hydrocarbon generation.[9,10] In general, maturation ratios either compare isomers (or groups of isomers) that either have different thermal stability or express proposed reactant-product relationships.

The abundance and composition of light hydrocarbons may be altered greatly by secondary physical, chemical, and/or microbial processes. Physical processes include gas-liquid phase separation[12], evaporative fractionation[13,14] (a process where an oil undergoes phase separation by the addition of methane to the reservoir), and water-washing or aqueous fractionation[15]. Phase separation processes are particular important in basins with predominantly vertical migration along faults and salt-and mud-diapirs. Thermochemical sulfate reduction (TSR), a chemical process whereby sulfate oxidizes hydrocarbons to produce H_2S, CO_2, and pyrobitumen, can greatly perturb both the molecular and isotopic light hydrocarbon ratios.[16,17] Microorganisms selectively catabolize normal and branched paraffins hydrocarbons; hence, oils that are biodegraded (common in reservoirs <80°C) have altered light hydrocarbon distribution that are depleted in these compounds.[18,19]

Light hydrocarbon composition and isotopic ratios have been used for many years for the correlation of oils and condensates.[1-5, 10,20,21] As with biomarkers, the light hydrocarbon can differentiate genetically related oils (oils derived from the same source facies, though possibly at different levels of thermal maturity). Such information is critical to defining migration pathways within the petroleum system. Because the light hydrocarbons are easily perturbed by secondary processes, they are particularly well suited for high-resolution oil-oil correlation used to distinguish reservoir continuity.[21-24]

2. METHODS OF ANALYSIS

Crude oil light hydrocarbons are characterized almost exclusively by gas chromatography (GC). Most crude oils are olefin-free as these compounds are readily hydrogenated in the subsurface environment. As such, nature crude oils resemble refinery reformates and alkylates. [A few, unusual oils contain light olefins that are formed by radiolysis.[25-26]] Although the resolving power of modern capillary gas chromatographic columns are considerable (>600,000 theoretical plates), they are not capable of resolving all light hydrocarbons that occur in petroleum. Consequently, the GC-methods employed by petroleum geochemists either analyze the entire range of hydrocarbons, compromising on separation, or analyze a small portion to resolve specific isomers.

2.1   Gas Chromatography of Light Hydrocarbons (C$_2$-C$_{9+}$)

ASTM D5134-98 specifies an industry standard procedure for light hydrocarbon analysis in crude oil (Figure 2).   Whole crude samples are introduced via a split injector and the light hydrocarbons are separated by a fused silica capillary column. The column is typically 50 meters or longer, with an internal diameter of <0.21 mm, and a thick-film (0.5μm), cross-linked methyl silicone stationary phase. Temperature and flow conditions are fixed and the retention times of hydrocarbon standards are determined to account for minor differences in column polarity. Even minor differences in the injection method, column dimensions and stationary phase, temperature programming, and flow conditions can change peak resolution or elution order.   Strict adherence to the ASTM method, is required to achieve the specified resolution, accuracy, and precision.

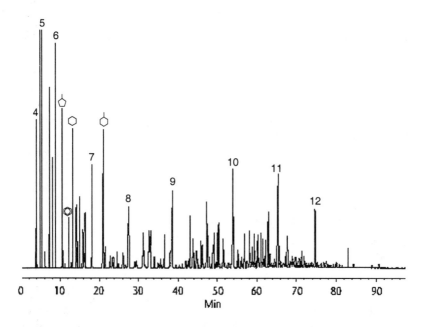

Figure 2. Typical crude oil chromatographic separation of C$_2$-C$_{9+}$ hydrocarbons following ASTM D5134-98. Normal alkanes are identified but carbon number, selected cycloalkanes and benzene are indicated by their structure. Separation is achieved using a 50m x 0.2 mm id, 0.5 μm film thickness capillary column with a100% methylsilicone cross-linked stationary phase.   Oven conditions are 35°C (30 min. initial hold) to 200°C at 2°C/min (10 min. final hold). Helium carrier gas is set at ~20 cm/sec at 35°C. The injector is set at 200°C and run with a 200:1 split. Detection is made with a 250°C FID supplied with ~30 mL/min H$_2$, ~250 mL/min air, and ~30 mL/min N$_2$ make-up gas.

Upstream-oriented geochemists tend not to use industry standards, relying more on proprietary methods. Nevertheless, most geochemical laboratories run a whole-oil gas chromatographic analysis similar, though usually not identical to

ASTM D5134-98. Proprietary procedures vary the method of injection, column selection, pressure and temperature programming, and data processing, usually to improve upon resolution in the light and mid-ranged hydrocarbons or to extend the analysis to higher molecular weights.

The major limitation in ASTM D5134-98 is its low injector and maximum oven temperature that limits crude oil analysis to $<C_{10}$ hydrocarbons. A simple modification to the method, increasing these temperatures from 200°C to 325°C, will allow separation up to $\sim n\text{-}C_{35}$. Using on-column injection, pressure programming, and capillary columns capable of withstanding programmed temperatures up to ~400°C, separations up to $\sim n\text{-}C_{60}$ are possible, while still preserving the separation of the light hydrocarbons (Figure 3). Flame ionization detectors are nearly universal in most methods, though other non-specific (e.g., pulsed helium photoionization), element-selective (e.g.., flame photometric and atomic emission), and specific (e.g, Fourier Transform IR and mass spectrometers) alternate detectors have been applied.

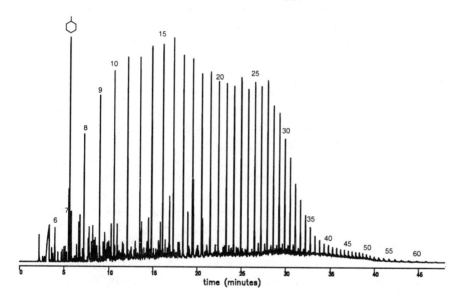

Figure 3. Proprietary chromatographic method for extended whole-oil analysis. Using on-column injection, pressure programming, and a high-temperature capillary column, separations from $n\text{-}C_3$ to $\sim n\text{-}C_{60}$ are possible. The sample is an oil from Far East Asia generated from an Oligocene lacustrine (algal-dominated) shale.

The major disadvantage of extending the chromatographic analysis to higher temperatures is column degradation. Column polarity changes rapidly due to loss of stationary phase, creation of active sites, and retention of non-volatile compounds, preventing the easy maintenance of the rigorous conditions established by ASTM D5134-98.

2.2 C_6-C_7 **Chromatographic Separations**

Most light hydrocarbon molecular parameters utilized by upstream geochemists involve the isomers of the C_6 and C_7 hydrocarbons. There is nothing inherently special about these compounds, except that they are large enough that all hydrocarbon classes (*n*-, iso- and cyclo-paraffins, and aromatic) exist, and all geological isomers can be separated and quantified by gas chromatographic methods. The many C_{8+} hydrocarbon isomers cannot be fully resolved. Distributions of these larger light hydrocarbons are used mostly in correlation studies where the relative ratio of even partially resolved peaks can be compared.[23,24]

C_7 is the highest carbon-number where all non-olefinic hydrocarbon isomers (seventeen) can be fully resolved. Baseline separation of all seventeen C_7 isomers, along with potentially co-eluting C_6 and C_8 hydrocarbons, cannot be achieved using standard gas chromatographic methods and commercially available capillary columns with chemically-bound stationary phases. For example, using a 100% methylsilicone column, 3-ethylpentane is only partially separated from the 1-*trans*-3-dimethylcyclopentane and 1-*cis*-2-dimethylcyclopentane is a shoulder on methylcyclohexane (Figure 4). Resolution of the latter compounds can be improved using conditions optimized for their separation, but at the loss of resolution of other compounds. Standard methods for the analysis of petroleum light hydrocarbons used by the American Society for Testing Materials and the Gas Processors Association (e.g., ASTM D 5134-98 and GPA 2186-95) resolve even fewer of the C_7 isomers.

Separation of all C_7 isomers can be achieved using highly apolar stationary phases, such as squalane or hexadecane-hexadecene. These phases are not available as chemical bonded capillary columns. Consequently they are limited to low temperature use (< 100°C), requiring only the volatile hydrocarbon fraction be introduced onto the column. With care, the gasoline-range fraction can be distilled and collected without alteration of the C_7 hydrocarbons.[4,5] An alternative approach uses a pre-column or temperature-programmed injector so that only light hydrocarbons are introduced onto the temperature-sensitive columns.

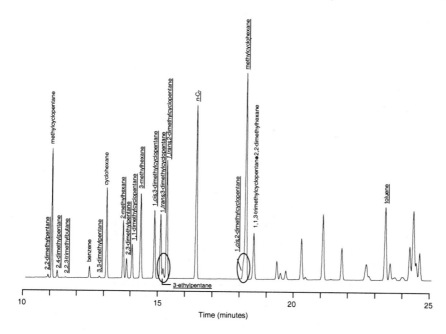

Figure 4. Optimized gas chromatographic separation of C7 hydrocarbons on a single-column gas chromatographic system, a Hewlett Packard 5890 Series II Plus gas chromatograph equipped with a 100-m 100% methylsiloxane capillary column (J & W DB-1, 0.25 mm ID x 0.5 μm film thickness). Hydrogen was used as the carrier gas and set with a linear velocity of ~18 cm/sec. Injector temperature was 270°C and run in split mode (~50:1). Flame ionization detector temperature was 350°C. The optimized program begins at 30°C and immediately ramps at 1.0°C/min to a final temperature of 107°C. The oven is then heated rapidly to 325°C and held for 10 min. to flush heavier hydrocarbons. This method achieves baseline resolution for all C7 isomers, except for 3-ethylpentane (partially separated from 1-*trans*-3-dimethylcyclopentane) and 1-*cis*-2-dimethylcyclopentane (shoulder on methylcyclohexane). C7 hydrocarbons are underlined.

We developed a multi-dimensional, gas chromatographic method for C7 hydrocarbon analysis that achieves baseline resolution of all isomers[27]. The method requires little sample preparation beyond the addition of an internal standard, uses commercially available capillary columns, minimal amounts of cryogenic cooling liquids, and produces rapid and reproducible quantitative results (Figure 5).

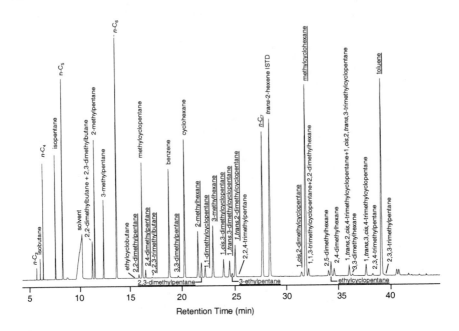

Figure 5. Multi-dimensional gas chromatographic separation of C$_7$ hydrocarbons (underlined) provides baseline resolution. A Siemens Sichromat 2-8 dual-oven GC with dual FID is fitted with a 100-m 100% methylsiloxane (J&W DB-1, 0.25 mm ID x 0.5 μm film thickness) pre-column and a 60-m 50% methyl-50% octylsiloxane (Supelco SPB Octyl 25 mm ID x 0.25 μm film thickness) column. The injector temperature is 250°C. Hydrogen is used as a carrier gas with a linear velocity of ~17 cm/sec in the pre-column and ~43 cm/sec in the analytical column. The temperature program for the pre-column begins with an 18 min. hold at 29°C, ramps at 3°C/min to 53°C, and is held for 15 min. The precolumn is then backflushed and rapidly heated to 230°C to remove heavier hydrocarbons. The temperature program for the analytical column begins with a 35 min hold at 28°C, ramped at 0.5°C/min to 31°C, then quickly heated to 130°C to remove heavier hydrocarbons. FIDs are set at 330°C for both ovens. Oil samples are prepared by dilution in CS$_2$ spiked with *trans*-2-heptene, which serves as an internal standard.

2.3 Compound Specific Isotopic Analysis (CSIA)

Determining the stable carbon isotopic composition of individual hydrocarbons in petroleum used to involve laborious laboratory procedures. Samples were separated into components by preparative gas chromatography, each isolated hydrocarbon was then combusted to CO$_2$, and then analyzed with an mass spectrometer designed specifically for determining the ^{12}C/^{13}C ratio to a high degree of accuracy. Great care was needed to assure that the hydrocarbon was not isotopically fractionated during sample preparation. Such analyses were restricted typically to C$_1$-C$_4$ hydrocarbon gases, as semi-volatile liquids were much more difficult to isolate from oils.

Development of a gas chromatograph coupled directly to an isotope ratio mass spectrometer (GC-IRMS) began in the late 1970's and became commercially available in the late 1980's.[28,29] The hybrid instrument required the complete combustion of individual hydrocarbons as they eluted from a GC column, separation of the product CO_2 from H_2O, and rapid determination of the $^{12}C/^{13}C$ ratio (Figure 6). To achieve complete combustion, eluting hydrocarbons flow through a ceramic capillary tube that contains strands of copper wire previous reacted with pure oxygen.   At temperatures of ~900°C, all hydrocarbons, including methane, can be completely combusted before they exit the furnace. Evolved H_2O is removed from CO_2 using either a cryogenic trap, or more commonly by a Nafion (semi-permeable membrane) filter. Further innovations in electronic instrument control and data acquisition/processing were necessary to allow for the rapid cycling of the mass spectrometer to obtain a profile of the $^{12}C/^{13}C$ ratio across a single eluting peak. The resulting hybrid instrument is frequently abbreviated as GC-IRMS (also GC-C-IRMS); and, the entire analytical procedure is termed *continuous flow* IRMS, or more commonly, compound specific isotopic analysis (CSIA).

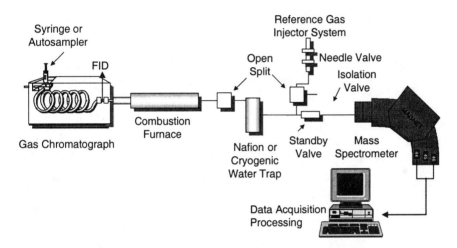

Figure 6. Diagram of a gas chromatograph-isotope ratio mass spectrometer (GC- IRMS) for compound-specific isotope analysis (CSIA).

CSIA was adapted quickly by petroleum geochemists and is now a routine technique. The stable carbon isotopic compositions of precursor biomolecules (biomarkers) are preserved during diagenesis and can be used to identify specific sources of individual hydrocarbons.[29] From such information, sedimentary carbon cycles can be reconstructed. CSIA is a valuable method in oil-oil and oil-source rock correlation. For example, the $\delta^{13}C$ values of individual *n*-alkanes easily differentiates bitumens from immature source rocks from in Brazil basins (Figure 7).[30]

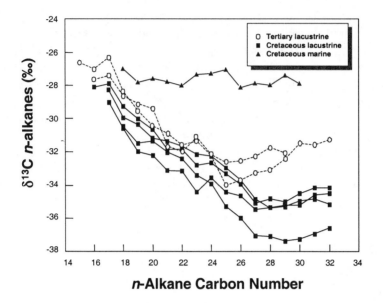

Figure 7. Compound-specific isotope analyses of *n*-alkanes differentiate extracts from different organic-rich source rocks in Brazil.[30] Similar analyses can be used to differentiate crude oils derived from these and other source rocks.

CSIA has proved particularly well suited for light hydrocarbons. Baseline chromatographic separation, coupled with little or no column bleed, allows measurements with high precise and accuracy of C_1-C_7 hydrocarbons (typically < ±0.2‰). As wide variations in $\delta^{13}C$ values may occur between individual light hydrocarbons, CSIA provides ideal criteria for oil-oil and oil-source correlation and well as monitoring a variety of alteration processes.[31-33]

A recent development in GC-IRMS is the determination of hydrogen isotope abundance of individual compounds (H/D or δD).[34,35] The combustion furnace is replaced by a pyrolysis unit that heats the eluting compounds to ~1400°C, converting hydrocarbons to atomic carbon and hydrogen that then recombines to molecular hydrogen. Preliminary compound specific hydrogen isotopic analyses of hydrocarbons in petroleum and source rocks suggest interesting future applications.[36-38]

3. APPLICATIONS OF LIGHT HYDROCARBONS TO PETROLEUM SYSTEMS ANALYSIS

The goal of this paper is not to be a comprehensive review of uses of light hydrocarbons in petroleum geochemistry, but to illustrate a few applications. The selected examples highlight the use of light hydrocarbons in determining thermal maturity, oil-condensate correlation, and the reservoir alteration process of thermochemical sulfate reduction. As noted above, other applications include determination of source characteristics (organic facies and depositional setting), temperature of oil generation and expulsion, and extent of biodegradation and water-washing. Such studies are critical to defining source rocks, geothermal history, migration pathways, and alteration processes – knowledge that is needed for successful development of exploration plays and prospects, for accurate pre-drill prediction of oil and gas quality, and is becoming increasingly important in reservoir development and production.

3.1   Thermal Maturity

Thermal maturation, either within the source rock during generation or within the reservoir rocks, changes the molecular and isotopic composition of crude oils. A variety of compound ratios have been proposed to quantify these differences.[4-8] One of approach was developed by Keith Thompson while at ARCO.[6] He proposed several ratios that empirically were found to be sensitive to changes in organic matter type, thermal maturity, and biodegradation. The two most widely used parameters are termed the Heptane (H) and Isoheptane (I) ratios. These parameters are defined as:

$$\text{Heptane Ratio} = \frac{100*(n\text{-heptane})}{\Sigma(\text{cyclohexane} + C_7 \text{ hydrocarbons})}$$

and

$$\text{Isoheptane Ratio} \quad \frac{(2\text{-methylhexane} + 3\text{-methylhexane})}{\Sigma \text{ (dimethylcyclopentanes)}}$$

The Heptane ratio was defined originally as the percent *n*-heptane relative to the sum of [cyclochexane + 2-methylhexane + 1,1-DMCP + 3-methylhexane + 1-cis-3-DMCP + 1-trans-3-DMCP + 1-trans-2-DMCP + *n*-heptane + methylcyclohexane], where DMCP = dimethylcyclopentane. The gas chromatographic separations then in use failed to resolve all isomers. The denominator of the Heptane ratio, as calculated in this early work, therefore also contains 2,3-dimethylpentane, 3-ethylpentane, and 1-cis-2-DMCP.

By examining the light hydrocarbons extracted from cuttings and outcrop samples of known maturity, Thompson showed that both the Heptane and Isoheptane ratios increase with increasing thermal stress. The relative rates of change differ from each other and are dependent on the type of kerogen. Thompson proposed that two general maturation curves could account for most of the sample variance. He termed these trend lines "aliphatic" and "aromatic", which roughly correspond to input from algal marine and terrestrial higher land plant biota, respectively. For oils following the "aliphatic" curve, those with Heptane ratios between 18 to 22 were termed *normal*, between 22 and 30 were named *mature*, while those over 30 were called *supermature*.

We have conducted a study to calibrate Thompson's C$_7$ ratios against rock-based thermal maturity parameters. Shown in Figure 7 is a set of samples of the

Kimmeridge Clay Formation, the major source formation of the North Sea region. These rock samples were collected at varying degrees of thermal maturity from wells in the Central North Sea (Table 1). The principle zone of oil formation in the basin occurs between 3500-4800 m. Data-points for the three highest maturity levels (%Ro_{eqv}.1.1 - 2.0; LOM 11-13.5) are from Thompson[6] based on graptolite, chitinozoan, scolecodont and pyrobitumen reflectance curves developed by Bertrand and Heroux [39]. This calibration proved to be in very good agreement with the curve calculated by Thompson (Ref. 6, p. 314) for Type II kerogens.

Table 1. Geochemical screening data for the rock samples used in the calibration of Heptane and Isoheptane ratios to thermal maturity. Samples are from wells in the Central North Sea, UK sector, Quadrants 22 and 29.

No	Well	Depth	%TOC	Rock-Eval				°C	%R_o	LOM	C_7 ratios	
				T_{max}	S_1	S_2	HI				H	I
1	22/14-3RE	3771	6.5	438	0.26	29.2	450	149	0.65	8.5-9	11.5	0.78
2	22/14-3RE	3770	2.5	440	0.21	6.26	250	149	0.65	9	13.5	0.57
3	22/14-3RE	3769	2.0	441	0.32	4.27	213	149	0.65	9	14.8	0.72
4	29/3B-4	4507	5.3	448	0.53	4.15	78	177	0.9	10.5	22.1	1.52
5	29/3B-4	4515	4.2	453	1.14	4.22	100	178	1.0	11	25.6	2.31
6	29/3B-4	4520	3.9	453	1.61	3.46	89	179	1.1	11	27	2.49

Depths are in meters sub-mud.   %TOC = %total organic carbon
S_1, S_2 = mg hydrocarbons/g Rock, T_{max} in °C, HI = Hydrogen Index (mg hydrocarbons/g TOC
Temperatures and vitrinite reflectance values (%R_o) are derived from regional down-hole trends.
LOM = Level of Organic Maturity
H= Heptane Ratio = 100*n-C_7/Σ(cyclohexane through methylcyclohexane)
I= Isoheptane Ratio = (2-methylhexane + 3 methylhexane)/ Σ (dimethylcyclopentanes)

C_7 ratios measured for oils, volatile oils, and condensates from the UK sector of the Central North Sea also are plotted in Figure 8. These petroleum fluids are produced from Jurassic and Triassic sands within the so-called 'high-pressure and high-temperature (HPHT) area of UK quadrants 22, 23, 29, and 30. By definition, the HPHT areas in the CNS have pressure gradients greater than 0.8 psi/ft and temperatures greater than 150°C. This region includes the deep, southern portion of the Forties-Montrose High, and portions of both the Eastern and Western Graben areas.

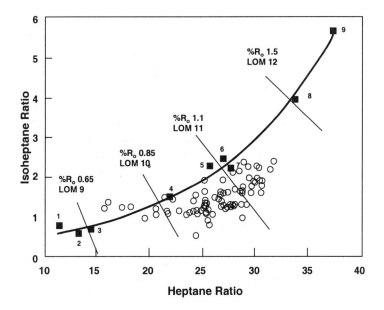

Figure 8. Gasoline-range hydrocarbons from rock extracts (■) . Calibrations have been made to Rock Eval T_{max}, vitrinite reflectance, and down-hole temperatures. Sample numbers 1 through 6 are identified by Table 1. Samples 7 through 9 are from Thompson's 1983 publication.[6] Values are plotted for liquid petroleum samples from the Central North Sea (○).

The primary source rocks of these oils are the Upper Jurassic Kimmeridge Clay (predominately algal marine) and underlying Heather Shales (mixed marine, algal and herbaceous organic matter).[40] The Middle Jurassic coals and coaly shales of the Pentland Formation are less likely to have contributed to the pools of liquid hydrocarbons as seismic facies and well-log analyses suggest that the coals have only a very limited areal distribution in this part of the CNS. These coals and coaly shales have, however, contributed to certain hydrocarbon accumulations in the Western Graben (e.g. block 29/7).

The overlay of the oil/condensate data onto the rock-based calibration curve suggests that only a few oils have maturities in the LOM 9.5 to 10 (%Ro_{eqv} 0.75-0.85), with the bulk of the oils, not surprisingly, at LOM's between 10 (%Ro_{eqv} 0.85) and 11 (%Ro_{eqv} 1.1). Volatile oils and condensates from the HPHT area grade upwards to LOM's of 11.5 (%Ro_{eqv} 1.2). The oils and condensates with Heptane ratios between 22 and 30, by far the largest population within the study area, are classified by Thompsons as *mature,* implying that the oil is '*undergoing continued heating with considerable thermal transformation (ring-opening and chain shortening)*'. The volatile oils and condensates above LOM 11 (%Ro_{eqv} 1.1) are undergoing more extensive in-reservoir thermal cracking, which has the effect of increasing both the Heptane and Isoheptane ratios as n-C_7 and iso-alkanes are preferentially generated. Examples of such *super-mature* oils/condensates are those of the Heron, Egret, and Puffin fields with reservoir temperatures between 163-177°C. In-reservoir thermal cracking at the Elgin and Franklin fields is evidenced by the formation of pyrobitumen in the pore volumes.

Compared to the rock data, many of the oils lie to the right and below the calibration curve. The most likely explanation is that the source facies for these oils are much more aliphatic in nature than the calibration samples. Oils and rock extracts from carbonate and algal-rich facies yield enhanced Heptane ratios, and the Kimmeridge Clay Formation is known to have highly varying organic-rich source facies.[41] As the algal-facies tend to be more oil-prone and high higher generative potential, they are likely to have a greater contribution of light hydrocarbons to the oil accumulations.

3.2   Oil-Condensate Correlations

Several different approaches have been published using light hydrocarbons. Most rely on comparing the ratios of close-eluting compounds, rather than on absolute abundance as the latter values are easily perturbed by numerous secondary processes (e.g., sample evaporation and fractionation at the wellhead separator). Comparison of the $\delta^{13}C$ values of individual hydrocarbons is particularly useful in correlation studies.

One of the more established approaches was proposed by Halpern based on five C_7 ratios.[42] These five correlation parameters are the ratios of individual C_7 alkylated pentanes to the sum of these compounds. Collectively, these five ratios are plotted using polar coordinates on what Halpern terms the C_7-OCSD (C_7 oil correlation star diagram). As these light hydrocarbons have approximately the same solubility in water, susceptibility to microbial alteration, and thermal stability, subtle differences in the distribution of these compounds should mostly reflect their source. Migrational and phase behavior effects may come into play, as there are small differences in volatility between these compounds.

We used the Halpern method in a study of DST samples from the Sable Island E-48 well from the Scotian Shelf, offshore Eastern Canada (Figure 9). Fifteen individual zones were tested between depths of 1460 and 2285 m. Using the Halpern correlation parameters, most of the DST samples cluster tightly within a narrow band. This group includes both degraded and non-degraded fluids. DST 1 and 9 define a second petroleum group. Numerically, the ratios are not very different, but the variance is evident when plotted on the star diagram. DST 1 and 9 correlate very well even though DST 1 lost appreciable volatile hydrocarbons during sampling and/or storage.

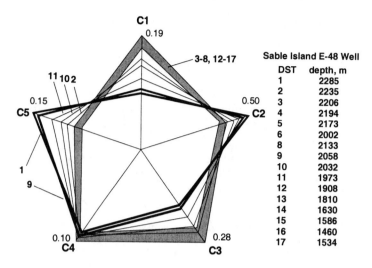

Figure 9. Halpern C_7 oil correlation star diagram for DST samples from the Sable Island E-48 well, Scotian Shelf, Canada. The numbers at the end of each axis are endpoint values for individual ratios. Two end-member oil groups are apparent. One group includes DST from depths of 1460-1908 m and 2133-2206 m (DST 12-17, 3-8). The other end member consists of DST 1 and 9. DST 2, 10, and 11 are mixtures of the two end member fluids. C1 = 2,2-DMP/P$_□$, C2 = 2,3-DMP/P$_□$, C3 = 2,4-DMP/P$_□$, C4 = 3,3-DMP/P$_□$,,C2 = 3-ethylpentane/P$_□$, where DMP=dimethylpentane and P3= □DMP's+3-ethylpentane.

DST 2, 10, and 11 appear to be mixtures of the two end-member fluid groups based on the correlation parameters. By taking the averaged, normalized values for each Halpern correlation ratio, we can estimate the percent contribution of each end-member to the mixed samples by minimizing the combined error. Following this procedure, DST samples with intermediate compositions can be expressed as: DST 11 = 67%, DST 10 = 38%, DST 2 = 23% of the end-member composition as defined by DST 1 and 9.

3.3   Thermochemical Sulfate Reduction (TSR)

Sulfate may oxidize hydrocarbon accumulation, forming H_2S, CO_2, and pryrobitumen. The exact reaction mechanisms for this process is not fully understood. Its impact on oil and gas quality can be substantial and the ability to predict its occurrence and extent prior to drilling is a critical element in successful risk assessment. The molecular and isotopic distributions of the light hydrocarbons are particularly diagnostic of the occurrence of TSR, even under incipient conditions.

CSIA of the hydrocarbons is a sensitive method to detect TSR in condensates.[33,43]. The change in $\delta^{13}C$ due to TSR appears to correlate with both molecular structure and reservoir temperature. Thermal cracking imparts up to a ~2-3‰ shift towards heavier $\delta^{13}C$ values (less negative) for individual hydrocarbons. Rooney [43] showed substantially greater isotopic shifts occur in the $\delta^{13}C$ of some light hydrocarbons in TSR-affected oils. The *n*-alkane and branched hydrocarbons in TSR-affected oils increase up to 22‰, whereas monoaromatics, such as toluene, show much smaller shifts in the range 3-6‰ (Figure 10). TSR may accelerate the destruction of some hydrocarbons compared to thermal cracking, with the remaining hydrocarbons becoming enriched in ^{13}C due to higher fractional conversion for each compound. This was supported by much lower concentrations of branched and normal alkanes with increasing reservoir temperature in TSR-affected oils relative to unaltered oils. Similar isotopic shifts, though not of the same order of magnitude, have been reported in TSR-altered oils from the Aquitaine Basin. [44]

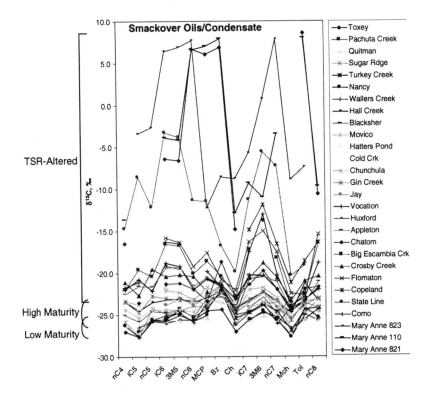

Figure 10. $\delta^{13}C$ of individual light hydrocarbons from Smackover oils and condensates.

Compositional changes in the proportion of light hydrocarbons also reflect the extent of alteration by TSR. One relationship that can be perturbed by TSR is the invariance isoheptanes for oils derived from the same source. [45] Mango showed that the sum of 2-methylhexane + 2,3-dimethylpentane co-varies with the sum of 3-methylhexane + 2,4-dimethylpentane (Figure 11). Although the cause of this invariance is debated, the relationship itself has been verified by us and others. [20] When oils and condensates derived from the Smackover Formation are plotted, non-altered samples that span a broad range of thermal maturity define a trend-line with a slope of 1.21. Oils that are altered by TSR deviate from the non-altered invariance, with the slope increasing with extent of TSR. Similar results have been reported for petroleums from the Western Canada Sedimentary Basin.[17]

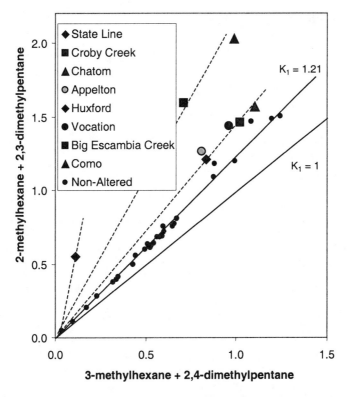

Figure 11. Mango showed oils from the same source exhibit invariance in specific isoheptane ratios. Non-altered oils from the Smackover Formation deviate from the ideal K1=1 trend line, and exhibit a slope of 1.21. Increasing deviation of the slope from the non-altered oils reflecting increasing alteration by TSR.

4.   FUTURE DIRECTIONS

To fully understand a petroleum system, it is necessary to characterize the complete molecular and isotopic composition of crude oil and to understand the factors and processes that influence this composition. There are considerable opportunities to increase our knowledge with a more complete analysis of the light and mid-range hydrocarbons. The use of comprehensive two-dimensional GC (GC x GC) techniques offers the promise of vastly improved separation.[46] Trace compounds, such as the adamantanes, have the potential to extent thermal maturation indicators well beyond the range where biomarkers are present.[47] Development is underway to expand compound specific analysis to include oxygen and sulfur isotopes. Petroleum geochemists are posed and anxious to take advantage of advances in petroleum analyses to address both fundamental questions concerning the origin and reactions of light hydrocarbons in the subsurface and to provide answers to practical problems facing upstream activities.

5.   ACKNOWLEDGEMENTS

Thanks are extended to Esso Exploration U.K., Shell Exploration, U.K., Exxon Exploration Company, and Exxon USA: Houston Production Office for permission to release the data herein.

6.   REFERENCES

1.   Hunt, J. M. *Petroleum Geochemistry and Geology*, 2nd ed., Freeman: San Francisco, 1995.
2.   Philippi, G. T. "On the depth, time and mechanism of petroleum generation", *Geochim. Cosmochim. Acta* **1965**, 29, 1021-1049.
3.   Hunt, J.M. "Generation and Migration of Light Hydrocarbons", *Science* **1984**, 226, 1265-1270.
4.   Mango, F.C. "The origin of light hydrocarbons in petroleum: A kinetic test of the steady-state catalytic hypothesis", *Geochim. Cosmochim. Acta* **1990,** 54, 1315-1323.
5.   Mango, F.C. "The origin of light hydrocarbons in petroleum: Ring preference in the closure of carbocylic rings", *Geochim. Cosmochim. Acta* **1994**, 58(2), 895-901.
6.   Thompson, K.F.M. "Classification and thermal history of petroleum based on light hydrocarbons", *Geochim. Cosmochim. Acta* **1983**, 47, 303-316.
7.   Hunt, J.M.; Miller, R.J.; Whelan, J.K. "Formation of C_4-C_7 hydrocarbons from bacterial degradation of naturally occurring terpenoids", *Nature* **1980**, 288, 577-578.
8.   Schaefer, R.G.; Littke, R. "Maturity-related compositional changes in the low-molecular-weight hydrocarbon fraction of Toarcian shales", *Org. Geochem.* **1988**, 13, 887-892.
9.   Mango, F.D. "The stability of hydrocarbons under the time-temperature conditions of petroleum genesis", *Nature* **1991**, 352, 146-148.
10. Mango, F.D. "The light hydrocarbons in petroleum: a critical review". *Org. Geochem.* **1997**, 26 41-44.
11. Barwise, A. J. G.; Goodwin, N. S. "Classification of petroleum using light hydrocarbons", Programs and Abstracts (Geochemistry Newsletter), Division of Geochemistry, American Chemical Society, 210th ACS National Meeting, Chicago, August, 1995.
12. Silverman S.R. "Migration and segregation of oil and gas" In: *Fluids in Subsurface Environments* (A. Young and G. E. Galley, eds.) American Association of Petroleum Geologists Memoir, **1965** Vol. 4, 53-65.
13. Thompson, K.F.M. "Fractionated aromatic petroleums and the generation of gas-condensates", *Org. Geochem..* **1987**, 11, 573-590.
14. Thompson, K. F. M. "Contrasting characteristics attributed to migration observed in petroleums reservoired" in *Petroleum Migration: Clastic and Carbonate Sequences in the Gulf of Mexico Region*, England, W. A.; Fleet, A. J. (Eds.), Geological Society Special Publication No. 59, 1991; pp. 191-205.
15. Palmer, S. E. "Effect of water washing on C_{15+} hydrocarbon fraction of crude oils from north-west Palawan, Philippines", *American Association of Petroleum Geologists Bulletin* **1984**, 68, 137-149.

16. Manzano, B.K.; Machel, H.G.; Fowler, M.G. "The influence of thermochemical sulphate reduction on hydrocarbon composition in Nisku reservoirs, Brazeau River Area, Alberta, Canada." *Org. Geochem.* **1997,** 27, 507-521.

17. Peters, K.E.; Fowler, M.G. Applications of petroleum geochemistry to exploration and reservoir management. *Org. Geochem..* **2002,** 33, 5-36.

18. Connan, J. "Biodegradation of crude oils in reservoirs", in *Advances in Petroleum Geochemistry volume 1,* Welte, D. H.; Brooks, J. M. (Eds.), Academic Press: New York, 1984; pp. 299-335.

19. James, A. T.; Burns, B. J. "Microbial alteration of subsurface natural gas accumulations", *American Association of Petroleum Geologists Bulletin* **1984,** 68, 957-960.

20. ten Haven, H.L. "Applications and limitations of Mango's light hydrocarbon parameters in petroleum correlation studies" *Org. Geochem.* **1996,** 24, 957-976

21. Halpern, H.I. "Development and applications of light-hydrocarbon-based star diagrams", *American Association of Petroleum Geologists Bulletin* **1995,** 79, 801-815.

22. Kaufman, R. L.; Ahmed, A. S.; Elsinger, R. J. "Gas chromatography as a development and production tool for fingerprinting oils from individual reservoirs: Applications in the Gulf of Mexico" in *Gulf Coast oils and gases: Their characteristics, origin, distribution, and exploration and production significance. Proceedings of the 9th Annual research Conference GCSSEPM,* Schumacher, D.; Perkins, B. F. (Eds.), Society of Economic Paleontologists and Minerologists Foundation, October, 1990; pp 263-282.

23. Beeunas, M.A.; Hudson, T.A.; Valley, J.A.; Clark, W.Y.; Baskin, D.K. "Reservoir continuity and architecture of the Genesis Field, Gulf of Mexico (Green Canyon 205): an integration of fluid geochemistry within the geologic and engineering framework". In: *Transactions of the 49th Annual Gulf Coast Association of Geological Societies & 46th Annual SEPM Gulf Coast Section* (Layfayette, LA, 9/15-17/1999) **1999,** 49, 90-95.

24. Nederlof, P.J.R.; Gijsen, M.A.; Doyle, M.A. "Application of reservoir geochemistry to field appraisal" In: *Geo '94; the Middle East petroleum geosciences; selected Middle East papers from the Middle East Geoscience Conference* (Ed. by M. I. Al Husseini), 1994 709-722. Gulf PetroLink, Manama, Bahrain.

25. Frolov, E.B.; Smirnov, M.B. "Unsaturated hydrocarbons in crude oils", *Org. Geochem.* **1994,** 21, 189-208.

26. Frolov, E.B.; Melikhov, V.A.; Smirnov, M.B. "Radiolytic nature of n-alkene/n-alkane distributions in Russian Precambrian and Palaeozoic oils" *Org. Geochem..***1996,** 24, 1061-1064.

27. Walters C.C.; Hellyer, C.L. "Multi-dimensional gas chromatographic separation of C_7 hydrocarbons" *Org. Geochem..* **1998,** 29, 1033-1041.

28. Matthews D. E.; Hayes, J. M. "Isotope-ratio monitoring gas chromatgraphy-mass spectrometry" *Anal. Chem.* **1978,** 50, 1465-1473.

29. Hayes J.M.; Freeman, K.H.; Popp, B.N.; Hoham, C.H. "Compound-specific isotopic analyses: A novel tool for reconstruction of ancient biogeochemical processes", *Org. Geochem.* **1990,** 16, 1115-1128.

30. Guthrie J.M.; Trindade, L.A.F.; Eckardt, C. B.; Takaki, T. "Molecular and carbon isotopic analysis of specific biological markers: Evidence for distinguishing between marine and lacustrine depositional environments in sedimentary basins of Brazil". American Association of Petroleum Geologists Annual Convention, San Diego, 1996. Abstract, p. A58.

31. Bjorøy M.; Hall, P.B.; Moe, R.P. "Variation in the isotopic composition of single components in the C_4-C_{20} fraction of oils and condensates", *Org. Geochem.* **1994,** 21, 761-776.

32. Chung H.M.; Walters, C.C.; Buck, S.; Bingham, G. "Mixed signals of the source and thermal maturity for petroleum accumulations from light hydrocarbons: An example of the Beryl field" *Org. Geochem.* **1998,** 29, 381-396.

33. Whiticar M.J.; Snowdon, L.R. "Geochemical characterization of selected Western Canada oils by C_5-C_8 Compound Specific Isotope Correlation (CSIC)" *Org. Geochem.***1999,** 30, 1127-1161.

34. Sessions A.L.; Burgoyne, T.W.; Schimmelmann, A.; and Hayes, J.M. "Fractionation of hydrogen isotopes in lipid biosynthesis", *Org. Geochem.* **1999**, 30, 1193-1200.

35. Scrimgeour C.M.; Begley, I.S.; Thomason, M.L. "Measurements of deuterium incorporation into fatty acids by gas chromatography/isotope ratio mass spectrometry" *Rapid Commun. Mass Spectrom.* **1999**, 13 271-274.

36. Schimmelmann A.; Lewan, M. D.; Wintsch, R. P. "D/H isotope ratios of kerogen, bitumen, oil and water in hydrous pyrolysis of source rocks containing kerogen types I, II, IIS, and III", *Geochim.Cosmochim. Acta* **1999**, 63, 3751-3766.

37. Andersen, N.; Schoell, M.; Carlson, R.M.K.; Schaeffer, P.; Albrecht, P.; Bernasconi, S.M.; Paul, H.A.; Klaas, C.; Hunkeler, D. "Compound-specific hydrogen isotope analysis as a novel tool in petroleum geology: Examples of source-rock paleoenvironment, diagenetic alteration, hydrogen donor, and biodegradation of petroleum hydrocarbons". In: American Chemical Society 222nd National Meeting (Chicago, August 26-30, 2001), pp. GEOC 7

38. Huang, Y.; Obermajer, M.; Jiang, C.; Snowdon, L.R.; Fowler, M.G.;Li, M. "Hydrogen isotopic compositions of individual alkanes as a new approach to petroleum correlation: case studies from the Western Canada Sedimentary Basin" *Org. Geochem..***2001**, 32, 1387-1399.

39. Bertrand, R.; Heroux, Y. "Chitinozoan, graptolite and scolecodont reflectance as an alternative to vitrinite reflectance in Ordovician and Silurian strata, Anticosti Island, Quebec, Canada", *American Association of Petroleum Geologists Bulletin* **1987**, 71, 951-957.

40. Cayley, G. T. "Hydrocarbon migration in the central North Sea" in *Petroleum Geology of North West Europe*, Brooks, J.; Glennie, K. (Eds.), Graham and Trotman, 1987; pp. 549-555.

41. Huc, A.Y., Irwin, H. and Schoell, M. 1985. Organic matter quality changes in an Upper Jurassic shale sequence from the Viking Graben. Petroleum Geochemistry in Exploration of the Norwegian Shelf, Norwegian Petroleum Society. Eds. B.M. Thomas et al. Graham and Trotman, p. 179-183.

42. Halpern H.I. "Development and applications of light-hydrocarbon-based star diagrams" *American Association of Petroleum Geologists Bulletin* **1995**, 79, 801-815.

43. Rooney, M.A. "Carbon isotope ratios of light hydrocarbons as indicators of thermochemical sulfate reduction". In: *Organic Geochemistry: Developments and Applications to Energy, Climate, Environment, and Human History*, 17th International Meeting on Organic Geochemistry (Ed. by J.O. Grimalt, C. Dorronsoro), 1995, pp. 523-525. European Association of Organic Geochemistry, Donostia-San Sebestián, Spain

44. Connan, J.; Lacrampe-Couloume, G.; Magot, M., "Origin of gases in reservoirs: In: *Proceedings of the 1995 International Gas Research Conference* (Ed. by D.A. Dolec), 1996, pp. 21-61. Government Institutes, Inc No. 1.

45. Mango, F.D. "Invariance in the isoheptanes of petroleum", *Science* **1987**, 247, p. 514-517.

46. Marriott, P.; Ong, R.; & Shellie, R. "Modulated multidimensional gas chromatography" *Amer. Lab.* **2001**, 33, 44-47.

47. Dahl, J.E.; Moldowan, J.M.; Peters, K.E.; Claypool, G.E.; Rooney, M.A.; Michael, G.E.; Mello, M.R.; Kohnen, M.L. "Diamondoid hydrocarbons as indicators of natural oil cracking", *Nature* **1999**, 399, 54-57.

Chapter 11

COUPLING MASS SPECTROMETRY WITH LIQUID CHROMATOGRAPHY (LC-MS) FOR HYDROCARBON RESEARCH

Chang Samuel Hsu
ExxonMobil Research and Engineering Co.,Process Research Laboratories,
Baton Rouge, LA 70821

1.   INTRODUCTION

Petroleum and other fossil fuels, such as coals and shale oils, are complex hydrocarbon mixtures that contain thousands of components.[1]   Numerous analytical techniques have been developed to characterize bulk properties and composition of petroleum and its products.[2]   However, much more detailed molecular-level compositional analysis of petroleum and its fractions is often needed for the understanding of the chemistry involved in refining processes and the correlation and prediction of the properties and performance of products.[3]   It is often desirable to separate petroleum crude oils into fractions to facilitate the analysis in manageable complexity.   The most common approach has been to distill petroleum into fractions (distillates) corresponding to the product streams in refineries, such naphtha, middle distillates, gas oils and residua (resids).   The other approach is to fractionate petroleum and its distillates according to the polarity of various compound types in the mixture.   In this aspect, supercritical fluid chromatography (SFC) has been developed for more volatile fractions, such as naphtha and middle distillates (650°F-), and liquid chromatography (LC) for less volatile fractions including gas oils and resids (650°F+).

Petroleum and its fractions are generally fractionated into saturated, aromatic and polar fractions using open columns or normal-phase high-performance liquid chromatography (HPLC) although reversed-phase HPLC has also been used to separate more polar fractions, such as polycyclic aromatic hydrocarbons.   The LC fractions are then analyzed by various analytical techniques for characterization, including mass spectrometry (MS).   However, samples containing relatively high concentrations of olefins are difficult to analyze by MS alone for the differentiation between olefins and naphthenes that have identical molecular formulae. Chromatographic separation of olefins from saturates and aromatics can also prove difficult. Hence, specialized techniques would be needed for olefin analysis, which are beyond the scope of this chapter.

Mass spectrometry (MS) is well suited for detailed molecular-level analysis because its ability to generate ions as surrogates for individual components in a complex mixture with sensitivity, selectivity and specificity. Its combination with chromatography, gas chromatography (GC), LC and SFC, further expands its capabilities of complex mixture analysis by increasing separation power and specificity. The most well known example is combined gas chromatography-mass spectrometry (GC-MS) that provides unprecedented capabilities in organic mixture analysis.[4-6] However, GC-MS is limited to volatile and thermally stable samples. For high boiling, polar and thermally labile samples the coupling of HPLC with MS (LC-MS) has become a logical choice to extend the GC-MS capabilities. Compared with GC, LC peaks are more likely to contain unresolved components. It is therefore often necessary to use ultraviolet (UV) detection in addition to MS for isomer differentiation. LC-MS can also be used for samples suitable for GC-MS when front-end separations by compound classes is preferred to boiling points.

Comparing to off-line LC separation followed by MS, LC-MS reduces sample size requirement, eliminates time-consuming solvent removal procedures, maintains peak integrity of chromatographic separations, and minimizes sample exposure to air for sensitive components. Early development of LC-MS was based on reserved phase LC using a polar solvent, such as water or methanol, as a mobile phase. The bulk of petroleum mixtures, however, are hydrocarbons that are not soluble in such polar solvents. Hence, different strategies and methods need to be developed for LC-MS based on normal phase HPLC where a non-polar solvent, such as hexane, is used as a mobile phase. There is no universal interface that can be used reliably to analyze all hydrocarbon samples. Various LC-MS interfaces have been investigated for their suitability to hydrocarbon analysis. Some of the LC-MS interfaces, such as thermospray (TSP)[7], electrospray (ESP)[8, 9] and atmospheric pressure chemical ionization (APCI)[10], also provide unique ionization capabilities. Only a limited number of interfaces, including moving belt (MB)[11] and particle beam (PB)[12], can be used for coupling LC with conventional MS ionization techniques, including electron-impact ionization (EI), chemical ionization (CI) and field ionization (FI). Table 1 summaries commonly used LC-MS interfaces and their compatible ionization techniques that will be discussed in more details later.

Table 1. Commonly Used LC-MS Interface and Compatible Ionization Techniques

LC-MS Interface	Compatible Ionization Techniques
Moving belt (MB) or particle beam (PB)	EI, CI and FI
Thermospray (TSP)	Filament or Discharge ionization
Electrospray (ESP)	Ionization by high voltage at nebulizer tip
Atmospheric Pressure Chemical Ionization (APCI)	Discharge ionization (solvent CI)

2. MASS SPECTROMETRY REVIEW

The key components of a mass spectrometer include sample introduction system, ion source, mass analyzer, and detection/recording system. In LC-MS, the sample is introduced into HPLC for separation, then through an interface to the ion source of the mass spectrometer. When a thermospray, electrospray or APCI interface is used, ions are produced inside the interface via ion-molecule reactions between solvent ions and sample molecules. Hence, the ionization methods are limited to chemical ionization using solvent ions as reagent gas ions. For applying a wide range of ionization techniques that have been developed for MS, a moving belt or particle beam interface would be the choice. The most commonly used ionization techniques in MS are electron-impact ionization (EI), chemical ionization (CI) and field ionization (FI).[13]

In EI sample molecules are bombarded by an electron beam generated from a hot filament to produce ions by losing valence electrons from the molecules. For highly reproducible spectra the filament voltage is generally set at 50-70 volts. At these voltages, fragmentation of molecular ions would occur due to excess internal energy. Although it can provide useful structural information, in a complex mixture the overlaps of fragment and molecular ions can make analysis very difficult. To eliminate or minimize fragmentation the filament voltage is reduced to 8-12 volts, slightly higher than the ionization potential of molecules, for compounds containing aromatic and polar functional groups.[4] However, the low voltage EI technique can not be applied to saturated hydrocarbons except multiple-ring condensed naphthenes because the ionization potentials of the molecules are near the appearance potentials of the fragment ions. It would be difficult to control the electron voltage for obtaining only molecular ions without significant amounts of fragment ions being present.

In chemical ionization (CI) an excess amount of reagent gas, such as methane, isobutane or ammonia, is introduced into the mass spectrometer's ion source to be ionized by an electron beam as in the EI process. The reactant ions formed from the reagent gas are then react with sample molecules via proton or hydride transfer. The internal energies of the protonated (MH^+) or hydride-abstracted (($M-H)^+$) molecular ions come from exothermometry of the reaction, which is usually a few tenth of an eV. Fragmentation is minimal due to low internal energies. Molecular weight information can therefore be obtained from these pseudo-molecular ions. When deuterated ammonia (ND_3) is used as a reagent gas, additional information, such as functionality, can be obtained via hydrogen-deuterium exchange.[14]

In field ionization (FI) molecules are ionized in the vicinity of high electric field generated by high voltage on a fine wire or at the edge of a razor blade.[15] Since ions are formed near or below ionization potential of molecules through quantum tunnelling, the internal energies of molecular ions are very low with no or little fragmentation. Hence, FI has been used for the determination of molecular weight or carbon number distribution of complex hydrocarbon mixtures, including saturated hydrocarbons.

There are many different types of mass analyzers available commercially. Some of them can only be operated in low-resolution mode where only nominal masses can be measured. A typical example is a quadrupole mass spectrometer. Others can be operated in high-resolution mode where the ions with the same nominal mass but difference in accurate masses can be resolved and measured. Mass spectrometers that can be operated in this mode include double-focusing magnetic sector [16], Fourier-transform ion cyclotron resonance (FT-ICR) [17], or reflectron time-of-flight (r-TOF) [18,19] mass spectrometers.

3.   LC-MS INTERFACES

3.1   Moving Belt (MB) Interface

Figure 1 shows the configuration of a moving belt (MB) interface. A polyimide belt is continuously moving around by the motion of two pulleys. The LC effluent is sprayed onto the belt, which is transported into mass spectrometer's ion source through two stages of differential pumping, at 5-50 and 0.1-0.5 torr, to evaporate off most of the LC mobile phase solvent. Hence, the moving belt interface is recommended to use for samples boiling above 600°F to avoid significant loss of volatile components during the transport stages. The spot on the belt containing the sample enters into the ion source at 10^{-6} torrs, where the sample is flashed off the belt by a nose heater at the tip of the interface. Fractions with boiling points greater than 1050°F can not be evaporated by the nose heater at a temperature around 350°C under the mass spectrometer vacuum to avoid breaking the polyimide belt by heat. The sample spot is continuously moving toward a wash zone to wash off residues before it moves back to the spray zone to deposit LC effluent.

The ion sources available commercially with a MB interface are EI and CI sources. Both ionization methods use a filament to generate an electron beam for ionization. An EI source is generally an open source in the 10^{-6} torr range that provides large mean free path between molecules to avoid interaction of the ions with molecules. A CI source, on the other hand, is a closed source where the pressure inside the source can reach 1 torr with reagent gas. Sample ions are produced from ion-molecule reaction between the reagent gas (reactant) ions and sample molecules. For FI source modification would be needed for it adaption with a moving belt interface.[20]

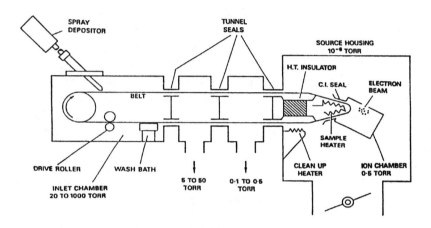

Figure 1. Moving belt LC-MS interface (Reproduced with permission from Micromass). An ion source for electron-impact ionization (EI) and chemical ionization (CI) is shown. For field ionization (FI) a source modification is described in Ref. 20.

A particle beam (PB) interface is ideal replacement for a MB interface, especially for more volatile samples where the front-end loss occurring in the MB interface can be avoided. A PB interface consists of concentric pneumatic nebulizer with temperature controlled desolvation chamber and a two- or three-stage momentum separator.[21] Mao, et al [22] used PB LC-MS to identify benzoquinolines and acridines in a Brazilian diesel oil. However, the sensitivity obtainable from a PB source is generally lower than a MB interface. The signals are not intense and stable enough for quantification purposes, especially at a high-resolution MS mode.

A MB interface described above is limited to hydrocarbon fractions with boiling points below 1050°F because the solute needs to be thermally desorbed from the surface of the belt made of a polymeric material under the mass spectrometer's vacuum. A LC-MS interface, including thermospray (TSP), electrospray (ESP) and atmospheric pressure chemical ionization (APCI), that sprays LC effluent directly into an ionization chamber without a thermal desorption process would be a logical choice to extend hydrocarbon characterization to high boiling fractions.

3.2   Thermospray (TSP)

Figure 2 is a schematic diagram of a thermospray LC-MS interface. The LC effluent is vaporized at the end of a heated nebulizer. In reversed phase HPLC, ammonium acetate can be added in an aqueous solution to produce ions without an external ionization device.[23] However, for mobile phases that are suitable for hydrocarbons, an external ionization source, such as a heavy-duty filament made of thoriated iridium or a discharge electrode, is placed upstream from the ion sampling aperture to produce ions.[24]   Since solvent molecules are present predominantly than sample molecules, solvent ions are formed initially which in turn ionize the sample molecules via ion-molecule reactions.   Hence, TSP is mainly a chemical ionization technique using solvent vapor as reagent gas that selectively ionizes molecules with high proton affinity such as aromatic hydrocarbons. The ions formed are repelled out of the vaporized stream in the spray chamber and enter the acceleration region of the mass spectrometer through a sampling cone.

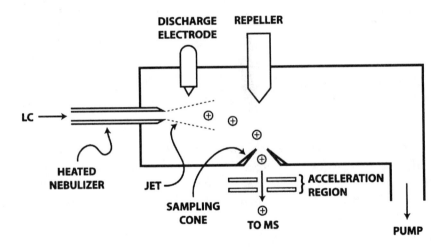

Figure 2. Simplified schematic diagram of a thermospray LC-MS interface.

3.3   Electrospray (ESP)

In ESP, LC effluent is sprayed out of a nebulizer (a capillary needle) at a high voltage (typically 3-5 kV) to produce multiply charged droplets of a solution containing analyte.   Solvent evaporation in an inert atmosphere (typically heated nitrogen) reduces the droplet into a smaller size and further breaks down the droplet by electrostatic repulsive force between charges to yield gas-phase ions.[8] The sample and ions passed through a counter electrode prior to being expanded through a sample cone and skimmer assembly into the mass spectrometer. When ESP is used as an ionization method, it's also referred to as electrospray ionization (ESI).

3.4 Atmospheric Pressure Chemical Ionization (APCI)

In APCI, the LC effluent containing the sample was pneumatically converted into an aerosol and rapidly heated into vapor phase at probe tip.[10] A corona discharge (pin) electrode is used to ionize the vapor from the nebulizer. As in ESP, the sample and reagent ions passed through a counter electrode prior to being expanded through a sample cone and skimmer assembly into the mass spectrometer. Most of the modern LC-MS instruments provide a single interface assembly for ESP and APCI that can be easily replaced from each other.

4. HOMOLOGOUS Z-SERIES FOR ELEMENTAL COMPOSITION DETERMINATION

As petroleum and its fractions contain hundreds and thousands of components, it is desirable to group them into homologous series that represent different families of compound types. For a homologous series, the group can be represented by a general formula of $C_nH_{2n+z}X$, where z represent hydrogen deficiency and X represent heteroatoms.[25] For example, all benzenes have z = -6 with no X and n ≥ 6. For thiophenes, z = -4 with X = S and n ≥ 4. These two classes of compounds share the same nominal mass series and require a high-resolution mass spectrometer to resolve them.

In the 1960's, Kendrick[26] introduced a concept of representing a homologous series by defining the exact mass of a methylene group as 14.000000, instead of 12.01565 on the ^{12}C = 12.000000 scale. Using his mass scale, all members in a homologous series would have the same mass defect. However, the Kendrick mass scale has found limited usage due to inherent accurate mass measurement errors with a high-resolution mass spectrometer. Later, Hsu, et al[25] introduced the use of nominal mass series in combination with the Kendrick mass scale to compensate uncertainties in accurate mass measurements.

5. LC-MS FOR PETROLEUM FRACTIONS

As mentioned earlier, normal phase LC generally separates hydrocarbon mixtures into saturate, aromatic and polar fractions. The applications of LC-MS for these fractions are discussed in the following sections. For more polar hydrocarbon fractions including aromatic fractions, reserved phase LC coupled with MS has also been used.

5.1   Saturates

A saturates fraction normally contain only paraffins ($z = +2$) and naphthenes (cycloparaffins, $z = 0$, -2, -4, etc.), particularly when sulfides are removed or isolated prior to the LC separation. Paraffins and naphthenes with a various number of rings differ in z numbers, thus, molar masses and molecular formulas. Field ionization (FI) MS that produces only molecular ions is an ideal ionization technique for saturates analysis. However, if the sample contains olefins the differentiation between naphthenes and olefins can prove difficult.

An on-line LC-MS has been developed for saturated hydrocarbon analysis using a moving belt interface coupled with FIMS.[20] It takes advantage of FI that generates predominantly molecular ions for normal paraffins and naphthenes but fragment ions for isoparaffins. Using these ionization characteristics, the split between normal paraffins and isoparaffins can be directly measured without the need of using gas chromatography for measuring normal paraffins. Traditionally, isoparaffins are determined by the difference of total paraffins measured by MS and normal paraffins measured by GC. In LC-FIMS, in addition to the differentiation and measurement of levels of naphthenes, normal paraffins and isoparaffins in a single analysis, the carbon number distributions of normal paraffins and naphthenes are also obtained.

5.2   Aromatics

Both normal phase and reverse phase HPLC methods have been developed for aromatic hydrocarbon analysis. The most often analyzed aromatic compounds by LC-MS are polycyclic aromatic hydrocarbons (PAHs) that are of environmental concerns due to many of them are known mutagens or carcinogens. Perreault et al [27] applied moving-belt LC-MS to analyze high molecular weight PAHs in a complex sample derived from coal tar. He found good preservation of chromatographic integrity with useful EI spectra for PAHs of low to moderate volatility up to molecular masses of at least 580 daltons (Da).

Singh et al [21] reported the use of reversed-phase HPLC and particle beam LC-MS for the determination of relatively high molecular weight PAHs between 228 and 300 Da. Pace and Betowski [28] used reversed-phase HPLC and particle beam LC-MS to measure PAHs in a contaminated soil. A group of PAHs ranging from 276 to 448 Da were identified.

Neutral PAHs are generally not amenable to electrospray ionization. However, Van Berkel et al [29,30] demonstrated electrospray methods for PAHs by forming molecular radical ions via either charge-transfer complexation or electrochemical oxidation. The analyte was ionized in solution via reaction with electron-transfer reagents trifluoacetic acid (TFA), 2,3-dichloro-5,6-dicyano-1,4-benzoquinone, or antimony pentafluoride. They suggest the method be used on-line as post-column derivatization. Airiau et al [31] demonstrated the use of ESP LC-MS for PAH analysis after post-column derivatization with tropylium cation to produce positive ions. They also employed tandem mass spectrometry (MS-MS) to obtain information on fragment ions derived from collision induced dissociation. Rudzinski et al [32] developed a novel method of adding Pd^{2+} in methanol to the ESP ionization source of an ion trap mass spectrometer to analyze sulfur-containing PAHs in the aromatic fraction of Maya crude oil for biodesulfurization studies.

Marvin et al [33] employed APCI LC-MS to analyze high molecular weight PAHs from organic extracts of a variety of sediments including coal tar-contaminated sediments and freshwater mussels from Hamilton Harbor in Western Lake Ontario. Lafleur et al. [34] characterized flame-generated C_{10} to C_{100} PAHs by atmospheric pressure chemical ionization (APCI) with LC coupling for PAHs up to C_{36} and with direct liquid injection for PAHs $>C_{36}$. The largest molecules identified at 1792 Da are comparable to nanoparticles of soot.

Anacleto et al. [35] compared various LC-MS interfaces for PAH analysis. They found that 17 target PAHs in the coal tar reference material could be detected and quantified within satisfactory agreement with certified values using APCI LC-MS when perdeuterated internal standards were employed. LC-MS total ion chromatogram of a carbon black extract is consistent with the chromatogram obtained from UV detection at 236-500 nm. Both moving belt (MB) and particle beam (PB) interfaces can provide library searchable EI spectra, but lose smaller and more volatile PAHs. Larger PAHs are difficult to analyze by PB, but form a thin layer on a moving belt and readily volatilize in the ion source by flash heating. MB is mechanically awkward, and PB has poor detection limits and nonlinear calibration curves in quantification. The ESI derivatization methods developed by Van Berkel et al, as described above, have limited choice of solvents that are incompatible with gradient elution to achieve efficient LC separations.

In normal phase HPLC aromatic hydrocarbons are separated into monoaromatics (or 1-ring aromatics, 1RA), diaromatics (2RA), triaromatics (3RA), tetraaromatics (aromatics containing 4 or more aromatic rings, 4+RA) and "polars" for components recovered by a polar solvent after the elution of "tetraaromatics". One advantage of combining LC with MS is the mass spectral characterization of LC peaks for detailed compound distribution. On the other hand, LC separations help resolve ambiguities in MS characterization. In the aromatic analysis, ambiguities of structural assignments are found in (1) the differentiation of alkylaromatics and naphthenoaromatics and (2) the distinction between aromatic hydrocarbon and thiophenes.[36]   Alkylaromatics and naphthenoaromatics can be isomers having identical molecular formulas but different number of aromatic rings. For example, compounds 1-4, shown in Figure 3, all have the same nominal mass at 380 Da. Compounds 1 and 2 have identical molecular formula, $C_{28}H_{44}$, with the same accurate molar mass of 380.344.   Hence, MS alone can not differentiate these two compounds. However, Compound 1 elutes off LC in the monoaromatic while Compound 2 in the diaromatic regions. Compound 2, an alkylnaphthalene, and Compound 3, an alkylbenzothiophene, have different molecular formulas and accurate molar masses.   The differentiation of these two compounds only requires 4200 resolving power that is achievable by many high-resolution mass spectrometers, including double-focusing magnetic sector and reflectron time-of-flight (TOF) mass spectrometers. However, Compounds 3 and 4 only differ in 3.4 milli-daltons. They would require over 110,000 resolving power that can only be achievable using a Fourier-transform ion cyclotron resonance (FT-ICR) mass spectrometer [17], but difficult or even impossible by other high resolution mass spectrometers. Again, these two compounds elute off the LC in different regions (triaromatic and tetraaromatic), as shown in Figure 4.   Hence, the combination of LC with MS reduces the resolution requirements for overlapping compound series and resolves overlapping isomers that share the same molecular formula.

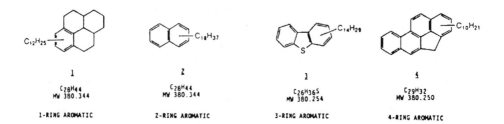

Figure 3. Compounds with the same nominal mass at 380 but difference in LC characteristics. (From Ref. 36. Reproduced with permission from American Chemical Society.)

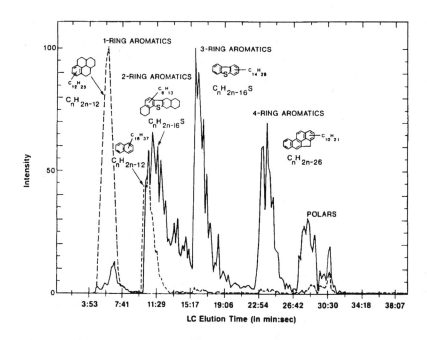

Figure 4. Overlapping compound series and isomers are resolved by LC separation coupled with high resolution MS, as shown in the selected ion chromatograms of m/z 380.25 (solid line) and 380.34 (dotted line) with a resolution window of 200 ppm. (From Ref. 36. Reproduced with permission from American Chemical Society.)

A comprehensive LC-MS analysis of aromatic compounds was reported by Qian and Hsu [37] using normal-phase HPLC and a MB interface. As mentioned previously, a MB interface has limitations for analyzing components of high and low volatility due to differential pumping and heat tolerance of polyimide belt. Vacuum gas oils (VGO) that boil between 650°F and 1050°F would be suitable for MB LC-MS analysis. A 650-1050°F distillates before and after hydrotreating were analyzed by HPLC using an evaporative mass detector (EMD) with their chromatograms shown in Figures 5. With low-voltage EI (LVEI), the mass spectra of each LC fractions essentially show molecular ions of the components. Figure 6 shows the LVEI spectra of 3RA before and after hydrotreating as an example. The difference in compound distributions of these two 3RA fractions, shown in Figure 7, are obtained from their LVEI spectra. The z = -18 series represent phenanthrenes, while the series with a z number less than -18 are naphthenophenanthrenes with the number of naphthene rings equals to (-18 - z)/2. The amounts of naphthenophenanthrenes are shown to be greater in the hydrotreated product than in the hydrotreating feed. The additional naphthenophenanthrenes apparently are derived from polyaromatic hydrocarbons with greater than 3 aromatic rings present in tetraaromatics and polars fractions of the feed.   From the molecular ions the carbon-number distribution of individual compound series before and after hydrotreating can also be obtained. An example is shown in Figure 8 for the phenanthrene series. It indicates the catalyst possesses hydrocracking activities because more low carbon number components are found in the products than in the feed.

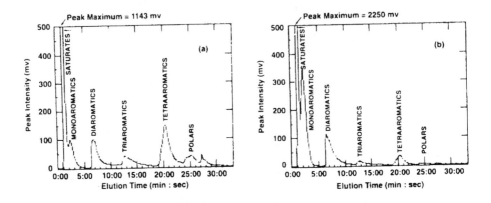

Figure 5. HPLC EMD (evaporative mass detector) chromatograms of the 650-1050 F distillates before (a) and after (b) hydrotreating. (From Ref. 37. Reproduced with permission from American Chemical Society.)

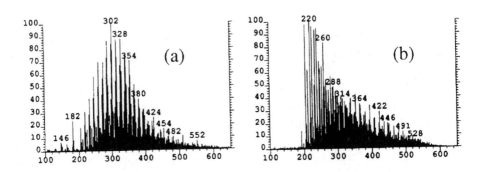

Figure 6. LVEI spectra of Triaromatics (3RA) before (a) and after (b) hydrotreating. (From Ref. 37. Reproduced with permission from American Chemical Society.)

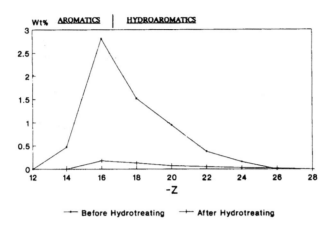

Figure 7. Compound type distribution of triaromatics before and after hydrotreating. (From Ref. 37. Reproduced with permission from American Chemical Society.)

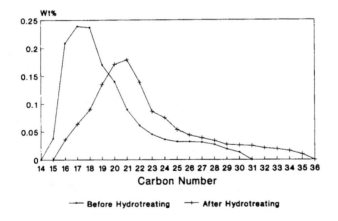

Figure 8. Carbon number distribution of z=-18 (phenanthrene) series before and after hydrotreating. (From Ref. 37. Reproduced with permission from American Chemical Society.)

For analyzing complex hydrocarbon mixtures by MS it is often desirable to obtain only molecular ions without significant amounts of fragment ions. LVEI mentioned above is one of the low energy ionization techniques for obtaining mass spectra of molecular ions only. However, it has several disadvantages: (1) effective electron energy is sensitive to ion source conditions and (2) sensitivity is relatively low due to low electron density and small ionization cross section.

Hsu and Qian [38] employed CS_2 charge exchange (CE) as a low-energy ionization technique to improve sensitivity, especially for high-resolution mass spectrometric measurement in the elemental composition determination of individual components. Unlike LVEI, small variation in electron beam energy does not affect the relative sensitivity in CS_2 CE. CS_2 CE also provides more uniform molar sensitivity for quantitative measurement of hydrocarbon molecules with different aromatic structure than LVEI. Both LVEI and CS_2 CE leave saturated hydrocarbon largely unionized; thus, providing selectivity for aromatic and polar compounds if saturates are also present in a hydrocarbon mixture. Field ionization can also be used to obtain molecular-ion only spectra for aromatic and polar compounds, but with lower sensitivity and limited resolution.

5.3   Polars

Polar compounds in petroleum contain heteroatoms, such as sulfur, nitrogen, oxygen and even metals. They are believed to be responsible for a variety of problems encountered in refining processes and petroleum products, such as catalyst poisoning, formation of deposits and sediments, etc. Sulfides can be separated by silver- or palladium-impregnated silica prior to the separation of other hydrocarbon types. Polar fractions, however, are generally collected after removing saturates and aromatics. Many HPLC methods have been developed to separate polar compounds into acidic, neutral nitrogen and basic nitrogen fractions. Although many other polar compounds are also of concern to petroleum processing, recent characterization focuses have been on naphthenic acids, basic nitrogen compounds and metal-containing organics.

Naphthenic acids are of concern to refinery units due to their corrosivity.[39] They are also of interest to biodegradation in geochemical studies [40] and refinery wastewater treatment for environmental compliance [41]. Hsu, et al. [42] pioneered the LC-MS studies of naphthenic acids using APCI and ESI. They found that negative-ion APCI using acetonitrile as a solvent and mobile phase yielded the cleanest spectra among the various mass spectrometric techniques evaluated. From the distribution of acid compound type and carbon number distributions, the presence of aromatic acids in acidic California crude oils was confirmed. However, due to limited resolving power of the MS used, no elemental composition of acidic species was determined. Using high-resolution MS naphthenic acids containing sulfur, nitrogen and more than two oxygen atoms were recently reported.[43,44]

Basic nitrogen compounds are of concern to petroleum processing due to their poisoning of active acidic sites of catalysts. The polars in an offshore California crude oil were fractionated by Biosol to collect "neutral" and "basic" nitrogen fractions that were subjected to moving belt LC-MS analysis.[14] In addition to low voltage EI for elemental composition of molecular ions, CI using deuterated ammonia as a reagent gas was also applied to determine the number of exchangeable hydrogen atoms in the molecule. Unlike the neutral nitrogen fraction that contained mainly carbazoles and their benzologs, the basic nitrogen fraction was much more complex. The predominant compound classes contain both nitrogen and oxygen atoms rather than pyridine benzologs. The presence of cyclic amides was confirmed by infrared analysis. Also present are unidentified sulfur-containing compounds that would need MS with higher resolving power, such as FT-ICR. However, a large variation in the composition of polars and basic nitrogen fractions among crude oils is expected. For oxygen-containing compounds, careful distinction between nascent and artificial (e.g., air exposure during sample storage) species need to made.

Metalloporphyrins and other metal-containing organics in crude oils present processing and environmental problems. Fukuda et al [45] utilized APCI LC-MS to determine low level (down to 200 pg) of metalloporphyrins, using nickel and vanadyl octaethyl prophyrins as model compounds. Both normal phase and reverse phase HPLC were used. By varying the cone voltage, the extent of fragmentation of molecular ions for obtaining possible structural information can be adjusted. A Cerro Negro heavy crude oil was recently analyzed by ESP ionization associated with FT-ICR for the determination of nickel and vanadyl porphyrin types, mostly etio and deoxophyllerythro-etio (DPEP), and the carbon number distribution of each type.[46]

The analysis of polar compounds in petroleum is a fertile field. Both normal- and reverse-phase HPLC are suitable for use. Bayona, et al [47] applied reversed phase LC-MS with "filament-on" TSP interface to characterize polar substituted polycyclic aromatic hydrocarbons (PAHs) in diesel exhaust particulates. Most of the polar compounds found were nitro-substituted. These compounds are likely formed from un-combusted PAH from diesel fuels reacting with oxides of nitrogen in the exhaust.[4]

5.4 Resids

Few attempts have been made to characterize resids and heavy oils by on-line LC-MS.[48,50] TSP was used to extend on-line LC-MS characterization of petroleum and synthetic fuel fractions with atmospheric equivalent boiling points (AEBP) beyond 1050°F. A DISTACT (short path distillation) fraction with AEBP of 1120-1305°F was analyzed by TSP LC-MS using a 2,4-dintroanilinoproply (DNAP) column gradiented with hexane, methylene chloride and isopropyl alcohol.[48] The overall TSP results were found to be in good agreement with those obtained from field desorption mass spectrometry (FDMS) with higher sensitivity. Post column addition of methylene chloride was found to enhance TSP signals. The integrated mass spectra over monoaromatic, diaromatic, triaromatic and tetraaromatic regions are shown in Figure 9. It can been seen that in a finite boiling range the average molar mass decreases steadily with higher aromaticity. For high boiling fractions the overlaps between sulfur (mostly thiophenic) compounds and aromatic hydrocarbons having a mass difference of only 3.4 milli-daltons can not be resolved by a double-focusing magnetic sector instrument.[36] It would require the use of ultrahigh resolution FT-ICR with resolving power >50,000 to resolve overlapping compound types. Miyabayashi et al [49] employed FT-ICR equipped with electrospray ionization to determine molecular formulas of components in Arabian mix vacuum residue by measuring accurate masses of molecular ions. Recently, Qian, et al [50] demonstrated the use of electrospray ionization coupled with FT-ICR for the analysis of nitrogen-containing compounds of a heavy crude oil.

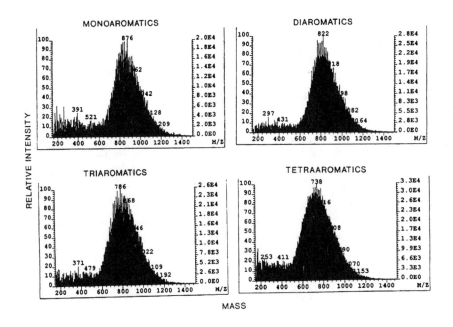

Figure 9. Integrated mass spectra over monoaromatic, diaromatic, triaromatic and tetraaromatic LC regions. (From Ref. 48. Reproduced with permission from American Chemical Society.)

6.   FUTURE TRENDS

In the previous sections, recent developments in the LC-MS analysis for various petroleum fractions were described. There are continuous needs for improving chromatographic separation for better defining various compound classes to facilitate mass spectrometric characterization. New ionization techniques to selectively ionize specific classes of compounds would provide new opportunities for compound identification. HPLC coupling with ultrahigh resolution FT-ICR would greatly improve the determination of elemental composition of the components. Alternatively, elemental composition can be determined in fair accuracy by accurate mass measurement of LC-reflectron TOF. The determination of functional groups may be achievable by pre- and post-column derivatization. The use of LC-MS-MS for structure determination remains to be demonstrated.

7. REFERENCES

1. Robbins, W. K.; Hsu, C. S. In *Kirk-Othmer Encyclopedia of Chemical Technology*, 4th ed., John Wiley & Sons: New York, 1996; pp. 352-370.
2. Hsu, C. S. In *Encyclopaedia of Analytical Science,* premiere edition, Academic Press: London, 1995; pp. 2028-2034.
3. Hsu, C. S. In *Encyclopaedia of Analytical Chemistry*, John Wiley & Sons: New York, 2000; pp. 6613-6622.
4. Hsu, C. S.; Drinkwater, D. in *Current Practice in GC-MS (Chromatogr. Sci. Series, Vol. 86)*, W. W. A. Niessen, ed. Dekker Marcel: New York, 2001; pp. 55-94.
5. Hsu, C. S.; Genovitz, M. W.; Dechert, G. J.; Abbott, D. J.; Barbour, R. in *Chemistry of Diesel Fuels*, C. Song; C. S. Hsu; I. Mochida (Eds.), Taylor & Francis: New York, 2000; pp. 61-76.
6. Malhatra, R.; Coggiola, M. J.; Young, S. E.; Hsu, C. S.; Dechert, G. J.; Rahimi, P. M.; Briker, Y. in *Chemistry of Diesel Fuels*, C. Song; C. S. Hsu; I. Mochida (Eds.), Taylor & Francis: New York, 2000; pp. 77-92.
7. Bakeley, C. R.; McAdams, M. J.; Vestal, M. L. *J. Chromatogr.* **1978**, 158, 261-276.
8. Whitehouse, C. M.; Dreyer, R. N.; Yamashita, M.; Fenn, J. B. *Anal. Chem.* **1985**, 57, 675-679.
9 Bruins, A. P.; Covey, T. R.; Henion, J. D. *Anal. Chem.* **1987**, 59, 2642-2646.
10. Horning, E. C.; Carroll, D. I.; Dzidic, I.; Lin, S. N.; Stillwell, R. N.; Thenot, J. P. *J. Chromatogr.* **1977**, 142, 481-495.
11. McFadden, W. H.; Bradford, D. C.; Games, D. E.; Gower, J. L. *Amer. Lab.* **1977,** 10, 55-64.
12. Winkler, P. C.; Perkins, D. D.; Williams, W. K.; Browner, R. F. *Anal. Chem.* **1988,** 60, 489-493.
13. Watson, J. T. *Introduction to Mass Spectrometry*, 3rd ed., Lippincott-Raven: Philadelphia, 1997.
14. Hsu, C. S.; Qian, K., Robbins, W. K. *High Resolution Chromatography* **1994**, 17, 271-276.
15. Beckey, H.-D. *Principles of Field Ionization and Field Desorption Mass Spectrometry*, William Clowes & Sons: London, 1977.
16. White, F. A.; Wood, G. M. *Mass Spectrometry: Applications in Science and Engineering*, Wiley-Interscience: New York, 1986.
17. Hsu, C. S.; Liang, Z.; Campana, J. E. *Anal. Chem.* **1994**, 66, 850-855.
18. Cotter, R. J. *Time-of-Flight Mass Spectrometry: Instrumentation and Applications in Biological Research*, American Chemical Society: Washington, D.C., 1997.
19. Hsu, C. S.; Green, M. *Rapid Comm. Mass Spectrom.* **2001,** 15, 236-239.
20. Liang, Z.; Hsu, C. S. *Energy Fuels* **1998**, 12, 637-643.
21. Singh, R.P.; Brindle, I. D.; Jones, T. R. B.; Miller J. M.; Chiba M. *J. Am. Soc. Mass Spectrom.* **1993**, 4, 898-905.

22. Mao J.; Pacheco C. R.; Traficante D. D.; Rosen W. *J. Chromatogr.* **1995,** 18, 903-916.
23. Vestal, M. L. *Anal. Chem.* **1984,** 56, 2590-2592.
24. Arpino, P. *Mass Spectrom. Rev.* **1990,** 9, 631-669.
25. Hsu, C. S.; Qian, K.; Chen, Y. C. *Anal. Chim. Acta* **1992,** 264, 79-89.
26. Kendrick, E. *Anal. Chem.* **1963,** 35, 2146-2154.
27. Perreault, H.; Ramaley, L.; Sim, P. G.; Benoit, F. M. *Rapid Comm. Mass Spectrom.* **1991,** 5, 604-610.
28. Pace, C. M.; Betwoski, L. D. *J. Am. Soc. Mass Spectrom.* **1995,** 6, 597-607.
29. Van Berkel, G. J.; McLuckey, S. A.; Glish G. L. *Anal. Chem.* **1991,** 63, 2064-2068.
30. Van Berkel, G. J.; Asano, K. G. *Anal. Chem.* **1994,** 66, 2096-2102.
31 Airiau, C. Y.; Brereton, R. G.; Crosby, J. *Rapid Comm. Mass Spectrom.* **2001,** 15, 135-140.
32 Rudzinski, W. E.; Sassman, S. A.; Watkins, L. M. *Prepr. - Am. Chem. Soc., Div. Pet. Chem.* **2000,** 45, 564-566.
33 Marvin, C. H.; McCarry, B. E.; Villella, J.; Bryant, D. W.; Smith, R. W. *Polycyclic Aromat. Compd.* **1996,** 9, 193-200.
34. Lafleur, A. L.; Taghizadeh, K.; Howard, J. B.; Anacleto, J. F.; Quilliam, M. A. *J. Am. Soc. Mass Spectrom.* **1996,** 7, 276-286.
35. Anacleto, J. F.; Ramaley, L.; Benoit, F. M.; Boyd, R. K.; Quilliam, M. A. *Anal. Chem.* **1995,** 67, 4145-4154.
36. Hsu, C. S.; Qian, K.; McLean, M. A.; Aczel, T.; Blum, S. C.; Olmstead, W. N.; Kaplan, L. H.; Robbins, W. K.; Schulz, W. W. *Energy Fuels* **1991,** 5, 395-398.
37. Qian, K.; Hsu, C. S. *Anal. Chem.* **1992,** 64, 2327-2333.
38. Hsu, C. S.; Qian, K. *Anal. Chem.* **1993,** 65, 767-771.
39. Piehl, R. L. NACE Conference, *Corrosion/87* **1987,** Paper No. 579.
40. Ahsan, A.; Kalsen, D. A.; Patience, R. L. *Mar. Pet. Geol.* **1997,** 14, 55-64.
41. Lai, J. W. S.; Pinto, L. J.; Kiehlmann, E.; Bendell-Young, L. I.; Moore, M. M. *Environ. Toxicol. Chem.* 1996; 15: 1482-1491.
42. Hsu, C. S.; Dechert, G. J.; Robbins, W. K.; Fukuda E. K. *Energy Fuels* **2000,** 14, 217-223.
43. Tomczyk, N. A.; Winans, R. E.; Shinn, J. H.; Robinson, R. C. *Energy Fuels* **2001,** 15, 1498-1504.
44. Qian, K.; Robbins, W. K.; Hughey, C. A.; Copper, H. J.; Rodgers, R. P.; Marshall, A. G. *Energy Fuels* **2001,** 15, 1505-1511.
45. Fukuda, E.; Wang, Y.; Hsu, C. S. *Proc. 44th ASMS Conf. on Mass Spectrom. and Allied Top.,* Portland, OR, May 12-16, 1996; p. 1273.
46. Rodges, R. P.; Hendrickson, C. L.; Emmett, M. R.; Marshall, A. G.; Greaney, M. G.; Qian, K. *Can. J. Chem.* **2001,** 79, 546-551.
47. Bayona, J. M.; Barcelo, D.; Albaiges J. *Biomed. Environ. Mass Spectrom.* **1988,** 16, 461-467.
48. Hsu, C. S.; Qian, K. *Energy Fuels* **1993,** 7, 268-272.
49 Miyabayashi, K.; Suzuki, K.; Teranishi, T.; Naito, Y.; Tsujimoto, K.; Miyake, M. *Chem. Lett.* **2000,** 172-173.
50. Qian, K.; Rodgers, R. P.; Hendrickson, C. L.; Emmett, M. R.; Marshall, A. G. *Energy Fuels* **2001,** 15, 492-498.

Chapter 12

ADVANCED MOLECULAR CHARACTERIZATION BY MASS SPECTROMETRY: APPLICATIONS FOR PETROLEUM AND PETROCHEMICALS

S. G. Roussis, J. W. Fedora, W. P. Fitzgerald, A. S. Cameron and R. Proulx
Research Department, Products and Chemicals Division
Imperial Oil
Sarnia, Ontario, Canada, N7T 8C8

1. INTRODUCTION

The availability of detailed molecular information about hydrocarbon feeds and products is important to many areas of the petroleum and petrochemical industries. Detailed compositional knowledge is needed for the development of novel raw material processing methods, the optimization of refinery processes, and the assurance of product quality. Compositional characterization across the boiling range is needed for property determination and overall valuation of crude oils. Development of control measures against corrosion requires knowledge about the nature and amounts of reactive sulfur and naphthenic acid compound types. Determination of the root-cause of refinery and chemical plant upsets depends on the availability of molecular information about the bulk hydrocarbon components, the additives, and the contaminants in the problematic streams.

Characterization of petroleum and petrochemical streams at the molecular level is very challenging due to the complexity of the streams, which increases very rapidly with boiling range.[1,2] Mass spectrometry (MS) was early recognized as a powerful method for the analysis of complex petroleum mixtures.[3,4] Low-resolution mass spectrometric (LRMS) methods were initially developed that used the characteristic fragment ion patterns of the different hydrocarbon compound types to determine their relative concentrations in the hydrocarbon streams.[3-5] Later, high-resolution mass spectrometry (HRMS) was introduced to characterize hydrocarbon mixtures based on the accurately measured masses of the fragment or molecular ions of the mixture

285

components.[6,7] HRMS permits the analysis of ions with the same nominal but different exact masses (isobars). The coupling of gas chromatography (GC) and liquid chromatography (LC) to mass spectrometry[8,9] can significantly reduce the complexity of the sample by separating the individual components as a function of boiling point (GC) or polarity/ionic strength (LC). The combined chromatography-mass spectrometry methods permit the characterization of compounds with the same exact masses but with different chemical structures (isomers).

In the current work, advanced methods of mass spectrometry used alone or in combination with chromatographic methods of separation are presented for the detailed characterization of complex petroleum and petrochemical samples. The methods are organized in sections addressing the requirements of the different application areas at the Sarnia Research Center. The majority of applications given here, illustrate the use of methods developed in our laboratory over the last several years and serve as examples to demonstrate the capabilities of advanced methods of mass spectrometry that can solve challenging analytical problems in the petroleum industry. Although it is not the intent of this work to thoroughly review the literature, whenever appropriate, the work of other researchers is discussed and the merits of pertinent methodologies are presented.

2.   APPLICATION AREAS

The majority of analytical requirements at the Sarnia Research Center originate from the application areas summarized in Table 1. Crude oil assays relate to the elaborate characterization of whole crude oils and distillation fractions to determine their physical and chemical properties. Knowledge of these properties is critical for crude oil purchasing and processing decisions. Historically, the majority of analytical tests performed as part of the crude assays evaluated the physical properties of crude oils and their fractions.[10] It was originally difficult to obtain detailed molecular information. More recently, GC[11,12] and GC-MS[13,14] methods have allowed the determination of the chemical composition of low-boiling petroleum fractions at great molecular detail, but similar levels of information are difficult to obtain for the heavier petroleum fractions.[15]

Table 1. Application Areas of Mass Spectrometry and Analytical Requirements

Area	Analytical Requirement
Crude oil assays	Detailed analysis across the boiling range
Refinery products	Molecular composition
Chemical plant products	Molecular composition
Corrosion control	Reactive sulfur species, naphthenic acids
Upstream, process, and products additives	Qualitative and quantitative analysis
Contaminants in refineries, chemical plants	Qualitative and quantitative analysis

The analysis of most refinery and chemical plant products predominantly involves the characterization of the hydrocarbon components in the mixtures. Compositional information is needed for process optimization and product quality control. The management of corrosion is another area of high economic importance to the refinery that requires molecular characterization tools to determine the nature and quantities of the corrosive species in crude oil feeds and products.[16]

An area of significant importance to the petroleum and petrochemical industries involves the characterization of trace non-indigenous components in the mixtures. These components can be purposely-added chemicals (additives) that improve certain qualities of the products[17] or undesirable chemical contaminants that adversely affect the properties of products. Contaminants are typically chemicals produced from unexpected events or secondary reactions. In many instances, improper usage of additives, or extreme operating conditions, may denature additive components and lead to contamination. Alternatively, indigenous petroleum compounds may unexpectedly react with denatured additives or other incompatible components to produce contaminants. Any mitigation actions would require knowledge of the chemical nature of the contaminant species.

3.   CRUDE ASSAYS

3.1   Unseparated Fractions

It was first recognized by Gallegos et al.[18] in 1967 that HRMS (5000 resolving power, 70 eV EI) could be used to analyze high boiling fractions (500°-950° F) without the need for lengthy sample fractionation via elution chromatography methods (e.g., silica gel). Their HRMS method permitted the determination of 19 hydrocarbon compound types (7 saturated, 9 aromatic, and 3 thioaromatic hydrocarbon types). An improved HRMS (10,000 resolving power) method was introduced later by Bouquet and Brument,[19] that provided compositional information about 33 compound types for samples boiling in the 500°-1200° F range. Further improvements were recently made to this method for the characterization of heavy petroleum fractions. [20]

A simplified method using LRMS was developed by Robinson[21] in 1971 that did not require prior fractionation and was applicable over a wide boiling range (200°-1100°F). The Robinson method accounted for the entire sample composition in terms of 4 saturated hydrocarbon types, 12 aromatic hydrocarbon types, 3 thioaromatic types, and 6 unidentified aromatic groups. The accuracy of the method was equivalent to the HRMS methods.[6,21] There are several advantages however, associated with the use of LRMS instrumentation over the use of HRMS instrumentation; among them: (1) lower cost, (2) smaller size, (3) lower maintenance, (4) simpler operation, and (5) easier automation.

3.2   Whole Crude Oils

The advantages of LRMS have been coupled in our laboratory with the separation capabilities of GC to provide a combined method of gas chromatographic simulated distillation-mass spectrometry (GCD-MS).[22] The method provides both weight and volume boiling point distributions for crude oils from hydrocarbon compound type information obtained across the boiling range (retention time). The boiling point distribution is one of the critical properties needed for the valuation and processing of crude oils. Most importantly, from the same analysis and in less than 60 min, the method provides compositional information about a crude oil as a function of boiling point and aromaticity (Z-series number).[22] This distributed compositional information can be used to differentiate crude oils and predict their physicochemical properties.[23] Figures 1A and 1B show the total ion currents obtained for a light and a heavy crude, respectively. Figures 2A and 2B show the corresponding compound type distributions obtained for the two crude oils as a function of boiling point. The plots show the weight percent distributions of summaries of hydrocarbon compound types across the boiling range of the crude oils. The wealth of information provided by GCD-MS and the simplicity of the instrumentation used make it one of the most powerful analytical tools in the petroleum industry.

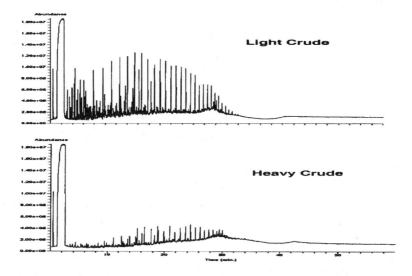

Figure 1. GCD-MS analysis of whole crude oils: (A) total ion current of a light crude oil in CS_2 solvent (~2 wt%); (B) total ion current of a heavy crude oil in CS_2 solvent.

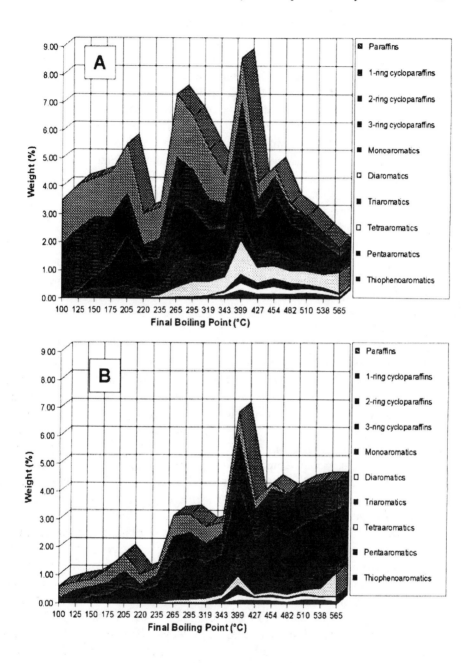

Figure 2. GCD-MS analysis of whole crude oils: (A) weight percent distributions of hydrocarbon compound types across the boiling range of a light crude oil; (B) weight percent distributions of hydrocarbon compound types across the boiling range of a heavy crude oil. Reproduced with permission from *Anal. Chem.* **2000**, 72, 1400-1409.

3.3   **Saturated Hydrocarbon Fractions**

ASTM (American Society for Testing and Materials) method D 2786 is the most commonly used method for the characterization of saturated hydrocarbon fractions. It is based on widely available LRMS instrumentation. However, the method does not allow for the determination of the molecular weight (MW) distributions of the different saturated hydrocarbon types or the structures of individual molecules. Field ionization mass spectrometry (FIMS) is a soft ionization method that does not transfer excess amounts of internal energy upon ionization and has successfully been used for the determination of the MW distributions of thermally labile saturated hydrocarbons.[2,24,25] The method can be combined with on-line GC[26,27] and LC[28] to provide distributions of isomeric structures. Liang and Hsu[28] found that FIMS generates predominantly molecular ions for normal paraffins and naphthenes but fragment ions for isoparaffins. In that fashion, it is possible to separate the three saturated hydrocarbon types by use of spectral editing software without the need for gas chromatography. The approach provides for the determination of the bulk amounts of the saturated hydrocarbon types but does not permit the characterization of the structures of individual molecules.

Work in our laboratory has shown that by using low-energy ionization (i.e., low-voltage electron ionization and CS_2 charge exchange chemical ionization methods) it is possible to obtain mass spectra for substituted saturated hydrocarbons with abundant molecular ions and characteristic fragment ions corresponding to the site of substitution.[29] Optimization of the low-energy ionization conditions is done by using suitable tuning compounds (e.g., n-heptadecane). Higher ionization cross-sections are obtained by the charge exchange experiments. Supersonic molecular beam (SMB)[30] and metastable atom bombardment (MAB)[31] are other techniques that have shown good potential for the analysis of saturated hydrocarbons by generating mass spectra with abundant molecular ions and limited fragmentation.

We have found that it is possible to determine the relative concentrations of target molecules of interest such as normal alkanes in petroleum fractions by applying the AMDIS deconvolution method[32] to samples analyzed by GC-MS.[33] Figure 3 shows the target n-alkane compounds determined in a middle distillate fraction. The determination of individual isoparaffin molecules is significantly more challenging due to the enormous complexity of the heavier fractions. However, improved determination of individual isoparaffin molecules is possible by using low-energy ionization methods to generate mass spectra with abundant molecular ions and fragment ions characteristic of the site of substitution. The ability to characterize the structures of individual isoparaffin molecules is of high importance to the lubricant oil business.

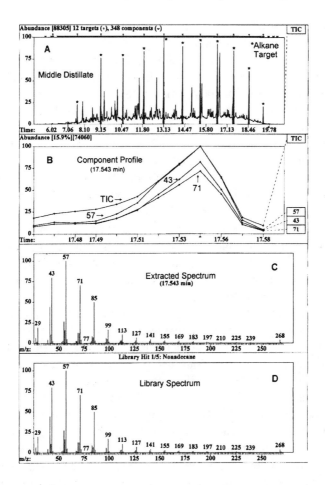

Figure 3. Target compound determination by AMDIS: (A) total ion current of a middle distillate sample; (B) profiles of most abundant ions; (C) extracted mass spectrum at 17.54 min; (D) library spectrum of best match. Reproduced with permission from *Energy Fuels* **2001**, 15, 477-486.

3.4 Aromatic Hydrocarbon Fractions

ASTM method D 3239 is a simple and extensively used method that determines the percent amounts of 18 aromatic hydrocarbons and 3 thiophenoaromatic compound types.[34] Similar to the use of other mass spectrometric ASTM methods, any inlet system may be used for sample introduction provided the inlet does not introduce contamination or compositional changes. The Brunfeldt all-glass heated system (AGHIS)[35] is a widely used system for the introduction of hydrocarbon fractions into a mass spectrometer. A dynamic batch inlet system (DBIS) is simpler to use and permits automated sample introduction with an autosampler.[36] We have recently used a GC oven for the introduction of hydrocarbon mixtures in the mass spectrometer.[33] The DBIS is more suitable for HRMS experiments since the sample can reside in the inlet for a long period of time, but the GC-MS permits the determination of individual target molecules.

The lower ionization potentials of aromatic and heteroatom-containing hydrocarbons in comparison to the ionization potentials of saturated hydrocarbons allow for the selective analysis of these compounds by low-energy ionization techniques. In that fashion, no prior separation of the samples into saturated and aromatic fractions is required. The use of low-voltage electron ionization (LV/EI) coupled with HRMS to extensively characterize aromatic fractions was pioneered by Aczel and co-workers.[7,37,38] Further developments have been reported since the original developments relating to the use of benzene and CS_2 as reagent compounds in chemical ionization charge exchange (CE) experiments for low-energy ionization of aromatic fractions.[39-42] The main benefits of CE ionization over LV ionization are: (1) higher sensitivities, and (2) approximately unity response factors.

The combination of low-energy ionization and high-resolution mass spectrometry has allowed for the extensive characterization of aromatic fractions. The reduction of the chemical composition information from the raw mass spectrometric data was initially done using calculations based on the fractional parts of the nuclidic masses of the atoms considered to be present in the hydrocarbon mixtures.[37,38] Later, a simplified method was developed by Kendrick based on $CH_2 = 14.0000$ to rapidly assign chemical formulas to the measured masses.[43] The Kendrick approach has the advantage of rapidity over the first formula assignment methods, however this advantage has become insignificant with modern high-speed computers. In addition, the use of the Kendrick method requires the pre-generation of mass tables relating the mass defects of exact masses to chemical formulas.[43,44] We have developed a simpler approach that exhaustively determines all hydrocarbon compound type distributions by the sequential comparison of the theoretical masses of compounds against those measured in the mass spectrum.[42] All compound types with the general hydrocarbon formula $C_nH_{2n+z}S_aO_bN_cY_d$ are exhaustively determined by the method. Y represents any additional heteroatom that may be part of the hydrocarbon molecule and a, b, c, d are the numbers of the different heteroatoms in the formula. Most importantly, the method determines the relative amounts of the ion species in overlapping mass doublets by consideration of the unresolved peak abundance and the difference between the measured and theoretical masses.

Figure 4 shows a summary of the compound class results obtained from the analysis of a hydrocracker feedstock and the corresponding product. All heteroatom-containing compound classes are reduced in concentration due to hydrocracking, while the relative concentration of the heteroatom-free hydrocarbons is increased. The Z-series distributions of the $C_nH_{2n+z}S$ class, which is the most abundant heteroatom-containing compound class is shown in Figure 5. The MW distributions for the dibenzothiophenes are shown in Figure 6. It is important to notice that the method[42] at a nominal resolving power of ~15,000 (10% valley criterion) using a magnetic sector mass spectrometer permits the separation of mass doublets (e.g., C_3/SH_4) that would require ~150,000 resolving power for separation at m/z 500. The changes in the Z-series distributions of the compound class containing a single oxygen atom are shown in Figure 7. The results in Figures 5 and 7 show that it is more difficult to remove the oxygen atom from the dibenzofuran structure that it is to remove the sulfur atom from the dibenzothiophene structure. The method provides an enormous amount of information, which plays a critical role to the understanding of the mechanisms of hydrocracking and the other raw material processing methods used at the refinery.

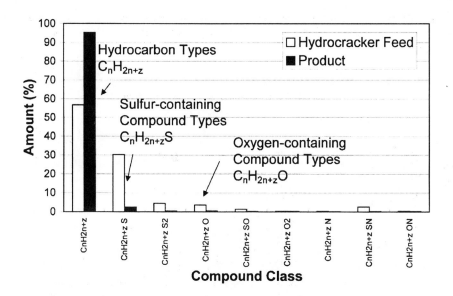

Figure 4. Summary of compound class results obtained from the analysis of a hydrocracker feedstock and the corresponding product. Reproduced with permission from *Rapid Commun. Mass Spectrom.* **1999**, 13, 1031-1051.

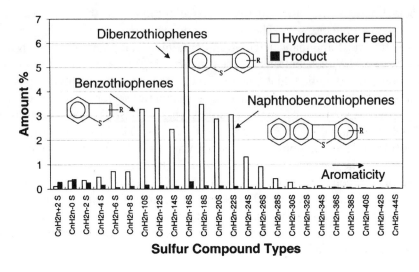

Figure 5. The analysis of a petroleum feedstock and its hydrocracked product by HRMS: Z-series distributions of compound class $C_nH_{2n+Z}S$. Reproduced with permission from *Rapid Commun. Mass Spectrom.* **1999**, 13, 1031-1051.

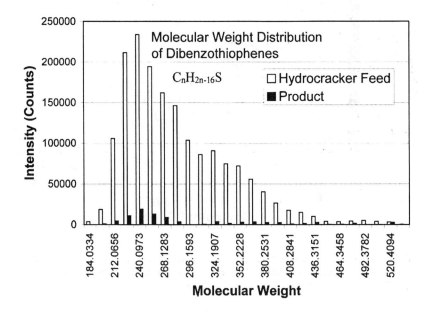

Figure 6. Molecular weight distributions of compound types $C_nH_{2n-16}S$ in a petroleum feedstock and its hydrocracked product. Reproduced with permission from *Rapid Commun. Mass Spectrom.* **1999**, 13, 1031-1051.

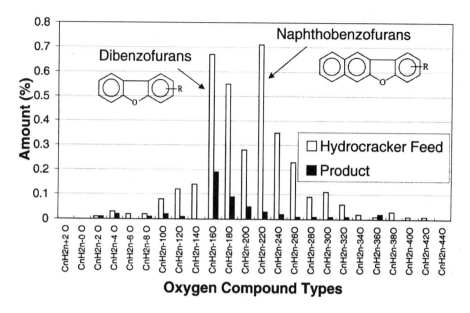

Figure 7. The analysis of a petroleum feedstock and its hydrocracked product by HRMS: Z-series distributions of compound class $C_nH_{2n+z}O$. Reproduced with permission from *Rapid Commun. Mass Spectrom.* **1999**, 13, 1031-1051.

The analysis of aromatic fractions has been reported using Fourier tranform ion cyclotron resonance (FT-ICR) MS instrumentation capable of ultrahigh resolution (>300,000) and high mass accuracy (<1ppm).[45] Although these instruments were originally associated with narrow mass and dynamic range limitations,[46] instrument re-designs and implementation of higher magnetic fields have overcome these barriers, allowing the successful characterization of complex petroleum mixtures.[47,48,70]

3.5   Olefins

Olefins are reactive compounds produced in the various processing stages at the chemical plants and refineries. Their reactive nature is often responsible for product quality issues and operational upsets due to high molecular compounds formed from polymerization reactions. The availability of analytical methods to determine the nature and amounts of olefins in hydrocarbon mixtures is therefore important. GC-MS has the capability to measure individual olefin molecules in the light boiling fractions (e.g., gasoline) but their determination in heavier fractions becomes problematic due to the increasing complexity of the fractions. Characterization from conventional 70 eV EI mass spectra is difficult because olefins and cycloalkanes (naphthenes) have the same molecular weight and very similar fragmentation patterns. An additional challenge is the concentration of the olefins, which is typically much lower than that of the naphthenes. We have developed a chemical ionization (CI) method based on the use of acetone as a reagent compound to selectively analyze olefins in the presence of naphthenes. Intense acetyl adducts are formed with the olefins but not with the naphthenes.[49] The capability of the method to selectively analyze olefins is shown in Figure 8. Figure 8A shows a part of the total ion current corresponding to the C_7 alkene isomer distribution obtained from the analysis of a whole gasoline by conventional 70 eV EI. No positive identification can be made for the olefins in the gasoline from the 70 eV EI results. Figure 8B shows the corresponding C_7 olefin distribution (m/z 144) for an olefin extract of the gasoline ionized by acetone-d_6. The olefin extract was obtained by modifying ASTM method D 1319-89 to permit the collection of the olefin fraction. Figure 8C shows the whole gasoline sample analyzed by acetone-d_6 CI. The traces in Figures 8B and 8C are almost identical illustrating the selectivity of the method. Similar results are obtained for olefins in heavier fractions.

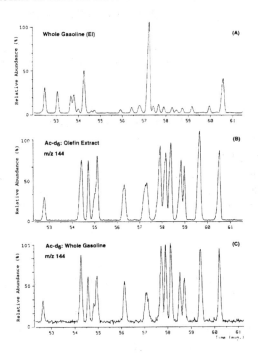

Figure 8. Analysis of C_7 olefins in gasoline: (A) part of total ion chromatogram obtained from the analysis of a whole gasoline by EI GC-MS; (B) part of selected ion chromatogram obtained from the analysis of the corresponding gasoline olefin extract by acetone-d_6 CI MS; (C) part of selected ion chromatogram obtained from the analysis of whole gasoline by acetone-d_6 CI MS. Reproduced with permission from *Anal. Chem.* **1997**, 69, 1550-1556.

4. CORROSION

4.1 Sulfur Compound Types

Reactive sulfur compound types such as sulfides, polysulfides, mercaptans and H_2S contribute to the corrosion of pipelines and refinery vessels. Knowledge of the types and amounts of the different sulfur compound types in petroleum feeds and products is therefore required for the management of corrosion. Additionally, the detection of the different sulfur compound types is important for the development of new processing methods for their removal from refinery products. The determination of the most abundant sulfur types in petroleum (thioaromatics and aliphatic sulfides) is challenging because these are neutral compounds, very similar in polarity to the bulk hydrocarbons. Development of separation methods to selectively extract them from the hydrocarbon matrix is generally difficult, involving several chemical treatment steps.[50,51] Moreover, the extraction efficiencies may depend on the molecular weights and structures of the individual sulfur compounds. Advanced spectroscopic techniques[52] have been successful in differentiating the relative amounts of thiophenic and sulfidic sulfur types but they cannot provide direct information on the MW or boiling range properties of individual compounds.

The distributions of aromatic sulfur compound types have been obtained by isobutane chemical ionization GC-MS.[53] GC/HRMS has been used to determine the distributions of aromatic sulfur types in oil fingerprinting studies.[54] In earlier work, Gallegos,[55] used the CSH^+ fragment ion in GC/HRMS experiment as characteristic fingerprint to monitor the distributions of sulfur compound types in fluid catalytically cracked naphtha. We have found that characteristic fragment ions are produced by the different sulfur compound types in 70 eV EI experiments.[56] Table 2 summarizes the fragment ions and their corresponding compound types. Figure 9 shows the distribution of the CSH^+ ion in a gas oil before and after hydrofining, and in Lavan crude oil acquired by GC/HRMS. Hydrofining severely reduces the amount of total sulfur. The method can be simplified by using LRMS instrumentation and multiple linear regression methods.[56] Tandem mass spectrometry (MS-MS) can also be used to selectively monitor the sulfur species in petroleum. Table 3 summarizes the most common MS-MS transitions for the different sulfur compound types (70 eV EI). Utilizing electrospray ionization, Rudzinski et al.[57] have reported the selective ionization of polyaromatic sulfur heterocyclic compounds by complexation with Pd^{+2}. The main benefits of the MS methods over the chemical methods are rapidity and simplicity, with no need for chemical separation of the sulfur fractions. A combination of the two can lead to extensive molecular information about the individual sulfur compound types.[51]

Table 2. Characteristic Fragment Ions in the 70 eV EI Mass Spectra of Different Sulfur Compound Types

Fragment Ion	Mass/charge	Sulfur Compound Type
$C_2H_5S^+$	61.0112	RSR/RSH
$H_2S_2^+$	65.9598	RSSR
S_2^+	63.9441	S_x, RSSR
CH_3S^+	46.9955	RSR/RSH/RSSR
CSH^+	44.9799	RSR/RSH/RSSR, Thiophenes
SH^+	32.9799	H_2S, RSR

Table 3. Characteristic MS-MS Transitions of Different Sulfur Compound Types

Precursor Ion	Product Ion	Compound Type	Reaction
CH_3S^+ 47 m/z	CHS^+ 45 m/z	RSR/RSH/RSSR	(loss of H_2)
$C_2H_5S^+$ 61 m/z	CHS^+ 45 m/z	RSR/RSH	(loss of CH_4)
$H_2S_2^+$ 66 m/z	S_2^+ 64 m/z	RSSR	(loss of H_2)

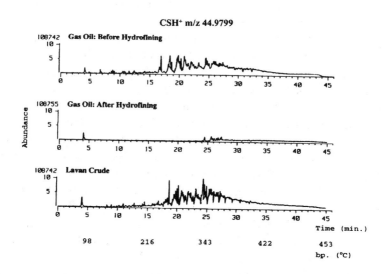

Figure 9. The distribution of m/z 44.9799 (CSH^+) monitored by GC/HRMS for a gas oil before hydrofining, a gas oil after hydrofining, and for Lavan crude oil.

4.2 Organic Acids

Due to their ionic character, it is much simpler to extract organic acids from hydrocarbon matrices. Seifert and Teeter originally used HRMS and other analytical methods to obtain detailed information about the nature of organic acids in petroleum.[58] The ionic character of the acids makes them amenable to several mass spectrometric ionization methods, among them: chemical ionization (CI),[59] fast atom bombardment (FAB),[60,61] atmospheric pressure chemical ionization (APCI),[61] and electrospray ionization (ESI).[47,61,62] We have used ESI MS to analyze commercially available acidic extracts and acidic fractions of whole crude oils.[62] The MW distributions of two commercially available naphthenic acid extracts are shown in Figure 10. Figure 11 shows the MW distributions of two crude oil acidic extracts obtained by ESI MS.

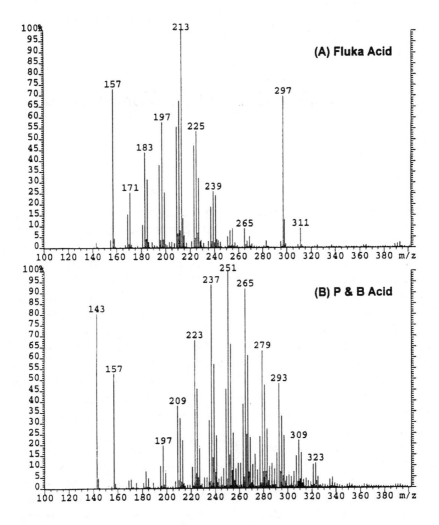

Figure 10. The MW distributions of commercial naphthenic acid extracts by ESI MS: (A) Fluka naphthenic acid extract; (B) P & B naphthenic acid extract.

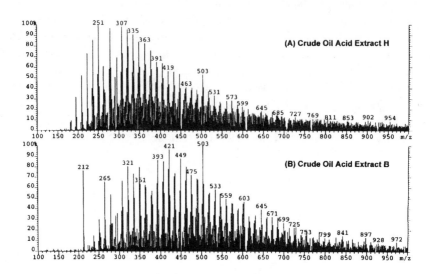

Figure 11. The MW distributions of two crude oil acid extracts analyzed by ESI MS.

4.3   Nitrogen Compounds

The characterization of basic nitrogen-containing compounds in crude oils and fractions is highly desired because these compounds are associated with the deactivation of zeolite-based acidic catalysts used in fluid catalytic and hydrocracking processes.[63] Basic nitrogen compounds also, although not directly associated to corrosion, they may affect the determination of the acids in the petroleum streams and may in fact reduce their effective corrosivity. The storage stability of fuel products depends on the presence of some non-basic nitrogen compound types.[64] Due to the polar nature of nitrogen compounds, it has been possible in the past to develop methods for their extraction from hydrocarbon matrices.[65] Chemical ionization with basic reagent gases also, has been used successfully for the selective determination of nitrogen compounds without the need for prior separation. This is an important advantage of the mass spectrometric techniques because lengthy and laborious fraction separation procedures are avoided.

Dzidic et al.[66] used ammonia CI GC-MS to determine carbazoles and benzocarbazoles in slurry oils. Creaser et al.[67] used ammonia CI in a ion trap mass spectrometer to selectively ionize nitrogen and sulfur heterocycles in petroleum fractions. Hsu et al.[68] introduced deuterated ammonia to determine the number of exchangeable hydrogen atoms on the structures containing the nitrogen atom. Veloski et al.[69] demonstrated the selective ionization of basic nitrogen compounds by ammonia CI and extensively characterized coal- and petroleum-derived products by HRMS. More recently, Marshall and co-workers demonstrated that electrospray ionization can be used in the positive ion mode, under ultrahigh FT-ICR MS resolving power conditions, to selectively ionize basic nitrogen compound types without the need for fractionation.[48,70]

5.   ADDITIVES AND CONTAMINANTS

The majority of petroleum additives are organic compounds that may contain several polar or ionic functional groups. These compounds are typically thermally unstable and analysis by conventional ionization methods that involve sample vaporization by heating (e.g., EI, CI) is unsuitable since it leads to decomposition prior to volatilization. Fast atom bombardment (FAB) MS was one of the most successful early ionization techniques that permitted the characterization of polar and ionic thermally labile additives.[71-73] Electrospray ionization (ESI) is readily compatible with on-line liquid chromatography and therefore is more suitable for the analysis of thermally unstable polar and ionic additives. Gough and Langley[74] recently used ESI to analyze several important classes of oilfield production chemicals. Oilfield chemical residues in water discharges and marine sediments were analyzed by Grigson et al.[75] by ESI-MS-MS.

We have used ESI, APCI MS, and MS-MS for the determination of indigenous and synthetic polar and ionic compounds in petroleum samples.[62,76] The APCI ionization process is more complex than that of ESI, producing mass spectra that may be difficult to interpret. ESI mass spectra are simpler and easier to interpret. However, in ESI absolute ion signal does not increase linearly with analyte concentration over a wide dynamic range, and there may be signal suppression effects depending on the nature of the mixture components. Nevertheless, it is possible to extend the dynamic range of ESI by using an internal standard, and reduce the possibilities for signal suppression by using LC to separate the mixture components. The dynamic range is wider for APCI (e.g., $> 10^6$:1) than for ESI (e.g., $< 10^4$:1). The sensitivity of ESI for the analysis of representative ionic compounds was found to be 10 to 100 times higher than that of APCI.[76]

Figure 12 shows selected ion chromatograms obtained from an LC/ESI MS experiment for the analysis of a diesel lubricity additive (m/z 390), a gasoline corrosion inhibitor (m/z 613), and an internal standard (m/z 308). The concentration of the additives in the fuels was 1 ppm each, and the concentration of the internal standard was 5 ppm. For the analysis of the diesel lubricity additive a polar extract was prepared by liquid-liquid extraction (70:30 fuel:methanol). A diluted sample (~10:1) in methanol was used for the analysis of the gasoline corrosion inhibitor. Figure 13 shows the calibration curves obtained for the two additives by normalizing the measured ion signal of the additives to that of the internal standard, over the 1 to 100 ppm concentration range.

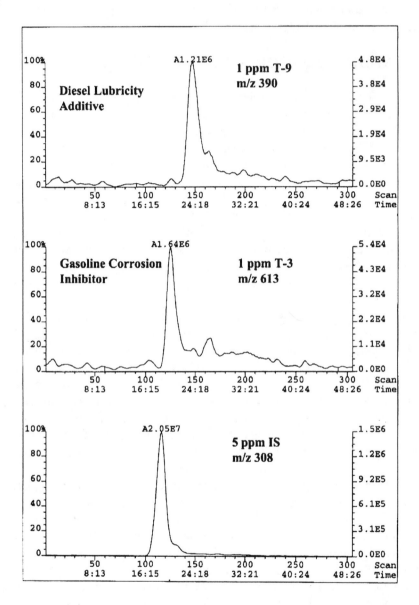

Figure 12 The analysis of additives by LC/ESI MS: selected ion chromatograms of (A) a diesel lubricity additive (1 ppm), (B) a gasoline corrosion inhibitor (1 ppm), and (C) an internal standard (5 ppm).

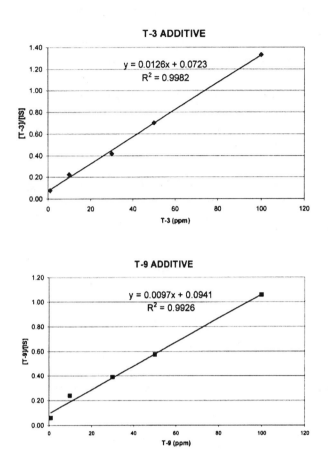

Figure 13. The analysis of additives by ESI MS: calibration curves obtained for (A) a gasoline corrosion inhibitor, and (B) a diesel lubricity additive.

Figures 14 and 15 illustrate the complementary nature of high-resolution mass spectrometry and MS-MS for the identification of unknown contaminants in petroleum samples. Figure 14 shows the molecular ion region of the high resolution (~10,000) CI (iso-butane) mass spectrum of a gasoline contaminant. The contaminant was collected from a filter in a gasoline stream that could not meet product specifications. The measured mass of the contaminant peak was consistent with the $C_8H_{19}O_5$ chemical formula (-9 ppm mass error). The MS-MS spectrum shown in Figure 15 produced an excellent match with the 70 eV EI

mass spectrum of tetraethylene glycol, which was identified as the most likely
contaminant structure. We have found that the tandem mass spectra of certain
compounds are very similar to those in standard 70 eV EI libraries of spectra,
which can be used for the rapid identification of unknowns.[77] In the absence of
good matches with existing libraries of spectra (MS or MS-MS), the most likely
structure can be validated by the analysis of model compounds or by
interpretation of the MS-MS spectra from first principles.

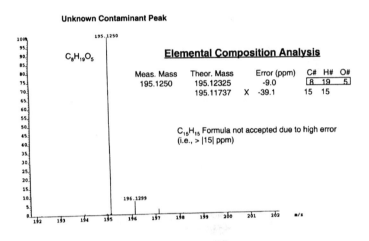

Figure 14. The analysis of unknown gasoline contaminant by HR-CI (iso-butane) MS.

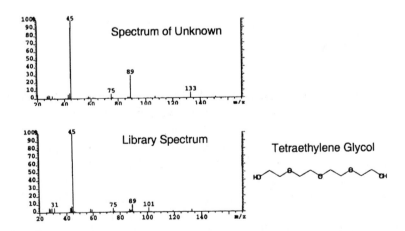

Figure 15. MS-MS spectrum of unknown gasoline contaminant and the corresponding 70 eV EI
library match (tetraethylene glycol).

A very powerful automated method has been developed for the characterization of contaminants and petroleum mixtures that combines the complementary capabilities of high resolving power mass spectrometry and MS-MS.[77] The method permits the automated acquisition of the product ion spectra of all precursor ions in a selected mass range by using a magnetic sector/orthogonal acceleration time-of-flight (oa-TOF) tandem mass spectrometer. High resolution MS produces a list of the most likely chemical formulas of the compounds in a sample and MS-MS provides structural information and validation of the precursor ion species identified in the MS experiment. The method is most useful for the analysis of complex spectra containing peaks at each nominal mass and peak distributions extending over a wide mass range.

6. ASPHALTS AND NON-BOILING FRACTIONS

Field desorption (FD MS) has been the most suitable method for the characterization of neutral heavy and non-boiling petroleum fractions.[78-80] Typically, the same ionization source is used as that for field ionization (FI) and the sample is deposited directly on the emitter. The ionization mechanism is related to that of field ionization with additional effects from the sample matrix. FD typically produces mass spectra with abundant molecular ions and negligible fragment ions. In that fashion, the method permits the acquisition of the molecular weight distributions of thermally labile and non-volatile molecules that cannot be obtained with other ionization methods. Unfortunately, due to the weak ion current signal generated by the method and the need for higher resolving power as a function of mass, it has been difficult in the past to perform high resolution experiments for the determination of the chemical composition of the molecules in heavy and non-boiling petroleum fractions. We have found that 70 eV EI or low-energy ionization (~10 eV) using a heated direct insertion probe provides adequate ion current signal for routine high resolving power experiments (~30,000). The direct insertion probe does not permit the vaporization of the entire heavy sample but it allows for the characterization of the volatile components under the experimental conditions (~10^{-6} Torr, 650°C maximum probe temperature). In that fashion, it has been possible to obtain detailed molecular information about asphalts generated from different crude oils.

Laser desorption (LD) MS has been used for the determination of the molecular weight distributions of heavy petroleum fractions.[81,82] A major question with the use of the laser ionization techniques relates to their ability to desorb all non-boiling components in the heavy sample. We have examined the possibility of using ESI MS for the analysis of non-boiling petroleum fractions.[83]

Stable adduct ions are formed between the silver ion and aromatic structures under conventional ESI conditions. Molecular weight distributions of vacuum residues obtained by Ag⁺ ESI MS are similar to those obtained by FD MS (Figure 16). The Ag⁺ ESI MS distribution is shifted to higher masses compared to the FD MS distribution due to adduct formation with the silver ion (m/z 107, 109). The Ag⁺ ESI method permits the generation of fragment ions (Fig. 16A), useful for the interpretation of spectra, under higher cone voltage conditions (e.g., >50 V). Fragment ions are not commonly generated under FD MS conditions (Fig. 16B). Two of the most significant features of the Ag⁺ ESI method are its high sensitivity and direct compatibility with LC methods of separation.

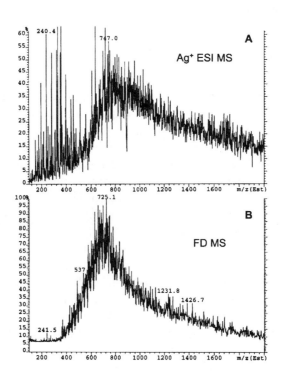

Figure 16. Distributions of petroleum vacuum residue obtained by: (A) Ag⁺ ESI MS (70 V cone voltage); (B) field desorption MS.   Reproduced with permission from *Anal. Chem.* **2002**, 74, 1408-1414.

7.   POLYMERS AND RESIDUES

Metal cationization laser desorption MS methods have been used for the characterization of non-volatile commercially important hydrocarbon polymers.[84-86] The use of mass spectrometry is desired for its ability to directly measure the MW distributions of polymers rather that the indirect determinations obtained by techniques such as light scattering, vapor phase osmometry, and gel permeation chromatography. The metal cationization laser desorption techniques compare favourably with the conventional FD MS methods of polymer analysis by mass spectrometry having the advantages of simplicity and rapidity. However, there are difficulties associated with the analysis of the more inert hydrocarbon polymers such as polyethylene with the laser desorption methods.[87] Our work with Ag^+ ESI MS indicates that the analysis of hydrocarbon polymers by metal cationization ESI MS is possible, in a fashion analogous to the laser desorption methods. ESI MS offers the additional advantages of direct compatibility with LC separation methods and on-line reagent ion complexation options for selective ionization.

In addition to the MW information, information is also required about the structures of the polymers. Structural information can be obtained from MS-MS experiments. Ionization techniques such as FD, LD or MALDI (matrix-assisted laser desorption/ionization) are limited from the sample amount available for consumption during MS-MS experiments. If the sample amount expires before all desired MS-MS experiments are performed, then a new sample has to be loaded for analysis. On the other side, ionization techniques such as FI, EI/CI, ESI, and APCI are readily interfaced with continuous introduction systems for the performance of long MS-MS experiments. Ultra-fast MS-MS experiments are expected to be possible with the tandem coupling of time-of-flight mass spectrometers.

Pyrolysis GC-MS has been a simple and reliable method for the structural characterization of polymers and the identification of impurities.[88-90] The method does not permit the determination of the MW distributions of polymers but it can be used for the fingerprinting of polymers and the determination of structural information. Pyrolysis GC-MS libraries can be created for the rapid identification of polymers.[91] We have introduced the use of a thermal extraction unit for the furnace-type pyrolysis of polymers. The system offers several advantages for the routine analysis of polymers and unknown chemical plant residues, among them: good reproducibility, minimum secondary reactions, capability for quantitative analysis, and minimum sample handling. A unique feature of the system is its ability to determine the extent of pyrolysis from the weight of the sample that did not undergo pyrolysis. In that fashion, it is possible to obtain thermogravimentric analysis (TGA)-type of information and collect the pyrolysis residue (typically inorganic components or coke) for further analysis (e.g., elemental composition analysis, microscopy, etc). The powerful capabilities of the thermal extraction system used as a furnace type pyrolyzer are shown in Figure 17. The figure shows the total ion current obtained from the analysis of an unknown residue from the chemical plant. For the analysis, the pyrocell was heated from ambient temperature to 600°C at a rate of 50°C/min. Although complete identification of all sample components was not possible due to the complexity of the sample (and perhaps not needed), identification of the most abundant sample components provided sufficient information for identification of the origin of the residue.

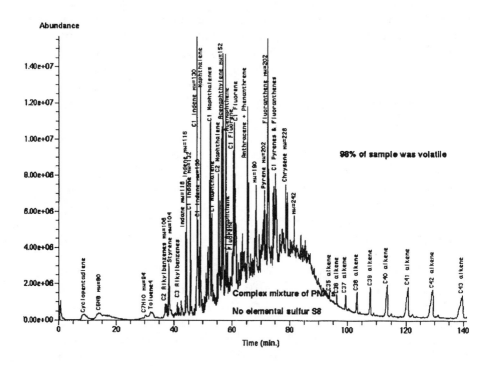

Figure 17. Pyrolysis/GC-MS total ion chromatogram obtained from the analysis of an unknown chemical plant polymer residue.

8.   CONCLUSION AND FUTURE CHALLENGES

The analysis of petroleum and petrochemical samples is challenging due to the nature of the compounds in the mixtures which can range from simple gases (e.g., CH_4) to non-volatile solids (e.g., asphaltenes, polymers). To address the analytical challenges posed by these compounds several mass spectrometric methods with different capabilities have been developed. Table 4 contains a summary of sample types from the most common application areas and the corresponding most suitable mass spectrometric methods. LRMS methods are simple and widely available which makes them attractive for many relatively simple applications (e.g., qualitative and quantitative analysis of simple mixtures). HRMS methods are more specialized but they can provide unique information about hydrocarbon and heteroatom-containing compounds (e.g., separate isobaric ion species and provide elemental composition information) in mixtures too complex to be separated by chromatographic methods.

Table 4. Summary of Sample Types and Mass Spectrometric Methods

Sample Type	Method
Whole Crude Oils	LRMS, GCD/MS, HRMS
Saturated Fractions	D 2786, GC-MS, FI/FD
Aromatic Fractions	D 3239, HRMS
Polar Fractions	HRMS, GC-MS, ESI, APCI
Olefins	GC-MS, Acetone CI
Sulfur Compound Types	GC-MS, HRMS
Acids	HRMS, ESI, FAB, APCI
Bases	HRMS, NH_3 CI, ESI
Ionics/Salts	FAB, ESI, APCI, LD/MALDI
Asphalts	Probe HRMS, Ag^+ ESI, LD/MALDI, Py-GC-M
Polymers	Py-GC-MS, ESI, FD, LD/MALDI
Insolubles/Solids	Py-GC-MS, Probe HR EI/CI, LD/MALDI

EI and FI/FD are general ionization methods that can analyze compounds independently of their chemical nature. Unfortunately, the boiling properties of the compounds affect the analysis. Non-boiling compounds cannot readily be analyzed by EI and FI and volatile compounds are not suitable for analysis by FD under typical experimental conditions. Due to the high vacuum requirements for these ionization methods, compatibility is restricted mainly to gas chromatography. CI, FAB, LD, APCI, and ESI are selective ionization methods, which can analyze specific compound types in complex mixtures. Depending on the needs of the applications, the ability to detect only the compound types of interest can be very useful. High-pressure chromatographic methods such as LC are most conveniently interfaced with mass spectrometers via ESI and APCI interfaces. Unfortunately, ESI cannot analyze neutral hydrocarbons under conventional experimental conditions. The use of APCI is also most suitable for polar and ionic compounds.

An ideal ionization method would be directly compatible with high-pressure chromatographic methods and would permit the analysis of compounds independently of their boiling properties, their polarity or ionic character. Such a method is not currently available. Efforts to that end have been reported by the development of atmospheric pressure photoionization (APPI) reported by Bruins and co-workers[93] and cap-EI by Cappiello et al.[94] Although there are limitations with both of these ionization methods, work in that direction is expected to lead to the development of more capable and versatile ionization methods.

FT-ICR MS is expected to play a major future role in the characterization of petroleum mixtures. In addition to the ultrahigh resolution requirements however, petroleum and petrochemical applications require fast chromatography and wide dynamic range capabilities. Quadrupole instruments provide rapid analysis and wide dynamic range but at low resolving powers. TOF technology has shown the capabilities for fast and accurate mass analysis [95] but with limited dynamic range and resolving power. High-resolution magnetic sector instruments still remain the most versatile instruments for the needs of petroleum and petrochemical applications. The ideal mass spectrometer should have ultrahigh resolving power capabilities, wide mass and dynamic ranges, and acquisition rates adequate for fast chromatography experiments.

9.   REFERENCES

1. Boduszynski, M. M. *Energy Fuels* **1987**, 1, 2-11.
2. Boduszynski, M. M. *Energy Fuels* **1988**, 2, 597-613.
3. Brown, R. A.; Taylor, R. C.; Melpolder, F. W.; Young, W. S. *Anal. Chem.* **1948**, 20, 5-9.
4. Brown, R. A. *Anal. Chem.* **1951**, 23, 430-437.
5. O'Neal, M. J.; Wier, T. P. *Anal. Chem.* **1951**, 23, 830-843.
6. Teeter, R. M. *Mass Spectrom. Rev.* **1985**, 4, 123-143.
7. Chasey, K. L.; Aczel, T. *Energy Fuels* **1991**, 5, 386-394.
8. Chapman, J. R. *Practical Organic Mass Spectrometry: A Guide for Chemical and Biochemical Analysis*; John Wiley & Sons: Chichester, 1993.
9. Ashcroft, A. E. *Ionization Methods in Organic Mass Spectrometry*; The Royal Society of Chemistry: Cambridge, 1997.
10. Gruse, W. A.; Stevens, D. R. *Chemical Technology of Petroleum*; McGraw-Hill: New York, 1960.
11. Canadian General Standards Board (CGSB) Method Can/CGSB-3.0 No. 14.3-99: "Standard Test Method for the Identification of Hydrocarbon Components in Automotive Gasoline Using Gas Chromatography", Canadian General Standards Board, Ottawa, Canada, K1A 1G6.
12. ASTM Method D 2427-87. *Annual Book of ASTM Standards*; American Society for Testing and Materials: Philadelphia, PA.
13. ASTM Method D 5769-95. *Annual Book of ASTM Standards*; American Society for Testing and Materials: Philadelphia, PA.
14. Teng, S. T.; Williams, A. D.; Urdal, K.; *J. High Resol. Chromatogr.* **1994**, 17, 469-475.
15. Altgelt, K. H.; Boduszynski, M. M. *Composition and Analysis of Heavy Petroleum Fractions*; Marcel Dekker: New York, 1994.
16. Tebbal, S.; Kane, R. D. *Prepr.-Am. Chem. Soc., Div. Pet. Chem.* **1998**, 43, 1, 111-113.
17. Rizvi, S. Q. A. *Lubr. Eng.* **1999**, 55, 4, 33-39.
18. Gallegos, E. J.; Green, J. W.; Lindeman, L. P.; LeTourneau, R. L.; Teeter, R. M. *Anal. Chem.* **1967**, 14, 1833-1838.
19. Bouquet, M.; Brument, J. *Fuel Sci. Technol. Int.* **1990**, 8, 961-986.
20. Fafet, A.; Bonnard, J.; Prigent, F. *Oil Gas Sci. Technolog.* **1999**, 54, 453-462.
21. Robinson, C. J. *Anal. Chem.* **1971**, 43, 1425-1434.
22. Roussis, S. G.; Fitzgerald, W. P. *Anal. Chem.* **2000**, 72, 1400-1409.
23. Ashe, T. R.; Roussis, S. G.; Fedora, J. W.; Felsky, G.; Fitzgerald, W. P. *U.S. Patent* 5,699,269 1997.
24. Scheppele, S. E.; Hsu, C. S.; Marriott, T. D.; Benson, P. A.; Detwilder, K. N.; Perreira, N. B. *Int. J. Mass Spectrom. Ion Phys.* **1978**, 28, 335-346.
25. Kasamatsu, K.; Noda, M. *Sekiyu Gakkaishi* **1983**, 26, 369-376.
26. Gallegos, E. J.; Fetzer, J. C.; Carlson, R. M.; Pena, M. M. *Energy Fuels* **1991**, 5, 376-381.

27. Briker, Y.; Ring, Z.; Iacchelli, A.; McLean, N.; Fairbridge, C.; Malhotra, R.; Coggiola, M. A.; Young, S. E.; *Energy Fuels* **2001**, 15, 996-1002.
28. Liang, Z.; Hsu, C. S. *Energy Fuels* **1998**, 12, 637-643.
29. Roussis, S. G.; Fedora, J. W. *Proceedings of the 46th ASMS Conference on Mass Spectrometry and Allied Topics*, Orlando, Florida, May 31 - June 4, 1998; 1078.
30. Dagan, S.; Amirav, A. *J. Am. Soc. Mass Spectrom.* **1995**, 6, 120-131.
31. Vuica, A.; Faubert, D.; Evans, M.; Bertrand, M. J. *Proceedings of the 46th ASMS Conference on Mass Spectrometry and Allied Topics*, Orlando, Florida, May 31 - June 4, 1998; 1498.
32. Stein, S. E. *J. Am. Soc. Mass Spectrom.* **1999**, 10, 770-781.
33. Roussis, S. G.; Fitzgerald, W. P. *Energy Fuels* **2001**, 15, 477-486.
34. Robinson, C. J.; Cook, G. L. *Anal. Chem.* **1969**, 41, 1549-1554.
35. Stafford, C.; Morgan, T. D.; Brunfeldt, R. J. *Int. J. Mass Spectrom. Ion Phys.* **1968**, 1, 87-92.
36. Roussis, S. G.; Cameron, A. S. *Energy Fuels* **1997**, 11, 879-886.
37. Johnson, B. H.; Aczel, T. *Anal. Chem.* **1967**, 39, 682-685.
38. Aczel, T.; Allan, D. E.; Harding, J. H.; Knipp, E. A. *Anal. Chem.* **1970**, 341-347.
39. Allgood, C.; Ma, Y. C.; Munson, B. *Anal. Chem.* **1991**, 63, 721-725.
40. Hsu, C. S.; Qian, K.; *Anal. Chem.* **1993**, 65, 767-771.
41. Roussis, S. G.; Fitzgerald, W. P.; Cameron, A. S. *Rapid Commun. Mass Spectrom.* **1998**, 12, 373-381.
42. Roussis, S. G. *Rapid Commun. Mass Spectrom.* **1999**, 13, 1031-1051.
43. Kendrick, E. *Anal. Chem.* **1963**, 35, 2146-2154.
44. Hsu, C. S.; Qian, K.; Chen, Y. C. *Anal. Chim. Acta* **1992**, 264, 79-89.
45. Guan, S.; Marshall, Scheppele, *Anal. Chem.* **1996**, 68, 46-71. Rodgers, R. P.; White, F. M.; Hendrickson, C. L.; Marshall, A. G.; Andersen, K. V. *Anal. Chem.* **1998**, 70, 4743-4750.
46. Hsu, C. S.; Liang, Z.; Campana, J. E. *Anal. Chem.* **1994**, 66, 850-855.
47. Qian, K.; Robbins, W. K.; Hughey, C. A.; Cooper, H. J.; Rodgers, R. P.; Marshall, A. G. *Energy Fuels* **2001**, 15, 1505-1511.
48. Qian, K.; Rodgers, R. P.; Hendrickson, C. L.; Emmett, M. R.; Marshall, A. G. *Energy Fuels* **2001**, 15, 492-498.
49. Roussis, S. G.; Fedora, J. W. *Anal. Chem.* **1997**, 69, 1550-1556.
50. Snyder, L. R. *Anal. Chem.* **1961**, 33, 1538-1543.
51. *Analysis of Heavy Oils: Method Development and Application to Cerro Negro Heavy Petroleum*; NIPER-452, U.S. Department of Energy: Bartleville, Oklahoma, 1989.
52. George, G. N.; Gorbaty, M. L. *J. Am. Chem. Soc.* **1989**, 111, 3182-3186.
53. Dzidic, I.; Balicki, M. D.; Rhodes, I. A. L.; Hart, H. V. *J. Chromatogr. Sci.* **1988**, 26, 236-240.
54. Tibbetts, P. J. C.; Large, R.in *Petrolanal. '87: Dev. Anal. Chem. Pet. Ind.*, 45-57, Crump, J. B., Ed.; Wiley: Chichester, UK, 1988.
55. Gallegos, E. J. *Anal. Chem.* **1975**, 47, 1150-1154.
56. Roussis, S. G.; Fedora, J. W.; Cameron, A. S. *U.S. Patent* 5, 744, 702 1998.
57. Rudzinski, W. E.; Aminabhavi, T. M.; Tarbox, T.; Sassman, S.; Whitney, K.; Watkins,L. *Prepr.-Am. Chem. Soc., Div. Pet. Chem.* **2000**, 45, 60-63.
58. Seifert, W. K.; Teetter, R. M. *Anal. Chem.* **1970**, 42, 180-189. Seifert, W. K.; Teeter, R. M. *Anal. Chem.* **1970**, 42, 750-758.
59. Dzidic, I.; Somerville, A. C.; Raia, J. C.; Hart, H. V. *Anal. Chem.* **1988**, 1318-1323.
60. Fan, T. -P. *Anal. Chem.* **1991**, 65, 371-375.
61. Hsu, C. S.; Dechert, G. J.; Robbins, W. K.; Fukuda, E. K. *Energy Fuels* **2000**, 14, 217-223.
62. Roussis, S. G.; Fedora, J. W. *Proceedings of the 47th ASMS Conference on Mass Spectrometry and Allied Topics*, Dallas, TX, June 13-17, 1999, MPG 158.
63. Scherzer, J.; Gruia, A. J. *Hydrocracking Science and Technology*; Marcel Dekker: New York, Chapter 8, 1996.

64. Dorbon, M.; Bigeard, P. H.; Denis, J.; Bernasconi, C. *Fuel Sci. Technol. Int.* **1992**, 10, 1313-1341.
65. Later, D. W.; Milton, L. L.; Bartle, K. D.; Kong, R. C.; Vassilaros, D. L. *Anal. Chem.* **1981**, 53, 1612-1620.
66. Dzidic, I.; Balicki, M. D.; Hart, H. V. *Fuel* **1988**, 67, 1155-1158.
67. Creaser, C. S.; Krokos, F.; O'Neill, K. E.; Smith, M. J. C.; McDowell, P. G. *J. Am. Soc. Mass Spectrom.* **1993**, 4, 322-326.
68. Hsu, C. S.; Qian, K.; Robbins, W. K. *J. High Resol. Chromatogr.* **1994**, 17, 271-276.
69. Veloski, G. A.; Lynn, R. J.; Sprecher, R. F. *Energy Fuels* **1997**, 11, 137-143.
70. Hughey, C. A.; Hendrickson, C. L.; Rodgers, R. P.; Marschall, A. G. *Energy Fuels* **2001**, 15, 1186-1193.
71. Lyon, P. A.; Stebbings, W. L.; Crow, F. W.; Tomer, K. B.; Lippstreu, D. L.; Gross, M. L. *Anal. Chem.* **1984**, 56, 8-13.
72. Freas, R. B.; Campana, J. E. *Anal. Chem.* **1986**, 58, 2434-2438.
73. Hodson, J.; Pidduck, A. J. Report (1985), MQAD-TP-905, BR98163; Order No. N86-28708/3/GAR. From: Gov. Rep. Announce. Index (U.S.) 1986, 86, Abstr. No. 650,648.
74. Gough, M. A.; Langley, G. J. *Rapid Commun. Mass Spectrom.* **1999**, 13, 227-236.
75. Grigson, S. J. W.; Wlkinson, A.; Johnson, P.; Moffat, C. F.; McIntosh, A. D. *Rapid. Commun. Mass Spectrom.* **2000**, 14, 2210-2219.
76. Roussis, S. G.; Fedora, J. W. *Proceedings of the 48th ASMS Conference on Mass Spectrometry and Allied Topics*, Long Beach, California, June 11-15, 2000, WOB pm03:40.
77. Roussis, S. G. *Anal. Chem.* **2001**, 73, 3611-3623.
78. Prokai, L. *Field Desorption Mass Spectrometry*; Marcel Dekker, Inc.: New York, 1990.
79. Larsen, B. S.; Fenselau, C. C.; Whitehurst, D. D.; Angelini, M. M. *Anal. Chem.* **1986**, 58, 1088-1091.
80. Aczel, T.; Dennis, L. W.; Reynolds, S. D. *Proceedings of the 35th ASMS Conference on Mass Spectrometry and Allied Topics*, Denver, CO, May 24-29, 1987; pp 1066-1067.
81. Seki, H.; Kumata, F. *Energy Fuels* **2000**, 14, 980-985.
82. Suelves, I.; Islas, C. A.; Herod, A. A.; Candiyoti, R. *Energy Fuels* **2001**, 15, 429-437.
83. Roussis, S. G.; Proulx, R. *Anal. Chem.* **2002**, 74, 1408-1414.
84. Kahr, M. S.; Wilkins, C. L. *J. Am. Soc. Mass Spectrom.* **1993**, 4, 453-460.
85. Dean, P. A.; O'Malley, R. M. *Rapid Commun. Mass Spectrom.* **1993**, 7, 53-57.
86. Keki, S. Deak, G.; Zuga, M. *Rapid Commun. Mass Spectrom.* **2001**, 15, 675-678.
87. Chen, R.; Yalcin, T.; Wallace, W. E.; Guttman, C. M.; Li, L. *J. Am. Soc. Mass Spectrom.* **2001**, 12, 1186-1192.
88. Hummel, D. O.; Dussel, H.-J.; Czybulka, G,; Wenzel, N.; Holl, G. *Spectrochimica Acta,* **1985**, 41A, 279-290.
89. Blazso, M. *Rapid Commun. Mass Spectrom.* **1991**, 5, 507-511.
90. Mundy, S. A. J. *J. Anal. Appl. Pyrolysis* **1993**, 25, 317-324.
91. Matheson, M. J.; Wampler, T. P.; Johnson, L.; Atherly, L.; Smucker, L. *Amer. Lab.* May **1997**, 29, 24C-24F.
92. Roussis, S. G.; Fedora, J. W. *Rapid Commun. Mass Spectrom.* **1996**, 10, 82-90.
93. Robb, D. B.; Covey, T. R.; Bruins, A. P. *Anal. Chem.* **2000**, 72, 3653-3659.
94. Cappiello, A.; Balogh, M.; Famiglini, G.; Mangani, F.; Palma, P. *Anal. Chem.* **2000**, 72, 3841-3846.
95. Hsu, C. S.; Green, M. *Rapid Commun. Mass Spectrom.* **2001**, 15, 236-239.

Chapter 13

Chromatographic Separation and Atmospheric Pressure Ionization/Mass Spectrometric Analysis of Nitrogen, Sulfur and Oxygen Containing Compounds in Crude Oils

Walter E. Rudzinski
Department of Chemistry and Institute for Environmental and Industrial Science
Southwest Texas State University
San Marcos, Texas 78666

1. NSO COMPOUNDS IN CRUDE OIL

Crude oil is a naturally occurring, flammable liquid whose chemical composition varies depending upon the source. Crude oil and products in refinery streams primarily consist of hydrocarbons (50-90%) with the remainder composed of compounds containing N, S, O, and trace amounts of organometallics. Oxygen containing compounds usually are alcohols, phenols, furans and acids. The acids include alkanoic, naphthenic and aromatic species. Sulfur containing compounds include thiols, sulfides, disulfides and thiophene type compounds as well as inorganic sulfur. The nitrogen containing compounds usually are anilines, pyrroles, pyridines, indoles, quinolines and carbazoles. In addition to compounds which contain only one functional group based on NSO, recent results indicate that up to 50% of the compounds in crude contain two or more functional groups per molecule.[1,2] These multifunctional compounds have yet to be elucidated.

The naphthenic acids are of interest because they can provide information about the maturity and extent of biodegradation of a crude oil. Weathered oils tend to have a lower concentration of alkanoic acids which are generally the first to be metabolized by soil organisms. The naphthenic acids also have higher water solubility than hydrocarbons that can lead to environmental challenges associated with the remediation of refinery wastewater.[3] Finally, in

313

refineries the naphthenic acids distill into the gas oil and vacuum gas oil fractions and can cause corrosion at process temperatures of 250-400 °C.

Nitrogen compounds in the form of pyrrole benzologues and pyridine benzologues are known to play a role in catalyst deactivation through coke formation on the catalyst surface.[4-6] Nitrogen compounds are also present as porphyrin compounds which contribute to heavy metal binding and catalyst deactivation.

Sulfur compounds (0.025 to 11% by weight) in petroleum are known to poison catalysts used in processing. In addition, several sulfur heterocycles are suspected mutagens or carcinogens. Upon combustion of fuels rich in sulfur, SO_2 is liberated which contributes to acid rain upon reaction with atmospheric oxygen and water.[7]

In the following chapter, for the sake of completeness, I will briefly review some of the classical, generic methods used for preliminary fractionation of the crude, i.e., distillation and open column or adsorption chromatography. I will then focus on high performance liquid chromatographic methods that further fractionate and reduce the complexity of the sample, and are compatible with atmospheric pressure ionisation / mass spectrometry. I will then review, specific methods for the fractionation of N,S,O compounds; in particular, naphthenic acids and organosulfur compounds. Finally, I will review the application of mass spectrometric methods based on atmospheric pressure ionisation to the analysis of these NSO compounds.

2.   GENERAL SEPARATION METHODS FOR CRUDE OIL AND RELATED PRODUCTS

2.1   Distillation

In order to evaluate crude, the petroleum generally is separated first by distillation. Distillation provides a preliminary fractionation based on the relative volatility of the constituents in the complex petroleum mixture. The fractions produced consist of straight run gasoline, middle distillate, heavy distillate, and residual oil. The gasoline fraction boiling up to 200°C will include hydrocarbons having 4 to 12 carbon atoms, subdivided into five main classes: (1) straight-chain paraffins, (2) branched-chain paraffins, (3) alkylcyclopentanes, (4) alkylcyclohexanes, and (5) alkylbenzenes. The gasoline fraction also contains aromatics such as toluene, xylenes, and *p*-cumene in small amounts, dienes, as well as volatile nitrogen, sulfur and oxygen compounds.   Relative amounts of trace constituents may vary depending upon the composition of original petroleum feedstock. The middle distillate fraction will include hydrocarbons having 12 to 20 carbon atoms, boiling at about 185-345 °C, from which are obtained kerosene, diesel, heating oils, jet, rocket, and gas turbine fuels. These higher boiling fractions may contain alkylnaphthalenes, biphenyls, and other polynuclear aromatic hydrocarbons (PAHs) as well as condensed naphtheno-aromatics, aromatic olefins, and compounds containing sulfur (thiophenes), nitrogen (pyrroles, pyridines, indoles and carbazoles) and oxygen (furans). The heavy distillate, having 20 to 44 carbon atoms and boiling at about 345-540°C, is the source of waxes, lubricating oils, and feedstock for catalytic cracking to produce gasoline. This fraction and the residue are usually enriched in polyaromatic heterocycles, and porphyrin–type compounds. The residue usually contains asphalt that is a dark-brown to black solid or semisolid. Asphalt consists of three components: (1) asphaltene, a hard, friable infusible powder, (2) resin, a semisolid ductile and adhesive material, and (3) oil which is similar to the lubricating oil fraction from which it is derived.

2.2   Adsorption Chromatography

Once the crude oil has been distilled, several open column liquid chromatography (OCLC) and medium pressure liquid chromatography (MPLC) fractionations may be employed. These chromatographic methods are based on the relative affinity of constituents in the petroleum fraction for a polar surface, usually silica or alumina. By coupling adsorption chromatography which exploits the intermolecular attraction of a solute for a polar surface with distillation which is based on volatility, two distinct separation mechanisms are employed which lead to petroleum fractions differing in polarity with a limited molecular weight range.

A number of chromatographic methods have been validated for the separation of species after distillation. ASTM Method D1319 has been employed for light distillates,[8] ASTM D2549 for middle and heavy distillates,[9] and ASTM Method D2007 for species which boil above 260°C.[10] The proper application of ASTM Method D2007 depends upon a prior removal of insoluble materials such as asphaltene with *n*-pentane.

In an early paper, Corbett [11] determined the composition of an asphalt by first precipitating the asphaltene fraction with *n*-heptane. The soluble fraction was then fractionated on activated alumina by sequential elution with heptane, benzene, then (50:50) benzene: methanol and trichloroethylene to yield the saturate, naphthene-aromatic and polar aromatic fractions. As a result four generic compound classes were produced which were used to determine the effect of vacuum distillation and crude source.[11] Variant approaches based on chemical class have been widely used since then for the separation of crude oils and are collectively known as SARA method.[12-15]   See Figure 1. The method takes its name from the fractions produced; namely, saturates (S), aromatics (A), resins (R), and asphaltenes (A). Altgelt et al.[16] Altgelt and Boduszynski [17], Leontaritis [18], Mansfield et al.[19] and Barman et al.[20] have reviewed the literature in this area.

As an example of the applicability of the SARA method, Ali et al.[21] separated an Arabian heavy crude oil. Each fraction was characterized by elemental analysis, infrared (IR), nuclear magnetic resonance (NMR), and high-resolution mass spectrometry. Their structural characterization led to the conclusion that the asphaltene fraction is the most hydrogen deficient (i.e., it has the lowest H/C ratio of 1.11) followed by resin (1.38), aromatic (1.60), and saturate (1.95) fractions. The results confirmed that the asphaltene fraction has a much higher degree of ring condensation than the resin and aromatic fractions. The resin fraction also has a higher degree of oxygen content (4.9 %) that indicates the presence of carboxylic acids, phenols, and other oxy compounds in this fraction.

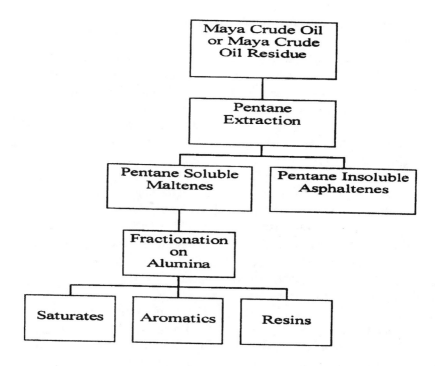

Figure 1. Scheme for the SARA Fractionation of a Crude Oil. (Reprinted from Ref. 60)

Akhlaq[22] also used the SARA method to separate crude oils and other petroleum fractions into various classes of aliphatics, one-, two-, and three-ring aromatics, polar compounds, resins, and asphaltenes. IR, NMR, and gas chromatography/mass spectrometry (GC-MS) were used for characterization.

2.3   High Performance Liquid Chromatography

After preliminary OCLC or MPLC fractionation, further separation is achieved using high performance liquid chromatography (HPLC). Silica, and -amino, -cyano, or -diol bonded silica columns have been extensively employed in the normal phase mode.[23,24] The HPLC separation of light, middle and gas oils isolates n-paraffins, isoparafins, naphthenes, olefins and aromatics. A study of eight different normal-phase columns showed that aminosilane and nitrophenyl columns provided the best selectivity for aromatic compounds.[25]

Hayes and Anderson [26-31] have developed several methods for hydrocarbon type analysis based on HPLC. In a culminating paper, they separated paraffins, olefins, naphthenes, and aromatics using on-line column switching HPLC with dielectric constant detection.[31] The separation is achieved using olefin-selective, unsaturate-selective and three naphthene-selective columns, two six-port valves and 2,2-dichloro-1,1,1-trifluoroethane as the mobile phase. They analyzed complex solutions of hydrocarbons and found that the accuracy for each structural group type was within 2% absolute, with a limit of detection of 4.0 vol % for naphthenes in a kerosene fuel. Robbins reported a fully automated HPLC system for the separation of six hydrocarbon types (saturate, 1-4 ring aromatics and polars) in a wide variety of heavy distillates.[32] Separation was carried out using two aminopropyl bonded phase columns to separate the saturate and monoaromatic species, while a dinitroanilino-propyl (DNAP) bonded phase column was used for the separation of aromatics. A ternary gradient of n-hexane, dichloromethane and isopropanol eluted the species. The separation efficiencies were verified by alternate separations and mass spectroscopy, and the aromaticities with ^{13}C NMR. The relative precision was determined to be 10%.

Although the aforementioned procedures have been applied primarily to saturate and aromatic species, they are also applicable to the analysis of NSO compounds which are also a mixture of saturates, mono-, di-, tri- and poly-aromatic and fused ring systems.

2.4   Mass Spectrometry

Although traditional thermal vaporization followed by electron impact or chemical ionization is effective for generating gas phase ions, other ionization methods are needed to extend MS to the high boiling and/or more polar molecules. With GC, the upper limit is about 60 carbon atoms, that can be extended to 100 carbon atoms, after which compound decomposition and stationary phase bleeding become significant.

Field desorption/field ionization (FD/FI) has been widely used for the ionization of high boiling, non-polar hydrocarbons.[33-37] The technique is based on introducing the sample to an emitting surface that consists of an array of sharp tips (cathode) which is very close to an anode. Upon application of a high electric field, quantum mechanical tunnelling of an electron is induced and the sample ionised. There is little sample fragmentation, and molecular ions are produced. Unfortunately, preparing the emitter is an art, and the sample loaded is quite small (less than a few ug) which makes it difficult to optimise the mass spectrometer for sample   analysis. In one of the more ambitious efforts, Bodusznski [37] analyzed a distilled vacuum oil for which he generated narrow boiling range fractions. These were then further separated by HPLC, followed by FD/FI.

LC-MS coupled with thermospray (TSP) ionization can produce protonated, molecular ions with a mass profile similar to FI/FD. Thermospray utilizes heat to nebulize the LC effluent and does not deposit excess energy into sample molecules; the ionisation process reduces fragmentation. Hsu and coworkers [38] used LC-MS coupled with thermospray to generate detailed hydrocarbon types from a vacuum residue.

More recently atmospheric pressure ionization/MS (API/MS) has been employed in the characterization of crude oils. API can either take the form of atmospheric pressure chemical ionization (APCI) or electrospray ionization (ESI) in either the positive or negative ion modes.

APCI is a soft ionization technique in which the sample is introduced as a fine mist of droplets that are vaporized in a high temperature tube. A high voltage is applied to a needle located near the exit end of the tube. The high voltage on the needle creates a corona discharge that forms reagent ions through a series of chemical reactions with solvent molecules and sheath gas (usually nitrogen). The reagent ions react with the sample molecules to form sample ions. Suitable solvents for APCI include hexane, cyclohexane and toluene.[39] We have used up to 50% dichloromethane with no deleterious effects. APCI is well suited for petroleum samples because of the limited solubility of oil in aqueous or alcohol solutions.

ESI is a soft ionization technique based on the nebulization of a sample by the application of a strong electric field. Generally ESI requires a more polar solvent, e.g., water, methanol, or acetonitrile, and sometimes the addition of an acid or base in order to promote ionization. Though rarely, acetone [40] and dichloromethane have also been employed. Spray formation is usually assisted pneumatically by means of a coaxial gas stream (Figure 2).

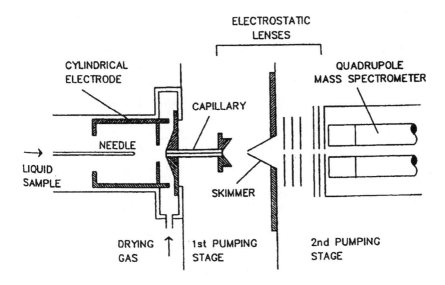

Figure 2. Fenn-Whitehouse ESI source design. Taken from J.B. Fenn et al. Mass Spectrom. Rev. **1990**, *9, 37.* (Reprinted by permission of John Wiley & Sons)

Ionization is achieved by either protonation of basic groups or deprotonation of acidic groups depending upon the polarity of the applied electric field. Though ESI is useful for easily-protonated peptides and proteins and for oligonucleotides that are easily deprotonated, the approach often fails for nonpolar substances and weakly acidic or basic species.

Both the APCI and ESI sources have been coupled to a number of different mass analyzers including: quadrupole, triple quadrupole, ion trap, double focusing magnetic sector, time-of flight and FT-ICR. The quadrupole is in essence a mass filter with low mass resolution, while the triple quadrupole is a sequence of three quadrupoles. The first and third are used for mass selection, while the second is ordinarily employed for in situ collisions of ions with inert gas for collision activated dissociation (CAD) experiments. The triple quadrupole allows for the determination of daughter and parent ions. Daughter ions are fragments obtained from the dissociation of a selected ion in the first quadrupole, while parent ions are molecular antecedents that produce a particular fragment in the third quadrupole. The triple quadrupole offers excellent sensitivity coupled with information about the molecular composition. The ion trap is a mass filter with limited capacity, but robust, and with the proper electronics, the ability to provide MS^n capability. MS^n refers to experiments in which the mass spectrum of a peak in a mass spectrum is obtained. Mass spectra can be obtained n times sequentially in order to obtain structural information about MS ions or fragments. Double focusing magnetic sector instruments (Double sector) have extremely high mass resolution, but are slow in obtaining wide range mass scans. Time-of-flight instruments are usually associated with matrix assisted laser desorption ionization (MALDI) and in principle have an infinite upper mass range. High field FT-ICR furnishes ultrahigh-mass resolving power that is needed to resolve mass overlaps in the high mass region; however, these systems are expensive.

Table 1. Analytical Figures of Merit for Various Mass Analyzers[41]

Type	Resolution[a]	Accuracy[b]	m/z limit[c]	Range[d]	Pressure[e]
Quadrupole	10^3	0.1%	10^4	10^5	10^{-5}
Ion trap	10^3-10^4	0.1%	10^5	10^5	10^{-3}
Double sector	10^5	5 ppm	10^4	10^7	10^{-7}
Time-of flight	10^3-10^4	0.01%-0.1%	10^6	10^5	10^{-7}
FTICR	10^6	1 ppm	10^4	10^5	10^{-9}

a) Resolution is defined as the ratio between an m/z of 1000 and the peak width in m/z units.
b) Accuracy is defined as the relative error at an m/z of 1000.
c) m/z limit is the highest mass/charge ratio observable.
d) Range is the concentration interval over which the measured signal abundance varies linearly. It is expressed as a ratio between the largest and smallest concentration.
e) Pressure is the operating pressure expressed in torr.

A further description of mass analyzer systems is beyond the scope of this chapter, but, in principle, all mass analyzers can be coupled with API sample introduction systems. See Table 1. for some of the analytical figures of merit for the various mass analyzers.

3.   METHODS FOR NSO COMPOUNDS

3.1   Separation of Acids

Crude oils are considered acidic if their total acid number (TAN) exceeds 0.5 mg KOH/g by nonaqueous titration. The TAN though useful does not provide any significant compositional information. In addition there is often little agreement between the TAN and the corrosivity of a crude oil. It is, therefore, imperative that further information be obtained with respect to the so-called naphthenic acids.

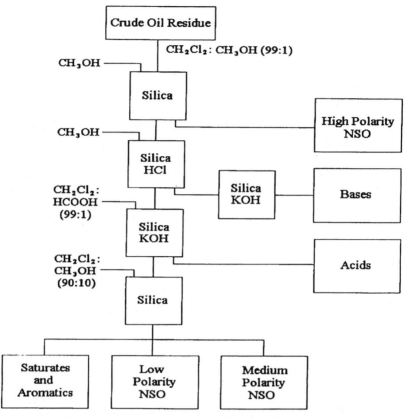

Figure 3. Scheme for the Acid-Base-Neutral Fractionation of a Crude Oil Residue. (Reprinted from Ref. 60)

Methods of obtaining carboxylic acid fractions include simple base extraction[42], or adsorption onto KOH impregnated silica gel.[43] A nonaqueous ion exchange method was developed by Green et al. [44] to separate liquid fossil fuels into acid, base and neutral concentrates. The acid concentrate obtained in this way was then separated into subfractions using in situ tetralkylammonium hydroxide modified silica HPLC.[45] Robbins [46] has pointed out some of the difficulties in extracting naphthenic acids. For example, aqueous caustic extraction cannot be used with acids >C_{20} because they form emulsions. Isolation efficiency is also difficult to assess because of the formation of salts in solution and the co-extraction of weak acids like phenols and carbazoles.

Hertz et al.[47] separated shale oil samples into three fractions (acids, bases and neutrals) using an extraction procedure adapted from Schmeltz [48] for the isolation of individual organic compounds from shale oil fractions. Willsch and co-workers [49] then further developed the method and extended its applicability to other samples: sediment, rock, coal extracts, and crude oils. The method resulted in a separation of the sample into five hetero-compound fractions: acids, bases, high-polarity, medium-polarity, and low-polarity nitrogen-sulfur-oxygen (NSO) compounds. See Figure 3 for the separation scheme. Acids and bases were retained according to their affinity for modified basic and acidic silica, respectively. Saturates and aromatics, and low-polarity NSO compounds were then fractionated using a medium pressure liquid chromatography (MPLC) technique. The isolated fractions were quantified by gas chromatography/flame ionisation detection (GC/FID) as well as by GC-MS.

Recently, we used the Willsch approach to isolate and characterize the components of a Maya crude oil vacuum residue into high polarity, acidic, and basic as well as saturates and aromatics (Sat + Ar), low polarity (LP) NSO, and medium polarity (MP) NSO compounds. See Figure 3. The nitrogen and sulfur content increased with the polarity of the unretained neutrals. See Table 2.

Table 2. Elemental Analysis Data for a Maya Crude Oil Residue

Fraction	% Recov.	%C	%H	%N	%S	H/C	Mol. Wt.
Sat + Ar	23	82.19	10.79	0.24	2.85	1.56	1073
LP	57	83.89	11.03	0.34	4.02	1.56	1000
MP	9	78.93	9.77	0.94	4.18	1.47	877

These fractions, also characterized by liquid chromatography coupled with atmospheric pressure chemical ionization/mass spectrometry (LC/APCI/MS), yielded a number average molecular weight (Mol. Wt.) which decreased from 1073 to 877 with increasing polarity of the fraction. The decrease in average molecular weight with increasing polarity of the fraction was in good agreement with the results of Miller et al.[50] who analyzed the residuum maltene fractions of a Maya crude oil.

Solid phase extraction (SPE) is a fairly recent approach for the efficient extraction of naphthenic acids. SPE is attractive because of stationary phase uniformity, low solvent consumption and the potential to exploit specific interactions in order to separate petroleum hydrocarbons from non-hydrocarbons. Naphthenic acids have been separated using SPE on a cyano column,[51] aminopropyl-functionalized silica gel,[2] and a quaternary amine ion exchange column.[52] Generally, the sample is adsorbed onto the column, then washed with hexane to remove the hydrocarbons. The aromatics and polars are then eluted with solvents or a mixture of solvents of higher eluotropic strength, e.g., toluene, dichloromethane or acetone. The adsorbed acids are then removed by washing with dilute acid (e.g., formic or acetic acid), and then reconstituted in an appropriate solvent. For GC analysis, the samples are reacted with a derivatizing reagent in order to increase volatility. Methylation using BF₃/methanol has been used successfully followed by GC analysis.[42] Reaction with diazomethane and esterification with fluoroalcohols have also been employed.[53]

3.2 Atmospheric Pressure Ionization/ Mass Spectrometry of Naphthenic Acids

API/MS has been used in order to characterize the naphthenic acid fraction in crude oils. Hsu and coworkers evaluated chemical ionisation (CI), liquid secondary ion mass spectrometry (LSIMS), APCI and ESI in both positive and negative ion modes for the determination of the molecular distribution of acids.[54,55] They found that APCI and LSIMS gave very similar mass spectra, while the CI mass spectrum yielded a lower average molecular mass distribution. The authors attributed the lower value obtained to more fragmentation, and an incomplete volatilization of high molecular mass components. Although the results were not presented, ESI in the negative ion mode was an order of magnitude less sensitive than APCI. APCI using acetonitrile as a mobile phase yielded the cleanest spectra.

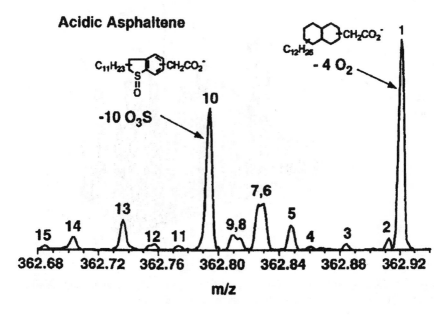

Figure 4. Ultrahigh Resolution mass spectral segment (m/z = 363) of an asphaltene sample. All 15 peaks resolved in the 0.26 mass window have been identified. (Reprinted with permission from Ref. 2)

Marshall and coworkers [2] used negative-ion microelectrospray high field Fourier Transform Ion Cyclotron Resonance/MS (FTICR/MS) for the analysis of heavy petroleums. They obtained an average mass resolving power, $m/\Delta m_{50\%} > 80,000$ where m is the nominal mass and $m_{50\%}$ is the mass spectral peak full width at half maximum peak height. They identified -SO_3, -SO_2 and -O_2 type acids present in a 45-fold concentrated extract as well as over 3,000 acids in a crude oil over the mass range $200 < m/z < 1000$ amu. Figure 4 illustrates the mass spectrum of a 0.26 mass window.

Siefert and coworkers [56,57] used exhaustive extraction of carboxylic acids and selective reduction to the parent hydrocarbons combined with high resolution MS to identify 3-ring and 4-ring steroid carboxylic acids.

Headley and coworkers [58] determined naphthenic acids in natural waters by using negative ion electrospray mass spectrometry. They preconcentrated their sample using a solid phase extraction procedure using a crosslinked polystyrene-based polymer with acetonitrile as the eluent. The recovery was highly pH dependent giving 100% recovery at a pH of 3. The limit of detection was 0.01 mg/L.

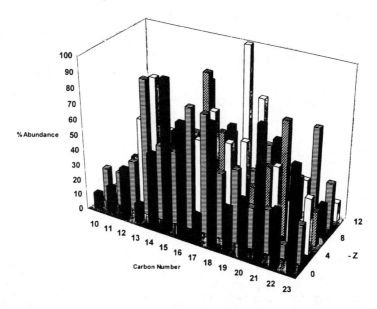

Figure 5. Relative abundance as a function of the Number of Carbon atoms and z-series (extent of hydrogen deficiency) for a commercial naphthenic acid mixture.

We studied the distribution of naphthenic acids in commercially available naphthenic acid samples (Figure 5). We found that ESI coupled with MS/MS experiments is extremely valuable in the determination of molecular weight distribution as well as confirmation of the identity of acid extracts. Carboxylic acids generate neutral losses of 44, 28 and 18 amu that can be attributed to M-CO_2, M-CO and M-H_2O.[59]

We used HPLC/ESI/MS (see Figure 6) to study the methanol extract obtained from the asphaltene fraction of a Maya crude oil.[60] Using HPLC/ESI/MS, we obtained partial separation of the mixture on a Supelco LC-18 bonded phase with pure methanol as the mobile phase. MS of the first eluted peak yielded an envelope in the range 273-350 amu (Figure 6). Careful inspection of the mass spectrum indicates that there are many peaks separated by 14 amu (-CH_2 groups) with two major homologous series: one starting at m/z = 273 and the other at m/z = 295. ESI/MS/MS revealed the presence of sulfonic acids as well as naphthenic acids based on fragmentation patterns.[59]

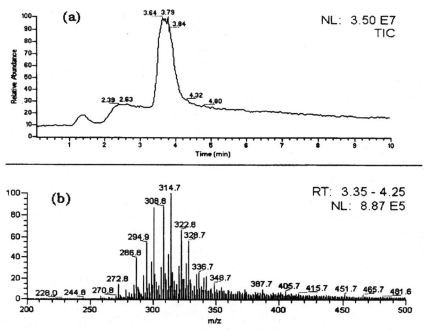

Figure 6. HPLC/ESI/MS of a Methanol Extract Obtained from the Asphaltene Fraction of a Maya Crude Oil.: (a) Total Ion Chromatogram (negative ion mode), (b) Mass Spectrum of peak at 3.8 min. (Reprinted from Ref. 60)

3.3 Separation of Nitrogen and Oxygen Compounds

In an early series of papers, Snyder and coworkers [61,62] determined the nitrogen and oxygen compound types in a 700-850°C distillate from a California Crude Oil. After OCLC separation, they determined compound types using low ionising voltage, electron impact, high resolution mass spectrometry which provided a mass accuracy of 5 ppm and a resolution of

10,000. They determined both the amount and the types of nitrogen and oxygen compounds. Major compound types (> 0.1% wt) included the indoles, carbazoles, benzcarbazoles, pyridines, quinolines, phenanthridines, hydroxypyridines, hydroxyquinolines, dibenzofurans, naphthobenzofurans, phenols, aliphatic ketones, carboxylic acids, and sulfoxides.

Yoshida and coworkers [63] developed a simple method for the trace analysis of volatile nitrogen and phenolic compounds in a synthetic naphtha (boiling point < 200°C) by a combined use of alumina adsorption chromatography followed by GC. The saturates on the alumina were eluted with pentane, the nitrogen compounds removed with methylene chloride, and the phenolic compounds removed with (60:40) methanol: chloroform. GC/MS with electron impact ionisation was then used to determine the compound types. The amount of nitrogen was less than 2% and predominately pyrrole-type, while oxygen compounds other than phenols were not found.

Ion exchange chromatography has been used to isolate nitrogen compounds in crude oil followed by analysis using spectroscopic and mass spectrometric techniques.[64-65] 3-(2,4-Dinitrobenzenesulphonamido) propylsilica and 3-(2,4-dinitroanilino) propylsilica (DNAP silica) columns with dichloromethane as the mobile phase were used for the HPLC determination of 22 nitrogen bases in gas, kerosene and diesel.[66] The approach uses chemically bound electron acceptors (nitroaromatic sorbents) in order to selectively attract nitrogen bases (azaarenes) through their pi and n electron donating ability. The selectivity allows for a group type separation of three and four ring PAH's, weak nitrogen bases (e.g., anilines, indoles, pyrroles, and carbazoles) and stronger bases containing a pyridine ring. The limit of detection for pyridine was between 50 –100 ppb. Azarenes were also fractionated from a crude oil by a selective extraction procedure followed by reverse phase chromatography. The fractions were collected then analyzed using GC-MS.[67] Nitrogen compounds in a heavy distillate (370-535°C) were also fractionated using anion and cation exchange resins, followed by HPLC separation using a neutral stationary phase and a cyclohexane/dichlomethane gradient as the mobile phase. GC-MS confirmed the presence of pyridines, pyrroles and amides.[68]

3.4   Atmospheric Pressure Ionization/ Mass Spectrometry of Nitrogen-containing Compounds

Marshall and coworkers [69] evaluated the relative electrospray efficiencies of selected nitrogen-containing compounds and found that the efficiency of positive-ion ESI strongly favors more basic molecules (pyridine benzologues) over pyrrole-type benzologues. This is not altogether surprising since basic compounds would have a higher tendency to be protonated and therefore form within the electrospray environment.

Marshall and coworkers also investigated the presence of high molecular weight peaks that appear in the mass spectra generated from a variety of ionization techniques (field desorption, MALDI and ESI). These have often been interpreted as noncovalent complexes. They observed that petroporphyrin molecules formed dimers since infrared multiple photon dissociation (IRMPD) experiments dissociated the complexes without fragmenting molecular ions.[70] With nitrogen-containing compounds, IRMPD had little effect on the mass distribution of the extra heavy crude oil. The results suggest that in the case of nitrogen-containing compounds in a heavy crude, high molecular weight species can be attributed to single molecules.

3.5 Separation of Organosulfur Compounds

Inorganic sulfur can be successfully removed from crude oil by a variety of physical separation methods, but organosulfur compounds are much more recalcitrant and more difficult to eliminate. New legislation in Japan and Europe will limit the sulfur content in light oil to 50 parts-per-million (ppm) maximum by 2005. The United States will limit the sulfur content to 15 ppm by the middle of 2006. Even though a variety of techniques, have been used to isolate and remove the sulfur, further developments are necessary in order to meet the new regulatory requirements.[71] Since the properties of polyaromatic sulfur heterocycles (PASH) are very similar to those of polyaromatic hydrocarbons (PAH), finding a suitable isolation procedure is difficult. Miller et al. have estimated that up to 75% of sulfur compounds in a Mayan asphaltene are sulfurs present in an aromatic core along with N and O heteroatoms.[50] From a different perspective, Tomczyk et al. [1] surmise that approximately ½ of the compounds which possess oxygen moieties also contain nitrogen, while ¼ also contain sulfur.

The nucleophilic interaction of sulfur compounds with $PdCl_2$ to form complexes has been well documented in the literature.[13,72-75] Nishioka et al.[72,76-79] successfully exploited this effect and developed a unique separation method based on the SARA approach followed by ligand exchange chromatography (LEC) using $PdCl_2$ and $CuCl_2$ on silica gel columns. $CuCl_2$ can be used to isolate aliphatic sulfur compounds like thiophenols and sufides,[78] while the $PdCl_2$ can be used to separate PASH compounds.[76] Andersson [80] disclosed some deficiencies in the Nishioka method; i.e., the elution of organosulfur compounds as $PdCl_2$ complexes, incomplete recovery of benzothiophenes, and the early elution of constituents with a terminal (as opposed to an internal) thiophene ring. In addition, the selectivity of the $PdCl_2$

varies with the organosulfur ring size. Milenkovic et al.[73] and Rudzinski and coworkers [81] found that the selectivity decreased in the order 3-ring> 2-ring> 1-ring PASH compounds.

Ma and coworkers [82] determined the sulfur compounds in the non-polar fraction of a vacuum gas oil (boiling range from 340 to >530^{0}C). The sulfur compounds were isolated from the non-polar fraction by ligand exchange chromatography, followed by ring-number separation using HPLC. The resulting sulfur compounds were identified by capillary GC with MS detection, and quantified by capillary GC with sulfur specific detection. Alkylbenzothiophenes, dibenzothiophenes, benzonaphtho- thiophenes and phenanthrothiophenes were found to be some of the major sulfur species in the non-polar fraction.

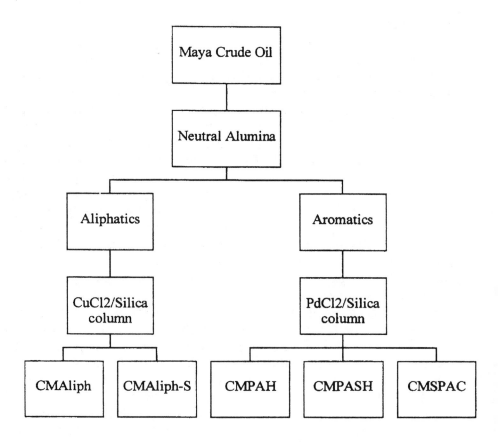

Figure 7. Scheme for the LEC Separation of a Crude Oil. (Reprinted from Ref. 60)

In an effort to isolate and analyze different fractions of a Maya crude oil and in view of the paucity of such published data, we isolated and characterized sulfur compounds in the aliphatic and aromatic fractions of a crude oil [81,83-85] according to the Nishioka Method [76,78] (Figure 7). The CM designation in front of the final five fraction names refers to "crude Mayan". In the protocol, the SARA method was used to separate the crude oil into saturate, aromatic, resin, and asphaltene fractions. $CuCl_2$ silica was then used to isolate the sulfur containing aliphatics (CMAliph-S) while $PdCl_2$ silica was used to separate the aromatic fraction into polyaromatic hydrocarbons (CMPAH), polyaromatic sulfur heterocycles (CMPASH), and sulfur-polar aromatic compounds (CMSPAC). Although the separation of sulfur containing compounds was not completely selective, the amount of sulfur in the CMAliph-S fraction was nine times as large as in the aliphatic CMAliph fraction, while the amount of sulfur in the CMPASH and CMSPAC fractions was 1.5 and 1.9 times as much as in the CMPAH fraction (Table 3). The $PdCl_2$ silica though offering an enrichment in aromatic sulfur is not nearly as selective as the $CuCl_2$ silica for the separation of aliphatic sulfur compounds. The fractions collected were analyzed by a combination of elemental analysis, Fourier transform infrared, ^1H and ^{13}C nuclear magnetic resonance (NMR), gel permeation chromatography (GPC), and atmospheric pressure chemical ionization /mass spectrometry (APCI/MS).[85]

Table 3. Elemental Analysis Data for LEC fractions from a crude oil

Fraction	%C	%H	%S
CMAliph	86.0	13.4	0.18
CMAliph-S	78.3	10.9	1.81
CMPAH	83.0	10.0	4.25
CMPASH	75.1	9.78	6.34
CMSPAC	75.2	9.52	8.12

Andersson and Schmid [86] used the perchlorobenzoic acid oxidation method for two-ring and the Nishioka method for three ring organosulfur compounds. Using gas chromatography with atomic emission detection (AED), which is specific for sulfur, and GC-MS, they were able to separate 69 organosulfur compounds and obtain quantitative data for 15 organosulfur compounds including alkylated benzothiophenes, dibenzothiophenes, and naphthothiophenes in an Austrian shale oil. The concentrations ranged from 65 ppm naphtho [2,3-b] thiophene to 974 ppm 2-methylbenzothiophene.

Dibenzothiophene (DBT) in crude oil was separated from other constituents using an aminosilane column and 1% dichloromethane in hexane as the eluent. Gas chromatography using the selective sulfur chemiluminescence detector (GC-SCD) was used to identify the DBT.[87]

Palmentier et al [88] separated parent and alkylated PAH compounds as well as sulfur-containing species into saturate and aromatic fractions using silica adsorption chromatography. After a further, preparative separation on an amino-bonded phase based on the number of aromatic rings, high resolution GC-MS and GC/FID were used to characterize the collected HPLC fractions. The authors were able to identify 22 alkylated dibenzothiophene and naphthothiophene type compounds ranging from 184-240 molecular weight at the part-per-billion (ppb) level. Wise and co-workers [89] extended the analysis of PASH's up to 80 compounds using GC-MS and a variety of selective stationary phases. Though both approaches exhibited unparalleled resolution of the components of the sample, the approach was limited by the volatility of the thiophenes.

3.6   Atmospheric Pressure Ionization/ Mass Spectrometry of Organosulfur Compounds

Since the properties of polyaromatic sulfur compounds are very similar to those of polyaromatic hydrocarbons, developing a selective mass spectrometric procedure is difficult. Development of a satisfactory mass spectrometric analysis for sulfur compounds in the presence of aromatic hydrocarbons requires that a combination of ionization conditions, ion-molecule reactions and/or derivatization steps be employed to direct fragmentation of the sample along a pathway which involves either elimination of a neutral containing sulfur or formation of a charged species characteristic of all or specific types of sulfur-containing compounds. Hunt and Shabanowitz [90] used collision activated dissociation (CAD) in a triple quadrupole instrument to follow the neutral loss of a 33 amu SH radical. See Figure 8. By setting quadrupole number 1 and quadrupole 3 to scan repetitively at a fixed mass separation, the detector monitored organosulfur compounds exclusively. A major pitfall of the approach was the inability to detect highly alkylated sulfur heterocycles and aliphatic sulfides.

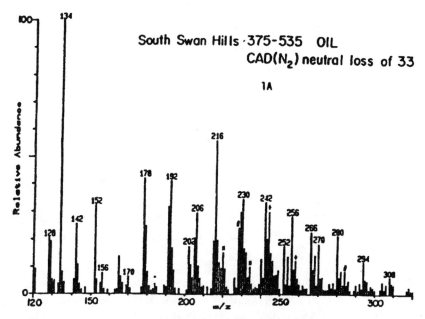

Figure 8. Collision activated dissociation (CAD) mass spectrum obtained in the 33 amu neutral loss mode on (m^{+1}) ions from organosulfur compounds in a South Swan Hills crude oil. Reprinted with permission from Hunt and Shabanowitz, <u>Anal. Chem.</u>, *54*, **1982**, *575*.

HPLC-MS in single ion monitoring (SIM) mode was used for the determination of dibenzothiophene (DBT) in coal liquids and crude oils. For this, M$^+$ with m/z 184 and M$^+$ with m/z 152 were monitored simultaneously. Only those species with a characteristic loss of an M-32 fragment were detected.[91]

Another approach for the analysis of PASH compounds involves the oxidation of the sulfur to the sulfone with m-chloroperbenzoic acid. Rudzinski et al [81] evaluated the oxidation approach for the analysis of DBT and found a sensitivity enhancement of 500 for the sulfone over the parent DBT (Figure 9).

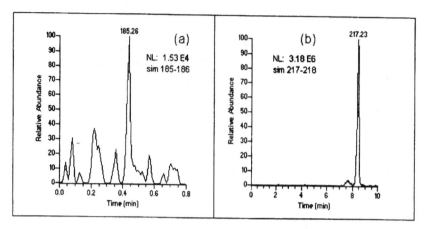

Figure 9. APCI MS of (a) DBT (1420 ppm) and (b) DBT sulfone (80 ppm). (Reprinted from Ref. 60)

A relatively new approach that has been developed for the selective determination of difficult to ionize species is coordination-ion spray.[92] In this approach, positively or negatively charged metal complexes are formed by the addition of a suitable central atom to an analyte, and the complexes are then detected by mass spectrometry. Neither an electric field nor the formation of a reagent gas plasma by corona discharge is necessary for ionization. Since pneumatic nebulization in the absence of an electric field is often insufficient to obtain a suitable spray, a supporting voltage is usually needed. Generally elements from the first and eighth transition groups are employed, e.g., Cu^+, Ni^{+2}, Pd^{+2}, Pt and Ag^{+1}. These form highly stable pi bonding complexes with both alkenes and polynuclear aromatic compounds.

One application of coordination ion-spray though not directly related to NSO compounds is the recent work by Roussis and Proux [93] who used Ag^+ to determine the molecular weight distribution of heavy petroleum fractions by electrospray ionisation/ mass spectrometry. They found that it was possible to complex aromatic compounds with silver ion leading to the formation of abundant adduct ions such as $[M+ Ag]^+$ and $[2M+ Ag]^+$. The concentration of the latter dimer could be reduced by increasing the sampling cone voltage.

Table 4. LC/ESI/MS LOD and Sensitivity Using Pd+2 as a Sheath Liquid (SIM mode)[a]

Compound	LOD	Sensitivity
Dibenzothiophene	46	12
4,6-dimethyl DBT	340	0.35
Thianthrene	10	130
Benzonaphthothiophene	22	28

[a] presented by Rudzinski et al. at the 220th National ACS Meeting, Washington, D.C., Aug., 2000.

Recently we have developed a variant of coordination ion spray in which Pd^{+2} in methanol is added to the electrospray ionization source of an ion trap mass spectrometer.[81,83] The addition of the Pd^{+2} produces charge transfer complexes with 3- and 4- ring organosulfur compounds which can dissociate to form radical cations $[M]^+$. The approach was used for the separation and analysis of several polyaromatic sulfur heterocycles (PASH) normally found in the aromatic fraction of a Maya crude oil. The sensitivity, and limit of detection (LOD) were evaluated. Thianthrene had the highest sensitivity and the lowest LOD among the PASH compounds tested. See Table 4. *Rhodococcus strain sp.* IGTS8 bacteria were then used to desulfurize the aromatic fraction from a Maya crude oil. A good correlation for a limited sample set was obtained between the elemental analysis data and the results obtained from LC/ESI/MS. See Table 5.

Table 5. Comparison of Sulfur Elemental Analysis and Mass Spectrometry for a Biodesulfurized Sample[a]

Fraction	%S from elemental analysis	Total Ion Area $(x10^{-3})^b$
Aromatic	4.30	3.50
Aromatics + cells	2.53	1.98
Aromatics + Cells + SO_3^-	3.76	3.13

[a] presented by Rudzinski et al. at the 220th National ACS Meeting, Washington, D.C., Aug., 2000. [b] Areas represent an average of three runs

Rudzinski and Zhang also explored the feasibility of using MS-MS methods for the identification of organosulfur compounds[94]. The addition of Pd^{+2} produced radical cations $[M]^+$ from 3- and 4- ring organosulfur compounds which were not alkylated, while 2-methyl-dibenzothiophene, and 4,6-dimethyldibenzothiophene decomposed by loss of either H^-, CH_3^- or HS^-.

4. ACKNOWLEDGEMENTS

WER would like to thank the Institute for Industrial and Environmental Science of Southwest Texas State University, the Environmental Protection Agency (Grant Application R825505) as well as the Texas Higher Education Coordinating Board (ATP Energy 2001) for support.

5. REFERENCES

1. Tomczyk, N.A., Winans, R.E., Shinn, J.H., and Robinson, R.C.,*Energy Fuels* **2001**,*15*, 1498-1504.
2. Qian, K., Robbins, W.K., Hughey, C.A.: Cooper, H.J.; Rodgers, R.P.; Marshall, A.G., Energy Fuels **2001**,*15*, 1505-1511.

3. Lai, J.W.S., Pinto, L.J., Kiehlmann, E., BendellYoung, L.I., Moore, M.M., *Environ. Toxicol. Chem.* **1996**, *15*, 1482-1491.
4. Guisnet, M., Magnoux, P. , *Appl. Catal.* **1989**, *54*, 1-27.
5. Qian, K., Tomczak, D.C., Rakiewicz, E.F., Harding, R.H., Yaluris, G. , Cheng, W.-C. Zhao, X., Peters, A.W., *Energy & Fuels* **1997**, *11*, 596-601.
6. Harding, R.H., Zhao, X., Qian, K., Rajagopalan, K., Cheng, W.-C. *Ind. Eng. Chem. Res.* **1996**, *35*, 2561-2569.
7. Speight, J.G. *Fuel Science and Technology Handbook*, Marcel Dekker, New York, **1990**.
8. ASTM D1319-95a. Annual Book of ASTM Standards; American Society for Testing and Materials; Philadelphia, PA, **1998**; Vol 5.01, pp. 465-470.
9. ASTM D2549-91. Annual Book of ASTM Standards; American Society for Testing and Materials; Philadelphia, PA, **1998**; Vol 5.01, pp. 892-897.
10. ASTM D2007-93. Annual Book of ASTM Standards; American Society for Testing and Materials; Philadelphia, PA, **1998**; Vol 5.01, pp. 655-661.
11. Corbett, L.W., *Anal. Chem.* **1969**, *41*, 576-579.
12. Campbell, R.M.; Lee, M.L. *Anal.Chem.* **1986**, *58*, 2247-2251.
13. Later, D.W.; Lee, M.L.; Bartle, K.D.; Kong, R.C.; Vassilaros, D.L. *Anal. Chem.* **1981**, *53*, 1612-1620.
14. Fan, T.P., *Energy & Fuels*, **1991**, *5*, 371-375.
15. Nali, M.; Fabbi, M.; Scilingo, A. *Petroleum Science and Technology* **1997**, *15*, 307-322.
16. Altgelt, K.H., Jewell, D.M. Latham, D.L., Selucky, M.L., in *Chromatography in Petroleum Analysis*, K.H. Altgelt and T.H. Gouw, Eds; *Chromatogr. Sci. Vol. 11*; Marcel Dekker: New York, **1979**; pp. 185-214.
17. Altgelt, K.H., Boduszynski, M.M., *Composition and Analysis of Heavy Petroleum Fractions* Marcel Dekker: New York, **1994**; pp. 203-255.
18. Leontaritis, K.J. *Proc. Int. Symp. Oilfield Chem.* **1997**, 421-440.
19. Mansfield, C.T.; Barman, B.N.; Thomas, J.V.; Mehrotra, A.K.; McCann, J.M. *Anal. Chem.* **1999**, 71, 81R-107R.
20. Barman, B.N., Cebolla, V.L., Membrado, L., *Crit. Rev. Anal. Chem.*, **2000**, *30*, 75-120.
21. Ali, M.F.; Bukhari, A.; Hassan, M. *Fuel Sci. Technol.* **1989**, *7*, 1179-1208.
22. Akhlaq, M.S. *J. Chromatogr.* **1993**, *644*, 253-258.
23. Fetzer, J.C., in *Chemical Analysis of Polycyclic Aromatic Compounds*, T.Vo-Dinh, Ed; Wiley: New York, **1989**; Chap. 3, pp. 59-109.
24. McKerrell, E.H., *Fuel*, **1993**, *72*, 1403-1409.
25. Matsunaga, A., *Anal. Chem.* , **1983**, *55*, 1375-1379.
26. Hayes, P.C., Jr., Anderson, S.D., *Anal. Chem.* **1985**, *57*, 2094-2098.
27. Hayes, P.C., Jr., Anderson, S.D., *Anal. Chem.* **1986**, *58*, 2384-2388.
28. Hayes, P.C., Jr., Anderson, S.D., *J. Chromatogr.* **1987**, *387*, 333-344.
29. Hayes, P.C., Jr., Anderson, S.D., *J. Chromatogr. Sci.*, **1988**, *26*, 210-217.
30. Hayes, P.C., Jr., Anderson, S.D., *J. Chromatogr. Sci.*, **1988**, *26*, 250-257.
31. Hayes, P.C., Jr., Anderson, S.D., *J. Chromatogr.* **1988**, *437*, 365-377.
32. Robbins, W.K., *J. Chromatogr. Sci.*, **1998**, *36*, 457-466.
33. Heine, C.E., Geddes, M.M., *Org. Mass Spectrom.* **1994**, *29*, 277-282.
34. Pfeifer, S., Beckey, H.D., Schulten, H.R., Fresenius Z. *Anal. Chem.* **1977**, *284*, 193-195.
35. Malhotra, R. , McMillen, D.F., Tse, D.S. , St. John, G.A., Coggiola, M.J., Matsui, H., P. , *Prepr.- Am. Chem. Soc. , Div. Petr. Chem.* **1989**, *34*, 330-338.
36. Rahimi, P.M., Fouda, S.A., Kelly, J.F., Malhotra, R., McMillen, D.F., *Fuel* **1989**, *68*, 422-429.
37. Boduszynski, M.M., *Energy & Fuels,* **1987**, *2*, 697.
38. Hsu, C.S., Qian, K., *Energy & Fuels*, **1993**, *7*, 268-272.

39. Cole, R.B., Electrospray Ionization Mass Spectrometry, John Wiley & Sons, New York, **1997**.
40. Zhan, D., Fenn, J.B., *Int. J. Mass Spec.* **2000**, *194*, 197-208.
41. Lambert, J.B., Shurvell , H.F. ,Lightner, D.A., Cooks, R.G., *Organic Structural Spectroscopy*, Prentice Hall, Upper Saddle River NJ, **1998**, p.375.
42. Tomczyk, N.A., Winans, R.E. Shinn, J.H., Robinson, R.C. *Prepr. -Am. Chem. Soc. Div. Pet. Chem.* **1998**, *43*, 123-125
43. McCarthy, R., Duthie, A., *J. Lipid Res.* **1962**, *3*, 117-119.
44. Green, J.B., Hoff, R.J., Woodward, P.W., Stevens, L.L., *Fuel* **1984**, *63*, 1290-1301.
45. Green, B.J., *J. Chromatogr.* **1986**, *358*, 65-75.
46. Robbins, W.K. Prepr.-Am. Chem. Soc. , Div. Pet. Chem. **1998** 43 137-140.
47. Hertz, H.S.; Brown, J.M.; Chesler, S.N.; Guenther, F.R.; Hilpert, L.R.; May, W.E.; Parris, R.M.; Wise, S.A. *Anal. Chem.* **1980**, *52*, 1650-1657.
48. Schmeltz, I. *Phytochemistry* **1967**, *6*, 33-38.
49. Willsch, H.; Clegg, H.; Horsfield, B.; Radke, M.; Wilkes, H. *Anal. Chem.* **1997**, *69*, 4203-4209.
50. Miller, J.T.; Fisher, R.B.; Thiyagarajan, P.; Winans, R.E.; Hunt, J.E. *Energy Fuels*, **1998**, *12*, 1290-1298.18. McCarthy,R; Duthie, A. *J. Lipid Res.* **1962**, *3*, 117-119.
51. Yepez, O., Lorenzo, R. Callaroti, R, Vera J, <u>Prepr. -Am. Chem. Soc. Div. Pet. Chem.</u> **1998**, *43*, 114-122.
52. Jones, D.M. Watson, J.S. , Meredith, W. , Chen, M., Bennett, B. *Anal. Chem.,* **2001**, *73*, 703-707.
53. Green , J.B., Yu, S., K-T; Vrana, R.P., *J. High Res. Chromatogr.* **1994**, *17*, 427-438.
54. Hsu, C.S., Dechert, G.J., Robbins, W.K., Fukuda, E., Roussis, S.G., *Prepr.--Am. Chem. Soc. Div. Petrol. Chem.* **1998**, *43*, 127-130.
55. Hsu, C.S.; Dechert, G.J.; Robbins, W.K.; Fukuda E.K., *Energy Fuels* **2000**, *14*, 217-223.
56. Seifert, W., Teeter, R., Howells, W., Cantow, M., *Anal. Chem.*, **1969**, *41*,1639-1646.
57. Seifert, W. , *Fortzchr. Chem. Org. Naturst.*, **1975**, *32*, 1-49
58. Headley, J.V., Peru, K.M., McMartin, D.W.,Winkler, M., *J. AOAC*, **2002**, 85, 182-187.
59. Rudzinski, W.E., Oehlers, L., Zhang, Y., Najera, B. *Energy & Fuels,* **2002** (submitted for publication).
60. Sassman, S., *Fractionation and Analysis of Maya Crude Oil*, M.S. Thesis, **1999**, Southwest Texas State U., San Marcos, TX, p.51.
61. Snyder, L.R., Buell, B.E., *Anal. Chem.* **1968**, *40*, 1295-1302.
62. Snyder, L.R., Buell, B.E., Howard, H.E., *Anal. Chem.* **1968**, *40*, 1303-1317.
63. Yoshida, T., Chantal, P.D., Sawatzky, H., *Energy & Fuels*, **1991**, *5*, 299-303.
64. Hsu, C.S., Qian, K., Robbins, W.K., *J. High Resolut. Chromatogr.*, **1994**, *17*, 271-276.
65. McKay J.F., Weber, J.H., Latham, D.R., *Anal.Chem.*, **1976**, *48*, 891-898.
66. Nondek, L., Chvalovsky, V., *J. Chromatogr.* **1984**, *312*, 303-312.
67. Schmitter, J.M., Colin, H., Excoffler, J. L., Arpino, P., Guichon, G.., *Anal. Chem.* **1982**, *54*, 769-772.
68. Ali, M.F., Ali, M.A., *Fuel Sci. Technol. Int.* **1988**, *6*, 259-290.
69. Qian, K., Rodgers, R.P.; Hendrickson, C.L., Emmett, M.R.; Marshall, A.G. *Energy Fuels* **2001**, *15*, 492-498.
70. Rodgers, R.P., Hendrickson, C.L., Emmett, M.R., Marshall, A.G., Greaney, M., Qian, K., *Can. J. Chem.* in press
71. Whitehurst, D.; Isoda, T.; Mochida, I. *Adv. Catal.* **1998**, *42*, 345-471.
72. Nishioka, M. *Energy Fuels* **1988**, *2*, 214-219.
73. Milenkovic, A.; Schultz, E.; Meille, V.; Loffreda, D.; Forissier, M.; Vrinat, M.; Sautet, P.; Lemaire, M. *Energy Fuels,* **1999**, *13*, 881-887.
74. Wright, B.W.; Peaden, P.A.; Lee, M.L.; Stark, T. *J. Chromatogr.* **1982**, *248*, 17-34.

75. Grang, B.Y. *Anal. Lett.* **1985,** *18,* 193-202
76. Nishioka, M.; Campbell, R.M.; Lee, M.L.; Castle, R.N. *Fuel* **1986,** *65,* 270 - 273.
77. Nishioka M.; Whiting, D.G.; Campbell, D.G.; Lee, M.L. *Anal. Chem,* **1986,** *58,* 2251-2255.
78. Nishioka, M.; Tomich, R.S. *Fuel* **1993,** *72,* 1007- 1010.
79. Nishioka, M.; Lee, M.L.; Castle, R.N. *Fuel* **1986,** *65,* 390-395.
80. Andersson, J.T. *Anal. Chem.* **1987,** *59,* 2207-2209.
81. Rudzinski, W.E., Sassman, Watkins, L.M., *Prepr.- Am. Chem. Soc. , Div. Petr. Chem.* **2000,** *45(4),* 564-566.
82. Ma, X., Sakanishi, K., Isoda, T. Mochida, I., *Fuel,* **1997,** *76,* 329-339.
83. Rudzinski, W.E.; Sassman, S. Proceedings of the 47th ASMS Conference on Mass Spectrometry and Allied Topics, Dallas, TX, June 13th, **1999.**
84. Rudzinski, W.E.; Rodriguez, R.; Sassman, S.; Sheedy, M., Smith, T,; Watkins, L.M. *Prepr.- Am. Chem. Soc. , Div. Fuel. Chem.* **1999,** *44*(1), 28-31.
85. Rudzinski, W.E., Aminabhavi, T.M., Spencer, L., Sassman, S., and Watkins, L., *Energy Fuels* **2000,** *14,* 839-844.
86. Andersson, J.T.; Schmid, B. *J. Chromatogr.* **1995,** *693,* 325-338.
87. Rebbert, R.E., Chesler, S.N., Guenther, F.R. , Parris, R.M., *J. Chromatogr.* **1984,** *284,* 211-217.
88. Palmentier, J.-P.F., Britten, A.J., Charbonneau, G., M., Karasek, F.W., *J. Chromatogr.* **1989,** *469,* 241-251.
89. Mossner, S.G., Lopez de Alda, M.J., Sander, L.C., Lee, M.L., Wise, S.A. , *J. Chromatogr.* **1999,** *841,* 207-228.
90. Hunt, D.F. Shabanowitz, J., *Anal. Chem.* **1982,** *54,* 574-578
91. Christensen, R.G., White, E. , *J. Chromatogr.* **1985,** *323,* 33-36.
92. Bayer, E., Gfrorer, P., Rental, C., *Angew. Chem. Int. Ed.* **1998,** *38,* 992-995.
93. Roussis, S.G., Prouix, R., Anal. Chem. **2002,** 74, 1408.
94. Rudzinski, W.E.; Zhang, Y.. Proceedings of the 50th ASMS Conference on Mass Spectrometry and Allied Topics, Orlando, FL, June 6th, **2002.**

Chapter 14

CHARACTERIZATION OF HEAVY OILS AND HEAVY ENDS

Lante Carbognani, Joussef Espidel and Silvia Colaiocco
PDVSA-INTEVEP. P.O. Box 76343. Caracas 1070 A. Venezuela

1.   INTRODUCTION

Current global reserves of conventional crude oils are estimated at approximately 1,000 billion barrels (1 bn bbls = 10^9), while extra heavy crude oil (XHC) plus tar sand in place volumes are estimated at around 7,000 bn bbls.[1] From the former estimation, Canada and Venezuela respectively, appears to be the leading countries in tar sand and XHC oil sources. Another study showed that three geographical regions in the world possess the majority of these resources.[2] The combined tar sands from Canada plus the former Soviet Republics and XHC from Venezuela plus the former Soviet Republics account for more than 90% of Earth´s reserves.

There is not a general agreement on the definitions of heavy and extra-heavy crude (HC/XHC) and bitumen. Boduszynski and Altgelt [3] provided useful insights on these definitions, as well as a review of common misconceptions. In this chapter, the proposed definitions from the UNITAR Center for Heavy Crude and Tar Sands are as follows; HC comprises the API° range from 10-20 (specific gravity range from 1.00-0.93), XHC crude represents less than 10 API°, and natural bitumen is the one that shows a viscosity greater than 10000 cP in-reservoir.[4] However, from the current production figures of nearly 70 million barrels/day (b/d), only 10% correspond to HC and only 1.4% to XHC.[1]

Despite the actual low production figure of HC/XHC, the importance of these resources cannot be neglected. Population is expected to increase from 5.8 to 8 billion people by the year 2020.[5] This population increase will demand an additional oil consumption of 20 million b/d. However, an important point is that light and medium crude oil reserves (>20 API°) are declining and the void will have to be filled by HC and XHC. It has been pointed out that by the year 2050, estimate world consumption will reach 50% XHC.[6] Therefore, by the middle of the 21st century at least half of the oil production is expected to be XHC, demanding optimized production, transportation, oil upgrading, and new environmental schemes to be fulfilled.

Cost-effective production of HC/XHC is the first aspect that requires close attention. Understanding their production mechanisms is a key aspect for operational improvements.[7] Technology has been shown to lie at the core of cost reductions and has been exemplified by the success achieved during the past decade.[1,8] Steam is probably the most widespread technology applied during enhanced oil recovery (EOR).[7] Steam assisted gravity drainage (SAGD) is a related technique that has recently gained wide spread acceptance.[9] New alternatives like solvent extraction with simultaneous deasphalting [10] and microbial enhanced recovery [11] are also currently used. The cold production or "foamy oil" mechanism is far from understood, but is currently applied and investigated.[12,13] Well known technologies like in-situ combustion, continue to find their application niches under the right geological conditions.[14]

Transportation of highly viscous HC/XHC is the ensuing issue once the oil has been brought to surface facilities. A review on heating, dilution, emulsion, core annular-flow and partial field upgrading summarizes the main technologies applied for HC/XHC transportation.[15] The emulsion technology has received particular attention, particularly in Venezuela, creating a new fuel known as Orimulsion^{TM}. Cost reduction by natural surfactant activation in Orimulsion^{TM} production [16] and microbial enhanced emulsification [17] are two aspects currently under investigation for improvement. Partial upgrading at the wellhead with a novel homogeneous catalytic process appears to be a breakthrough in XHC transportation.[18]

Upgrading of HC/XHC has received a continuous attention during the past three decades. Low demanding technology schemes like asphalt production, intermediate technologies like carbon rejection and high technology approaches based on catalytic conversion (with/without hydrogen addition) have been reviewed.[19,20] However, the substantial cost involved in such processes made cheaper alternatives appear viable, as in the case for asphalt production. Venezuela is such an example, with its primacy in the above business and an ever-increasing refinery diet of HC/XHC.[21]

Technological issues related to HC/XHC were discussed in previous paragraphs. Several alternatives were identified to cope with practical issues imposed by the particular nature of these crudes. However, in order to optimize operations, knowledge of the raw material is one of the key aspects that need to be investigated. Distillation can be considered as the fundamental benchmark for the oil industry.[22] Unfortunately, in spite of great efforts conducted to date, characterization is still an area on which more information is deemed necessary. HC, and particularly XHC, contain large amounts of heavy ends, vacuum residua in particular. The term heavy end refers to hydrocarbon mixtures boiling above

343°C. Cerro Negro crude oil from the Orinoco Belt in eastern Venezuela illustrates very well the abundance of vacuum residua present in XHC (Fig. 1).[23] Cerro Negro 550°C+ and 700°C+ residua represent 59.2 and 45.8 wt% of the total crude. The estimated average molecular weight (AMW) for the 700°C+ residue lay around the range of 1500 Daltons (107 carbon atoms in average). Reportedly, the complexity of this kind of hydrocarbon materials is overwhelming. A calculated number of possible isomers for paraffin molecules with 100 carbon atoms lie in the order of 10^{39}, according to Altgelt and Boduszynski [3]. Indeed, the detailed report by Green et al.[23] on Cerro Negro showed that thousands of compounds were tentatively identified within the whole crude. This complexity is the reason that leads other authors to assert that "crude oils are Ultra-Complex Fluids within the branch of physics called Complex Fluids".[24]

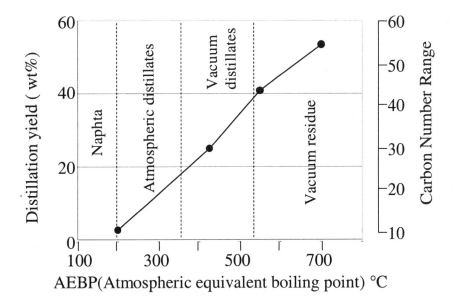

Figure 1. Cerro Negro XHC distillation cuts. Typical commercial refinery cuts ranges are shown. The "Equivalent Carbon Number Scale" is based on n-Paraffins. The high vacuum distillate (550-700°C) was obtained with a short path distillation unit.

This chapter will cover some of the strategies that have been pursued in order to cope with the characterization of extremely complex materials, like HC, XHC and distillation residua. Some of their weaknesses and advantages will be discussed, and some emerging alternatives will also be described.

2.   HEAVY OILS / HEAVY ENDS SEPARATION AND CHARACTERIZATION SCHEMES

As it has been stated in the previous section, one of the most important factors limiting the characterization of HC/XHC systems is their complex molecular composition. Another constraint has been the limited funds available for carrying out detailed analyses of heavy hydrocarbon fractions, which are the less valuable oil components. Therefore, simplified strategies are commonly employed, and will continue to be applied and optimized. Ever increasing computer capabilities enables the development of new strategies that delivers the required information in a timely manner, with the least possible effort and cost. Some of these strategies are presented on Figure 2.

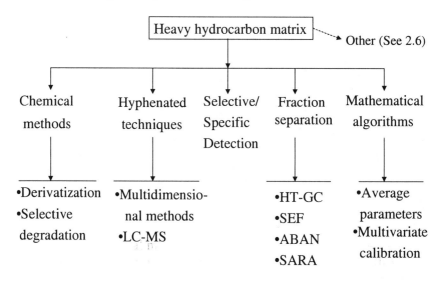

Figure 2. Heavy oil/heavy ends characterization schemes. Abbreviations: LC-MS (liquid chromatography-mass spectrometry); HT-GC (High temperature gas chromatography); SEF (sequential elution fractionation); ABAN (acids, bases, amphoterics and neutrals); SARA (saturates, aromatics, resins and asphaltenes).

The remaining of this section will be devoted to comments on what appears to be the most interesting aspects and trends regarding the cited characterization strategies. In the last section of this chapter, applications from the authors´ laboratory will illustrate some of the discussed aspects concerning these strategies.

2.1 Chemical Methods

Chemical fractionation methods for complex hydrocarbon mixtures were formerly very common, as reported by Speight [25]. Extensive use of selective chemical functional derivatization, combined with spectrometric techniques allowed a detailed compound distribution identification for Cerro Negro XHC.[23] Despite the fact that chemical methods are generally labor intensive, simple separation schemes proposed many years ago are still useful. Speight and coworkers [26,27] developed acid/base partitioning of carbonyl, carboxyl and ester functionalities from resin and asphaltenes. This technique was recently adopted for the characterization of organic compounds co-precipitated with mineral matter, during Canadian [28] and Venezuelan [29] oil production operations. NaOH and HCl modified silica sorbents are essential in a methodology developed for the isolation of basic and neutral nitrogen compounds extracted from the Venezuela Boscan HC.[30] Further adoption of commercial sulfonic modified silica allowed the detailed determination of mass distributions of neutral and basic nitrogenates.[31] The Strausz research group in Canada can probably be credited for the most involved group in chemical separation of hydrocarbon fractions. The sulfide separation via their oxidation to sulfoxides, retention on polar sorbents and reduction to the original form using $LiAlH_4$ is a classic example.[32] The sulfide contents on Canadian Athabasca and Venezuelan Cerro Negro XHC have been determined following this scheme.[33] Similarly, sulfur type compounds were differentiated in asphalts derived from Venezuelan, Canadian and USA HC.[34]

Selective complexation of heterocyclic compounds is commonly employed for their isolation. Sulfur heterocycles were selectively retained on $PdCl_2$-Silica [35], nitrogen species were removed from shale oil using $CuCl_2$ [36], and basic nitrogenates were isolated with the aid of $TiCl_4$ [37]. Asphaltenes isolated from Cerro Negro and Boscan Venezuelan crudes were separated into soluble and colloidal phases by complexation with p-nitrophenol.[38] A comprehensive monograph covering complexation separation is available in the literature.[39] Removal of sulfur heterocycles from aromatic fractions was suggested as an improvement for the ring number speciation of aromatic types [40], as found in open literature [3].

Selective degradation of complex hydrocarbon matrices is another useful technique for their characterization. Oxidative studies [41] and Ruthenium ion catalyzed oxidation as independently described by two research groups [42,43] are very interesting approaches. The Ru methodology is now commonly called RICO (ruthenium-ion catalyzed oxidation). Under RICO reaction conditions, carbons from aromatic moieties are selectively oxidized at the junction points, generating carboxylic derivatives. Alkyl appendages give origin to monocarboxylic acids, alkyl links between aromatic rings as well as fused cycloalkane rings generate diacids, and polyaromatic compounds give rise to aromatic polyacids. In this way, the RICO reaction is a powerful tool to characterize aromatic moities present in complex hydrocarbon mixtures, as recently illustrated with Boscan asphaltenes.[44] However, its weakness is the way the identified building blocks are connected in the original sample. Improvements in RICO material balances were recently described in the published literature.[45] Currently, the RICO reaction is used for characterization of asphaltenes isolated from stable HC and unstable light crudes, in order to shed light on the intrinsic stability of these crude oil components.[46]

2.2   Hyphenated Techniques

The term Hyphenated, or coupled techniques, describes those analytical methodologies that combine chromatographic separation plus spectroscopic techniques. In this way, analysis time is reduced and sample integrity is better maintained due to less manipulation. Davis et. al. [47] discussed many of the advantages achievable by hyphenation in a review. These advantages guarantee future applications for hyphenated techniques. The most successful hyphenated technique to date is GC-MS (gas chromatography-mass spectrometry). GC-FTIR (GC coupled to Fourier-transform infrared spectroscopy) [48], LC-NMR (liquid chromatography coupled to nuclear magnetic resonance spectroscopy) [49], LC-LC [50] and LC-GC [51] have also been described. Despite of many hyphenated methodologies developed to date that involve GC, there is one implicit limitation for their application to petroleum heavy ends. The drawback is the limited volatility of high molar mass (MM) hydrocarbons. It has been stated that nearly 20% of the whole population of known organic molecules can be analyzed by GC, whereas LC is suited for the analysis of the remaining 80%.[52]

LC-MS (liquid chromatography-mass spectrometry) has been demonstrated to be an excellent technique for large MM hydrocarbons. There are some reviews that cover diverse LC-MS approaches [53-55] and a monograph has also been published on this topic [56]. High-pressure ionization techniques like API (Atmospheric Pressure Ionization) [53] seem very attractive from an instrumental point of view. Recently, the analysis of fullerenes and polycyclic aromatic hydrocarbons (PAH) up to C_{75} was reported by HPLC-APCI-MS (high performance liquid chromatography-atmospheric pressure chemical ionization).[57] Hsu and coworkers from Exxon published many LC-MS applications on heavy distillates and oil residua in the past decade. By using a moving belt interface and low energy electron impact ionization, LC-MS allowed them to distinguish between pure and sulfur aromatic compounds in a vacuum distillate, with low resolution requirements.[58] Furthermore, this system proved useful for understanding the molecular transformations occurring during the hydrotreating of a vacuum distillate.[59] A better alternative than the formerly described was subsequently developed, achieving ionization with CS_2.[60] By using a thermospray LC-MS system, aromatic hydrocarbons in a high vacuum distillate (606-707°C) were selectively ionized.[61] A patented modification of a field ionization system allowed quantifying n-paraffins, cycloparaffins and iso-paraffins in a single LC-MS analysis.[62] Advantages of API were demonstrated in a recent paper in which commercial or extracted (California oil) naphthenic acids were analyzed by diverse LC-MS approaches where APCI was the best-identified alternative.[63] To facilitate data handling, a procedure based on the Kendrick mass scale (the mass for a methylene group $-CH_2-$ is defined as 14.0000 Daltons) was developed by Hsu and coworkers [64]. Based on these successful achievements, future LC-MS applications for heavy hydrocarbon ends seem promising.

The selected examples covered in the ensuing discussion, point out what is believed to be the next step in LC-hyphenation i.e., improving the first dimension provided by the LC separation by the addition of multiple spectral dimensions. One example that can be considered as a classic is the use of HPLC-DAD-MS (DAD: diode array detection) for the characterization of PAH, allowing for simultaneous determination of Uv-Vis spectra and mass fragmentograms for each of the chromatographic separated signals.[65] These combined results led, in many cases, to unambiguous identification of single isomers and provided information for coeluting peaks. For a similar HPLC-DAD application, the interpretation of the information through multivariate statistical methods allowed the authors to determine aromatic groups in lube oil and VGO's.[66,67] Deconvolution of overlapping signals and quantitative analysis of hydrocarbon groups were shown by this chemometric approach. In a final example involving nitrogen compounds, the separation of pyridines/pyrrol benzologs in middle distillates was demonstrated by on-line coupling of HPLC-DAD-MS in a single analysis, performed in 1.5 hours.[68] In this way, a remarkable reduction in analysis time was achieved in the detailed characterization of carbazole and indole species. Previous reported open column approaches (section 2.1) required days before their completion.[30] Off-line HPLC separation followed by GC-MS and GC-N-selective detection, was an improvement compared with open column, but still demanding two separation systems plus two successive separation sequences, as reported in a study conducted with rock extracts and a California HC.[69] The discussed examples with nitrogen compounds illustrate many of the advantages achievable through hyphenation.

The ultimate trend in hyphenation can be credited to Giddings [70], one of the great theoreticians in separation science. He discussed the so called "multidimensional separations", operations which require that at least two independent and selective solutes displacements are performed and the separated species must remain resolved across the separation. A broad spectrum of LC-GC separations have been reported, many of which cover some crude oil aspects.[47,51] Analyses of complex PAH mixtures with up to 7 rings were shown in some of the examples presented in these reviews. Another nice example of the power of LC-GC was presented in one monograph devoted to this topic.[71] Here, the detection of benzo[a]pyrene isomer in a liquefied shale oil was achieved with a combination of normal and reverse phase HPLC with simultaneous ultraviolet and fluorescence detection. Detailed information on the naphtalene fraction from Venezuela´s Cerro Negro XHC was provided by coupled LC-GC-MS.[72] A promising tool for petroleum heavy end characterization is the multidimensional coupling of supercritical fluid chromatography (SFC-SFC), enabling the analysis of PAH with up to 7 rings.[73]

It is the belief of the authors of this chapter, that the analysis of target compounds within complex petroleum heavy ends will continue to be studied by multidimensional methodologies like those already discussed. Finally, the development of hyphenated techniques for heavy ends that do not depend on sample volatility such as LC-FTIR and LC-NMR, is most interesting. Sensitivity limitations of these techniques are recognized, but the information that can be obtained could be very valuable because the evaluated species never leave the liquid phase.

2.3   Selective/Specific Element Detection

Selectivity for hydrocarbon group types and/or single compounds in petroleum heavy ends was achieved in many cases by multidimensional techniques, like those described in the previous section. However, the determination of elemental profiles is another area covered during the analysis of large MM hydrocarbons. Generally, the goal is to determine the distribution of heteroatomic species like N, S, O and metal compounds, which are deleterious from the upgrading and environmental points of view.

Vanadium and nickel are particularly abundant in HC, XHC and heavy ends. Venezuelan XHCs' generally have a combined total of 400-500 ppm of both metals, where Boscan HC is well known worldwide for its high V content (1200 ppm). Metal distribution is as important as total content. Krull has covered specific strategies for the determination of these types of distributions in a monograph.[74] Among these, the molecular size distributions of metal compounds have been investigated by size exclusion chromatography with plasma detectors, SEC-ICP [75,76] and SEC-DCP [77]. These studies showed that high MM metallorganic compounds are difficult to convert and survive most upgrading processes. The ICP technique appears to be better suited for low metal content.

Sulfur and Nitrogen detectors for HPLC have been described in the literature.[78,79] However, applications for large MM hydrocarbons are lacking. A comprehensive review of sulfur speciation in HC, coal and rock extracts is available.[80] The X-ray photoelectron spectroscopy (XPS) and X-ray absorption near-edge structure spectroscopy (XANES) are derived from studies on S, N compounds in coal. These techniques are particularly useful to get insights about the chemical functionality distribution within the sample. Extensive research has been conducted at Exxon laboratories with these techniques, summarized in a recent article.[81] This article stated that XPS and XANES are better suited than other techniques that are easier to carry out from the experimental point of view, safeguarding their future application despite their complexity.

Mitra and coworkers[82] reported the application of XANES to petroleum asphaltenes derived from crudes of diverse origin. They found that pyrrolic nitrogen is by far the most abundant functionality. Results reported by Gorbaty and Kelemen [81] using coal samples are in agreement with the above findings.

2.4   Fraction Separation

Separation has been demonstrated as a valid strategy since ancient Cesar's Rome. Recently, Boduszynski and coworkers [83] quoted the principle of "divide and conquer". In the preceding section on hyphenated techniques, fast separation methodologies at analytical scale were described. However, isolation of preparative amounts of fractions is convenient in many instances for further testing of their properties. The ensuing discussion will cover four of the most widespread separation methodologies applied to heavy hydrocarbons. Three are commonly operated at preparative scale.

The first to be discussed is gas chromatography simulated distillation (GC-SimDist), which is run at analytical scale. Since the 80´s, this technique has been routinely used instead of labor intensive physical distillation and standardized under ASTM protocols. The development of high temperature capillary columns widens the operating range up to 700-800°C atmospheric equivalent boiling points.[84,85] The chromatographic technique allows decreasing response time from one week to a couple of hours, guaranteeing its continuous use in the future. High temperature GC-SimDist appears suitable for analysis of molecules up to 120 carbon atoms.[86] The analysis of high MM alkane molecules isolated from bottom sediments from stored oil was demonstrated with this approach.[87]

The next technique is sequential elution fractionation (SEF). SEF is a solubility fractionation technique proposed for the "truly nondistillable resid" (700°C+). Details of the solvent sequence and deposition of the sample onto an inert solid support have been published by Altgelt and Boduszynski [3]. What is important, is to take into account the fact that crude oils are continuous mixtures of hydrocarbon and heterocyclic compounds, being some of their properties deduced from an extended atmospheric equivalent boiling point curve (AEBP). The idea is that intermolecular forces govern molecular interactions; the kinetic energy provided by temperature overcomes such forces along the distillation process, and the release of such forces is brought by solubilization for the non-volatile portion of the sample. The extended AEBP scale as proposed by Altgelt and Boduszynski, allows the comparison of whole crude oils on a common rationale basis, providing information on high vacuum gas oils (550-750°C), distillates that will probably be produced by the industry in the near future.[88] Despite the important aspects derived from this proposal, wide acceptance from the global petroleum community is still pending.

The hydrocarbon group-type separations widely known as SARA (saturates, aromatics, resins and asphaltenes), comprise many different in-house developed methodologies routinely applied at each petroleum laboratory. Gulf investigators introduced the concept.[89] Several reviews have appeared afterwards.[3,90-92] Commonly, the most polar part of the sample (asphaltene) is separated by precipitation with an alkane solvent. The remainder (maltene fraction), is separated by liquid chromatography over polar sorbents with an appropriate solvent gradient. In the following section, we will address some of the drawbacks affecting SARA separations. It is worth mentioning here, that a faster alternative has been proposed, based on thin layer chromatography with flame ionization detection (TLC-FID). The fundamental work of Poirier and George [93,94] gave rise to many published approaches, like the one followed in our laboratories [95]. A common problem faced in TLC-FID is quantitation, requiring careful and continuous calibration in order to obtain useful results, as reported by Vela and coworkers [96]. Another drawback to be considered, is the loss of the volatile C_{13}-fraction (< 220°C) when light or medium crude oils are analyzed by TLC-FID. A final aspect commonly disregarded is controlling of ambient conditions where the separation is carried out, since these were shown to affect the outcome of the analysis.[97] Despite all the quantitative issues mentioned, the technique is very powerful for assessment of the quantity and properties of the organic matter present in reservoir cores.[98]

A few closing comments regarding fraction separation and on a technique for which the term ABAN is proposed (acid, bases, amphoterics and neutral hydrocarbons) follows. Inconveniences in aqueous base/acid extractions carried out with large MM hydrocarbons, led the U.S.A. Bureau of Mines to develop a non-aqueous ion exchange separation of ABN fractions, employing ion exchange resins. One article by Green and coworkers is believed to summarize all this effort.[99] The detailed work carried out with Wilmington HC from California and Cerro Negro XHC from Venezuela is available from the U.S.A-Department of Energy.[23]

The original ABN scheme was not suited for separating amphoteric compounds, which were retained in the first ion exchange column of the series, commonly the anionic one. Subsequently, amphoterics were isolated together with acidic compounds. The scheme was further developed at Western Research Institute (Wyoming), allowing to separate the four fractions, achieving the extra benefit of placing the most rugged cationic column as the first one of the series.[100] Recent work conducted in our laboratories with vacuum residua failed to provide quantitative recovery of isolated ABAN fractions. The failure to replicate the technique remains an unknown, but a solution has been found by extracting the strongly adsorbed amphoterics from the anionic resin with a solvent system comprising $CH_2Cl_2/CS_2/MeOH$ 6/3/1 by volume.[101] The significance of ABAN separations will be discussed in greater detail in the final section of this article. The schematic for ABAN separations and the recoveries for some vacuum residual samples with the mixed solvent extraction are presented on Fig. 3.

2.5   Mathematical Algorithms

One approach to simplify handling of complex hydrocarbon mixtures relies on the concept of average structural properties. To this end, Neurock-Klein [102] and Diallo et. al. [103] have recently presented different approaches in regards to this subject. Another possible alternative is to look for structural trends in the different fractions by means of average molecular parameters (AMP), using the appropriate molecular representations to visualize such parameters.

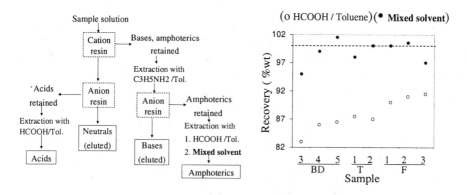

Figure 3. Schematic of ABAN separations with non-aqueous ion exchange resins. Fraction recoveries are compared when amphoterics are extracted following the routine protocol or by using $CS_2/CH_2Cl_2/MeOH$ mixed solvent.

The AMP method of analysis was developed in the early days of NMR to satisfy the needs to characterize very complex molecular systems like HC and heavy ends.[104] Proton [105] and carbon-13 [106,107] NMR can be used to extract the spectral information for the AMP determination, where proton is the fastest and the least time consuming technique. It is important to be aware that, as its name indicates, they are average values of some wide ranging molecular properties, which provide structures that may or may not really exist.[25] They provide general trends and are useful for comparison among samples subject to diverse studies or processes. In the last section of the chapter, one application involving Venezuelan HXC will illustrate the benefits derived from AMP usage.

The increase use of computers allows gathering huge amounts of data from complex materials under analysis, such as HC/XHC and heavy ends. However, all this information needs to be interpreted and processed. For that reason, data analysis methods that can handle large quantities of data, such as multivariate statistics are mandatory.[108] Chemometric approaches are very useful techniques to reduce sample preparation and to simplify separation schemes prior to chemical characterization. They can also allow samples classification according to some input variables, and to generate calibration models to predict some quality parameters of interest.

Multivariate calibration is a mean to determine how to use simultaneously measured variables x_1, x_2, ..., x_n for quantification of a target variable y.[109] With multivariate calibration, it is possible to generate calibration models using results obtained from relatively rapid, precise but non-specific analyses, in order to predict information normally obtained from slower, more expensive, but more relevant analyses [110]. Once the models are developed, the reference method is only used for periodical checking of model performance. Selective input variables are no longer needed, it is the output result that must be selective.[109] The input variables can be physical, chemical, spectroscopic, chromatographic, etc. Mostly, the type of data handled is spectroscopic and the multivariate statistics are useful in spectral analyses, because the simultaneous inclusion of multiple spectral intensities from many samples greatly improves the precision and applicability of quantitative spectral analyses.[111]

The three most used multivariate calibration methods are multiple linear regression (MLR), principal component regression (PCR) and partial least squares (PLS). Some reviews and tutorials have been published regarding these multivariate calibration techniques.[108,109,112,113] Multivariate calibration methods such as MLR, PCR and PLS are implicit modeling techniques, which means that no fundamental theory about expected behavior is imposed on the data.[114]

One of the first areas of application of the multivariate calibration was near infrared spectroscopy (NIR), due to the low resolution and selectivity of the higher overtones and combination bands, and the simplicity of the instrumentation. In the petroleum industry, the initial applications of multivariate calibration were focused on the prediction of fuel quality parameters in refining streams. Octane number determination (RON and MON) was the most common application.[115,116] Other applications of multivariate calibration are: (i) the determination of hydrocarbon type distributions [116,117], (ii) the determination of methanol and methyl tert-butyl ether (MTBE) in gasoline [118],

(iii) the determination of parameters for diesel fuel [119] and (iv) BTU content in natural gas [120].

Some recent publications described the estimation of crude oil and heavy cuts quality parameters through linear multivariate calibration techniques. Aske et al. [121] evaluated the estimation of SARA group-types in crude oil by infrared and near-infrared spectroscopy. They used PLS to establish the relationship between the spectral data and the HPLC results. An analytical method was proposed to predict the asphaltene content in crude oils by Fourier-transform infrared spectroscopy (FTIR) through PLS regression.[122] Chung and Ku [123] compared the use of near-infrared, infrared and Raman spectroscopy to predict the specific gravity (API°) of atmospheric residua with the use of PLS regression.

One of the limitations of PLS and PCR regression is that even though they can handle some non-linearities by including additional terms in the model, it is unsatisfactory for reasons that have been already addressed.[113] Methods such as neural network algorithms (NN), projection pursuit regression (PPR) and multivariate adaptive regression splines (MARS) are alternatives for fitting non-linear data.[124]

Artificial neural networks (ANN) attempt to simulate the mechanism of human learning. They also try to simulate the execution of typically human activities such as memorization, pattern recognition, generalization or modeling, and decision-making. Both linear and non-linear functions may be defined by configuring the neural network. Detailed information about NN has already been published.[125-127] ANN have also been used to predict quality parameters such as fuel octane number from near-infrared absorbance data [128] and from gas chromatographic analysis [129]. More recently, some applications of ANN to crude oil and heavy ends have been published, such as asphaltene content estimation in crude oil [130], asphalt rheological properties [131,132] and asphalt performance grade determination [131]. Details of these three applications will be provided in the final section within this chapter.

2.6 Other Characterization Schemes for HC, XHC and Heavy Ends

Within this section, we will address some important aspects concerning heavy ends, particularly those arising from HC and XHC. The first aspect is related to the propensity of such materials to give rise to carbonaceous phases during upgrading operations. These solid phases are defined as "petroleum cokes". Standardized analytical methods (ASTM D-189, D-524 and D-4530) are currently used for determining the coking propensity of the sample. It has been shown that the main parameter governing coke generation is the hydrogen content of the feed to be upgraded.[24,133,134] However, minor structural effects have been cited in a study covering HC from California, Mexico and Venezuela.[134] Thermogravimetry under inert atmosphere has been recently proposed for determination of cracking propensity and coke yields.[135] A thermogravimetry method carried out under vacuum was also proposed for assessment of distillate yields, however this pioneering work was not further developed to date.[85]

Solids generation during crude oil blending, XHC dilution for transportation and heavy ends upgrading has been addressed during recent symposiums focused on "fouling" aspects. A recent monograph covers this topic in detail.[136] Sample titration with precipitant streams is commonly carried out for the determination of crude oil stability and compatibility.[137-139] The amount of precipitated foulant is generally determined by the temperature gradient caused in reactor walls by the layered deposit.[140]

A decade long controversy on the state of polar fractions inside the crude oil medium seems to have reached a final point with the development of analytical techniques suited for studying whole samples under varying concentration, pressure and temperature conditions. It appears that colloidal dispersions better describe the physicochemical state of crude oils and particularly, of heavy ends. Particle scattering techniques like small angle neutron or x-ray scattering spectroscopy (SANS and SAXS) put into evidence the sample aggregation level, its particle size distribution as a function of experimental conditions, and the presence of natural dispersants (resins) or synthetic additives.[141-144] Less demanding analytical alternatives like rheology [145], laser light scattering [146] and surface tension measurement [147] provided many useful insights, which agreed with those shown by SANS or SAXS. A systematic decreasing aggregate size with increasing pressure or temperature was observed in these works. However, some studies showed that temperature affects sample behavior in a complex manner.[141,145] Reportedly, at 50-65°C, a liquid-solid transition was observed. Residua structures were observed to disappear in the 150-200 °C temperature interval, and the flocculation begins within this range. The smallest measured particle sizes were detected near 400°C, and irreversible effects were observed upon cooling.[145]

In all likelihood, the molecular mass (MM) can be considered as one of the most important parameters characterizing single hydrocarbon molecules. For complex mixtures, the situation is further complicated and their spanned MM ranges, plus the average value must be considered. However, despite extensive research conducted for determining these two parameters, it was recently recognized that the "true molecular weight distribution for any crude still can not be determined".[83] Depending on the chosen analytical technique for MM determination, orders of magnitude were reported for crude oil polar fractions (spanning from 600 to 300,000 Dalton, according to Speight et al. [148]). The reason for this drawback is the strong aggregative tendency shown by polar hydrocarbons, particularly by the asphaltene fraction. Thanks to the work of Speight, Boduszynski and coworkers [148,149], the recognition of the role played by intermolecular associations, was a breakthrough that showed the relatively small nature of hydrocarbons compounding HC, XHC and heavy ends. Recent developments of soft ionization MS techniques, size exclusion chromatography (SEC) of derivatized fractions, fluorescence spectroscopy and high temperature vapor pressure osmometry (VPO), confirmed the above authors findings. However, nowadays there is ample room for debate, since the trend in MM ranges displayed by polar hydrocarbons and heavy ends has been reversed. Very low MM values were published in recent reports, as illustrated in the asphaltene samples included in Table 1.

Despite the fact that asphaltenes from diverse sources were studied, the broad MM ranges reported (Table 1) suggests that MM determination for large hydrocarbon mixtures continues to pose challenges, demanding improvements for the analytical techniques commonly used. A recent article reports that VPO strongly depends on variables like concentration, temperature and solvent polarity.[150] SEC for virgin non-derivatized asphaltenes suffer from strong adsorption effects on chromatographic supports.[159] MS methods are subjected to discrimination effects imposed either by fraction volatility [149,154] or by ionization effects brought by the power of the ionizing source [153,157].

3. ILLUSTRATIVE EXAMPLES ON THE CHARAC-TERIZATION OF HC, XHC AND HEAVY ENDS

Recent studies conducted at our research laboratories will illustrate how some simplifying strategies are currently applied. There are some aspects that need to be considered for these methodologies: (1) which aspects demand corrective actions for their improvement, and (2) how they are expected to evolve with the aid of recent developments in data handling.

Table 1. Asphaltene Molecular Mass (MM) and Molecular Mass Distribution (MMD) Determined by Diverse Techniques.

Sample	Technique	MMD (Daltons)	MM average or "peak"(Dalton)	Reference
C_7-Athabasca and Cold Lake XHC	High Temp. VPO (o-diClϕ @130°C)	-	1800 ("Monomers")	150
C_6-Hamaca XHC	SEC (THF) for . Octylated sample	-	3900	151
C_7-Kuwait, France & California crudes	Fluorescence Spectroscopy	500-1000	750	152
C_7-Kuwait Atmospheric Resid	LD-MS	500-6000	1500	153
C_7-Maya Vacuum	LD-MS	200-600	300	154

Sample	Technique	MMD (Daltons)	MM average or "peak"(Dalton)	Reference
Resid				
C_7-Arab Atmospheric Resid	LD-MS	220-9000	1500	155
C_5-Arab Vacuum Resid	MALDI-TOF-MS	200-3500	450	156
C_7-Sumatra and Arab Vac. Resids	LD-MS	200-6000	1500	157
C_7- Ratawi Vacuum Resid	LVEI-MS	200-800	440	158
C_7-Russian Crude Vacuum Resid	FI-MS	300-1700	700	149

The term C_x in sample boxes refers to the n-Alkane used for asphaltene precipitation.
MS abbreviations: LD (laser desorption), MALDI/TOF (matrix assisted laser desorption ionization-time of flight), LV/EI (low voltage electron impact), FI (field ionization).

3.1 SARA Group-type Analysis

The preparative-scale SARA separations are widely practiced worldwide. However, there are some inconveniences that must be solved in order to improve their accuracy and most important, to be able to compare data within diverse laboratories. Results from a Round-Robin testing on Cerro Negro XHC showed that SARA distribution reports vary depending on the methodology employed [160]. On the one hand, small variations were observed when different separation media were adopted (Fig. 4). On the other hand, strong variations can also be appreciated, but other causes were responsible for the results from these outliers. Uncompleted recovery of resins is one of them (Fig. 4).

Improvements leading to better SARA fraction recoveries are needed for optimizing petroleum heavy end analysis. In our laboratories, the adoption of derivatized adsorbents has permitted quantitative recovery of vacuum residual samples.[161] Further standardization of analysis methods for heavy ends will enable various laboratories to have a common database and to share their results in a more direct way.

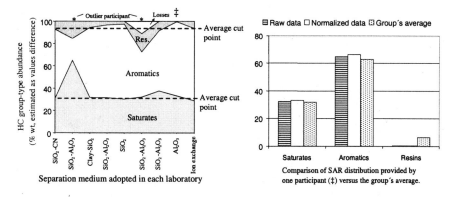

Figure 4. Round Robin results from nine participant laboratories providing SAR analysis of the 340-425°C cut from Cerro Negro XHC. Two outlier laboratories were identified (*). A detailed analysis from one participant laboratory (‡) illustrates the problem derived from resin losses.

One aspect commonly disregarded and poorly understood is fraction oxidation during preparative isolation processes. This applies to SARA analysis and to the other preparative schemes as well. Oxidation of hydrocarbon fractions under poorly controlled ambient conditions happens frequently, as reported in recent articles.[162,163] Despite detailed studies on oxidation mechanisms (for example, see Petersen [164]), preventive actions are not generally recommended within separation /characterization methodologies, in order to avoid the effects derived from the influence of atmospheric oxygen. It is believed that small variations in the abundance of polar compounds can bring noticeable changes to the colloidal structure of heavy ends.

Having mentioned some of the drawbacks affecting SARA analysis, it could be said that their future usage is secured, given the simplicity of the involved procedures. Extensive use of SARA has recently led us to the understanding of the intrinsic stability of crude oils, from a compositional point of view.[29,165] The "problematic components", i.e., asphaltenes with tendency to precipitate, are shown to be the most aromatic, pericondensed, more dense and hydrogen deficient portion among these polar fractions.[166] The existence of diverse asphaltene types is illustrated on Fig. 5.

Further application of SARA separations for large paraffin (waxy) fractions, led to the development of high temperature methodologies suited for isolation [167] and analysis of very large alkanes [168]. High temperature is a mandatory condition for the successful and quantitative separation of large alkane mixtures. Examples for HC/XHC isolated alkane fractions were presented before.[169,170] One illustrative example is included in Fig. 6. The sequential elution of a wax deposit following a step temperature gradient was observed to restrict the elution to low MM components, when the temperature was initially set at a low value. Setting the experimental temperature above the melting range of the mixture (75°C), allowed the elution of the highest MM components. By following this relatively simple experimental approach, very large alkane molecules can be analyzed. Their existence has been put into evidence before, relying on MS and high temperature GC.[86,87,149]

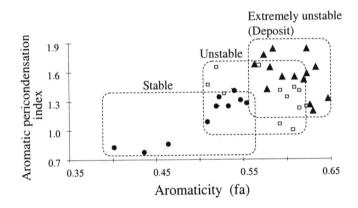

Figure 5 Average parameters for asphaltenes isolated from crudes with varying stability and from precipitated solid deposits.

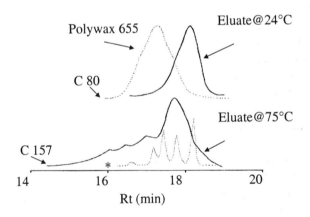

Figure 6. Elution of alkane components from a wax deposit with an step temperature gradient. Fractions were reanalyzed by high temperature size exclusion chromatography with evaporative light scattering detection. Reference standard n-alkanes*(n-C22-30-38-44-60) and Polywax-655 were overimposed on the chromatograms for estimating the molecular size ranges displayed by the eluted fractions.

3.2 Studies on XHC and Isolated ABAN Fractions. One Application of Average Molecular Representations

As it has been stated in previous sections, one of the most important factors limiting the characterization of HC/XHC is their complex molecular composition. The fraction separation method used is a determining step to minimize this complexity, specially the polar fraction i.e., resin and asphaltenes. They constitute 30-40% of the whole crude. One way to reduce this complexity is to avoid molecular aggregation of these fractions. It is very well documented in the literature that resins, and more particularly asphaltenes, are very prone to suffer this type of phenomena, which can disturb significantly any structural information.[148]

One possible parameter that can be used to monitor this effect is the fraction's average molecular weight (AMW) determined by vapor pressure osmometry (VPO), a method of analysis with problems of its own.[148,150] A comparison is presented on Table 2 for the polars obtained from SARA and ABAN separation of two Venezuelan XHC.

Table 2. Comparison of Average Molecular Weights (Daltons) for XHC fractions.

Crude oil	Resins	Asphaltenes	Acids	Bases	Amphoterics
Cerro Negro	1110	5758	670	867	1322
Hamaca	983	4596	734	827	1718

Experiments were carried out in CHCl$_3$ solvent at 45°C.

It is evident from the above table the inherent problem arising from the use of the SARA separation. One cannot totally overrule the presence of aggregates on the ABA fractions, but at least they are minimized, as can be seen from their lower AMW values. From this perspective, the ABAN fractionation scheme gives the best starting point (i.e., less aggregated fractions).

Once the method of separation has been decided, the next question is how to approach the characterization of the separated fractions. There are not too many possibilities in this direction, and the path one takes depends on the kind of information that is needed, as well as the costs and time involved in such a work. On one hand, it is possible to perform a detailed characterization with the aim of extracting either fractions or complete hydrocarbon families for their identification.[23,44] On the other hand, it is possible to model the system using known molecular structures that fit analytical data, like elemental analysis and NMR-AMP, with the purpose of predicting behavior and reactivity.[102,103] For the authors, a particular case of interest to study by means of AMP involving HC/XHC and their production behavior was the foamy oil characteristics that some of them possess. In these cases, oil production is higher than expected due to trapped gas bubbles within the oil or in the actual formation.[13] For gas bubbles to be trapped, some kind of flexible molecular network has to be present in the crude oil that limits molecular diffusion of the gas fast moving molecules. The initial study was conducted on three XHC, namely Cerro Negro, Hamaca and Zuata, that where fractionated into their respective ABAN components. The respective fraction distribution is presented on Table 3. It can be seen from these values, that the crudes are very similar in composition.

Table 3. Mass Balance (wt%) for ABAN fractions from XHC.

Crude	Acids	Bases	Amphoterics	Neutrals	Balance
Cerro Negro	8.0	7.1	15.3	66.3	96.7
Hamaca	7.9	8.2	15.5	64.1	95.8
Zuata	7.5	7.0	13.9	68.9	97.3

As far as the structural information is concerned, the main focus of interest was placed on the acid, base, and amphoteric fractions. Mainly, because of the fact that they have been identified as those responsible for the formation of the intermolecular network, specially the amphoteric one.[171] The authors of this chapter have investigated this point further by means of ^1H transverse relaxation measurements (T_2), done on some of the above crudes that have been doped with a known amount of the isolated fractions. T_2 relaxation is directly associated with the mobility (or viscosity) of the system under investigation.[12,172] The higher the molecular motion, the more isotropic the system is, and the higher the T_2 values associated to it. Experiments on the Hamaca oil at 60°C will illustrate these findings. 2, 4 and 10 wt% of each of the ABAN fractions were incorporated to the crude. The results are presented on Fig. 7.

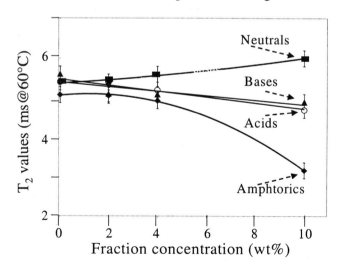

Figure 7. Transverse relaxation values (T2) vs. the percentage of the ABAN fraction incorporated on Hamaca XHC.

Below 4 wt%, changes in T_2 values, and therefore in mobility, induced by the addition of the different fractions is very small or within experimental error. But at 10 wt% addition, the changes are very significant, where the addition of the amphoteric fraction is the one that produces the stronger effect, in comparison with those observed from acid and bases. The neutral fraction (saturates and aromatics) can be considered as the solvent in these systems. One may conclude, that the functional groups present in the amphoteric fraction play an important role in the formation of the molecular network in these crudes.

These results suggest that one has to look for some significant structural characteristic on the amphoteric fraction that may produce a flexible structure that may hinder the gas molecular diffusion. From the calculated AMP, there are two parameters that are very important to describe how flexible the amphoteric species may be. They are the number of aromatic rings (#Ra), and more importantly, the number of aromatic carbons that may fuse four or more aromatic rings (Car,ar,ar).[173,174] There are two alternatives with these two parameters when constructing graphic representations from the AMP. On one hand, it is possible that all aromatic rings have to be condensed to satisfy the number of Car,ar,ar. On the other hand, the number of Car,ar,ar is such that not all aromatic rings are condensed, and in this case it is possible that two or more aromatic nuclei are linked by paraffin chains in this type of molecular representation.

In this particular case, it was possible to produce both types of graphic representations for the determined AMP. They are presented on Fig. 8 for the amphoteric fraction of each of the crudes. From the representations, those with two aromatic nuclei are the more flexible structures. The effect that they might have on the gas molecules movement has been formerly investigated employing asphaltene molecules.[175] Molecular simulation carried out in that study has shown that asphaltene molecules with three aromatic nucleuses are capable of trapping more gas molecules than those constituted with just one aromatic sheet.

So, it looks like that molecular structures with more than one nucleus, combined with the functional groups may produce the right molecular network capable of trapping gas molecules during crude oil depressurization. One could speculate that, as the pressure goes down, there might be gas over-saturation that will reduce the crude oil viscosity, and in this way produce a better flow.

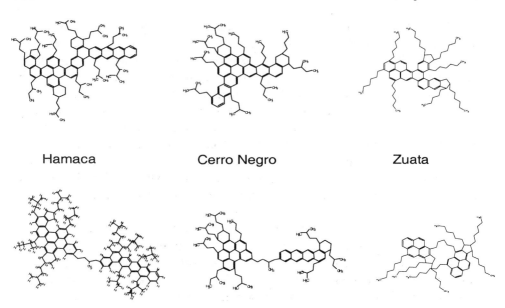

Hamaca   Cerro Negro   Zuata

Figure 8. Graphic representations of AMP for the amphoteric fractions from the studied XHC.

Further investigation into the crude oil structuring involved the use of electron microscopy images obtained from oil cryo-replicates.[176] In this technique, the oil sample is rapidly frozen using freon/liquid nitrogen, to prevent large ice crystal formation resulting in sample distortion. After the sample is frozen, it is fractured to reveal the interior features and the exposed surface is etched by sublimation under high vacuum conditions at -110°C. The surface is replicated with carbon and shadowed at 45° angle by evaporating platinum/carbon. The replica is cleaned with chloroform to remove all traces of adhering crude and then examined using a transmission electron microscope (TEM). Images taken using this technique are presented on Fig. 9 for two Venezuelan HC/XHC. The Tia Juana HC is introduced as an example where only one type of molecular representation can be produced from the determined AMP, i.e. a *peri*-condensed aromatic ring type of structure.

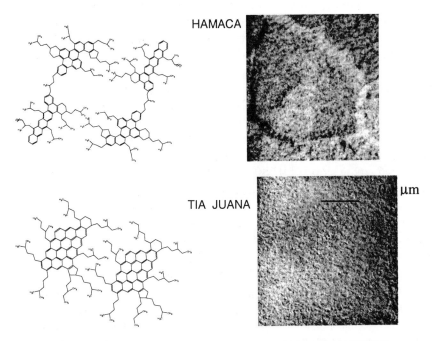

Figure 9. AMP representations and TEM images for two Venezuelan HC/XHC.

From these TEM images, it is possible to see that the surface morphology is significantly different between the two specimens. The one from Tia Juana is flat, whereas the one from Hamaca presents more structure and depth. The same type of images as in the ones Hamaca's, where obtained for the Cerro Negro and Zuata XHC. It is also important to mention that the graphic representations of the AMP do not show the functional groups that may be responsible for the formation of the network between the species present in the ABA fractions.

This section has illustrated one of the possible crude oil characterization pathways but by no means a complete one. As mentioned, there are two sensitive areas, the AMW and functional group determination. The ABAN separation scheme provides a method to reduce molecular aggregation of the polar fractions, providing a better starting point for any type of characterization.

3.3 Estimation of Crude Oil and Heavy Ends Quality Parameters Using Neural Networks Algorithms

As it was mentioned in section 2.5, multivariate calibrations can greatly simplify the determination of quality properties of crude oil, its fractions and products, generating estimated but quick characterization results. Two of such applications have been carried out in our laboratories. These are the determination of the n-C_7 asphaltene content in crude oil by FT-IR spectroscopy [130], and the estimation of asphalt's performance grades using ^{1}H-NMR [131,177]. To account for both, the mathematical algorithm employed was a three-layer back-propagation neural network (NN).[126]

FT-IR spectra of crude oils were taken using an attenuated total reflectance kit with a ZnSe crystal. Absorbances in the spectral range from 1800 to 685 cm^{-1}, baseline corrected at 2000 cm^{-1}, were the input patterns to the NN. The output pattern was the asphaltene concentration. Fig. 10 shows predicted versus actual asphaltene contents. There is a linear relationship between both results with an R^2 value of 0.996, and a slope of 0.980 ± 0.030. The standard error of calibration is 0.37 wt %. Fig. 10 also presents the absolute residuals plot, showing a random distribution with no bias, along the entire range of concentration. The trained NN allows the precise determination of asphaltene content in crude oils from FT-IR spectra, reducing the analysis time from hours to minutes and eliminating the need of preparative scale separation of the asphaltenic compounds.

In a similar fashion, for the asphalt, the input patterns to the NN were the temperature at which the rheological properties were measured, and 14 AMP. They were calculated from the ^{1}H-NMR integrals, carbon and hydrogen percentages and the sample and standard weight amounts. The output patterns were the following asphalt rheological properties: (i) permanent deformation susceptibility (G*/sinδ), (ii) G*/sinδ after RTFOT (rolling thin film oven test), (iii) Fatigue-cracking propensity (G*sinδ), (iv) creep stiffness (S), and (v) what is known as the m-value. These five properties are used to determine the performance grades (PG) with the aid of the SHRP specification tables.[171]

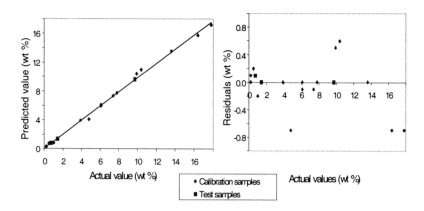

Figure 10. FT-IR-NN predicted vs actual n-C_7 asphaltene content (wt %) in crude oils. Absolute residuals vs actual asphaltene concentration values, in wt %.

Each rheological property was modeled independently and the NN configuration had only one neuron in the output layer. Fig. 11 shows the predicted and actual values for two of the above rheological properties. It also shows the *y=x* line and the error limits of the reference laboratory measurements. The estimated values fall within experimental error. Few exceptions were encountered at the low end of the scale, and they do not affect the PG determination.

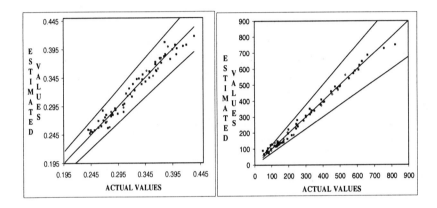

Figure 11. Predicted vs actual results for (a) S (MPa) and (b) m-value. (■ Calibration set, ○ Test set)

Table 4 presents a comparison between predicted and actual PG values for the asphalt´s test set, showing an excellent match. The discrepancy found for sample #4 was due to the fact that one of the determined rheological properties was near the boundary between two PG grades, for both the experimental and the estimated values. One of the important features of the described approach is the ability to predict the rheological properties at different temperatures from NMR data measured at room temperature. This is possible because the temperature is incorporated as an input pattern to the NN. Therefore, a considerable reduction in laboratory time (days to minutes) is achieved by avoiding the sample aging protocols.

Table 4. Laboratory and estimated performance grades (PG) for the asphalt´s test set.

Sample	Laboratory PG	Estimated PG
1	58-28	58-28
2	58-28	58-28
3	64-28	64-28
4	64-28	58-28

One of the advantages of applying chemometrics to spectroscopic data is the possibility of moving to flow-through cell configurations, taking the applications from the laboratory to the field. With this approach, the characterization results could be more accessible to the end user, allowing real time characterizations, better process control and easier decision-making.

4.   CONCLUSIONS

Heavy hydrocarbon feedstocks like heavy/extra-heavy crude and oil distillation residua, are expected to play an ever increasing global role on the energy scene during the first half of the 21st century. Their complexity imposes the adoption of simplified characterization strategies like those developed during the past two decades and reviewed in this chapter. However, some needs were identified for the improvement of such strategies, and ample room for their optimization still exists. Increase use of mathematical algorithms is expected thanks to the steady increase in computing capabilities.

5.   ACKNOWLEDGEMENTS

PDVSA-Intevep is acknowledged for funding and granting us permission to publish this work. Zoraida Hernandez and Dr. Lupe Marquez from PDVSA-Intevep, provided some of the experimental data presented and contributed with helpful discussions. Our sincerest gratitude goes to Dr. Chang S. Hsu from ExxonMobil for his kind invitation to participate in this project.

6.   GLOSSARY OF FREQUENT REFERRED TERMS

ABAN:	acids, bases, amphoterics and neutrals hydrocarbon group-types.
AEBP:	atmospheric equivalent boiling point.
AMP:	average molecular parameters.
AMW:	average molecular weight.
Asphaltenes:	solubility fraction defined by its insolubility in alkane solvents plus its solubility in aromatic solvents.
Bitumen:	hydrocarbon material possessing viscosities greater than 10,000 cP at the reservoir conditions.
Heavy end:	hydrocarbon mixture with boiling points > 343°C.
HC:	heavy crude.
Hyphenated:	analysis methodology that simultaneously combines separation plus spectroscopy techniques.
LC-MS:	liquid chromatography-mass spectrometry.

MM / MMD: molecular mass / molecular mass distribution.
NN / ANN: neural network / artificial neural network.
NMR: nuclear magnetic resonance spectrometry.
SARA: saturates, aromatics, resins and asphaltenes hydrocarbon group-types.
SEF: sequential elution fractionation.
VPO: vapor pressure osmometry.
XHC: extra-heavy crude.

7.　REFERENCES

1. Haaland, O.; Klovning, R.; Sem, T. *Proceeds. 7ᵗʰ UNITAR Int. Conf. Heavy Crude and Tar Sand.* **1998**, 1, 73-90; Paper 1998-008.
2. AOSTRA (Alberta Oil Sands Technology and Research Authority) Annual Report. 1992.
3. Altgelt, K. H.; Boduszynski, M. M. *Composition and Analysis of Heavy Petroleum Fractions;* Marcel Dekker: New York, 1994.
4. Meyer, R. F. *Proceeds. 15ᵗʰ World Petr. Cong.* **1998**, 2, 459-471.
5. Lindsay, I. *Proceeds. 15ᵗʰ World Petr. Cong.*, Vol 1, 85-88. 1998.
6. Stosur, G. J.; Waisley, S. L.; Reid, T. B.; Marchant, L. C. *Proceeds. 7ᵗʰ UNITAR Int. Conf. Heavy Crude and Tar Sand.* Vol.1, 1998; Paper No. 002, pp. 11-27.
7. Prats, M. *Proceeds. 15ᵗʰ World Petr. Cong.*, Vol 2, 1998; pp. 473-484.
8. Layrisse, I.; Chacin, J. *Proceeds. 7ᵗʰ UNITAR Int. Conf. Heavy Crude and Tar Sand.* Vol.1, 1998; Paper 001, pp. 1-9.
9. Renard, G.; Delamaide, E.; Morgan, R.; Fossey, J-P. *Proceeds. 15ᵗʰ World Petr. Cong.*, Vol 2, 1998; pp. 485-494.
10. Jha, K. N.; Butler, R. M.; Lin, G. B.; Oballa, V. *Proceeds. 6ᵗʰ UNITAR Int. Conf. Heavy Crude and Tar Sand.* Vol.1, 1995; pp. 759-774.
11. Bryant, R. S. *Proceeds. 7ᵗʰ UNITAR Int. Conf. Heavy Crude and Tar Sand.* Vol.1, 1998; Paper No. 110, pp. 1027-1033.
12. Fisher, D. B.; Espidel, J.; Huerta, M.; Randall, L.; Goldman, J. *J. Transport in Porous Media* **1999**, 35, 189-204.
13. Sheng, J. J.; Maini, B. B.; Hayes, R. E.; Tortike, W. S. *Transport in Porous Media,* **1999**, 35, 157-187.
14. Sarathi, P. *Proceeds. 7ᵗʰ UNITAR Int. Conf. Heavy Crude and Tar Sand.* Vol.2, 1998; Paper No. 124, pp. 1189-1200.
15. Guevara, E.; Gonzales, J.; Nuñez, G. *Proceeds. 15ᵗʰ World Petr. Cong.*, Vol 2, 1998; pp. 495-502.
16. Rivas, H.; Silva, F.; Gutierrez, X.; Nuñez, G. Paper 1998-058, *Proceeds. 7ᵗʰ UNITAR Int. Conf. Heavy Crude and Tar Sand.* Vol. 1, 1998; Paper No. 124 pp. 585-595.
17. Hayes, M. E.; Hrebenar, K. R.; Murphy, P L.; Bolden, P. L.; Deal, J. F.; Futch, L. E. Surfactant Compositions Containing Bioemulsifier for Reducing Oil Viscosity by Emulsification. Patent WO 8501889, 1985.
18. Pereira, P.; Flores, C.; Zbinden, H.; Guitian, J.; Solari, R. B. *Oil & Gas J.* **2001**, 99, 79-85.
19. Verlaan, J. P.; O´Connor, P.; Sonnemans, W. H.; Inoue, Y. *Proceeds. 15ᵗʰ World Petr. Cong.*, Vol 2, 1998; pp. 705-717.
20. Solari, R. B. Chapter 7 in *Asphaltenes and Asphalts, 2. Developments in Petroleum Science, 40B,* Yen, T. F.; Chilingarian, G. V. (Eds), Elsevier, 2000; pp. 149-171.
21. Perera, A.; Peñuela, N.; Rosales, J. L.; Scaglione, G. *Vision Tecnologica.* **2001**, 9(1), 11-16.
22. Yergin, D. *The Prize,* Simon & Schuster, 1991.

23. Green, J. A.; Green, J. B.; Grigsby, R. D.; Pearson, C. D.; Reynolds, J. W.; Shay, J. Y.; Sturm, G. P.; Thomson, J. S.; Vogh, J. W.; Vrana, R. P.; Yu, S. K. T.; Diehl, B. H.; Grizzle, P. L.; Hirsch, D. E.; Hornung, K. W.; Tang, S. Y.; Carbognani, L.; Hazos, M.; Sanchez, V. "Analysis of Heavy Oils: Method Development and Application to Cerro Negro Heavy Petroleum", Report NIPER-452, Vol. 1/2. US-DOE, Bartlesvile, OK. 1989.
24. Wiehe, I. A.; Liang, K. S. *Fluid Phase Equilibria* **1996**, 117, 201-210.
25. Speight, J. G. *The Chemistry and Technology of Petroleum,* Marcel Dekker: New York, 1999.
26. Moschopedis, S.; Fryer, J. F.; Speight, J. G. *Fuel* **1976**, 55, 184-186.
27. Speight, J. G.; Moschopedis, S. E. *Am. Chem. Soc. Prep. Div. Pet. Chem.* **1981**, 26, 907-911.
28. Majid, A.; Sparks, B. D.; Ripmeester. J. A. *F. Sci. Technol. Int'l.* **1993**, 11, 279-292.
29. Carbognani, L.; Espidel, J.; Izquierdo. A. Chapter 13 in *Asphaltenes and Asphalts, 2. Developments in Petroleum Science, 40B,* Yen, T. F.; Chilingarian, G. V. (Eds). Elsevier: Amsterdam, 2000; pp. 335-362.
30. Ignatiadis, I.; Schmitter, J. M.; Arpino, P. *J. Chromatogr.* **1985**, 324, 87-111.
31. Merdrignac, I.; Behar, F.; Albretcht, P.; Briot, P.; Vandenbroucke, M. *Energy Fuels* **1998**, 12, 1342-1355.
32. Payzant, J. D.; Mojelsky, T. W.; Strausz, O. P. *Energy Fuels* **1989**, 3, 449-454.
33. Payzant, J. D.; Montgomery, D. S.; Strausz, O. P. *Org. Geochem.* **1986**, 9, 357-369.
34. Green, J. B.; Yu, S. K. T.; Pearson, C. D.; Reynolds, J. W. *Energy Fuels* **1993**, 7, 119-126.
35. Nishioka, M.; Campbell, R. M.; Lee, M. L.; Castle, R. N. *Fuel* **1986**, 65, 270-273.
36. Nishioka, M.; Tomich, R. S. *Fuel* **1993**, 72, 1007-1010.
37. Zhang, H.; Que, G.; Li, P. *Am. Chem. Soc, Prep. Div. Pet. Chem.* **1998**, 43, 182-184.
38. Gutierrez, L. B.; Ranaudo, M. A.; Mendez, B.; Acevedo, S. *Energy Fuels* **2001**, 15, 624-628.
39. Cagniant, D. *Complexation Chromatography,* Marcel Dekker: New York, 1991.
40. Boduszynski, M. M., 1993 (personal communication).
41. Stock, L. M.; Obeng, M. *Energy Fuels* **1997**, 11, 987-997.
42. Stock, L. M.; Hsien Wang, S. *Fuel* **1987**, 66, 921-924.
43. Mojelski, T. W.; Ignasiak, T. M.; Frakman, Z.; McIntyre, D. D.; Lown, E. M.; Montgomery, D. S.; Strausz, O. P. *Energy Fuels* **1992**, 6, 83-96.
44. Strausz, O. P.; Mojelsky, T. W.; Lown, E. M.; Kowalewski, I.; Behar, F. *Energy Fuels* **1999**, 13, 228-247.
45. Zijun, W.; Wenjie, L.; Guohe, Q.; Jialin, Q. *Pet. Sci. Technol.* **1997**, 15, 559-577.
46. Orea, M.; Bruzual, J.; Carbognani, L. *Rev Soc. Venezolana de Quimica* **2001**, 24(2), (in press).
47. Davis, I, L.; Raynor, M. W.; Kithinji, J. P.; Bartle. K. D. *Anal. Chem.* **1988**, 60, 683A-702A.
48. Griffiths, P. R.; Pentoney, S. L.; Giorgetti, A.; Shafer, K. H. *Anal. Chem.* **1986**, 58, 1349A-1366A.
49. Albert, K.; Braumann, U.; Tseng, L-H.; Nicholson, G.; Bayer, E.; Spraul, M.; Hofmann, M. *Anal. Chem.* **1994**, 66, 3042-3046.
50. Raglione, T. V.; Sagliano, N.; Floyd, T. R.; Hartwick, R. A. *LC-GC* **1986**, 4, 328-338.
51. Mondello, L.; Dugo, P.; Dugo, G.; Lewis, A. C.; Bartle K. D. *J. Chromatogr. A* **1999**, 842, 373-390.
52. Snyder, L. R.; Kirkland, J. J. *Introduction to Modern Liquid Chromatography,* Wiley-Interscience: New York, 1979.
53. Huang, E. C.; Wachs, T.; Conboy, J. J.; Henion, J. D. *Anal. Chem.* **1990**, 62, 713A-725A.
54. McLuckey, S. A.; VanBerkel, G. J.; Goeringer, D. E.; Glish, G. L. *Anal. Chem.* **1994**, 66, 737A-743A.
55. Niessen, W. M. A. *J. Chromatogr. A* **1998**, 794, 407-435.
56. Yerguey, A. L.; Edmonds, C. G.; Lewis, I. A. S.; Vestal, M. L. *Liquid Chromatography/ Mass Spectrometry,* Plenum Press: New York, 1990.

57. Yuan Xie, S.; Liu Deng, S.; Jia Yu, L.; Huang, R-B.; Zheng, L-S. *J. Chromatogr. A* **2001,** 932, 43-53.
58. Hsu, C. S.; McLean, M. A.; Qian, K.; Aczel, T.; Blum, S. C.; Olmstead, W. N.; Kaplan, L. H.; Robbins, W. K.; Schulz, W. W. *Energy Fuels* **1991,** 5, 395-398.
59. Qian, K.; Hsu, C. S. *Anal. Chem.* **1992,** 64, 2327-2333.
60. Hsu, C. S.; Qian, K. *Anal. Chem.* **1993,** 65, 767-771.
61. Hsu, C. S.; Qian, K. *Energy Fuels* **1993,** 7, 268-272.
62. Liang, Z.; Hsu, C. S. *Energy Fuels* **1998,** 12, 637-643.
63. Hsu, C. S.; Dechert, G. J.; Robbins, W. K.; Fukuda, E .K. *Energy Fuels* **2000,** 14, 217-223.
64. Hsu, C. S.; Qian, K.; Chen, Y. C. *Anal. Chim. Acta* **1992,** 264, 79-89.
65. Quilliam, M. A.; Sims, P. G. *J. Chromatogr. Sci.* **1988,** 26, 160-167.
66. Varotsis, N.; Pasadakis, N.; Gaganis, V. *Fuel* **1998,** 77, 1495-1502.
67. Pasadakis, N.; Gaganis, V.; Varotsis, N. *Fuel* **2001,** 80, 147-153.
68. Mao, J.; Pacheco, C. R.; Traficante, D. D.; Rosen, W. *J. Chromatogr. A* **1994,** 684, 103-111.
69. Li, M.; Larter, S. R.; Stoddard, D. *Anal. Chem.* **1992,** 64, 1337-1344.
70. Giddings, J. C. *Anal. Chem.* **1984,** 56, 1258A-1270A.
71. Cortes, H. J.; Rothman, D. Chapter 6 in *Multidimensional Chromatography,* Cortes H. J. (Ed), Marcel Dekker: New York, 1990; pp. 219-250.
72. Welch, K. J.; Hoffman, N. E. *J. HRC & CC* **1992,** 15, 171-175.
73. Davis, I, L.; Xu, B.; Markides, K.; Bartle, K. D.; Lee, M, L. *J. Microcolumn Sepn.,* **1989,** 1, 71-84.
74. Krull, I. S. *Trace Metal Analysis and Speciation.* J. Chromatogr. Library, vol. 47. Elsevier: New York, 1991.
75. Biggs, W. R.; Brown, R. J.; Fetzer, J. C. *Energy Fuels* **1987,** 1, 257-262.
76. Reynolds, J. G. *Pet. Sci. & Technol.* **2001,** 19, 979-1007.
77. Izquierdo, A.; Carbognani, L.; Leon, V.; Parisi, A. *F. Sci. & Technol. Int'l.* **1989,** 7, 561-570.
78. Ryerson, T. B.; Dunham, A. J.; Barkley, R. M.; Sievers, R. E. *Anal. Chem.* **1994,** 66, 2841-2851.
79. Yan, X. *J. Chromatogr. A.* **1999,** 842, 267-308.
80. Snape, C. E.; Mitchell, S. C.; Ismail, K.; Garcia, R. "Speciation of Organic Sulfur Forms in Heavy Oils, Petroleum Source Rocks, and Coals" in *Reviews on Analytical Chemistry, Euroanalysis VIII.* Littlejohn, D.; Thornburn Bums D. (Eds.), The Royal Society of Chemistry: London, 1994.
81. Gorbaty, M. L.; Kelemen, S. R. *Fuel Process. Technol.* **2001,** 71, 71-78.
82. Mitra-Kirtley, S.; Mullins, O. C.; van Elp, J.; George, S. J.; Chen, J.; Cramer, S. P. *J. Am. Chem. Soc.* **1993,** 115, 252-258.
83. Boduszynski, M. M.; Rechsteiner, C. E.; Carlson, R. M. *Proceeds. 7th UNITAR Int. Conf. Heavy Crude and Tar Sand.* Vol.2, 1998; Paper No. 168, pp. 1579-1587.
84. Trestianu, S.; Zillioli, G.; Sironi, A.; Saravalle, C.; Munari, F.; Galli, M.; Gaspar, G.; Colin, J. M.; Jovelin, J. L. *JHRC & CC.* **1985,** 8, 771-781.
85. Schwartz, R. E.; Brownlee, R. G.; Boduszynski, M. M.; Su, F. *Anal. Chem.* **1987,** 57, 1393-1401.
86. Thomson, J. S.; Rynaski, A. F. *JHRC & CC.* **1992,** 15, 227-234.
87. Thomson, J. S.; Grigsby, R. D.; Doughty, D. A.; Woodward, P. W.; Giles, H. N. *Proceeds. 4th Int. Conf. Stability and Handling Liq. Fuels,* Giles, H. N. (Ed), 1, 65-78, 1992.
88. Boduszynski, M. M.; Grudoski, A.; Rechsteiner, C. E.; Iwamoto, J. D. *Oil & Gas J.* **1995,** 93(37), 39-45.
89. Jewell, D. M.; Albaugh, E. W.; Davis, B. E.; Ruberto, R. G. *Ind. Eng. Chem. Fundament.* **1974,** 13(3), 278-282.
90. Altgelt, K. H.; Gouw, T. H. *Chromatography in Petroleum Analysis.* Chromatographic Series, Vol. 11, Marcel Dekker: New York, 1979.

91. Lundanes, E.; Greibrokk, T. *JHRC & CC.* **1994,** 17, 197-202.
92. Carbognani, L. *Rev. Soc. Venezolana de Quimica* **2001,** 24(2), 21-29.
93. Poirier, M. A.; George, A. E. *Energy Sources* **1983,** 7, 151-164.
94. Poirier, M. A.; George, A. E. *J. Chromatogr. Sci.* **1983,** 21, 331-333.
95. Sol, B.; Romero, E.; Carbognani, L.; Sanchez, V.; Sucre, L. *Rev. Tecnica Intevep* **1985,** 5, 39-43.
96. Vela, J.; Cebolla, V. L.; Membrado, L.; Andres, J. M. *J. Chromatogr. Sci.* **1995,** 33, 417-425.
97. De Zeeuw.; Franke, J. P.; van Halem, M.; Schaapman, S.; Logawa, E.; Siregar, C. J. P. *J. Chromatogr. A* **1994,** 664, 263-270.
98. Orea, M.; Alberdi, M.; Ruggiero, A. *Rev. Soc. Venezolana de Quimica* **2001,** 24(2), 13-20.
99. Green, J. B.; Hoff, R. J.; Woodward, P. W.; Stevens, L. L. *Fuel* **1984,** 63, 1290-1301.
100. Kim, S.; Branthaver, J. F. *F. Sci. & Technol. Int'l.* **1996,** 14, 365-393.
101. Carbognani, L.; Espidel, J.; Salazar, P.; Cotte, E.; Oliveros, A. *Am. Chem. Soc. Prep. Div. Pet. Chem.* **2000,** 45(4), 659-660.
102. Neurock, M.; Klein, M. T. Chapter 4 in *Asphaltenes and Asphalts, 2. Developments in Petroleum Science, 40B,.* Yen, T. F.; Chilingarian, G. V. (Eds), Elsevier: Amsterdam, 2000; pp. 59-102.
103. Diallo, M. S.; Cagin, T.; Faulon, J. L.; Goddard, W. A. Chapter 5 in *Asphaltenes and Asphalts, 2. Developments in Petroleum Science, 40B.* Yen, T. F.; Chilingarian, G. V. (Eds), Elsevier: Amsterdam, 2000; pp. 103-125.
104. Williams, R. B.; Chamberlain, N. F. *Proceeds. 6th World Petr. Cong.,* Sec. V, 1963; paper no. 17, pp. 217-230.
105. Yen, T. F.; Wu, W. H.; Chilingar, G. V. *Energy Sources* **1984,** 7, 275-303.
106. Dickinson, E. M. *Fuel* **1980,** 59, 290-294.
107. Michon, L.; Martin, D.; Planche, J. P.; Hanquet, B. *Fuel* **1997,** 76, 9-15.
108. Beebe, K. R.; Kowalski, B. R. *Anal. Chem.* **1987,** 59, 1007A-1017A.
109. Martens, H.; Naes, T. *Multivariate Calibration,* John Wiley & Sons : New York, 1989.
110. Martens, H.; Naes, T. *Trends in Analytical Chemistry* **1984,** 3, 204-210.
111. Thomas, E. V.; Haaland, D. M. *Anal. Chem.* **1990,** 62, 1091-1099.
112. Geladi, P.; Kowalski, B. R. *Anal. Chim. Acta* **1986,** 185, 1-7.
113. Kowalski, B. R.; Seasholtz, M. B. *J. Chemometrics* **1991,** 5, 1290-145.
114. Seasholtz, M. B.; Kowalski, B. R. *Applied Spectroscopy* **1990,** 44, 1337-1348.
115. Kelly, J.; Barlow, C.; Jinguji, T.; Callis, J. *Anal. Chem.* **1989,** 61, 313-320.
116. Parisi, A.; Nogueiras, L.; Prieto, H. *Anal. Chim. Acta* **1990,** 238, 95-100.
117. Kelly, J.; Callis, J. *Anal. Chem.* **1990,** 62, 1444-1451.
118. García, F.; De Lima, L.; Medina, J. *Applied Spectroscopy* **1993,** 47, 1036-1039.
119. Foulk, S.; Desimas, B. *Process Control and Quality* **1992,** 2, 69-72.
120. Brown, C.; Lo, S. *Applied Spectroscopy* **1993,** 47, 812-815.
121. Aske, N.; Kallevik, H.; Sjöblom, J. *Energy Fuels* **2001,** 15, 1304-1312.
122. Wilt, B. K.; Welch, W. T. *Energy Fuels* **1998,** 12, 1008-1012.
123. Chung, H.; Ku M. *Applied Spectroscopy* **2000,** 54, 239-245.
124. Seasholtz, M. B.; Kowalski, B. R. *Anal. Chim. Acta* **1993,** 277, 165-177.
125. Smits, J. R. M.; Melssen, W. J.; Buydens, L. M. C.; Kateman, G. *Chemometrics and Intelligent Laboratory Systems* **1994,** 22, 165-189.
126. Wythoff, B. L. *Chemometrics and Intelligent Laboratory Systems* **1993,** 18, 115-155.
127. Melssen, W. J.; Smits, J. R. M.; Buydens, L. M. C.; Kateman, G. *Chemometrics and Intelligent Laboratory Systems* **1994,** 23, 267-291.
128. Crawfod, N. R.; Hellmuth, W. W.; Marcellus, D. H.; Chou, K. J. *Process Control and Quality* **1992,** 4, 13-20.

129. Van Leeuwen, J. A.; Jonker, R. J.; Gill, R. *Chemometrics and Intelligent Laboratory Systems* **1994,** 25, 325-340.
130. Colaiocco, S. R.; Farrera, M. "Determination of Asphaltene Content in Crude Oil by Attenuated Total Reflectance Infrared Spectroscopy", Submitted to *Journal of Process Analytical Chemistry,* 2001.
131. Colaiocco, S. R.; Espidel, J. *Vision Tecnologica* **2001,** 9, 17-24.
132. Michon, L.; Hanquet, B. *Energy Fuels* **1997,** 11, 1188-1193.
133. Roberts, I.. *Am. Chem. Soc. Prep. Div. Pet. Chem.* **1989,** 34, 251-254.
134. Green, J. B.; Shay, J. Y.; Reynolds, J. W.; Green, J. A.; Young, L. L.; White, M. E. *Energy Fuels* **1992,** 6, 836-844.
135. Laux, H.; Butz, T.; Rahimian, I. *Oil & Gas Science and Technology- Rev. IFP* **2000,** 55, 315-320.
136. Musrush, G. W.; Speight, J. G. *Petroleum Products: Instability and Incompatibility,* Taylor & Francis: Philadelphia, 1995.
137. Wiehe, I. A.; Kennedy, R. J. *Energy Fuels* **2000,** 14, 56-59.
138. Wiehe, I. A.; Kennedy, R. J. *Energy Fuels* **2000,** 14, 60-63.
139. Carbognani, L.; Contreras, E.; Guimerans, R.; Leon, O.; Flores, E.; Moya, S. SPE 64993, *Proceeds. SPE Int. Symp. Oilfield Chem.,* Houston, TX-USA. 2001.
140. Watkinson, A. P.; Navaneetha-Sundaraman, B.; Posarac, D. *Energy Fuels* **2000,** 14, 64-69.
141. Thiyagaran, P.; Hunt, J. E.; Winans, R, E.; Anderson, K. B.; Miller, J. T. *Energy Fuels* **1995,** 9, 829-833.
142. Barre, L.; Espinat, D.; Rosenberg, E.; Scarsella, M. *Rev. Inst. Fr. Pet.* **1997,** 52, 161-175.
143. Storm, D. A.; Barresi, R. J.; Sheu, E. Y.; Bhattacharya, A. K.; DeRosa, T. F. *Energy Fuels* **1998,** 12, 120-128.
144. Sheu, E. Y.; Acevedo, S. *Energy Fuels* **2001,** 15, 702-707.
145. Storm, D. A.; Barresi, R. J.; Sheu, E. Y. *Energy Fuels* **1995,** 9, 168-176.
146. Nielsen, B. B.; Svrcek, W. Y.; Mehrotra, A. K. *Ind. Eng. Chem. Res.* **1994,** 33, 1324-1330.
147. Leon, O.; Rogel, E.; Espidel, J.; Torres, G. *Energy Fuels* **2000,** 14, 6-10.
148. Speight, J. G.; Wernick, D. L.; Gould, K. A.; Overfield, R. E.; Rao, B. M. L.; Savage, D. W. *Rev. Inst. Fr. Pet.* **1985,** 40, 51-61.
149. Boduszynski, M. M.; McKay, J.. F.; Latham, D. R. *Proceeds. Association Asphalt Paving Technology* **1980,** 49, 123-143.
150. Yarranton, H. W.; Alboudwarej, H.; Jakher, R. *Ind. Eng. Chem. Res.* **2000,** 39, 2916-2924.
151. Acevedo, S.; Escobar, G.; Ranaudo.; M. A., Rizzo, A. *Fuel* **1997,** 77, 853-858.
152. Groenzin, H.; Mullins, O. C. *Pet. Sci. and Technol.* **2001,** 19, 219-230.
153. Seki, H.; Kumata, F. *Energy Fuels* **2000,** 14, 980-985.
154. Miller, J. T.; Fisher, R. B.; Thiyagaran, P.; Winans, R. E.; Hunt, J. E. *Energy Fuels* **1998,** 12, 1290-1298.
155. Bartholdy, J.; Lauridsen, R.; Mejlholm, M.; Andersen, S. I. *Energy Fuels* **2001,** 15, 1059-1062.
156. Artok, L.; Su, Y.; Hirose, Y.; Hosokawa, M.; Murata, S.; Nomura, M. *Energy Fuels* **1999,** 13, 287-296.
157. Fujii, M.; Sanada, Y.; Yoneda, T.; Sato, M. *Am. Chem. Soc. Prep. Div. Fuel. Chem.* **1998,** 43, 735-740.
158. DeCanio, S. J.; Nero, V. P.; DeTar, M.; Storm, D. A. *Fuel* **1990,** 69, 1233-1236.
159. Carbognani, L.; Espidel, J. "Preparative and Analytical Scale Size Exclusion Chromatography of Petroleum Resin and Asphaltene", *Proceeds. Int. Symp. On Asphaltene and Wax Deposition,* Cancun, MX, 2001.
160. Carbognani, L.; Espidel, J.; Garcia, X. *Analytical and Statistical Study on Cerro Negro Distillation Cuts,* Final report for the 4th Round Robin on Heavy Crudes and Bitumen. United

Nations Institute for Training and Research (UNITAR). Centre for Heavy Crudes and Tar Sands. Edmonton, Alberta, Canada. 1995.

161. Carbognani, L.; Izquierdo, A. *J. Chromatogr.* **1989,** 484, 399-408.
162. Carbognani, L.; Espidel, J.; Carbognani, N.; Albujas, L.; Rosquete, M.; Parra, L.; Mota, J.; Espidel, A.; Querales, N. *Pet. Sci, & Technol.* **2000,** 18, 671-689.
163. Boukir, A.; Aries, E.; Guiliano, M.; Asia, L.; Doumenq, P.; Mille, G. *Chemosphere* **2001,** 43, 279-286.
164. Petersen, J. C. *Pet. Sci. & Technol.* **1998,** 16, 1023-1059.
165. Carbognani, L.; Orea, M.; Fonseca, M. *Energy Fuels* **1999,** 13, 351-358.
166. Carbognani, L. *Energy Fuels* **2001,** 15, 1013-1020.
167. Carbognani, L; Orea, M. *Pet. Sci. & Technol.* **1999,** 17, 165-187.
168. Carbognani, L. *J. Chromatogr. A* **1997,** 788, 63-73.
169. Carbognani, L.; Duarte, D.; Rosales, J.; Villalobos, J. *Pet. Sci. & Technol.* **1998,** 16, 1085-1111.
170. Carbognani, L.; DeLima, L.; Orea, M.; Ehrmann, U. *Pet. Sci. & Technol.* **2000,** 18, 607-634.
171. Youtcheff, J. S.; Jones, D. R. *Guideline for Asphalt Refiners and Suppliers,* Report SHRP-A-686, Strategic Highway Research Program, National Rsearch Council, Washington DC, USA. 1994.
172. Morriss, C. E.; Friedman, R.; Straley, C.; Johnston, M.; Vinegar, H. J.; Tutunjian, P. B. *Proceeds. SPWL 35ᵗʰ Annual Logging Symposium,* 1994; paper C, pp. 1-24.
173. Qian, S-A.; Zhang, P-Z.; Li, B-L. *Fuel* **1985,** 64, 1085-1091.
174. Zhao, S.; Kotlyar, L. S.; Woods, J. R.; Sparks, B. D.; Hardcare, K.; Chung, K. H. *Fuel* **2001,** 80, 1155-1163.
175. Marquez, L. 1997, Unpublished results (Internal report on investigation INT-3703,97).
176. Espidel, J.; Salazar, P.; Hernandez, Z.; Carbognani, L. 2001, Unpublished results (Internal report on investigation INT-9108, 2001).
177. Colaiocco, S. R.; Espidel, J. "System and Method for Predicting Parameters of Hydrocarbon with Spectroscopy and Neural Networks". USA Patent pending. Application, May 2000.

Chapter 15

ADVANCES IN NMR TECHNIQUES FOR HYDROCARBON CHARACTERIZATION

Gordon J. Kennedy
ExxonMobil Research and Engineering Company, Corporate Strategic Research, 1545 Route 22 East, Annandale, NJ 08801

1. INTRODUCTION

It is clear from the amount of material in this book focused on various chromatographic techniques, that the analysis of complex hydrocarbon mixtures continues to be one of the biggest challenges facing analytical chemists in the pharmaceutical, biomedical, environmental, petroleum, polymer, and petrochemical arenas. Nuclear magnetic resonance (NMR) has long been recognised as the spectroscopic technique of choice for obtaining detailed structural, dynamic, and chemical information of organic compounds. However, the analysis of complex organic mixtures with traditional one-dimensional (1D) NMR techniques usually suffers from severe spectral overlap, making it virtually impossible to extract detailed compositional information. Many ingenious and elegant 1D, 2D, and multidimensional NMR spectral editing techniques have been developed that address various aspects of this resolution problem and are now routinely implemented on modern commercial spectrometers. Excellent texts describing both the theoretical aspects and the practical application of these techniques are available.[1-5]

NMR spectroscopy is a vibrant field and it is neither practical nor the intent of this chapter to capture all the recent advances in this multidisciplinary field. The intent of this chapter is to focus on some of the recent developments in NMR spectroscopy that have potential for improving our ability to analyse liquid phase or soluble hydrocarbons. Some of the improvements in NMR instrumentation that impact hydrocarbon characterisation include higher magnetic fields, improvements in probe technology, facile implementation of pulsed field gradients, and improved selectivity through coupling with separation techniques (e.g. HPLC-NMR and coupling with capillary electroseparation techniques).

By design, this chapter will highlight certain techniques and examples the author thinks will be of interest to the reader. The choice of material presented, examples included, and references cited are by no means exhaustive. The goal of this chapter is to give the reader an appreciation for some of the recent advances in NMR techniques for hydrocarbon characterisation and to point them to areas of research rather than enumerate the vast amount of work in NMR technique development published in the past few years.

2.   DISCUSSION

2.1   Availability of Higher Magnetic Field Strengths Provides Increased Sensitivity and Resolution

There have been continuous efforts to increase the field strengths (B_o) of NMR magnets ever since the first spectrometers were introduced. This is because NMR theory and experience have shown that the sensitivity increases in proportion to $B_o^{3/2}$ and chemical shift dispersion increases linearly with B_o. Thus, the information content and spectral interpretation are typically improved as the magnetic field strength is increased. The current state-of-the-art field strength is 18.8T (800 MHz for ^1H) with the first 21.1T (900MHz for ^1H) delivered in 2001.

In the case of polyolefins it is well accepted that many important physical properties are related to the branching microstructure, the type of structural information readily determined from NMR analysis. High resolution ^{13}C NMR has been used for many years in detailed structural studies of commercial polymers, and resonances for branches shorter than six carbons long have been successfully assigned. Rinaldi et. al.[6] recently described an interesting application of high magnetic fields (17.6T, 750 MHz for ^1H) to improved spectral resolution in the ^{13}C NMR of polyethylene where unique resonances from chain branches up to 10 carbons long were resolved. These higher magnetic field strengths are also finding important biomedical applications, such as in determining 3D structures of high molecular weight proteins.[7]

Although the need for increased sensitivity and chemical shift dispersion have been the driving forces for increasing magnetic field strengths these newer ultra high field magnets are prohibitively expensive and are not yet widely available for routine use. This has lead to increased focus on national labs and centers of excellence to defray cost while maintaining capabilities. It is anticipated that the increased availability of these newer high field systems in the coming years will have far reaching influence on our ability to analyse complex hydrocarbons. It is also expected that these improvements in superconducting magnet technology will provide benefits in other analytical techniques such as FT-ICR mass spectrometry.

2.2 Improvements in Sensitivity from Higher Magnetic Fields and New Probe Designs Facilitate Further Development of On-line Coupling with Separation Techniques

On-line coupling of HPLC and NMR is not a new concept.[8-10] NMR is attractive because it can provide a wealth of structural information not available with any other detector. However, the HPLC-NMR technique suffered for many years because of the inherent low sensitivity of NMR relative to the more traditional detectors. There have been significant recent improvements in NMR sensitivity realised from the availability of higher magnetic fields, the design of smaller receiver coils with higher dynamic ranges, and the optimisation of solvent suppression techniques.[11,12] These efforts have resulted in commercially available HPLC-NMR systems that are now being successfully applied in pharmaceutical and biomedical research to the study of drug metabolites and natural products.[12-14]

Even with these improvements HPLC-NMR is still limited by the large sample requirements and the high rate of solvent consumption. The high solvent consumption limits the use of expensive deuterated solvents thus making it necessary to use solvent suppression techniques. Solvent suppression techniques may obscure peaks of interest in the NMR spectrum and reduce the amount of structural information obtained from the system under investigation. One approach to deal with the solvent problems has been to use supercritical fluid chromatography (SFC) as the coupled separation technique. This allows the use of nonprotonated eluents, such as CO_2, and the whole ^1H NMR spectrum can be recorded without interference from the solvent and without the need for solvent suppression. Bayer et. al.[15] successfully demonstrated the feasibility of this coupled SFC-NMR technique by separating a mixture of five phthalates under SFC conditions with CO_2 as the eluent and monitoring the process with ^1H NMR. For NMR detection they designed a pressure proof continuous flow probehead.

Work in separation methods aimed at miniaturisation has resulted in the development of capillary electrophoresis (CE), capillary HPLC (CHPLC), and capillary electrochromatography (CEC).[16-21] These separation techniques have the advantages over conventional methods of shorter separation times, lower eluent consumption and lower sample weight requirements. Clearly, miniaturisation of both the separation and the NMR components would make the use of deuterated solvents cost effective and overcome the problems associated with solvent suppression.

The limiting factor in coupling these capillary separation techniques to NMR has been the sensitivity (S/N) of the NMR detection system. This limitation has motivated the development of new microcoil NMR probes designed to maximise the filling factor of the receiver coil. The continuous flow probes used in HPLC-NMR and SFC-NMR both use the saddle type rf coils similar to those used in conventional high resolution NMR spectroscopy (Figure 1 a,b).

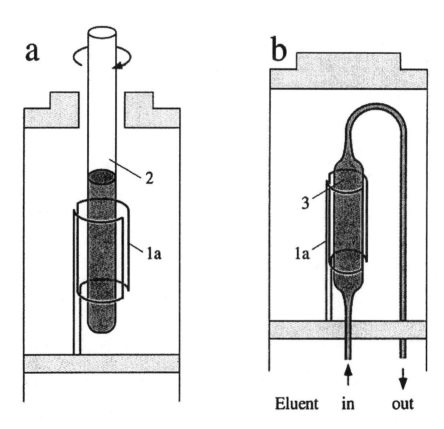

Figure 1. Different geometries of NMR probes. (a) Conventional probe with rotating NMR tube. (b) NMR probes with continuous-flow cells for HPLC-NMR and SFC-NMR coupling. 1a, Saddle-type rf coil; 2, NMR tube (5 mm diameter); 3, flow cell (reprinted with permission from Ref. 21, copyright (1996) American Chemical Society).

Two different microcoil designs, shown in Figure 2, have been developed and implemented for coupling capillary separation methods to NMR. The solenoidal geometry provides the best filling factor giving better sensitivity compared to saddle coils but is impractical because it requires the preparation of a new coil for every separation. Another drawback of the solenoid design is the difficulty in constructing such solenoid coils greater than 1mm in length for capillary columns without loss of resolution and sensitivity in NMR detection. Although the saddle configuration has a mass sensitivity of a factor of two lower than the solenoid type, separation columns can easily be changed and it accommodates a variation of the column dimensions within the range of the internal diameter of the rf coil. These practical advantages of the saddle rf coil design make it much more amenable to commercial development by instrument manufacturers and the future implementation beyond the specialist's laboratory.

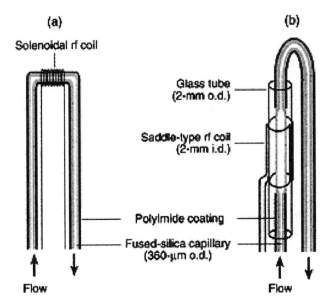

Figure 2. Solenoid rf coil (a) and saddle type rf coil (b) for coupling capillary separation techniques with NMR (reprinted with permission from Ref. 18).

Samples of natural products, amino acids, and drug metabolites in the range of 0.1 - 0.5 nmol have been successfully analysed with CHPLC-NMR, CE-NMR, and CEC-NMR.[16-21] This detection limit is approximately ten fold better than obtainable in conventional HPLC-NMR. These initial examples suggest tremendous potential for the use of capillary separation techniques coupled with NMR in the analysis of complicated mixtures.

Similar approaches have been demonstrated in the analysis of polymers. Size exclusion chromatography (SEC) has been successfully coupled to NMR for studying the molecular weight dependence of compositional features of polymers, such as tacticity and comonomer composition, that are readily detectable by NMR.[22,23] More recently, direct coupling of a SEC to a high field 750MHz ^{1}H NMR spectrometer was used to directly determine the molecular weight distributions of four different poly (methylmethacrylate) samples without the need for a conventional calibration procedure.[24] The accurate determination of the end groups relative to the repeat units required for this type of analysis was made possible by the high sensitivity of this magnetic field strength.

While two decades have elapsed since the initial demonstration of HPLC-NMR, the coupling of capillary separation techniques with NMR is a much more recent development. As a result, practitioners of NMR can purchase commercial HPLC-NMR systems but would have to work with home-built instrumentation for CE- or CEC-NMR. It is anticipated that ongoing development in rf coil and flow cell design along with the increased availability of high magnetic fields will result in further improvements in sensitivity and wider application of all of these emerging NMR-coupled separation techniques. This, in turn, will eventually enable the development and marketing of commercial instrumentation.

2.3   "Chromatography in a NMR Tube": - Spectral Editing with Pulsed Field Gradient (PFG) Techniques Improves Analysis of Hydrocarbon Mixtures

Of the many recent advances in NMR spectroscopy, the development and application of techniques based on the use of pulsed field gradients has tremendous potential in the study of hydrocarbons. This section presents the basic principles underlying these techniques and highlights some representative applications.

In conventional 1D NMR a single pulse (or pulse sequence) is applied and the resultant free induction decay (FID) is detected as a function of a single time variable. Fourier transformation of this FID is performed to give the familiar spectral representation of intensity as a function of frequency. In a two-dimensional (2D) experiment an array of 1D NMR experiments is recorded as a function of two time variables. Thus, a second dimension is introduced to the experiment that is used to separate two different effects and the result is easier to interpret than the corresponding 1D spectrum. Two-dimensional (2D) NMR is a general term for a class of experiments based on this principle.

Figure 3 compares the basic 1D and 2D NMR experiments.   A basic 2D NMR pulse sequence is characterised by three time intervals: preparation, evolution, and acquisition (or detection). In a number of 2D experiments a further interval is added, the mixing time. During the preparation time the spin system of interest is prepared for the experiment, for example by the generation of transverse magnetisation by applying a 90° pulse as in 1D NMR. In the evolution time (t_1) the spin systems develop under the influence of different factors, such as Larmor precession or spin-spin coupling, before it is detected during the acquisition time (t_2).

This sequence in Figure 3 does not constitute a 2D experiment by itself. In fact a 2D experiment is only possible after a series of 1D spectra have been measured. That is, in a 2D experiment this sequence is repeated with a systematic variation of the evolution time (t_1). Each FID, or t_2 signal, is then transformed to give a number of 1D spectra in the F2 frequency domain. This series of spectra show a phase and/or amplitude modulation with respect to the evolution time, t_1. This series of spectra can now be transformed with respect to the time axis, t_1. A frequency axis (F1) results which contains the frequencies of those mechanisms which have been effective during the evolution time, t_1, and caused the modulation of the signal phase or amplitude. For example, if spin-spin couplings were effective during t_1 and Larmor precession during t_2, F1 will now contain the coupling information and F2 the chemical shift information. The net result being the ability to separate the coupling and chemical shifts, that overlap and are indistinguishable in the 1D experiment, and now present them on two distinct frequency axes.

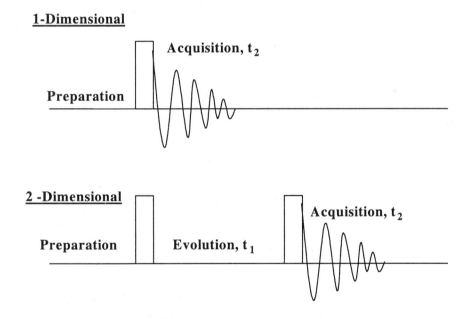

Figure 3. The basic principles of 1-Dimensional (1D) and 2-Dimensional (2D) NMR experiments. In 1D NMR a single pulse (or pulse sequence) is applied and the resultant free induction decay (FID) is detected as a function of a single time variable. In 2D NMR an array of 1D NMR experiments is recorded as a function of two time variables. Thus, introducing a second dimension that is used to separate two different effects.

The two main classes of 2D experiments are resolving experiments and correlation experiments. 2D-resolved experiments separate two different interactions along two frequency axes. This separation aids spectral interpretation in cases where the two effects, such as spin-spin coupling and chemical shifts, overlap in the conventional 1D. 2D-correlation spectroscopy correlates spins that interact though dipolar or scalar coupling. The key thing to remember about 2D NMR is that a response is detected as a function of two variables.

An extension of these 2D experiments is the introduction of different NMR dimensions that depend on properties such as size, shape, and mass. In NMR spectroscopy each spin in a magnetic field precesses along the field at a speed that is related to the external magnetic field strength. If a magnetic field gradient, whose field strength is position dependent, is superimposed on the main magnetic field then the precessing frequency of each spin will also be position dependent. This principle forms the basis of pulsed field gradient nuclear magnetic resonance spectroscopy, or PFG NMR. In PFG NMR a transient magnetic field is generated by passing current through a coil specifically wound to produce a regularly varying magnetic gradient, G, across the length of the sample. The length and amplitude of the gradient are controlled throughout the experiment. This results in spatial labelling of the spins that permits detection of translational movement of the molecules.

The first example of the use of PFGs was called the pulsed gradient spin echo (PGSE) method.[25] The pulse sequence and its effect on the magnetisation are shown in Figure 4. In this experiment a gradient pulse, G, is applied after the first $\pi/2$ pulse and a short delay, t_1. This gradient pulse spatially encodes the nuclear spins along the length of the sample. This spatial encoding occurs because the nuclei precess at frequencies that are dependent on their position relative to the applied gradient (G). Using gradients for spatially encoding nuclei forms the basis of magnetic resonance imaging (MRI) where the position of a nucleus is encoded in two or three dimensions with the use of gradient pulses along the x, y and z axes.

As shown in Figure 4c the magnetisation precesses in the xy plane during the period following the gradient pulse (4b). A π pulse (Figure 4d) is applied along the y axis to effectively reverse the direction of the transverse precession. A second gradient pulse is then applied that is equal in amplitude and phase to the first pulse. This second gradient pulse will spatially decode the spins in the sample. The free induction decay (FID) is recorded when the spin echo is formed at a time 2τ after the first $\pi/2$ pulse. The effect of a rapidly diffusing small molecule on the magnetisation is shown in Figure 4e. Application of the second gradient pulse will not effectively decode these spins because these encoded spins (denoted with the dashed arrow) have diffused to another region of the sample, resulting in some signal attenuation (4f). Also illustrated in Figure 4e and 4f is the effect that this experiment has on the larger, more slowly diffusing, molecules. Little attenuation of the signal (denoted with the solid arrow) occurs because of the larger molecules diffuse slowly. The basic PGSE sequence forms the foundation for all subsequent gradient techniques.[26-29] PFG NMR has developed into a very effective method for measuring molecular diffusion.[30]

Combining the 2D and PFG NMR methodology described above and illustrated in Figures 3 and 4 has tremendous potential in the analysis of hydrocarbon mixtures because it combines the detailed structural information traditionally associated with NMR and the size selectivity of diffusion coefficients. Thus setting the stage for development and application of "chromatography in a NMR tube" experimental techniques.

Diffusion-Ordered NMR SpectroscopY (DOSY) is the two-dimensional (2D) analogue of the PFG spin echo technique.[31,32] In the DOSY experiment the data are displayed as a pseudo-2D spectrum in which signals are dispersed according to chemical shift in one dimension and diffusion coefficient in the other.

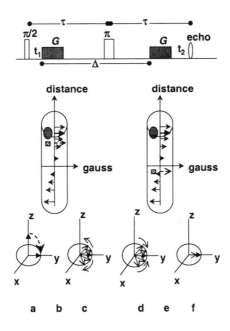

Figure 4. Outline of the PGSE experiment illustrating how signal intensity is dependent on diffusion. $\pi/2$ and π denote the 90° and 180° radio frequency pulses, τ the time between these two pulses, G the magnitude and the duration of the pulsed magnetic field gradients, and Δ the time between the onset of these two pulses. "Echo" denotes spin-echo formation. The experiment is described in detail in the text in terms of six components; (a) initial $\pi/2$ pulse, (b) gradient pulse, (c) spin evolution, (d) π pulse, (e) gradient pulse, and (f) signal detection.

An obvious application of PFG NMR is simplification of complex spectra of mixtures where resonances can be attenuated on the basis of the different diffusion coefficients of the components in the mixture. An interesting practical application of PFG NMR to the analysis of polymer additives was described by Larive et. al.[33], using PFG NMR to resolve a mixture of polymer additives with similar or overlapping chemical shifts but differing diffusion coefficients. They demonstrated by careful optimisation of the experimental parameters, amplitude and duration of the gradient pulses and delay times, that resonances of the different components of the mixture were attenuated on the basis of their rate of diffusion. The different diffusional properties that are manifested in the PFG experiment, effectively results in the generation of individual subspectra for each component that permits straightforward assignment of the ^{1}H NMR spectrum to the corresponding component in the mixture. In this work they used the PFG NMR experiments to determine diffusion coefficients of the individual components of the mixture in order to make peak assignments.

It is important to note that is not necessary to determine the diffusion coefficients of the individual components in order to apply this technique. Larive et. al.[33] and Müller et. al.[34] independently illustrated this point by demonstrating the application of this gradient technique to edit the NMR spectra of polymer solutions. The DOSY experiment can be used to selectively separate the resonances of fast diffusing components, such as residual solvent or low molecular weight additives, from the more slowly diffusing polymer. An example of the application of this concept using a mixture of n-decane and polypropylene (M.W. 44,000) is shown in Figure 5. This figure shows that even in this simple mixture while there is severe spectral overlap in the traditional 1D ^{1}H NMR spectrum, it is possible to separate them in a second diffusion dimension due the large differences in diffusion coefficients. Individual spectra of each component are obtained by extracting individual rows of the diffusion dimension.

Jerschow and Müller [34] have also extended this to the analysis of polymer mixtures. They have shown, with mixtures of various molecular weight polypropylenes and mixtures of polypropylene/polystyrene as examples, that it is possible to separate components of polymer mixtures in the diffusion dimension that severely overlap in the chemical shift dimension. Although the degree of separation achieved with DOSY is limited by differences in the diffusion coefficients, it has the potential to become a valuable complement to GPC in the analysis of polymer homogeneity by providing structural information not available from GPC.

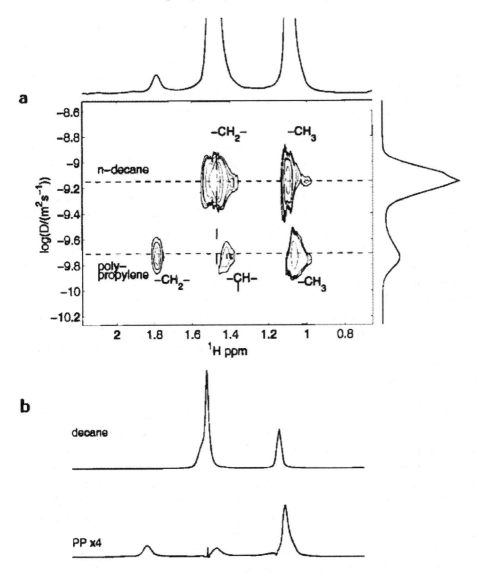

Figure 5. (a) ^{1}H - DOSY spectrum of a mixture of polypropylene (0.2%) and *n*-decane (0.8%) in tetrachloroethane-d_2. The x-axis is the ^{1}H chemical shift scale and the y-axis is the diffusion coefficient. (b) projections of individual rows with different diffusion coefficients give ^{1}H spectra of the individual components (reprinted with permission from Ref. 34, copyright (1998) American Chemical Society).

NMR spectroscopy has been widely used to provide average structural information about the composition of petroleum products, fractions, and process streams for many years.[35]   However, the ^{1}H and ^{13}C NMR spectra of these materials are not amenable to detailed interpretation due to overlapping signals from the different hydrocarbon classes that are present.   To address this resolution problem various separation techniques combined with both 1D and 2D editing and correlation techniques have been applied to provide more detailed structural information.[36,37]   DOSY offers the potential to provide similar structural detail without the need for physical separation.

Berger et al.[38] have recently described their initial attempts in applying the DOSY technique to complex petroleum products. Shown in Figure 6 is the ^{13}C - DOSY spectrum of a mixture of naphthalene, decalin, tetradecane, and 2,6,10,14-tetramethylpentadecane.   The traditional 1D ^{13}C NMR spectrum, shown at the top of the Figure, illustrates the difficulty in identifying individual components for even this relatively simple mixture. However, the signals of the different components show better resolution in the diffusion dimension of the DOSY spectrum, where signals have been filtered according to molecular mobility. The signals are well resolved in the diffusion dimension and can be correlated with the other signals of the individual component. Also shown in the figure is that the individual spectra of each component can be extracted from the different diffusion rows. Which leads to the important point that the DOSY experiment not only shows the presence of a number of components, but also leads to the structural identification of each component without the need for standard reference compounds.

A much more challenging example described by Berger et al.[38] is the analysis of the ^{1}H NMR spectrum of the aromatic fraction of a diesel fuel. Shown in Figure 7 are the ^{1}H - DOSY spectrum and traditional 1D ^{1}H NMR spectrum of this sample. Although the diffusion dimension is broad and highly overlapped, significant information can be extracted from this data set by examining different rows of the 2D plot. Also shown in the Figure are the individual subspectra extracted from three separate rows of the diffusion dimension. Scanning through the diffusion contour shows that the composition changes quite significantly. The spectrum along row A (the lowest diffusion coefficient row) shows mostly single ring aromatics with long aliphatic chains while the fastest diffusing component (Row C) is predominately methyl substituted polycyclic aromatics.   This illustrates that even in complex hydrocarbon mixtures it is possible to obtain a profile of molecular types by analysis of the DOSY subspectra without the need for prior physical separation of the entire sample.

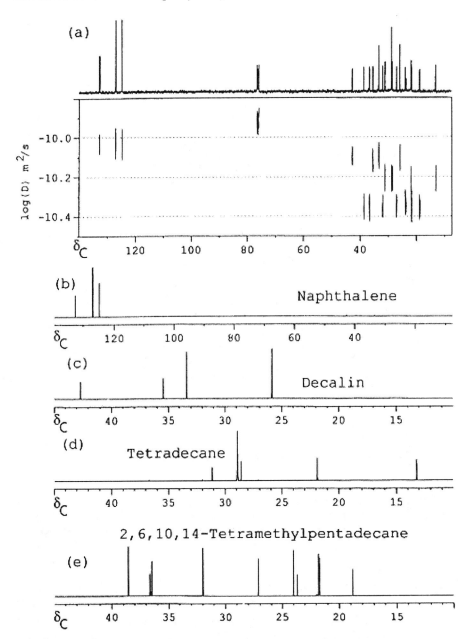

Figure 6. (a) ^{13}C-detected DOSY contour plot for a mixture of naphthalene, decalin, tetradecane, and 2,6,10,14-tetramethylpentadecane. The 1D ^{13}C NMR spectrum of the mixture is shown at the top. The x-axis is the ^{13}C chemical shift scale and the y-axis is the diffusion coefficient. (b)-(e) The bottom spectra are the extracted ^{13}C-spectra of the individual components from the DOSY spectrum. Only aliphatic region is shown for saturated components (reprinted from Ref. 38, copyright (2000) with permission from Elsevier Science).

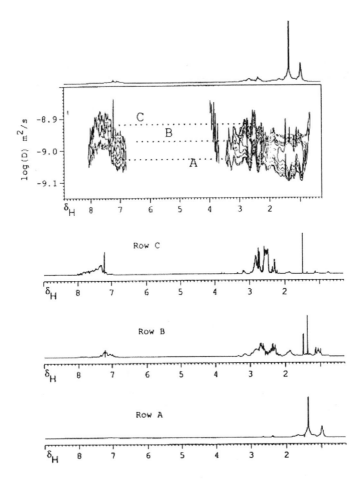

Figure 7. ¹H-DOSY contour plot for an aromatic fraction of diesel sample along with its 1D ¹H NMR spectrum at the top. The x-axis is the ¹H chemical shift scale and the y-axis is the diffusion coefficient. The bottom spectra are the extracted spectra of the average molecules corresponding to three diffusion rows (reprinted from Ref. 38, copyright (2000) with permission from Elsevier Science).

These diffusion-edited NMR techniques have recently been extended to the development of "affinity NMR" as an analytical screening tool in combinatorial drug discovery programs.[39,40]   As suggested by the name, affinity NMR is designed to distinguish active ligands that exhibit binding affinity to a particular receptor in a pool of nonactive ligands. Differences in the diffusional properties of the small unbound ligands and the much larger bound complexes form the basis of the spectroscopic separation. That is, the PFG NMR technique can selectively detect binding ligand signals in a complex mixture of other ligands because their diffusion coefficients will change when complexed with the receptor. With no need for physical separation affinity NMR has the potential to greatly increase the efficiency of high throughput drug screening.

These are just a few examples of how these PFG-based techniques can be applied to problems of practical significance. There is much activity in both the development and the use of these techniques and it is anticipated that applications to complex hydrocarbon mixtures will continue to grow at an ever increasing rate.

3. CONCLUSIONS AND FUTURE PROSPECTS

Some of the recent advances in NMR techniques that are of particular relevance to the analysis of liquid-phase or soluble hydrocarbons have been described. Work in superconducting magnet technology has increased the availability of higher magnetic field strengths for NMR spectroscopy that translates directly into increased sensitivity and resolution. These improvements in sensitivity from higher magnetic fields and new probe designs have facilitated the commercialisation of HPLC-NMR instruments. Miniaturisation of NMR detection systems has greatly improved the sensitivity of on-line coupling with separation techniques. It is expected that further improvement in coil design and instrument configurations will help in the separation and structural identification of hydrocarbon mixtures.

The concept of "chromatography in a NMR tube" using pulsed field gradient (PFG) techniques has been described and is a powerful tool for the analysis of hydrocarbon mixtures without the need for physical separation. Applications for this Diffusion-Ordered SpectroscopY (DOSY) are varied and range from the analysis of polymers and petroleum fractions, to uses in combinatorial drug discovery programs. Ongoing efforts in improving NMR hardware and software are expected to aid in the proliferation of new applications in the future.

4. ACKNOWLEDGEMENTS

The helpful comments of Steven Rucker, Debra Sysyn, Michael Lacey, Carrie Kennedy, Mobae Afeworki, C. Sam Hsu, and two anonymous reviewers are gratefully acknowledged.

5. REFERENCES

1.   Ernst, R. R.; Bodenhausen, G.; Wokaun, A. *Principles of Nuclear Magnetic Resonance in One and Two Dimensions*, Oxford University Press: Oxford, U.K., 1987.
2.   Freeman, R. *Spin Choreography*, Oxford University Press: Oxford, U.K., 1998.
3.   Günther, H. *NMR Spectroscopy; Basic Principles, Concepts, and Applications in Chemistry*, John Wiley and Sons: New York, 1994.

4.   Friebolin, H. *Basic One- and Two-Dimensional NMR Spectroscopy*, Wiley VCH: Weinheim, Germany, 1998.
5.   Braun, S.; Kalinowski, H. -O., Berger, S. *150 and More Basic NMR Experiments*, Wiley VCH: Weinheim, Germany, 1998.
6.   Liu, W.; Ray, D. G.; Rinaldi, P. L. *Macromolecules* **1999**, 32, 3817.
7.   Purvushin, K.; Riek, R.; Wider, G.; Wuthrich, K. *Proc. Natl. Acad. Sci. USA* **1997**, 94, 12366.
8.   Watanabe, N.; Niki, E. *Proc. Jpn. Acad., Ser. B* **1978**, 54, 194.
9.   Bayer, E.; Albert, K.; Neider, M.; Grom, E.; Keller, T. *J. Chromatogr.* **1979**, 186, 497.
10. Dorn, H. C. *Anal. Chem.* **1984**, 56, 747A.
11. Lindon, J. C.; Nicholson, J. K.; Wilson, I. D. *Adv. Chromatogr.* **1996**, 36, 315.
12. Albert, K. *J. Chromatogr. A.* **1995**, 703, 123.
13. Lindon, J. C.; Nicholson, J. K.; Sidelman, U. G.; Wilson, I. D. *Drug Metab. Rev.* **1997,** 29, 705.
14. Albert, K.; Bayer, E. *Anal. Methods Instrum.* **1995**, 2, 302.
15. Albert, K.; Braumann, U.; Tseng, L. -H.; Nicholson, G.; Bayer, E.; Spraul, M.; Hofmann, W.; Dowle, C.; Chippendale, M. *Anal. Chem.* **1994,** 66, 3042.
16. Wu, N.; Peck, T. L.; Webb, A. G.; Magin, R. L.; Sweedler, J. V. *J. Am. Chem. Soc.* **1994**, 116, 7929.
17. Wu, N.; Peck, T. L.; Webb, A. G.; Magin, R. L.; Sweedler, J. V. *Anal. Chem.* **1994**, 66, 3849.
18. Gfrorer, P.; Schewitz, J.; Pusecker, K.; Albert, K.; Bayer, E. *Anal. Chem.* **1999**, 71, 315A.
19. Pusecker, K.; Schewitz, J.; Gfrorer, P.; Tseng, L. -H.; Albert, K.; Bayer, E. *Anal. Chem.* **1998**, 70, 3280.
20. Gfrorer, P.; Schewitz, J.; Pusecker, K.; Tseng, L. -H.; Albert, K.; Bayer, E. *Electrophoresis* **1999**, 20, 3.
21. Behnke, B.; Schlotterbeck, G.; Tallarek, U.; Strohschein, S.; Tseng, L.-H.; Keller, T.; Albert, K.; Bayer, E. *Anal. Chem.* **1996,** 68, 1110.
22. Hatada, K.; Ute, K., Kitayama, T., Yamamoto, M.; Nishimura, T.; Kashiyama, M. *Polym. Bull.* **1989**, 21, 489.
23. Hatada, K.; Ute, K.; Kitayama, T.; Nishimura, T.; Kashiyama, M.; Fufimoto, N. *Polym. Bull.* **1990**, 22, 549.
24. Ute, K.; Nimi, R.; Hongo, S.; Hatada, K. *Polym. J.* **1998**, 30, 439.
25. Stejskal, E. O.; Tanner, J. E. *J. Chem. Phys.* **1965**, 42, 288.
26. Tanner, J. E. *J. Chem. Phys.* **1970**, 52, 2523.
27. Gibbs, S. J.; Johnson Jr., C. S. *J. Magn. Reson.* **1991**, 93, 395.
28. Wu, D.; Chen, A.; Johnson Jr., C. S. *J. Magn. Reson. A* **1995**, 115, 260.
29. Kärger, J.; Pfeiffer, H.; Heink, W. In *Advances in Magnetic Resonance Vol. 12,* Warren, W. S. (Ed.), Academic Press: New York, 1988; pp. 1.
30. Stilbs, P. *Prog. Nucl. Magn. Reson. Spectrosc.* 1987, 19, 1.
31. Morris, K. F.; Johnson Jr., C. S. *J. Am. Chem. Soc.* **1992,** 114, 3139.
32. Johnson Jr., C. S. *Prog. Nucl. Magn. Reson. Spectrosc.* **1999**, 34, 203.
33. Jayawickrawa, D. A.; Larive, C. K.; McCord, E. F.; Roe, D. C. *Magn. Reson. Chem.* **1998**, 36, 755.
34. Jerschow, A.; Müller, N. *Macromolecules* **1998**, 31, 6573.
35. Clutter, D. R.; Petrakis, L.; Stenger Jr., R. L.; Jensen, R. K. *Anal. Chem.* **1972,** 44, 1395.
36. Cookson, D. J.; Smith, B. E. *Energy Fuels* **1987**, 1, 111.
37. Sarpal, A. S.; Kapur, G. S., Chopra, A.; Jain, S. K.; Srivastava, S. P.; Bhatnagar, A. K. *Fuel* **1996**, 75, 483.
38. Kapur, G. S.; Findeisen, M.; Berger, S. *Fuel* **2000**, 79, 1347.
39. Gounarides, J. S.; Chen, A.; Shapiro, M. J. *J. Chromatogr.* **1999,** 725, 79.
40. Shapiro, M. J.; Gounarides, J. S. *Prog. Nucl. Magn. Reson.* **1999**, 35, 153.

Chapter 16

ANALYSIS OF POLYMERIC HYDROCARBON MATERIALS BY MATRIX-ASSISTED LASER DESORPTION/IONIZATION (MALDI) MASS SPECTROMETRY

Stephen F. Macha and Patrick A. Limbach
University of Cincinnati, Department of Chemistry, Cincinnati, OH 45221

1. INTRODUCTION

There are a number of analytical techniques available for the analysis of hydrocarbon polymeric materials. The most common and useful techniques include pyrolysis, chromatography, thermal analysis, spectroscopy (NMR, IR & Raman) and mass spectrometry.[1] The growth of mass spectrometry for the analysis of synthetic polymers has nearly doubled in eight years (as measured by ASMS abstracts) as noted by the editor of the journal of the American Society for Mass Spectrometry (ASMS) in 1998.[2] Among the many mass spectrometry techniques available, matrix-assisted laser desorption/ionization (MALDI) applications are constantly increasing. The MALDI technique was initially developed for biological molecules, mainly proteins and peptides,[3,4] but was eventually utilized for other analytes including synthetic polymers [5]. Although MALDI is quite successful for the analysis of polar polymers, it has not worked as well for nonpolar analytes such as hydrocarbon materials. This limitation is attributed to lack of proper understanding of some fundamental issues related to this technique such as the desorption/ionization mechanism and the role of the matrix in the MALDI process. In this chapter, MALDI-MS analytical approaches for the analysis of hydrocarbon polymeric materials are discussed.

2.   MALDI-MS

2.1   Overview

MALDI is a soft ionization technique used for the analysis of large nonvolatile molecules such as biopolymers and synthetic polymers. The MALDI process produces intact *pseudomolecular* ions of the analyte through irradiation by a pulsed laser beam on a solid mixture of a particular analyte dissolved in a suitable matrix compound. Typically, short pulse laser beams that range from 1 to 100 nanoseconds are used.[6]   Normally UV type lasers are used with the nitrogen laser ($\lambda \square \square 337$ nm) being the most common. More recently, IR lasers have also been used in MALDI effectively.[7,8]

2.2   Sample preparation

A MALDI experiment involves several steps. The first and most critical step is sample preparation. There are various sample preparation techniques that have been developed depending on the nature of sample to be analyzed. Among the common established techniques include the dried-droplet method [3], fast solvent drying method [9], two-layer method [10], and most recently, solid/solid compressing method used for hard to dissolve samples [11,12]. The dried-droplet method is the most common because of its simplicity. However, some disadvantages of using this method have been pointed out; most of all being creation of heterogeneous matrix crystals (hot spots) which result in poor spectral reproducibility.[13]

Application of the dried-droplet method involves mixing a very small amount of the analyte with an excess of matrix typically in a ratio of 1:1000, in a solvent of choice depending on the nature of the matrix and analyte used. Practically, analyte and matrix solutions are prepared separately then mixed together to obtain the required ratio. After mixing, a few microliters of the matrix/analyte mixture are deposited on the sample plate, which is typically a stainless steel or gold substrate of a particular shape and size depending on the instrument being used. The sample on the plate is left to dry, then introduced into the MALDI instrument vacuum chamber. Typically the pressure in the MALDI mass spectrometer varies between 10^{-7} torr to 10^{-10} torr depending on the instrument; with the higher vacuum allowing for better resolution.

2.3   Desorption/Ionization process

Once the required vacuum has been reached in the instrument the sample is then analyzed. Sample desorption and subsequent ionization is facilitated by the laser pulse. When the laser energy hits the sample, most of the photon energy is absorbed by the matrix. As a requirement, the matrix must have a stronger absorption at the wavelength of the laser used than the analyte, and it has to be in higher quantity than the analyte in the mixture. Imbedding the analyte in the matrix minimizes direct irradiation of the analyte that would otherwise result into fragmentation. Normally fragmentation is a problem when direct laser desorption techniques are used. Although the desorption and ionization mechanisms in MALDI are not well understood, ionization of the analyte is generally considered to occur in the condensed phase just above the sample surface in the plume of the desorbed material.[14] Analyte ionization results from collisions between analyte neutrals and various ions formed from the excited matrix molecules. The ions in the gas phase immediately after desorption include ions from the matrix molecules, (protonated, deprotonated or fragmented molecules), protons released from the matrix molecules, Na^+, K^+, or other metal ions inherent in the sample or added during the sample preparation to enhance ionization through metal cationization. Addition of metal cationization reagent is necessary for synthetic polymer analysis. Nevertheless, there has been a number of proposed desorption/ionization mechanisms in MALDI process.[15,16]

2.4   Mass analyzer

After the desorption/ionization event, the formed ions are introduced into the mass analyzer in which ions are separated according to their mass-to-charge (*m/z*) ratios. There are different mass analyzers used in mass spectrometers but only two are common for MALDI generated ions: time-of-flight (TOF), and Fourier transform ion cyclotron resonance (FT-ICR) mass analyzers. Of these two, TOF is the most commonly used for various reasons including cost effectiveness, simplicity in both design and use, high ion transmission, unlimited mass range (theoretically), ability to detect all the ion species at the same time, fast analysis, and good compatibility with the pulsed ionization sources.[14] In fact because of these advantages, continuous ionization sources like electrospray have been modified to be compatible with TOF mass analyzers.[17-19]

A TOF mass analyzer works on the basic principle that when a temporally and spatially well-defined packet of ions of different *m/z* ratios are subjected to the same applied electric field, then allowed to drift in a region of constant electric field, they will travel through this region in a time which depends on their *m/z* ratios, allowing the ions to separate on the basis of their masses. Practically, ions generated in the source region of the MALDI instrument are accelerated to a constant kinetic energy (eV) and then allowed to drift through a field free region to the detector placed at the end of the ion flight path. Normally the path length ranges from 1 m to 3 m depending on the instrumental design. The flight time *t* (seconds) of ions through a drift region of length *l* (meters) has been shown to be proportional to the square root of their *m/z* ratios as shown in equation 1.1 in which e is the fundamental charge (1.6×10^{-19} C), and V (volts) is the accelerating voltage of the ions in the source region.[14,20 21]

$$t = \sqrt{m/2zeV} \times L$$

(1.1)

Hence, the lightest ions reach the detector first followed by other ions of successive heavier mass.

The main disadvantage of TOF mass analyzers is its limited mass resolution. Mass resolution is referred to as the ability to resolve m/z values of two adjacent ion peaks of a slightly different mass; peak m and $m + \Delta m$ in the mass spectrum. For a TOF mass analyzer resolution (R) is proportional to the flight time t of the ion divided by twice the time interval (Δt) of ion arrival at the detector as shown in equation 1.2.

$$R = \frac{m}{\Delta m} = \frac{t}{2\Delta t}$$

(1.2)

To overcome this limitation, two main instrumental modification modes have been developed. These instrumental modes are referred to as time-lag-focusing or delayed extraction [19,20,22,23] and reflectron mode [14,21].

For ion detection in MALDI instruments typically multichannel plate (MCP) detectors are used. These are thin plate-like detectors with many cylindrical channels made of lead-doped glass material with fiber optics properties. The channels are 4 to 25 μm in diameter with a center-to-center distance ranging from 6 to 32 μm.[24] The functioning of these detectors depends on the direct ion-to-electron conversion that takes place at the detector surface. The electron signal produced is eventually amplified and collected. Depending on the instrument design, these detectors can be used individually or in combinations of two or three to enhance sensitivity. With MCP detectors, signal amplification can be as high as 10^8.[24]

Practically the number of emitted secondary electrons depends on the mass and velocity of the impacting analyte ions such that ions of different masses cause different numbers of secondary electrons to be emitted at the detector surface.[25] Ions of high molecular weight usually fail to produce adequate secondary ions when they impinge the detector surface. Hence detector efficiency is low for relatively high molecular weight samples.[25] Also, depending on the amount of ions produced per laser pulse, and their velocity, detectors can be saturated and hence give unreliable results. Two types of detector saturation effects have been noted: one is saturation due to a high output ion pulse and the other is due to limited detector dynamic range.[26] The saturation recovery time for a detector is longer than its normal response time; hence the timed events of ion formation and detection under such conditions are overlapped. Therefore, because of these noted detector problems, MCP detectors, though common in MALDI instruments, have not performed well for all types of MALDI amenable analytes. For example, MALDI has not provided reliable results for polydisperse polymers. Among the reasons suggested is the MALDI detector problem.[26-30]

2.5 Advantages of Using MALDI

From the very beginning of the MALDI technique, it was demonstrated that this technique has several advantages over many others used for similar analyses. Some of the MALDI advantages include low sample consumption, ease of sample preparation, short analysis times, and soft ionization which leads to negligible or no fragmentation of the analytes.[3] With minimal fragmentation, MALDI provides spectral simplicity characterized by singly charged ions for easy spectra interpretation. Another very important advantage of using MALDI is its ability to analyze analytes of very broad mass ranges when combined with the TOF mass analyzer.[31]

Application of MALDI particularly for analysis of synthetic polymeric materials has demonstrated several advantages over most of the traditional techniques such as gel permeation chromatography (GPC) and viscometry. MALDI gives more definitive answers about the composition and purity of the polymer. It provides absolute, fast and accurate molecular masses for polymers with narrow polydispersity as opposed to relative masses provided by other techniques. It provides masses for the entire polymer distribution instead of the average value, hence providing molecular mass information which can be used to obtain the mass of the end-groups, mass of the repeat unit (monomer), and chemical modifications on the polymer if oligomer resolution is attained.[31-33]

However, saturated hydrocarbon polymeric materials can be analyzed using a number of other analytical techniques. Such techniques include field desorption ionization (FD)[34], plasma desorption (PD)[35], and laser desorption (LD)[36]. These techniques have various limitations particularly analyte fragmentation, mass limit and difficult sample preparations. The ion currents in most of experiments with these techniques are very small and decrease rapidly with increasing molecular weight of the polymer.[37] There has been a great deal of effort to develop MALDI protocols for hydrocarbon material analysis mainly because this is a promising technique to overcome some of the current limitations encountered when using other techniques.

2.6   Matrix Requirements

MALDI evolved from the early experiments with LD when it was realized that instead of direct laser desorption of the analyte, mixing the analyte with a matrix compound for pulsed laser desorption mass spectrometry not only produces better results but also improves the desorption efficiency of non-volatile analytes into the gas phase.[6] The use of a matrix was found to extend the mass range for thermally labile biomolecules to a range not attainable by other techniques.[6,38]. Whether the matrix compound used for analysis is organic as initially proposed by Hillenkamp and Karas [6], or inorganic as initially proposed by Tanaka and co-workers [4], they all have a strong energy absorption capability at the wavelength of the laser used, vacuum stability, and solvent compatibility. The absorption characteristic allows a controlled energy deposition to the analyte that then results into a much softer ionization as compared to the LD technique. Although molar absorptivity values for MALDI matrices in the gas phase have not been determined, good matrix compounds have been found to have solution absorption coefficient values between 10^3 - 10^5 L mol^{-1}cm^{-1}.[39,40] For this reason, compounds that originally could meet this requirement were small aromatic acid compounds with phenyl rings on which electron-withdrawing groups are attached.[38,40-42] The gas-phase inorganic chemistry studies for some selected compounds and complexes have been shown to reflect the solution chemistry of some compounds.[43,44] In comparison, MALDI is a much softer ionization technique than FAB which is also a useful technique for fragile organic molecules since the introduction of liquid matrices.[45]

Although the role of the matrix is very crucial in MALDI, very little is known about what makes a particular compound a good matrix. *A priori* determination of whether a particular chemical compound will function as suitable matrix is not yet possible.   To date, matrix compounds are still discovered through trial and error because there has not been enough research to investigate what are the factors that affect the functioning of such compounds in the MALDI process.   Apart from absorbing the laser energy at the wavelength used, the matrix dilutes and isolates macromolecules in the sample preparation step, a phenomenon called matrix isolation, and plays a role in analyte desorption and ionization process.[21,39] The analyte molecules are evenly distributed within these highly absorbing matrix molecules forming a solid solution of both the matrix and analyte after the solvent has evaporated.[41] In this case, the matrix inhibits potential interactions among the analyte molecules and between the analyte and the target surface.[16] Practically, the ratio of the analyte to the matrix ranges from $1:10^3$ up to $1:10^5$ although for high molecular weight samples such polystyrene 10 kDa to 900 kDa much higher ratios, up to $1:\sim8 \times 10^6$, have been used.[31]

2.7 MALDI and Nonpolar Analytes

In MALDI-MS the primary means of forming gas-phase ions is by proton or metal ion attachment. Most MALDI amenable analyte molecules contain heteroatoms such as N or O in their molecular structures which provide lone pairs of electrons on which cationization can take place. Similarly, hydrocarbon polymers with unsaturated double bonds have polarizable π bonds on which cationization reagents, mainly metal ions, can be attached. For saturated hydrocarbon polymers like polyethylene, typical MALDI experiment does not yield reliable results due to the inert nature of such analytes which makes them lack the suitable sites for metal ion attachment.

Nevertheless, there are some challenges that limit MALDI application to hydrocarbon analytes. The main challenges include availability of proper matrices for specific analytes, proper cationization reagents, common solvents for both analyte and matrix, as well as challenges in sample preparation techniques. These limitations and various others inherent in MALDI instrumentation such as problems in ion formation (desorption/ionization), ion transmission and detection, and difficulties in reproducing spectra, have rendered this technique applicable for qualitative purposes only. MALDI is generally employed as a tool for obtaining the analyte mass but not the analyte concentration or even its relative amount from the measured ion abundance values. To date, application of MALDI remains an active research area where improved sample preparation methods are developed, new and better performance matrices are explored, and various new applications are discovered.

2.8 Analyte/Matrix Miscibility

The inability of polar matrices to interact well with nonpolar analytes causes solubility problem in the MALDI matrix isolation step. Although the matrix and analyte molecules are completely separated in the solid mixture,[21,29] the ability of the matrix molecules to separate the analyte molecules is facilitated by the chemical nature of the interacting molecules. Because of their similar chemical properties, polar matrix molecules interact well with polar analyte molecules. Through trial and error, some polar matrices such as 2-(4-hydroxyphenylazo)-benzoic acid (HABA), *all-trans* retinoic acid (RTA), 1,5-dihydroxy bezoic acid (DHB) and dithranol have been found to work for hydrocarbon polymers. For example, HABA and *all-trans*-RTA have been applied to polystyrene.[32,46-50] The matrix *all-trans*-RTA has also been applied to polybutadiene and polyisoprene.[48] DHB and *trans*-indole acrylic acid have also been used for hydrocarbon polymers.[32,51] The most commonly used matrix for hydrocarbon polymers so far is dithranol.[50,52-59] As mentioned before, MALDI sample preparation requires a uniform formation of a solid solution between the matrix and the analyte. In this case, the polar matrices and the hydrocarbon polymers will definitely not allow such uniformity. Juhasz and co workers [46] suggested that HABA does not work well for polystyrene because it is too polar to form a uniform solid-phase mixture with this nonpolar polymer. Also polystyrene samples mixed with DHB when observed under a microscope indicated that the two compounds do crystallize separately.[36] These observations have led to a conclusion that miscibility of the analyte and the matrix in the condensed phase is an essential requirement for a good matrix. Hence, matrices that work well for polar analytes such as biopolymers cannot provide the required optimal conditions for nonpolar hydrocarbon synthetic polymer analysis.

Other important characteristics of the matrix include non-reactivity with the analyte, as well as vacuum and photochemical stability.[40] Because the commonly used matrices are small organic acid compounds, it has been established in protein analyses that good results are obtained at pH values of the matrix compound itself or in a more acidic environment.[41]

2.9   Solvents

Universal application of MALDI to providing solutions for analytical problems has also been limited by solvent systems needed for different analytes during sample preparation. Since the inception of MALDI, all matrix/solvent systems were developed to work for biological materials that are soluble in aqueous media.[6,38] Although it has been shown that water-soluble synthetic polymers can be analyzed by following similar protocols used for the analysis of biopolymers,[5,36,60] analytes that are soluble in organic solvents have been difficult to analyze with water-soluble matrices. Danis and Karr [61] started the investigation of the possibility of using organic solvents for both the polymer and matrix as a sample preparation technique. In their work, acidic matrices and some polar polymers were dissolved in acetone instead of water. Since then, several organic solvents have been applied to solvate different synthetic polymers during MALDI sample preparation. The use of a single solvent system to prepare the polymer and matrix mixture is required for successful analysis.[39] However, there are cases where the use of a binary solvent system becomes inevitable. In such cases, various problems have been observed. For example, addition of 1-5% water to a PMMA/polar matrix sample prepared in organic solvents resulted into mass discrimination when the samples were MALDI analyzed.[62] Yalcin and coworkers [63] studied extensively the effect of solvent in polymer sample preparation. They showed that binary solvent systems can be used for preparing the samples as long as both solvents are compatible; both the polymer and the matrix must have the same solubility in each of the solvents, and both solvents should have the same volatility. Slight differences in solubility of the polymer and the matrix in the solvents leads to significant differences in the MALDI results. It was concluded from that work that in cases where a common solvent is not available, binary solvents to be used should allow co-crystallization of the polymer and the matrix, or crystallization of the matrix before polymer precipitation.

There are various polymeric materials that are difficult to dissolve in organic solvents under ordinary conditions. For such materials, sample preparation for MALDI analysis is very challenging. Examples of such polymers are polyamides, teflon and polyethylene. Zenobi's group [11] has developed a solvent-free MALDI sample preparation technique for these kinds of samples; a similar method like one used for KBr pellet sample preparation for infrared analysis was proposed. Another group has tried to make this solvent-free method applicable to various kinds of analytes including soluble synthetic polymers as well as biopolymers.[12]

2.10   Cationization of Polymers in MALDI

As was mentioned before, MALDI was developed originally for relatively large, non-volatile biomolecules. Since then this technique has found a tremendous application in biopolymers such as proteins, oligonucleotides and polysaccharides; compounds which are all polar and undergo protonation or deprotonation in the MALDI process. It was in the early 1990s when MALDI was applied to synthetic polymer analysis. Today, polar synthetic polymers of more than 200 kDa can be analyzed by following the same standard protein MALDI protocols.[5,36,61] Examples of such polymers include polyglycols [36], polystyrenesulphonic acid [5], and polymethylmethacrylate [26,64]. These polymers produce oligomer ions that are singly protonated in MALDI when analyzed with acidic matrices.

While ionization of biopolymers in MALDI occurs by protonation, the ionization of synthetic polymers is achieved through cationization by metal ions; group I metal salts are good cationization reagents for polar synthetic polymers. Polymers such as polyethyleneglycol (PEG) and polymethylmethacrylate (PMMA) which contain basic groups in their repeat unit interact well with alkali metal ions.[52,55,56,65] In general, MALDI-MS has been more successful for the analysis of polar synthetic polymers than for nonpolar synthetic polymers. Polar synthetic polymers have a higher chemical resemblance to biopolymers, such as proteins and peptides, for which the MALDI technique was originally developed, and contain heteroatoms such as N or O which provide sites for proton or metal cation attachment.[36,51,61] For synthetic polymers, formation of ions via cation attachment is much more favored than via protonation.[66,67] Water-soluble polymers, such as polyethylene glycol and polypropylene glycol, have been successfully analyzed with MALDI using DHB as a matrix with alkali-metal salt solutions to increase the yield of the cationized species.[66] Other water-soluble polymers, like polyacrylic acid and polystyrene sulfonic acid, have been analyzed using sinapinic acid as a matrix.[48] The organic soluble polar polymers, such as polymethylmethacrylate, polyvinylacetate and polyvinylchloride, have been analyzed using various matrices including DHB, t-3-indoleacrylic acid (IAA) [5,61] and 2-(4-hydroxyphenolazo)benzoic acid (HABA) [5].

It has been established, however, that analysis of hydrocarbon polymers such polystyrene, polybutadiene or polyisoprene requires a different approach; standard MALDI protein protocols do not work for these type of polymers. Hydrocarbon polymers require an additional cationization reagent in the sample preparation step. More interestingly, hydrocarbon polymers do not undergo protonation or metal cationization by monovalent Group I metal salts ions. To date, only silver and copper salts (Cu^+ or Cu^{2+}) have been found to work relatively well for nonpolar polymer cationization in MALDI. (Cu^{2+} is reduced to Cu^+ in the process). The performance of these two metal salts is matrix dependent. A few reports in the literature indicate that some low molecular weight polystyrene analytes can be cationized using Group I metal salts in MALDI, though not efficiently.[57] It should be recalled here that LD experiments have indicated that some monovalent gas-phase metal ions such as Al^+, Cr^+, Fe^+ and Cu^+ can efficiently cationize both PEG and polystyrene analytes, hence providing alternative cationization to the Na^+, K^+ and Ag^+ ions. However, LD works well only for polymers of less than 10 kDa because of fragmentation problem that occurs for larger ones.[6]

The work done by Reinhold et al. [68] has shed some light on why only specific metal ions work in MALDI. That work examined the binding energy between the metal ion and the analyte (polymer oligomer). It was observed that aromatic species, such as polystyrene, and polyolefins, such as polybutadiene and polyisoprene, containing double bonds are ionized less efficiently in MALDI using sodium ions because of the low binding energy between the sodium ion and the polymer oligomer.[68] This observation led to a general conclusion that the binding energy in such systems is not sufficient to prevent dissociation of the metal-oligomer complex during the multiple collisions which take place during the ion extraction in the MALDI process. Other reasons which have been given by some researchers suggest that for a metal ion to be an effective cationization reagent, it should have a strong affinity for unsaturated electrons present on the polymer and also it should be able to sustain an oxidation state of + 1 in the MALDI environment.[50] This condition limits the number of transition metals that can be effective cationization reagents. In fact, little research has been performed to find out why silver and copper ions in particular are the only metal ions capable of efficiently cationizing hydrocarbon polymers in MALDI.

3. SYNTHETIC POLYMERS AS MALDI ANALYTES

3.1 Polymer Distribution

Information provided by MALDI data for polymer analysis includes polymer average molecular weight, molecular weight distribution, mass of the end-groups, sample purity, and even initiation and polymerization mechanisms.[69,70] In general, the physical properties of the polymer depend on the polymer molecular weight. By using MALDI mass spectrometry, the number-average molecular weight (M_n) and the weight-average molecular weight (M_w) values for a polymer of a narrow distribution, (with polydispersity [PD] values < 1.2) can be obtained. Synthesis of synthetic polymers always results in the formation of polymer chains of various lengths that form a product that is a mixture of molecules of different molecular weights. Because of this molecular weight variability in the product, synthetic polymer molecular weights are always referred to as average values after considering all the individual polymer molecules (oligomers) in the total distribution. Equations 1.3 and 1.4 are standard equations used for polymer average molecular weight computation in mass spectrometry.

$$M_w = \frac{\sum N_i M_i^2}{\sum N_i M_i}$$

(1.3)

$$M_n = \frac{\sum N_i M_i}{\sum N_i}$$

(1.4)

N_i is the measured peak area or signal intensity (ion abundance) of oligomer i with mass M. The ratio of M_w to M_n is an indicator for how broad is the molecular weight distribution in the polymer sample; this ratio is called polydispersity index (*PD*).

$$PD = \frac{M_w}{M_n}$$

(1.5)

Normally M_w is greater than M_n unless the polymer sample consists of oligomers of the same weight.

It was noted in the early 1990s when MALDI was first applied to synthetic polymers that, unlike biomolecules, the total available charge produced during MALDI analysis of synthetic polymers is distributed over a large number of different molecular weight.[36] When MALDI is applied to polymers with broad molecular mass distributions (PD > 1.2),[6,28,71] the charge distribution leads to an upper mass limit above which individual oligomers cannot be distinguished from the noise. As a result, MALDI data have been found to underestimate the higher mass polymer distribution hence resulting in noticeably lower average molecular weight values.[29,47,71]

The mass discrimination effect in MALDI is very significant for polymers with high PD values. This effect is thought to be due to a combination of factors including mass discrimination in the ion detection and sample preparation.[26,27,29,47,64] To date, application of MALDI for polymer analysis is still limited to polymers with low PDs (< 1.2).[30] Comparative results have shown that polymer analysis results obtained by MALDI are in good agreement with those determined by conventional methods such as gel permeation chromatography (GPC) only if the polymer has a narrow distribution.[28,30,71,72] For polymers with broad distributions, a combination of GPC and MALDI does provide a useful method for determining molecular weight distributions. In the MALDI-GPC combined method, the polydisperse polymer is first fractionated by GPC, then the collected fractions are analyzed off-line by MALDI which provides data to calibrate the GPC chromatogram.[73] The GPC fractionation of a polydisperse polymer for MALDI analysis requires a separate optimization of the GPC method and determination of the number of fractions and fraction volumes required to provide adequately narrow distributions across the whole GPC chromatogram. Before MALDI analysis, sample concentration might be necessary. Recently a liquid chromatography (LC) interface to the GPC was developed and demonstrated to be a useful GPC-MALDI sample preparation for fractionated polymers.[74] The LC interface uses heated sheath gas and a capillary nozzle to remove most of the GPC mobile phase and deposit the required eluants on the precoated matrix on a moving MALDI plate.

4.   MATRICES FOR POLYMER ANALYSIS

Finding a suitable matrix that will work well for nonpolar polymers has traditionally been through trial and error. The most common MALDI matrices which have been applied to polystyrene include dithranol [53,57,64], DHB [58], IAA [51,61,75], *all-trans*-retinoic acid [29,31,32,49], HABA [67], and 2-nitrophenyl octyl ether [66]. These matrices have been used together with cationization reagents such as Ag^+, Cu^{2+}, Li^+ or Na^+.[29,31,32,47,49,51,64]

The presence of the phenyl functionality on polystyrene is considered to be responsible for its higher ionization probability compared to other nonpolar polymers such as polybutadiene, polyisoprene, polyethylene or polypropylene. To date, there have been no reports in the literature regarding the successful MALDI analysis of polyethylene or polypropylene. The presence of the phenyl group of styrene or the unsaturated double bond on butadiene provides a site of high metal affinity.[76] Such a site is not available on polyolefins, which probably accounts for them being the hardest to analyze by MALDI.[47,68] However, polyethylene samples have been analyzed using laser desorption Fourier transform mass spectrometry (LD/FTMS).[77]

There are fewer reports in the literature regarding the analysis of polybutadiene and polyisoprene with MALDI-MS than for polystyrene. Polybutadiene and polyisoprene have been analyzed by MALDI using the same or similar matrices as those that have been used for polystyrene. Examples of matrices which have been applied to analyze polybutadiene are azo compounds [67], IAA [51,75], 1,4-di-[2-(5-phenyl oxazolyl)]benzene [75], 2-nitrophenyl octyl ether [51], *all-trans*-retinoic acid [36] and DHB [58]. Matrices that have been applied to polyisoprene include IAA [51,75], DHB [58], *all-trans*-retinoic acid [36], and dithranol [53].

4.1 Nonpolar Matrices

Nonpolar matrices, such as anthracene, pyrene, acenaphthne and terthiophene, are suitable for MALDI-MS analysis of low molecular weight general nonpolar analytes.[78,79] Nonpolar analytes and the nonpolar matrices are soluble in a common organic solvent, which facilitates analyte:matrix co-crystallization upon spotting on the MALDI sample plate. The nonpolar matrices have low ionization energies and high molar extinction coefficients in the UV, facilitating the production of radical molecular cations of the matrix during the MALDI event; hence, they are not good sources of protons in the MALDI process.

As mentioned earlier, hydrocarbon synthetic polymers can be difficult to characterize using conventional MALDI-MS due to the lack of suitable matrices. Studies have shown that nonpolar matrices, such as anthracene, acenaphthene and pyrene, are amenable to MALDI-MS analysis of hydrocarbon analytes. The nonpolar matrices have the following general characteristics: high molar extinction coefficients at the wavelength of the UV laser being used, low enthalpies of sublimation, and low vertical ionization energies.[78,79]

These matrices have been shown to be effective as matrices for the analysis of low molecular weight hydrocarbon synthetic polymers. These matrices have been applied to hydrocarbon polymers such as polybutadiene 760, 1100 and

2940, polystyrene 1940, 2557, 2800 and 5120, as well as polyisoprene 2300 and
2600 (the numbers represent the mass-weighted average molecular weight of the
polymer). It should be mentioned here that Tang et al. [80] had previously reported
that polycyclic aromatic hydrocarbons, such as anthracene, benzo[a]pyrene,
chrysene, naphthalene, phenanthrene and pyrene, did not function as suitable
matrices for the analysis of synthetic polymers. However, that study focused
primarily on polar synthetic polymers, with the exception of a polybutadiene
sample.

Adequate isolation of the analyte by the matrix appears to be a prerequisite
for successful MALDI-MS experiments.[81] A survey of the common matrices
that are shown to be suitable for the analysis of hydrocarbon synthetic polymers
finds that each of these matrices is soluble (to some extent) in a suitable organic
solvent, although the extent of isolation of the analyte by the matrix is matrix
dependent. For example, a study of hydrocarbon polymers, mainly polystyrene
and polybutadiene, with DHB as a matrix showed that these polymers do not mix
well with this matrix.[66,67] Under microscopic observation, poly(styrene) crystals
were found to be separated from the DHB matrix crystals after mixing and
drying.[66] It should be noted, however, that Pastor et al. [58] have found that DHB
was superior to IAA and HABA for MALDI FT-ICR mass spectrometric
analysis of nonpolar polymers such as polystyrene and polybutadiene.

Nonpolar matrices improve the interaction between the polymer and matrix
by dissolving each in a common solvent. The polymer samples and nonpolar
matrices were dissolved in the same solvent (THF) to promote the interaction of
the analyte and matrix on the MALDI sample plate.[61] With the use of nonpolar
matrices, spectra reproducibility from spot to spot was found to be high. It was
also observed that finding a spot that gives a high quality mass spectrum was not
difficult. Although there are other factors that influence phase separation
between the analyte and matrix (e.g., the number of chain elements and the
Flory-Huggins interaction parameters), experimental results imply that the
sample and matrix interact well and uniformly and that uniform interaction is
retained upon solvent evaporation. The polybutadiene, polyisoprene and
polystyrene samples mentioned above did not generate representative molecular
ions when analyzed as neat samples under the experimental conditions used for
the MALDI investigation.

4.2   Nonpolar Matrices with Cationization Reagents

Nonpolar matrices are not suitable for the analysis of nonpolar polymers in
the absence of cationization reagents. Figure 1 is representative mass spectral
data of polystyrene 1940 which was analyzed using anthracene (Fig. 1a) and *all-
trans*-retinoic acid (Fig. 1b) as matrices. In both cases, silver trifluoroacetate
(AgTFA) was added as the cationization reagent. The oligomer distributions are

similar between the two matrices investigated. In general, it was found that, under similar experimental conditions, anthracene yielded a cleaner mass spectrum, as measured by the base-line signal, than *all-trans*-retinoic acid especially when Ag was used as a cationization reagent. As shown in Fig. 1b, the use of *all-trans*-retinoic acid generally resulted in a higher background level with several interfering ions at low *m/z* values. It has been noted that when Ag salts are used with RTA for MALDI-MS analysis, depending on experimental conditions, an interfering background signal due to silver clusters may be present up to around *m/z* 7000.[82] A similar background pattern appears on some of the spectra obtained by Yalcin et al.[44]

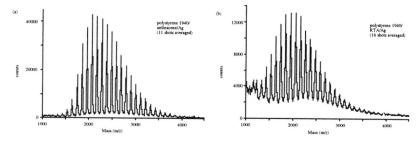

Figure 1. Linear-mode MALDI-TOF mass spectra of polystyrene 1940 with (a) anthracene and (b) all-trans-retinoic acid as matrices. AgTFA was used as the cationization reagent in both cases.

Similar results were obtained for the analysis of polystyrene 2557, 2800 and 5120. As with the polystyrene 1940 sample, addition of a cationization reagent to the nonpolar matrix results in the formation of oligomer-cation adducts. In general, anthracene yielded more abundant and reproducible results than did pyrene or acenaphthene. In general, the use of nonpolar matrices does not yield substantially different mass and number weighted distributions for the polystyrene samples.

Figure 2 represents the MALDI results obtained from polybutadiene 2940 using anthracene (Fig. 2a) and *all-trans*-retinoic acid (Fig. 2b) as matrices. In Figure 2a, AgTFA was included as a cationization reagent, and in Figure 2b, $Cu(NO_3)_2$ was included as a cationization reagent. In general, similar results were found using these different matrix preparations. It should be noted that the elevated baseline and background signal observed in Fig. 1b did not occur with *all-trans*-retinoic acid when Cu was used as a cationization reagent.

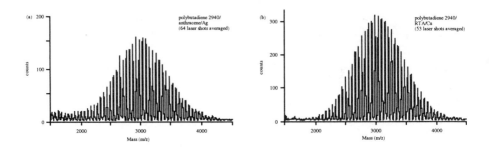

Figure 2. Reflectron-mode MALDI-TOF mass spectra of polybutadiene 2940 with anthracene as the matrix and AgTFA as the cationization reagent and (b) all-trans-retinoic acid as the matrix and $Cu(NO_3)_2$ as the cationization reagent.

Polyisoprene, similar to polybutadiene and polystyrene, is typically characterized in MALDI-MS by addition of a cationization reagent to the standard MALDI matrix. Figure 3 are representative mass spectra arising from the analysis of a polyisoprene sample with anthracene as the nonpolar matrix in the presence of AgTFA (Fig. 3a) and $Cu(NO_3)_2$ (Fig. 3b) as cationization reagents and *all-trans*-retinoic acid with $Cu(NO_3)_2$ (Fig. 3c).

Similar to the results found for polybutadiene, $Cu(NO_3)_2$ is a poorer cationization reagent for nonpolar matrices than is AgTFA. In Figure 3a, a well-defined polymer distribution can be detected when AgTFA is used as the cationization reagent. However, as seen in Figure 3b, the quality of the mass spectral results decreases dramatically when $Cu(NO_3)_2$ is incorporated as the cationization reagent.

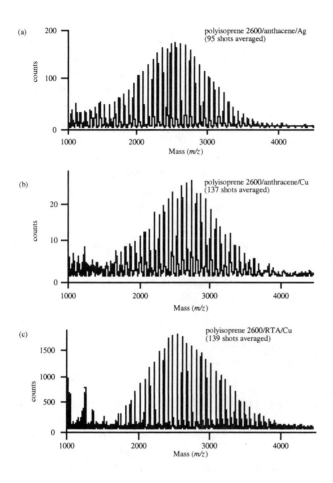

Figure 3. Reflectron-mode MALDI-TOF mass spectra of polyisoprene with (a) anthracene as the matrix and AgTFA as the cationization reagent, (b) anthracene as the matrix and $Cu(NO_3)_2$ as the cationization reagent and (c) all-trans-retinoic acid as the matrix and $Cu(NO_3)_2$ as the cationization reagent.

4.3 Ag vs. Cu Cationization Reagents

The inability of nonpolar matrices with $Cu(NO_3)_2$ as a cationization reagent to generate reasonable mass spectral data from the polybutadienes or polyisoprene is not yet understood. Yalcin et al. [36] have previously found that $Cu(NO_3)_2$ is the preferred cationization reagent (vs. Ag salts) for the analysis of high molecular weight polybutadienes using polar matrices. So far it is not clear why $Cu(NO_3)_2$ is not a suitable cationization reagent for the analysis of polybutadienes with nonpolar matrices such as anthracene. Possible explanations could include a higher interaction between the nonpolar matrix and Cu, poor mixing of the cationization reagent with the analyte/matrix solution upon crystallization,[76] or Cu salts are more reactive in acidic media.

Lehmann et al.[76] previously discussed the role of pre-formed ions in the analysis of polystyrene performed in the presence of a cationization reagent. In that work, the authors suggested that the use of nonpolar matrices favors a situation where the transition metal cation and its counterion crystallize separately from the polystyrene sample, which co-crystallizes with the matrix. Under those experimental conditions, no pre-formed ions of polystyrene and the transition metal cation are expected to be present; only gas-phase reactions can lead to the production of polystyrene-cation adducts. While both AgTFA and $Cu(NO_3)_2$ each yielded similar, high-quality results for the analysis of polystyrene samples with nonpolar matrices, a similar situation does not exist for the polybutadiene or polyisoprene samples. Although many Cu^+-alkene complexes are known,[83] the reason for the poorer interaction of copper salts, in the presence of nonpolar matrices, with these nonpolar polymers remains to be determined.

5.   CONCLUSIONS

In this chapter the application of nonpolar matrices for the analysis of low molecular weight nonpolar polymers using MALDI-MS was studied. These matrices, unlike the conventional MALDI matrices, do not have acidic or basic functional groups which are responsible for the feasibility of the MALDI analysis of polar synthetic polymers and biopolymers. Thus, these matrices may be suitable choices for the characterization of modified hydrocarbon polymers which may be sensitive to the acidic conditions present when conventional matrices are used. The matrices used in this study mix uniformly with nonpolar polymers resulting in a more homogeneous crystal formation on the sample plate which increases the reproducibility of the spot-to-spot signal. Analysis of low molecular weight hydrocarbon polymers is not feasible with nonpolar polymers without the addition of a cationization reagent. These matrices do not operate under a charge-transfer mechanism, most likely due to the unfavorable thermodynamics of charge-transfer reactions between the nonpolar matrices and polymers. The use of nonpolar matrices with cationization reagents yields similar results to those found using conventional matrices, although Ag is found to be the preferred cationization reagent relative to Cu with the nonpolar matrices.

6. REFERENCES

1. Smith, P.B., *et al.*, Anal. Chem., 1997. 69: p. 95R-121R.
2. Cook, K. *J. Am. Soc. Mass Spectrom.* **1998**, 9, 267.
3. Karas, M.; Hillenkamp, F. *Anal. Chem.* **1988**, 60, 2299-2301.
4. Tanaka, K., *et al.*, Rapid Commun. Mass Spectrom., 1988. 2(8): p. 151-153.
5. Danis, P.O., *et al.*, *Mass Spectrometry.* Org. Mass Spectrom., 1992. 27: p. 843-846.
6. Hillenkamp, F., *et al.*, Anal. Chem., 1991. 63(24): p. 1193A-1203A.
7. Niu, S.; Zhang, W.; Chait, B. T. *J. Am. Soc. Mass Spectrom.* **1998**, 9, 1-7.
8. Talrose, V.L., *et al.*, Rapid Commun. Mass Spectrom., 1999(13): p. 2191-2198.
9. Vorm, O.; Roepstorff, P.; Mann, M. *Anal. Chem.* **1994**, 66, 3281-3287.
10. Dai, Y.; Whittal, R. M.; Li, L. *Anal. Chem.* **1999**, 71, 1087-1091.
11. Skelton, R.; Dubois, F.; Zenobi, R. *Anal. Chem.* **2000**, 72(7), 1707-1710.
12. Trimpin, S., *et al.*, Rapid Commun. Mass Spectrom., 2001. 15: p. 1364-1373.
13. Gusev, A.I., *et al.*, Anal. Chem., 1995. 67: p. 1034-1041.
14. Muddiman, D.C., *et al.*, J. Chem. Educ., 1997. 74(11): p. 1288-1292.
15. Zenobi, R.; Knochenmuss, R. *Mass Spec. Rev.* **1998**, (17), 337-366.
16. Karas, M.; Gluckmann, M.; Schafer, J. *J. Mass Spectrom.* **2000**, 35(1), 1-12.
17. Muddiman, D.C., *et al.*, Anal. Chem., 1995. 67: p. 4371-4375.
18. Whittal, R. M.; Russon, L. M.; Li, L. *J. Chromatogr. A* **1998**, 794, 367-375.
19. Wiley, W. C.; McLaren, I. H. *J. Mass Spectrom.* **1997**, 32, 4-11.
20. Cotter, R. J. *Anal. Chem.* **1999**, 445A-451A.
21. Limbach, P. A. *Spectroscopy* **1998**, 13(10), 16-27.
22. Whittal, R. M.; Li, L. *Anal. Chem.* **1995**, 67, 1950-1954.
23. Whittal, R.M., *et al.*,Anal. Chem., 1997. 69: p. 2147-2153.
24. Hoffmann, E.D., et al. J. Charette, and V. Stroobant, *Mass Spectrometry Principles and Applications.* 1996, Masson, Paris: John Wiley & Sons Ltd. 91-93.
25. "Methods in Enzymology" *Mass Spectrometry,* Vol. 193. McCloskey, J.A. (Ed.), Academic Press: San Diego, CA 1990; pp. 61-86.
26. Rashidzadeh, H.; Guo, B. *Anal.Chem.* **1998**, 70, 131-135.
27. Axelsson, J., *et al.*, Macromolecules, 1996. 29: p. 8875-8882.
28. McEwen, C.N., et al. Int. J. Mass Spectrom. Ion Processes, 1997. 160: p. 387-394.
29. Schriemer, D. C.; Li, L. *Anal. Chem.* **1997**, 69, 4176-4183.
30. Byrd, H. C. M.; McEwen, C. N. *Anal. Chem.* **2000**, 72, 4568-4576.
31. Schriemer, D. C.; ; Li, L. *Anal. Chem.* **1996**, 68, 2721-2725.
32. Whittal, R. M.; Schriemer, D. C.; Li, L. *Anal. Chem.* **1997**, 69, 2734-2741.
33. Schriemer, D.C., R.M. Whittal, and L. Li, Macromolecules, 1997. 30: p. 1955-1963.
34. Craig, A.G., et al., Int. J. Mass Spectrom. Ion Phys., 1981. 38: p. 297-304.
35. Feld, H., *et al.*, Anal. Chem., 1991. 63: p. 903-910.
36. Bahr, U., *et al.*, Anal. Chem., 1992. 64: p. 2866-2869.
37. Fenn, J.B., *et al.*, Mass Spectrom. Rev., 1990. 9: p. 37-70.
38. Karas, M., *et al.*, Int. J. Mass Spectrom. Ion Processes, 1987(78): p. 53-68.
39. Williams, J. B.; Gusev, A. I.; Hercules, D. M. *Macromolecules* **1996**, 29, 8144-8150.
40. Beavis, R.C. *Org. Mass Spectrom.* **1992**, 27, 653-659.
41. Ehring, H.; Hillenkamp, F. *Rapid Commun. Mass Spectrom.* **1992**, 27, 472-480.
42. Krause, J., et al., Rapid Commun. Mass Spectrom., 1996. 10: p. 1927-1933.
43. Dubois, F., et al., Eur. Mass Spectrom., 1999. 5: p. 267-272.
44. Fisher, K.J., et al., Rapid Commun. Mass Spectrom., 1996. 10: p. 106-109.
45. Barber, M., *et al.*, Anal. Chem., 1982. 54(4): p. 645A-657A.
46. Juhasz, P.; Costello, C. E. *Rapid Commun. Mass Spectrom.* **1993**, 7, 343-351.
47. Schriemer, D. C.; Li, L. *Anal. Chem.* **1997**, 69, 4169-4175.

48. Yalcin, T.; Schriemer, D. C.; Li, L. *J. Am. Soc. Mass Spectrom.* **1997**, 8, 1220-1229.
49. Zhu, H.; Yalcin, T.; Li, L. *J. Am. Soc. Mass Spectrom.* **1998**, 9, 275-281.
50. Guo, B., et al., Polym. Preprints, 2000. 41(1): p. 650.
51. Belu, A.M., *et al.*, J. Am. Soc. Mass Spectrom., 1996. 7: p. 11-24.
52. Jackson, A.T., *et al.*, J. Am. Soc. Mass Spectrom., 1997. 8: p. 132-139.
53. Jackson, A.T., et al., J. Am. Soc. Mass Spectrom., 1997. 8: p. 76-85.
54. Zenobi, R. *Chimia* **1997**, 51, 801-803.
55. Mowat, I. A., et al., Rapid Commun. Mass Spectrom., 1997. 11: p. 89-98.
56. Scrivens, J.H., *et al.*, Int. J. Mass Spectrom. Ion Processes, 1997(165/166): p. 363-375.
57. Deery, M.J., *et al.*, Rapid Commun. Mass Spectrom., 1997. 11: p. 57-62.
58. Cornett, L., *et al.*, ASMS Proceedings: The 46th ASMS Conference on mass spectrometry and alllied topics May 31 - June 4, 1998 Orlando Florida, 1998.
59. Wesdemiotis, C., *et al.*, Polym. Preprints, 2000. 41(1): p. 629.
60. Rashidzadeh, H., et al., Rapid Commun. Mass Spectrom., 2000. 14: p. 439-443.
61. Danis, P. O.; Karr, D. E. *Org. Mass Spectrom.* **1993,** 28, 923-925.
62. Chen, H.; Guo, B. *Anal. Chem.* **1997,** 69, 4399-4404.
63. Yalcin, Y., et al., J. Am. Soc. Mass Spectrom., 1998. 9: p. 1303-1310.
64. Larsen, B.S., et al., J. Am. Soc. Mass Spectrom., 1996. 7: p. 287-292.
65. Wong, C. K. L.; Chan, T. W. D. *Rapid Commun. Mass Spectrom.* **1997**(11), 513-519.
66. Liu, H. M. D.; Schlunegger, U. P. *Rapid Commun. Mass Spectrom.* **1996**, 10, 483-489.
67. Pastor, S. J.; Wilkins, C. L. *J. Am. Soc. Mass Spectrom.* **1997**, 8, 225-233.
68. Reinhold, M.,et al., Rapid Commun. Mass Spectrom., 1998. 12(23): p. 1962-1966.
69. Dourges, M.A., *et al.*, Macromolecules, 1999. 32: p. 2495-2502.
70. Williams, J.B., et al., Macromolecules, 1997. 30: p. 3781-3787.
71. Wu, K. J.; Odom, R. W. *Anal. Chem.* **1998**(July 1), 456A-461A.
72. Jackson, C., et al., Anal. Chem., 1996. 68: p. 1303-1308.
73. Lou, X., et al., J. Chromatogr. A, 2000. 896: p. 19-30.
74. Hanton, S. D.; Liu, X. M. *Anal. Chem.* **2000**, 72, 4550-4554.
75. Danis, P.O., *et al.*, Rapid Commun. Mass Spectrom., 1996. 10: p. 862-868.
76. Lehmann, E., et al., Rapid Commun. Mass Spetrom., 1997. 11: p. 1483-1492.
77. Kahr, M. S.; Wilkins, C. L. *J. Am. Soc. Mass Spectrom.* **1993**, 4, 453-460.
78. McCarley, T.D., et al., Anal. Chim., 1998(70): p. 4376-4379.
79. Macha, S.F., et al., Anal. Chim. Acta., 1999. 397: p. 235-245.
80. Tang, X., et al., Rapid Commun. Mass Spectrom., 1995. 9: p. 1141-1147.
81. Strupat, K., et al., Int. J. Mass Spectron. Ion Processes, 1991. 111: p. 89-102.
82. Macha, S.F., *et al.*, J. Am. Soc. Mass Spetrom., 2001. 12: p. 732-743.
83. Cotton, F.A., *et al.*, *Advanced Inorganic Chemistry*. 6th ed. 1999: Wiley-Interscience Publication. 855-863.

Chapter 17

LASER DESORPTION/IONIZATION (LDI)- AND MALDI-FOURIER TRANSFORM ION CYCLOTRON RESONANCE MASS SPECTROMETRIC (FT/ICR/MS) ANALYSIS OF HYDROCARBON SAMPLES

Chad L. Robins[1], Stephen F. Macha[1], Victor E. Vandell[2]
and Patrick A. Limbach[1]
[1]*Department of Chemistry*
University of Cincinnati
Cincinnati, OH 45221
[2]*Analytical Science Division*
Hercules Incorporated
Wilmington, DE. 19808

1.   INTRODUCTION

There are several distinct advantages to utilizing Fourier transform ion cyclotron resonance mass spectrometry (FT/ICR/MS) for the analysis of hydrocarbon materials. These advantages include the inherently high mass accuracy and resolution afforded by this technique (e.g., to distinguish between ^{12}CH and ^{13}C). In this chapter, we describe investigations of different sample preparation techniques for achieving high-resolution laser desorption/ionization (LDI)-FT/ICR/MS of metalloporphyrins, and we discuss the use of nonpolar matrices for the analysis of low molecular weight hydrocarbon materials.

2.   FT/ICR/MS OVERVIEW

2.1   Fundamentals of Ion Motion

Mass spectrometry based on an ion's natural motion in the presence of a magnetic field is termed ion cyclotron resonance mass spectrometry. Fourier transform ion cyclotron resonance mass spectrometry (FT/ICR/MS) was developed by Melvin B. Comisarow and Alan G. Marshall in 1974.[1] The development of FT/ICR/MS was an extension of ion cyclotron resonance mass spectrometry used extensively for gas-phase ion molecule chemistry in the 1960's and 1970's. The motion of a charged particle (in our case we are primarily concerned with ions of relatively large mass) in a magnetic field is described by the Lorentz equation (eq 1).

$$F_{Lorentz} = m\mathbf{a} = q\,(\mathbf{v} \times \mathbf{B}) + q\,\mathbf{E} \qquad (1)$$

where q is the fundamental charge, $\mathbf{v}$ is the velocity of the ion of mass m, $\mathbf{E}$ is the electric field, and $\mathbf{B}$ is the magnetic induction. It is convenient to specify the magnetic induction in Cartesian coordinates as $\mathbf{B} = B_0\mathbf{k}$. We will also assume (for the moment) that there is no electric field, $\mathbf{E}$. It is immediately obvious from eq 1 that for a force to act on the ion it must have a velocity component perpendicular to the magnetic induction; the strength of this force is proportional to the magnetic induction.

The motion of an ion in a magnetic field can be found from the Lorentz force equation.

$$F_{Centripetal} = \frac{m(v_{xy})^2}{r} = qv_{xy}B_0 = F_{Lorentz} \qquad (2)$$

$$r = \frac{mv_{xy}}{qB_0} \qquad (3)$$

$$\omega_c = \frac{v_{xy}}{r} = \frac{q\,B_0}{m} \qquad (4)$$

We find that ions have a natural circular motion in the x-y plane with angular frequency ω_c. Providing the magnetic field is homogeneous to within ~10 ppm, the frequency of the ion may be measured quite precisely (to 1 part in 10^9 in favorable cases) and the determination of the frequency of the ion can be made which is independent of the ion's initial position and velocity.

Once ions are in the magnetic field, they are confined in the x-y plane to orbit about magnetic field lines at their cyclotron frequency. However, some physical device must be used to confine the ions z-motion in a specific volume so that they may be analyzed. The ion trap is used for just this task. An electrostatic trap used in a magnetic field is referred to as a "Penning" ion trap. Ion traps used for FT/ICR/MS can take a variety of configurations.

Whereas the motion of an ion in a magnetic field is described by eq 1, an electrostatic potential, V, applied to any of the plates on the ion trap generates an electric field given by

$$\mathbf{E} = -\nabla V \quad \text{where} \quad \nabla = \mathbf{i}\frac{\partial}{\partial x} + \mathbf{j}\frac{\partial}{\partial y} + \mathbf{k}\frac{\partial}{\partial z} \tag{5}$$

To confine ions in the z-direction (i.e., the direction parallel to the magnetic field), a quadrupolar trapping potential is usually applied to the plates, which are perpendicular to the direction of the magnetic field. The quadrupolar potential takes the form:

$$V(x,y,z) = V_T(\gamma - \frac{\alpha}{2a^2} [x^2 + y^2 - 2z^2]) \tag{6}$$

in which $\alpha = 2.77373$ and $\gamma = 1/3$ for a cubic trap. Solving eq 1 with this electrostatic potential we find that, instead of one natural frequency of motion per ion, there are three natural frequencies:

$$\omega_+ = \frac{\omega_c}{2} + \frac{\omega_c}{2}\sqrt{1 - \frac{m}{m_{crit}}} \tag{7}$$

$$\omega_- = \frac{\omega_c}{2} - \frac{\omega_c}{2}\sqrt{1 - \frac{m}{m_{crit}}} \tag{8}$$

$$\omega_t = \sqrt{\frac{4\beta q V_T}{ma^2}} \tag{9}$$

$$m_{crit} = \frac{q\, a^2\, B_o^2}{8\beta V_T} \tag{10}$$

where β is a trap geometry factor ($\beta = 0.72167$ for the cubic trap), V_T is the trapping potential, and a is the radius (cross-section) of the cell. Eq 7 defines the "reduced" cyclotron frequency i.e., the natural cyclotron frequency, ω_c, shifted by the electric field. Eq 8 defines the "magnetron" frequency, ω_-, which is a low frequency (force proportional to $\mathbf{E} \times \mathbf{B}$) drift motion of the ion. Eq 9 defines the harmonic trapping frequency of the ion. The fundamental motion of the ion, ω_+, is proportional to m/q as is the lower frequency, ω_-, drift motion, whereas the trapping motion, ω_t, is proportional to $\sqrt{m/q}$.

2.2   Experimental Sequence

Once ions have been placed in the ion trap and are confined, a detectable signal must be generated. Typically, a radiofrequency pulse is applied to the excitation plates. When the frequency of this pulse matches the natural frequency of an ion in the trap, the ion will absorb energy and be excited to a larger radius. Once the ion is excited to a sufficiently large radius, the excitation power is shut off. This excitation pulse results in a coherent ion packet of sufficient radius for efficient detection by the detection circuit (Figure 1).

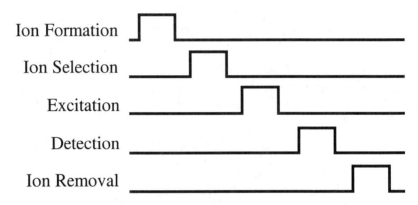

Figure 1. Experimental sequence in FT/ICR/MS.

A feature of FT/ICR/MS is that ions of different mass are excited to the same radius if the excitation is uniform in magnitude at each frequency. However, at this final radius ions of different mass will have different translational energies. With excitation profiles that can be well defined (e.g., square-waved inverse Fourier-transform (SWIFT) excitation), monoenergetic ion packets may be created or ion packets with well-specified final radii can easily be created.

One of the more widely used excitation methods is frequency sweep ("chirp") excitation.[2,3] "Chirp" excitation excites a broader range of frequencies compared to those forms based on single frequency excitation mentioned above. Unfortunately, "chirp" excitation suffers from excitation profiles with poor selectivity at low and high frequencies and has an uneven distribution of power to ions of different mass-to-charge ratio. Thus the final radius of the ion packet will vary depending on the cyclotron frequency of the ion.

In SWIFT excitation,[4-6] the frequency domain excitation profile is specified and an inverse Fourier transform of this excitation profile is applied to the excite plates. SWIFT excitation produces an equal response of ions to the excitation power over a large frequency range, allows for high-resolution isolation and ejection events, and permits the implementation of more complicated experiments.

As excitation of the ion packet can be performed with several different forms, detection in FT/ICR/MS also can take several different forms. The standard mode of detection is the differential dipole method. While the electrode configuration used in detection in FT/ICR/MS can vary, the signal model which relates the induced charge of the ion packet to the detected signal in the mass spectrum is more straightforward.[7] After the time domain spectrum has been acquired and digitized in FT/ICR/MS, the discrete Fourier transform is used to extract the correct frequencies and amplitudes from this spectrum. In typical FT/ICR/MS analysis, the magnitude-mode peak area is displayed.

3.   LDI-FT/ICR/MS ANALYSIS OF PORPHYRINS

3.1   Sample Preparation

Sample preparation for mass spectral analysis has always been an important factor when designing an experiment. The way a sample is prepared can favourably or adversely affect the quality of the data collected during the experiment (e.g. ion abundance, peak resolution, ion suppression). In laser desorption/ionization (LDI) mass spectrometry, sample preparation is critical and generally involves dissolving the sample in some solvent, spotting this solution onto a sample probe tip, and allowing the solvent to evaporate. Factors like the concentration of the sample in solution just prior to spotting, the type of solvent used to dissolve the sample and the technique used for actually spotting the sample are all relevant criteria important to the overall process. After the deposition of sample and subsequent evaporation of the solvent, the sample remains behind on the sample probe as crystals or a thin film. The mass spectral quality in laser desorption methods (MALDI or LDI) is directly dependent on the morphology of the sample that remains after evaporation of the solvent. In this section we demonstrate that samples deposited as crystals yield higher quality LDI-FT/ICR/MS data than did the samples deposited as a thin film on the sample probe.

3.2   Thin Films

To examine and demonstrate the influence of sample deposition on the quality of an LDI mass spectrum, we utilized metalloporphyrins as our model compounds. LDI mass spectrometry of metalloporphyrins has been demonstrated previously.[8-11]   Much of the success of LDI-MS of metalloporphyrins can be attributed to the high molar extinction coefficients of these analytes in the ultraviolet wavelength range. Researchers have demonstrated that, not only are metalloporphyrins amenable to the laser desorption process, but they can be successfully ionized using a variety of different lasers (i.e., wavelengths).

One drawback, however, to the use of laser desorption for metalloporphyrin analysis is the characteristic fragmentation that occurs. Stable metalloporphyrin ions are short lived in LDI experiments due to their tendencies to decay at extremely fast rates. Metastable decay has been found to be an inherent part of the metalloporphyrin laser desorption experiments. Given this fact, one should be able to probe the influence of sample preparation on the quality of the signal observed by monitoring the resolution, ion abundance and effective mass range of metalloporphyrins analyzed under LDI-MS conditions and correlated to state of the sample post-deposition. The internal and kinetic energy distributions for the analyte may also have some correlation to the state of the deposited sample (i.e. film vs. crystalline).

During the course of prior investigations into metalloporphyrin characterization by LDI-MS, we observed that the preparative method generally reported for metalloporphyrins yielded exceptionally low mass resolution on our time-of-flight (TOF) mass spectrometer and very low ion abundance on our FTICR mass spectrometer. A previous method for sample preparation involved dissolving the metalloporphyrin in a volatile solvent (e.g., methanol, acetonitrile). It was observed that spotting of the porphyrin solution on the sample plate and allowing the solvent to evaporate rapidly resulted in the formation of a thin film of the analyte.

A mass spectrum collected from the sample when spotted as a thin film is shown in Figure 2. Figure 2a shows the LDI-TOF mass spectrum obtained from the spotting of a 1 nmol/μL solution of vanadyl octaethylporphyrin (VOOEP) in dichloromethane on the sample plate. Similar results were observed for nickel octaethylporphyrin (NiOEP), as seen in Figure 2b. In both cases, very broad peaks are observed in the mass spectra. Two explanations for the occurrence of the broad peaks were proposed. The first explanation was that the broad peaks could be attributed to the TOF mass analyzer (a linear system), which has limited mass resolving capabilities. The second explanation was that the broad peaks were due to a broad kinetic energy distribution of the laser desorbed ions, a process which has a detrimental effect on TOF mass spectral data.

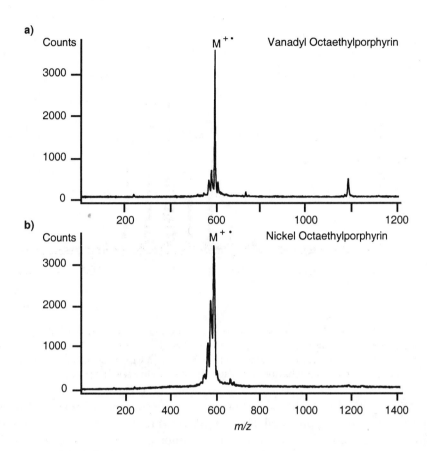

Figure 2. Positive-ion LDI-TOF mass spectra of (a) Vanadyl Octaethylporphyrin and (b) Nickel Octaethylporphyrin. The samples were prepared by spotting the analyte in dichloromethane onto the sample plate. A broad molecular ion and several low abundance fragment ions are detected.

However, when the same analysis was attempted using FT/ICR/MS, extremely low parent ion signals were obtained for all of the samples tested and the overall quality of the mass spectral results was poor. Figure 3 is the mass spectrum obtained during the analysis of a metalloporphyrin sample deposited as a thin film on the sample plate. In this example, the molecular ion for NiOEP is barely detectable, a varying distribution of fragment ions is seen and the relative abundances of the isotope peaks do not agree with the calculated isotope distributions. Further, the ion signal decayed away much quicker than would be expected based on the background pressure of the mass spectrometer resulting in a poor signal-to-noise ratio throughout the mass range.

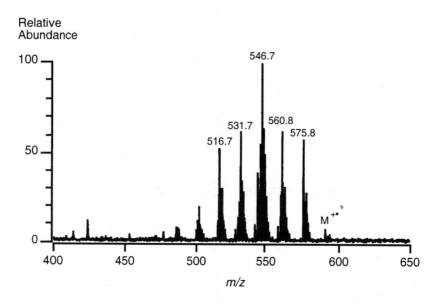

Figure 3. Positive-ion LDI-FTICR mass spectrum of Nickel Octaethylporphyrin. The sample was prepared by spotting the analyte in dichloromethane onto the sample plate. A minor abundance molecular ion signal is detected along with a number of high abundance fragment ions demonstrating the metastable behavior of this analyte under LDI-FT/ICR/MS conditions.

The FT/ICR/MS results, in conjunction with the TOFMS results, led us to suspect that samples prepared as thin films were being desorbed with high internal energy and kinetic energy distributions. The high kinetic energy distribution was suggested by the broad peaks found in the TOFMS spectra and due to the rapid decay of the transient in the FT/ICR/MS experiment. We observed that the transient decayed away at a rate higher than one would expect due to the loss of ions from the FTICR ion trap in the axial direction. Although gated trapping was employed in the FT/ICR/MS experiment, immediately after trapping the ions at high trapping potentials (8 V), the trapping plates were lowered to 1.5 V for the excitation and detection events. The signal-to-noise ratio in the FT/ICR/MS experiments increased if the trapping potential during excitation and detection were kept at higher values ($V_T > 4$ V).

The FT/ICR/MS data suggests that the metalloporphyrins are formed with an internal energy distribution significant enough to result in fragmentation of the molecular ions. If the fragment ions were formed directly during the laser desorption process, they would be much more abundant in the TOF mass spectrum (Figure 2). However, as suggested by the data presented here, if the fragment ions were being formed as a result of a metastable decay process (i.e., fragmentation occurs on the μsec or longer time-scale), then the ions would not be resolved in the linear-mode TOF instrument used for this investigation. Those fragment ions would be readily apparent in the FTICR mass spectrum due to the longer delay between ion formation and detection ($\sim$ 500–2000 msec).

3.3 Crystalline Sample Preparation

Attempts to improve the signal for the metalloporphryins involved spotting highly concentrated aliquots of the sample on the sample probe. This approach, however, did not change the mass spectral quality to any appreciable extent. It was believed that the laser power was too high and could be responsible for the poor mass spectral results. Attenuation of the laser power was performed and found to be unsuccessful at producing improved mass spectral results.

It was then thought that crystals of the analytes, rather than a thin film, might improve the quality of the mass spectral results. Thus, methods to improve the sample preparation were addressed as a potential solution. Cullen and Meyer had previously reported the growing of triclinic NiOEP crystals.[12] This procedure involved the recrystallization of NiOEP from a 1:1 solution of pyridine:1,4-dioxane. This approach was employed as a sample preparation technique in LDI-MS of metalloporphyrins. Long rod-shaped burgundy-red crystals were collected from the solution of pyridine:1,4-dioxane as it was cooled in a freezer. In addition, the recrystallization technique was applied to the VOOEP sample and found to work as well. Dark purple, fluffy, rod-shaped crystals were obtained.

A small portion of the crystals were spotted onto the sample probe tip in the following manner. Nanopure water was used to load the NiOEP crystals onto the probe tip. Loading was accomplished by pipetting 2 μL of water onto the probe tip. A small portion of the NiOEP crystals (enough to cover the tip of a microspatula) was added to the drop of water on the probe tip. Most metalloporphyrins are hydrophobic, so the NiOEP crystals did not dissolve in the water. With the microspatula, the crystals were stirred into the water until the crystals were evenly distributed into the droplet. The water was then evaporated by a continuous stream of air directed at an angle across the surface of the droplet. The NiOEP crystal were left adhered to the probe tip once all of the water was evaporated to dryness. A similar sample spotting technique was used for the VOOEP crystals and for a variety of other metalloporphyrin compounds.

The VOOEP and NiOEP samples were subsequently analyzed via LDI-FT/ICR/MS. Figures 4a and 4b show the porphyrin spectra obtained for NiOEP. The sample in Figure 4a was prepared by recrystallizing NiOEP in a 1:1 solution of pyridine:1,4-dioxane. The crystals were then crushed and applied to the sample probe as a powder. The mass spectrum was acquired under identical conditions to those used to acquire the mass spectrum in Figure 3. The recrystallization process alone results in an improvement in the quality of the mass spectral data: a more abundant molecular ion is now present and fragmentation is greatly reduced.

Figure 4b was obtained by taking the recrystallized NiOEP and applying the crystals directly to the sample probe in a water slurry. An even greater improvement in the quality of the mass spectral results is found. The molecular ion signal is more abundant in Figure 4b as compared to Figure 4a and fragmentation is reduced even further. Similar results were obtained for VOOEP, Copper Octaethylporphyrin (CuOEP), and Magnesium Octaethylporphyrin (MgOEP). A similar improvement in mass spectral quality during TOF mass analysis was obtained also for each of these samples when utilizing the recrystallization method for sample preparation.

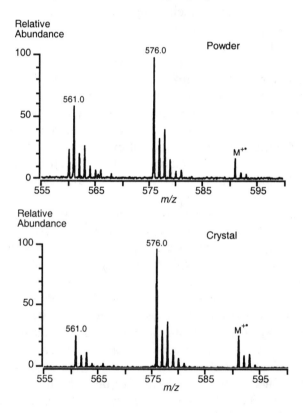

Figure 4. Positive-ion LDI-FTICR mass spectra of Nickel Octaethylporphyrin. In (a), the sample was prepared by spotting recrystallized NiOEP as a powder onto the sample plate. In (b), the sample was prepared by spotting recrystallized NiOEP directly onto the sample plate. Both mass spectra demonstrate the dramatic reduction in metastable fragmentation upon recrystallization of the sample. As seen in (b), an abundant molecular ion with minimal fragmentation is found when the recrystallized sample is analyzed directly.

Optimization of the recrystallization technique was investigated. The ratio of pyridine to 1,4-dioxane was varied and the amount of crystals collected was monitored. It was found that the optimal mixture ratio for pyridine and 1,4-dioxane was 3:1, respectively. This particular ratio produced the greatest yield for NiOEP and VOOEP, calculated as 65% and 51% respectively. The calculated yield for NiOEP and VOOEP in the previous 1:1 solution of pyridine:1,4-dioxane was 51% and 18%, respectively. Similar results were observed for CuOEP and MgOEP.

Sample preparation is very critical in LDI mass spectral analysis of metalloporphyrins. It is evident that ion abundances are directly proportional to the method used for sample preparation. Analysis of an analyte in its purest possible form is always the objective of any researcher prior to laser desorption analysis. The sample preparation method presented here yields dramatic improvements in molecular ion abundance compared to prior methods. Such sample preparation methods are directly transferable to MALDI-MS, where others have shown also that sample preparation techniques play an important role in the quality of the mass spectral results.[13-15]

4.   MALDI-FT/ICR/MS ANALYSIS OF NONPOLAR ANALYTES

Since the introduction of MALDI-MS, the primary utilization of this technique has been for the characterization of polar organic and bio-organic molecules. The matrices used for analysis of such compounds are acidic organic molecules such as 2,5-dihydroxybenzoic acid, α-cyano-4-hydroxycinnamic acid, and 3,5-dimethoxy-4-hydroxycinnamic acid (sinapinic acid). For these types of matrices and analytes, the predominant mechanism for ionisation appears to be either proton-transfer reactions between analyte and matrix or cation adduction. Although not nearly as common as reports on the production of protonated or cation-adducted molecular ions, there are several reports in the literature on the formation of analyte molecular radical cations during MALDI-MS experiments.[16-18] In addition, laser desorption studies of matrix materials reveal that many compounds that have been used as matrices form $M^{+\bullet}$ ions upon ionisation.[19, 20]

Juhasz and Costello presented the first and most complete description on the production of radical cations using MALDI.[16] They reported the successful use of four different matrices (2-(4-hydroxyphenylazo)benzoic acid (HABA), quinizarin, dithranol, and 9-nitroanthracene) for the characterization of ferrocene and ruthenocene oligomers and polymers up to m/z 10,000. These authors noted that the role of the matrix in the ionisation process could not be established: either charge transfer from the matrix to the analyte or photoionization of the

analyte directly could account for their results. Further, the chemical stability of
these matrixes was not reported in their work (although in another paper they
noted that HABA is characterized by a strong $(M+H)^+$ ion,[21]) and it is unclear
whether it was a matrix molecular radical cation or some other feature that
yielded the reported mass spectral results.

Michalak et al. have demonstrated successful MALDI mass spectra of several
proteins (insulin, cytochrome c, and serum albumin) using C_{60} as a matrix.[17] The
positive ion mass spectrum of C_{60} shows production of the molecular radical
cation and a large number of fragment ions while the negative ion mass spectrum
is composed primarily of the molecular radical anion. As C_{60} has no proton
sources for proton-transfer reactions to occur, these authors assign the molecular
ions of each protein analyzed as the $M^{+\bullet}$ or $M^{-\bullet}$ species. However, due to the
high molecular weight of these proteins ($M_r > 5,000$ u in all cases) and the low
resolution of the time-of-flight mass spectrometer used, these assignments are
tentative. Formation of radical molecular ions from these proteins would be
unique and in contrast to other studies of radical molecular ion formation. For
example, Juhasz and Costello noted that insulin could not be detected when 9-
nitroanthracene, which has no labile protons for proton-transfer reactions, was
used a matrix, but the $(M+H)^+$ ion of insulin was detected when other matrixes
that have labile protons, such as dithranol, were used.[21]

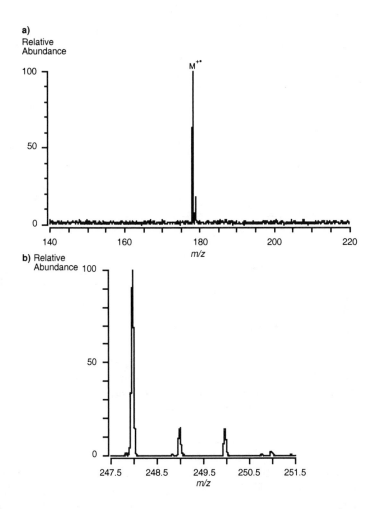

Figure 5. LDI-FTICR mass spectra of (a) anthracene and (b) terthiophene. Each of these potential charge-transfer matrices produces a radical molecular cation exclusively under LDI conditions.

We have been investigating a new class of non-polar compounds having suitable properties to function as charge-transfer matrices for the MALDI-MS analysis of non-polar analytes.[22,23] Our initial studies focused on examining the MALDI properties of such matrices. Figure 5 contains mass spectral results obtained by the direct laser desorption/ionisation of two potential charge-transfer matrices: anthracene (Figure 5a) and terthiophene (Figure 5b). Both of these potential matrices have strong molar extinction coefficients at 337 nm, the wavelength of the nitrogen laser used for these studies. As seen in Figure 5, upon direct LDI, only the radical molecular cation for each compound is detected. As illustrated in Figure 5b, the isotope distribution agrees with the predicted isotope distribution (predicted: M 100%, M+1 16%, M+2 15%; found: M 100%, M+1 15%, M+2 14%) and we have found no evidence of the formation of $(M+H)^+$ ions from these types of matrices.

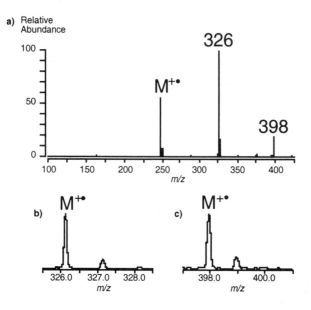

Figure 6. (a) MALDI-FTICR mass spectrum of an equimolar mixture of decamethylferrocene (*m/z* 326) and diferrocenylethane (*m/z* 398) with terthiophene used as the matrix. High abundance radical molecular cations for both analytes are detected as seen in (b) and (c).

We have utilized such compounds as matrices for the analysis of low molecular weight, nonpolar analytes. For example, in Figure 6 a mixture of decamethylferrocene and 1,2-diferrocenylethane were analyzed in the presence of terthiophene using MALDI-FT/ICR/MS. Under these experimental conditions we detect radical molecular cations for each analyte exclusively. We have investigated a number of other nonpolar analytes including ferrocene and 2,2'-methylenebis(6-tert-butyl-4-methylphenol).[23] For all analytes investigated, an abundant radical molecular cation was detected only in the presence of the nonpolar matrix. To establish that direct photoionization of the analyte was not responsible for the production of radical molecular cations, matrices having molar extinction coefficients five orders of magnitude larger than the analytes were used and the matrix-to-analyte mole ratio was varied from 100:1 to 10,000:1. Direct laser desorption/ionisation of the analytes in the absence of a matrix did not result in the production of a radical molecular cation.

In the majority of the cases studied, the presence of an analyte radical molecular cation can be predicted from the thermodynamics of the charge-transfer process: analytes with ionisation energies less than the ionisation energy of the matrix are detected while analytes with ionisation energies greater than the ionisation energy of the matrix cannot be detected. That work lends further evidence for the mechanism of ionisation using this relatively new approach to MALDI-MS. This technique should prove valuable for the analysis of electro-active polar and nonpolar molecules, such as those found in crude-oil materials (e.g., Figure 7).

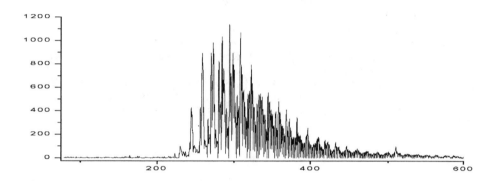

Figure 7. MALDI mass spectrum of a crude oil fraction analyzed using anthracene as the matrix.

5. ACKNOWLEDGEMENTS

Acknowledgement is made to the donors of the Petroleum Research Fund administered by the ACS, the American Society for Mass Spectrometry research award sponsored by Finnigan and the University of Cincinnati for partial financial support of our work in MALDI-MS of hydrocarbon materials.

6. REFERENCES

1. Comisarow, M. B.; Marshall, A. G. *Chem. Phys. Lett.* **1974**, *25*, 282-283.
2. Comisarow, M. B.; Marshall, A. G. *Chem. Phys. Lett.* **1974**, *26*, 489-490.
3. Marshall, A. G.; Roe, D. C. *J. Chem. Phys.* **1980**, *73*, 1581-1590.
4. Marshall, A. G.; Wang, T.-C. L.; Ricca, T. L. *J. Amer. Chem. Soc.* **1985**, *107*, 7893-7897.
5. Chen, L.; Wang, T. C. L.; Ricca, T. L.; Marshall, A. G. *Anal. Chem.* **1987**, *59*, 449-454.
6. Marshall, A. G.; Wang, T.-C. L.; Chen, L.; Ricca, T. L. In *Amer. Chem. Soc. Symposium Series*; Buchanan, M. V., Ed.; American Chemical Society: Washington, D. C., 1987; Vol. 359, pp 21-33.
7. Grosshans, P. B.; Shields, P. J.; Marshall, A. G. *J. Chem. Phys.* **1991**, *94*, 5341-5352.
8. Brown, R. S.; Wilkins, C. L. *J. Am. Chem. Soc.* **1985**, *108*, 2447-2448.
9. Brown, R. S.; Wilkins, C. L. *Anal. Chem.* **1986**, *58*, 3196-3199.

10. Forest, E.; Marchon, J.-C.; Wilkins, C. L.; Yang, L.-C. *Org. Mass Spectrom.* **1989**, *24*, 197–200.
11. Irikura, K. K.; Beauchamp, J. L. *J. Am. Chem. Soc.* **1991**, *113*, 2767-2768.
12. Cullen, D. L., Meyer, E.F. *J. Am. Chem. Soc.* **1974**, *96*, 2095-2102.
13. Lidgard, R. O.; McConnell, D. B.; Black, D. S. C.; Kumar, N.; Duncan, M. W. *J. Mass Spectrom.* **1996**, *31*, 1443–1445.
14. Dai, Y.; Whittal, R. M.; Li, L. *Anal. Chem.* **1996**, *68*, 2494–2500.
15. Hensel, R. R.; King, R. C.; Owens, K. G. *Rapid Commun. Mass Spectrom.* **1997**, *11*, 1785–1793.
16. Juhasz, P.; Costello, C. E. *Rapid Commun. Mass Spectrom.* **1993**, *7*, 343–351.
17. Michalak, L.; Fisher, K. J.; Alderdice, D. S.; Jardine, D. R.; Willett, G. D. *Org. Mass Spectrom.* **1994**, *29*, 512–515.
18. Lidgard, R.; Duncan, M. W. *Rapid Commun. Mass Spectrom.* **1995**, *9*, 128–132.
19. Ehring, H.; Karas, M.; Hillenkamp, F. *Org. Mass Spectrom.* **1992**, *27*, 472–480.
20. Gimon-Kinsel, M.; Preston-Schaffter, L. M.; Kinsel, G. R.; Russell, D. H. *J. Am. Chem. Soc.* **1997**, *119*, 2534–2540.
21. Juhasz, P.; Costello, C. E.; Biemann, K. *J. Am. Soc. Mass Spectrom.* **1993**, *4*, 399–409.
22. McCarley, T. D.; McCarley, R. L.; Limbach, P. A. *Anal. Chem.* **1998**, *70*, 4376–4379.
23. Macha, S. F.; McCarley, T. D.; Limbach, P. A. *Anal. Chim. Acta* **1999**, *397*, 235-245.

Chapter 18

X-RAY ABSORPTION SPECTROSCOPY FOR THE ANALYSIS OF HYDROCARBONS AND THEIR CHEMISTRY

Josef Hormes* and Hartwig Modrow[†]
* *Center for Advanced Microstructures and Devices (CAMD)*
  Louisiana State University
  6980 Jefferson Hwy., Baton Rouge, LA 70806
[†] *Institute of Physics*
  Bonn University
  Nussallee 12, D-53115 Bonn, Germany

1.   INTRODUCTION

In the X-ray region, the photoabsorption cross-section of all atoms shows the so called "absorption edges" corresponding to an X-ray photon having enough energy to excite an electron from an atomic inner shell (e.g. n = 1 (called K-shell) or n = 2 giving the three L - edges) into the continuum. The fine structure, which is observed in the cross section in the region of these absorption edges of atoms in molecules or solids, has been known for over 70 years. The first experimental detection of the fine structure past absorption edges was reported by Fricke [1] and Hertz [2]. A first unsuccessful theory to explain the fine structure which was observed up to several hundreds of electron volts beyond the absorption edge was developed by Kronig [3]. He tried to correlate the observed fine structure with the long-range order of the materials. However, the X-ray absorption fine structure remained a scientific curiosity until the 1970's when Sayers et al.[4] based on a theoretical formula pointed out that the Fourier transform of an extended X-ray absorption fine structure (EXAFS) spectrum should peak at distances corresponding to the radial distance of the excited atom to the neighboring coordination shells. This model opened the way to extract exact structural information directly from the measured spectra. However, as important as an applicable theory was the increasing availability of synchrotron radiation (SR) sources in the early seventies. The time required to take an

EXAFS spectrum of a metal foil dropped from several days with a standard X-ray source to some minutes using SR according to the corresponding intensities. An overview about the status of the understanding of the extended fine structure and the now available practical computational methods is given in a recent review article by Rehr and Albers [5] together with references to the older literature.

The X-ray absorption near edge structure (XANES) covers the energy range from the absorption threshold to the point where EXAFS begins, i.e. about 50 to 100 eV above the ionization potential at the energy where the wavelength of the outgoing photoelectron is equal to the distance between the absorbing atom and its nearest neighbors. The near edge structure can be divided into three parts. The spectral features below the ionization potential correspond to transitions of the excited electron into non- or antibinding molecular orbitals and unoccupied conducting bands, respectively. The region above the ionization potential is dominated by multiple scattering effects of the outgoing electron wave. The third region is the "absorption edge" itself, in general a sharp rise of the absorption cross section when the photon energy is sufficient to excite the electron from a localized inner shell into the continuum. All three regions are influenced by the geometric and/or electronic structure of the surrounding of the excited atom. The molecular orbitals into which the inner shell electron is excited reflect the chemical environment (symmetry, electronegativity of neighboring atoms, type of chemical bond etc.) of the excited atom; the energy position of the absorption edge is determined by the "chemical shift" of the initial orbital which is an atomic and highly localized orbital, and the chemical shift of the corresponding vacuum level. In favorable cases the valency of the excited atom can be derived from the energy position of the edge. The resonances above the ionization potential – often called "shape resonances – reflect the geometry of the surrounding of the excited atom qualitatively, but unfortunately so far no general approach had been developed to use those resonances for a quantitative determination of the geometrical structure. Thus, EXAFS and XANES are both sensitive to geometry and often complementary to each other: EXAFS provides mainly a tool to extract analytically geometric information whereas XANES can be used to extract information about the electronic structure and the "type" of coordination geometry.

Whereas EXAFS spectroscopy of high Z elements with suitable edges above ≈ 5 keV, i.e. for elements heavier than titanium (Z=22), developed rapidly into a valuable analytical tool already in the 70s, XANES spectra of these elements were in most cases ignored as there was no reliable theoretical understanding of these structures. X-ray absorption fine structure (XAFS) experiments for low Z elements were much more difficult until the early 1980s because in spite of availability of synchrotron radiation the necessary intense and tunable radiation was not available in the corresponding the energy range. The main reasons for that had been technical problems connected with beam line optics and monochromators. Though there are single crystals available that can be used in X-ray monochromators for monochromatizing also X-ray energies down to about 2 keV (for example InSb (111) and Beryll)), experiments are more difficult in this range as now vacuum (or at least a He atmosphere) is required in the beam line and the monochromator to avoid the strong absorption by air. Also the Be-windows that are used at most SR-facilities to separate and protect the accelerator vacuum from the vacuum in the beamline and the experimental stations, respectively, reduce the intensity so drastically that even today measurements at the K-edge of the low Z-elements such as Al, Si, P and S are still not "standard".

The situation gets even worse when we look at elements with K-edge below 1 keV such as C, N, O, and F for which measurements at the corresponding K-edges were hardly possible in the 70s. Radiation in the relevant energy range between 200 eV and 1 keV has to be monochromatized using grating monochromators under grazing incidence conditions. Monochromators that provided the necessary energy resolution had all been based on the "Rowland circle" principle and because of the additional "boundary conditions" of experiments carried out with synchrotron radiation (fixed light source and fixed position of the sample behind the exit slit) in most cases several optical elements had been required reducing the intensity of the monochromatic light to a level where experiments were extremely difficult.[6] In spite of these problems several fundamental experiments had been carried out by J. Stöhr and coworkers using the so-called "grasshopper" monochromator at the Stanford Synchrotron Radiation Laboratory (SSRL). The results of these experiments are summarized in a book by Stöhr (Ref. 33). The arrival of more suitable soft X-ray monochromators such as the so-called Dragon monochromator with a spherical grating [7] and the SX 700 using plane grating [8] and the availability of insertion devices (undulators) in dedicated SR sources gave the field a new boost. On the other hand, the near edge structure of low Z elements could and had been recorded already in the late 70s early 80s by electron energy loss spectroscopy (EELS) or to be more precise by inner shell electron energy loss spectroscopy (ISEELS). This technique has for example been used to study the C-K-spectra of several hundred carbonaceous samples and many of those results are summarized in Ref. 9.

Today, XAFS is a very important technique especially for the investigation of materials without long-range order and for the in situ observation of chemical and biological reactions. A broad variety of those applications are presented in textbooks [10-12] and in the proceedings of the corresponding conferences [13].

As compared to other techniques for structural analysis, X-ray absorption spectroscopy has some very specific advantages:

• no long range order is required in the sample;

- the technique is element specific so that the local surrounding of each type of atom can be investigated separately;
- the "complete" spectra provide information about the geometric (EXAFS) as well as about the electronic structure (XANES);
- the investigation does not destroy the samples; the radiation damage that occurs is in most cases negligible;
- due to the penetration strength of X-rays, XAFS measurements (at least in transmission and fluorescence mode) do not require a good vacuum and in many cases even in situ measurements are possible.

To characterize carbonaceous materials and especially polymers in a wider sense and to understand their properties, geometric <u>and</u> electronic information is necessary because there is a strong interdependence between these two properties. For hydrocarbon – rich materials one would like to have the following information:

- geometric structure on an atomic scale (type, distance and number of neighboring atoms)
- the spatial distribution of the various constituents of a heterogeneous sample on a nanometer scale
- electronic structure (valency of the excited atom, the nature of the chemical bond (for carbon, for example, single or multiple bonds), band structure when applicable)
- interaction of molecules with surfaces (coatings, catalysis)
- orientation of molecules at the surface of a sample or on the surface of a substrate.

All this information can be derived from XAFS measurements. In this review we will discuss not just the application of XAFS spectroscopy to "pure" hydrocarbon systems, i.e. measurements at the carbon K-edge, but we will also include examples from interaction with hetero-atoms (nitrogen, oxygen, sulfur) and finally we will also discuss some catalysts of relevance for the hydrocarbon chemistry. It is of course not the goal of this paper to present a "complete review" of these applications but we will discuss selected examples demonstrating the specific advantages of XAFS and we will provide starting references for those readers who want to learn more about a specific application. Several of these examples are taken from our own work carried out over the last years. For a more detailed discussion of the theoretical basis and the various experimental aspects of XAFS the reader is referred to corresponding textbooks.[11,12]

2. X-RAY ABSORPTION SPECTROSCOPY: THEORETICAL BACKGROUND

As a typical example of an X-ray absorption spectrum, the Pt - L - III XAFS absorption spectrum of a Pt-foil is shown in Figure 1.

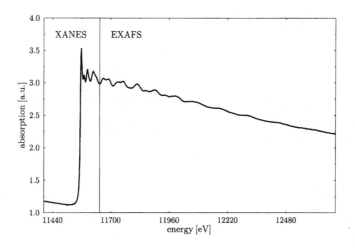

Figure 1. X-ray absorption spectrum of a platinum foil at the L-III edge measured at room temperature.

As marked in this figure also today two regions are distinguished in the spectra: the structure close to the absorption edge (XANES = X-ray absorption near edge structure) and the high energy structure (EXAFS). In the following the basic theory for both regions will be described and the information that can be extracted from the structure will be discussed.

2.1 Extended X-ray Absorption Fine Structure (EXAFS)

For photon energies E above the ionization potential E_0 the electron leaving the atom can be described as a spherical wave with the wavenumber:

$$k = \sqrt{\frac{8 \pi^2 m_e}{h^2} (E - E_0)} \qquad (1)$$

The extended fine structure (EXAFS) is now a result of the interference of this outgoing spherical wave with parts of the same wave that are backscattered from neighboring atoms. Assuming that there is not too much static disorder in the sample (which is of course not always true for very small particles!) the resulting interference pattern $\chi(k)$ can be described analytically by:

$$\chi(k) = \sum_{i=1}^{n} S_{0,i}^2(k)(\frac{N_i}{kR_i^2})f_i(k)e^{-2k^2\sigma_i^2} e^{\frac{-2R_i}{\lambda_i}} \sin(2kR_i + 2\delta_i(k) + \Phi_i(k)) \quad (2)$$

The sum in equation (2) is over coordination shells i with N_i identical backscattering atoms.

S_0^2 is less than 1 and describes the reduction of the EXAFS - amplitude due to multi - electron effects which are not included in the simple picture described above;

$f_i(k)$ is the backscattering amplitude describing the potential of an atom to "reflect" the electron wave. As the shape of f as a function of k depends on the type of atom it is in principle possible to determine the type of backscattering atom from the measured $\chi(k)$ function;

R_i is the radial distance from the excited atom to the neighboring atoms in coordination shell i;

λ_i is the mean free path describing the reduction in coherence of the electron wave due to inelastic scattering processes and the lifetime of the excited final state;

$e^{-2k^2\sigma_i^2}$ corresponds to the Debye - Waller factor known from X-ray diffraction . However, for EXAFS the mean square relative displacement between the central atom and the backscattering atoms has to be considered whereas in the diffraction formalism the mean square displacement of each type of atom has to be taken into account;

Φ_i is the phase - function of the respective backscattering atoms, δ_l is the phase shift of the central atom.

The analysis of the phase of the interference pattern provides very accurate information on radial distances R_i between the absorbing atom and the neighbor atoms in the first coordination shells. From the amplitude of the interference structure information on the type and the number of atoms N_i in the respective coordination shell can be derived. This latter information is much less accurate than the information about the radial distances. Details of the EXAFS - theory and the process of data analysis are discussed extensively in the literature.[11,12]

2.2   X-ray Absorption Near Edge Structure (XANES)

A theoretical description of the XANES region of an X-ray absorption spectrum is much more difficult than the EXAFS range. XANES incorporates the structure below the ionization potential (IP) as well as the structure extending about some tenths of eV above the IP. Spectral features below the ionization potential correspond to transitions of the excited electron into non- and antibonding molecular orbitals and unoccupied conducting bands respectively; the region above the IP is dominated by multiple scattering effects of the outgoing electron wave. The XANES part of the spectrum contains information on the geometrical arrangements of the atoms around the absorbing atom <u>and</u> on the electronic structure of this atom.[14]

As this region of the spectrum cannot be described analytically, the available information is in many cases extracted from the spectra by applying a "fingerprint" methods, i.e. by comparing the spectra of well characterized reference compounds with the spectrum of the sample under investigation. However, for the interpretation of XANES some useful empirical rules exist which have been derived by systematic investigations of well-known model compounds. Perhaps the most important is the observation that the edge in the absorption spectra (in most cases defined as the turning point in the rise of the absorption cross section to the continuum) is shifted towards higher energies as a function of increasing formal valency of the absorbing atom and also as a function of increasing electronegativity of the neighboring atoms.[15] In the last few years there has been some significant progress in the theoretical description of XANES spectra closely connected to computer codes like FEFF based on the development of the scattering series [16,17] or CONTINUUM [18].

Figure 2 highlights the chemical sensitivity of XANES spectra. Figure 2a shows the sulfur K-XANES spectra of 4 sulfur compounds (and the spectrum of an aged rubber sample) in which the sulfur valency varies between +2 for the dioctenylsulfide and +6 for zinc-sulfate. The corresponding white line, i.e. the first strong absorption line, changes its energy position by more than 10 eV so that it is in general very easy to determine the valency of sulfur in an unknown compound by just looking at the energy position of the white line. In the spectra shown in Figure 2a the white line corresponds in the molecular orbital picture to a transition from S 1s into a π^*(S-C) (for the dioctenylsulfide) and into a π^*(S-O) orbital for the other compounds.

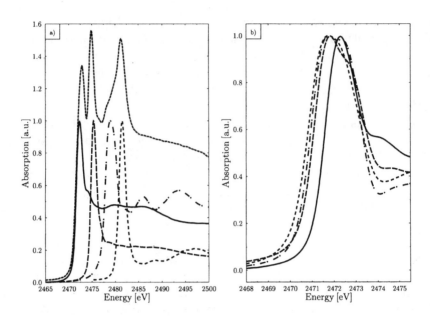

Figure 2. Sulfur K-XANES spectra of various sulfur containing molecules; a): dioctenylsulfide (solid line), dimethylsulfoxide (long dashes), dimethylsulfone (dash – dot), Zinc sulfate (short dashes), rubber sample aged for 160 min at 190°C (norrow short dashes): b): dioctenyl-n-sulfanes: n = 1 (solid line), n = 2 (long dashes), n = 3 (dash – dot), n = 4 (short dashes)

However, the sensitivity of XANES spectra goes much beyond the capability to determine "valency". Figure 2b shows the S-K-XANES spectra of 4 dioctenylsulfides (R-S_x-R with x = 1,2,3 and 4). For the monosulfide the white line has just one component corresponding to the transition S 1s $\rightarrow$ π*(S-C); for the disulfide the white line has two components: a high energy one that corresponds again to the S 1s $\rightarrow$ π*(S-C) transition and a low energy one that corresponds to the 1s $\rightarrow$ π*(S-S) transition. For longer sulfur-chains, mainly the intensity ratio changes in favor of the transition into the (S-S) orbital shifting the maximum of the white line to lower energies. This position sensitivity of the white line as a function of the sulfur-chain length offers, for example, the opportunity for a "semi-quantitative" analysis of the chain length distribution in rubber samples (see below) or for monitoring thermal oxidation processes of sulfur crosslinks.[19]

The success of the above mentioned fingerprint method is based on the "building block model". In this model one assumes that the XANES spectra of more complicated systems can be interpreted and also simulated quantitatively by using the spectra of "simple building blocks", for example the spectra for R-C-S_x-C-R (x = 1 -4) from a "suitable" molecule to simulate the spectrum of an aging rubber sample. This approach is made clear also in Figure 2a. From this figure it is obvious that the rather complicated spectrum of the rubber sample can be simulated by a weighted superposition of the spectra of (at least) the four reference compounds shown in the same figure.

For years, the generally accepted explanation for the above demonstrated sensitivity of the pre-edge part of the XANES towards changes of the local atomic environment has been based on two arguments: on symmetry considerations and, more often, on local electron density considerations. On the basis of this model, several properties could be derived e.g. the formal oxidation state of the absorber atom [19, 20, 21], electronegativity of the ligands [22, 23, 24], ionicity [25, 26], and type of bond [27, 28], as well as effective atomic charges [29, 30, 31]. As local charge density is mainly influenced by the nearest neighbors of the absorber atom, the idea that the XANES spectrum can be explained using local charge density considerations leads in a natural way to a "building block" concept of a complex molecule, [32,33] i.e. the idea that the direct environment of the absorber atom exerts the dominant influence on a XANES spectrum and therefore allows the (de)composition of the spectrum of the complex molecule from the spectra of simple molecules with identical direct environment.

However, the building block model and thus also the fingerprint method have limitations. In most cases an identical direct environment is not sufficient for a "suitable" building block and larger "units" i.e. molecules with identical second (or even higher) coordination shell have to be used to come to reliable interpretation of more complex spectra. In some cases especially when strong electron correlation effects are observed in a molecule, the simple building block models fails completely. This is demonstrated in Figure 3 where the S-K-XANES spectra of 4 molecules are shown. These molecules have identical first and second coordination shell as can be seen from the structures in Figure 4. However, there is a very strong influence from the third coordination shell leading to significant differences in the spectra.

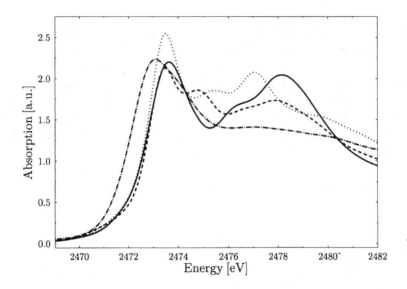

Figure 3. Sulfur K-XANES spectra of 4 "monosulfanes": — t-butynyl monosulfane,····
silylethynyl nonosulfane, ---- phenylethynyl monosulfane, ·–·–·– vinyl monosulfane.

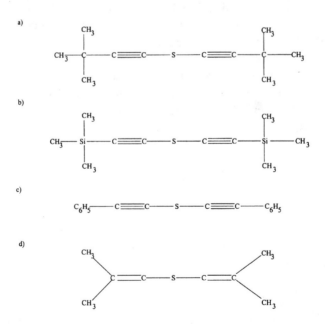

Figure 4. Molecular structure of the compounds from which the S-K-XANES spectra are shown
in Figure 3: a) t-butynyl monosulfane, b) silylethynyl nonosulfane, c) phenylethynyl monosulfane,
d) vinyl monosulfane.

Thus, it is extremely important to choose reference compounds with structures as similar as possible as compared to the sample to be analyzed, especially when spectra should be analyzed quantitatively. An example for such the quantitative analyzes of complex spectra of rubber is given in a paper by Modrow et al.[19]

3. EXPERIMENTAL TECHNIQUES

A typical experimental set up for measuring X-ray absorption spectra in the transmission mode as it is used with sometimes slight modifications in all synhrotron radiation laboratories is shown in Figure 5. The white light emitted from a storage ring enters a double crystal monochromator employing two perfect single crystals. Changing the Bragg angle in the monochromator varies the photon energy. The primary intensity I_0 falling on the sample and the intensity I transmitted through the sample are measured by ionization chambers. The preamplified currents of the ionization chambers are digitized and stored in a microcomputer system. By dividing the two signals and taking the logarithm of the result, a value directly proportional to the desired absorption coefficient is obtained.

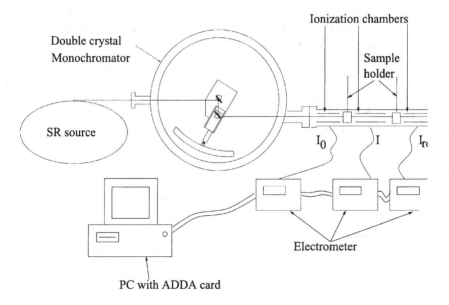

Figure 5. Scheme for the experimental set-up for measuring X-ray absorption spectra in transmission mode.

To eliminate the influences of beam instabilities on the energy calibration of the monochromator, very often a third ionization chamber is used and a reference sample is put between the second and the third chamber. The transmission spectrum of this reference sample is now recorded simultaneously with the spectrum of the actual sample. Thus, it is possible to determine e.g. the energy position of absorption edges relative to each other to better than 0.1 eV. Because of the penetration strength of X-rays in the harder X-ray range (above ≈ 5 keV) no vacuum is required around the sample so that measurements can also be carried out "in situ", for example, under high pressure. Samples can be solids, liquids or even in the gas phase.

An electron from a higher shell fills up the inner shell hole created by the outgoing electron. This process is accompanied by the emittance of either a characteristic photon (fluorescence photon) or an electron (Auger electron). The intensity of both processes recorded as a function of the excitation energy shows the same fine structure in the absorption cross section as the transmitted intensity. Fluorescence XAFS spectra are recorded for example for diluted samples where the concentration of the atoms of interest is below 1 at%.

Grazing incidence grating monochromators are used in the energy range below 1000 eV for measuring XAFS spectra of low Z elements.[7,8] To make fully use of the chemical sensitivity of XANES spectroscopy at the C-K-edge for the analysis of hydrocarbons, these monochromators should provide an energy resolution of at least 100 meV. Recently even vibronic effects have been observed at a resolution of about 50 meV in the XANES spectra of polystyrene isotopomers.[34] At energies below 1 keV, the penetration strength of X-rays is very limited and thus transmission experiments are difficult and require very thin samples (several hundred Å). Thus, spectra are very often recorded in fluorescence or electron yield mode.[33] The electrons produced by the absorption process that actually leave the sample, originate from a thin layer at the surface of the sample. This is due to the limited range of electrons with energies between some and several thousand eV in solids. Thus, XAFS measurements in the so-called electron yield mode are surface sensitive. The effective escape depth of photoelectrons for hydrocarbon films from C-K-XANES measurements and thus the information depth of such an experiment have recently been determined by Ohara et al.[36] For n-alkane the escape depth is about 35 Å.

Over the past decade analytical soft X-ray microspectroscopy has developed into a powerful tool for characterizing polymers [36, 37] but also environmental and biological samples [38, 39]. Spectromicroscopy combines the superior chemical sensitivity of XANES with adequate spatial resolution of about 50 nm to address a lot of questions in materials sciences relating for example to the composition of multicomponent thin polymer films or to polymer fibers. The resolution obtained so far is at least an order of magnitude better than what can be obtained by complementary techniques such as IR, Raman and NMR spectroscopy. For the investigation of carbonaceous materials mainly scanning transmission X-ray microscopes (STXM) are used. Here monochromatized synchrotron radiation – in most cases coming from an undulator to provide higher intensity – is focused by a Fresnel zone plate. An image at a given wavelength/energy of the sample is generated by raster-scanning the sample at the focus of the zone plate and measuring the transmitted intensity. Detail of this technique are published elsewhere [38, 40] and the operational facilities at the NSLS (Brookhaven) and at the ALS (Berkeley) are described in recent publications [41, 42, 43].

As mentioned earlier, inner shell excitation spectra for low Z elements can also be recorded by electron energy loss spectroscopy (EELS). In EELS monoenergetic electrons with energies up to 3 keV are used in forward scattering geometry to study inner shell dipole excitation processes by recording the energy loss spectrum of the electrons.[44] Dedicated EELS apparatuses can have energy resolution equal or higher than what can be achieved using X-rays for XANES measurements. Today, EELS is very often an integral part of analytical transmission and scanning transmission electron microscopes (TEM/STEM) and an energy filtered TEM image is equivalent to a STXM image. These instruments have per se an extremely good spatial resolution, but in most cases not a very good energy resolution so that their sensitivity to variations of the chemical environment is not as good as XANES microscopy at a high resolution synchrotron beam line. Using photons instead of electrons also reduces the radiation damage for sensitive samples such as biological and polymeric materials. This was quantitatively analyzed for polyethylene terephthalate by Ade et al.[45]

4.   XANES SPECTROSCOPY AND MICRO-SPECTROSCOPY AT THE CARBON K-EDGE

For low Z elements the lifetime of a K–shell hole is relatively long (as compared to high Z elements) so that the natural linewidth is relatively small: less than 0.1 eV for carbon and about 0.6 eV for sulfur.[46] Thus, with a good experimental energy resolution also overlapping lines can be distinguished, making a detailed interpretation of spectra and especially any quantitative analysis more reliable. This sensitivity to the various chemical states of carbon is the basis for the success of C-K-XANES spectroscopy and microspectroscopy. However, this sensitivity is also an absolute necessity keeping in mind the "complex" chemistry of carbon, based on the essentially covalent chemical bonds and the numerous combinations of single and multiple bonds between the carbon atoms and other low Z heteroatoms, such as O, N, F, S, and Cl.

From the C-K-XANES spectra the bond type (single, double, or triple) can easily be determined e.g. for C-O but also for C-C bonds.[33] This is demonstrated for the C-O bond for simple molecules in Figure 6 where the (ISEELS) spectra of C≡O, $H_2C=O$ and H_3C-OH are shown. For the molecules with π-bonding, a strong π* - resonance corresponding to a transition C 1s → π* (C-O) is observed whereby the energy position depends on the bond type. For the single bonded methanol, no such resonance is observed. The sensitivity to the bond type has, for example, also be used to characterize different solid carbon modification: natural diamond, tetrahedral amorphous carbon and highly oriented pyrolytic graphite.[47] The historical development until the late 80s/early 90s of XANES (NEXAFS), the status concerning the experimental data (especially for low Z-elements) and the theoretical approaches for describing the spectra are summarized in the "classical book" by J. Stöhr[33]. In this book also the early basis of the above discussed building block model are presented. Here, it is assumed that the spectra of polyatomic molecules can be viewed as the sum of corresponding diatomic or quasi diatomic building block contributions. A comprehensive overview about NEXAFS (XANES) spectroscopy and spectromicroscopy of polymers is given in recent articles by Kikuma and Tonner [48], Unger et al.[49], Urquhart et al.[50], and by Ade and Urquhart[51].

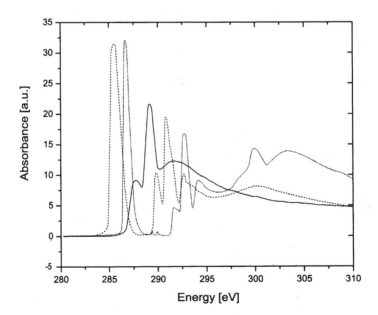

Figure 6. C-K-edge inner shell electron energy loss (ISEELS) spectrum of carbon monoxide (dots), formaldehyde (short dashes), methanol (solid line); for a better comparison the spectrum of CO was divided by 10. (spectrum based on Figure 4.5 in Ref. 33)

As an example for the sensitivity of C-K-edge XANES, the spectra of some polymers containing unsaturated functional groups will now be discussed in some detail. In Figure 7 the C-K-XANES spectra of polyurea and polyurethane polymers (and some other unsaturated polymers) are shown.[51] As to be expected, the spectra of these systems are dominated by a strong π* resonance at about 285 eV corresponding to the C 1s → π* (C=C) transition. However, this resonance clearly shows some "fine structure" caused obviously by the overlap of two transitions. According to the two major types of "chemically different" carbon atoms in the molecules, these transitions can be assigned to carbon bond to H (C-H) and carbon bond to the "rest molecule" (C-R). In the spectra of TDI polyurea and TDI polyurethane the $C_{(C-H)}$ 1s → π* (C=C) transition is split again into two peaks that are however hardly resolved in the spectra. Comparing the intensity of the π* - resonances of all the samples, one finds that the intensity for MDI polyurea and polyurethane is about half of the intensity for TDI polyurea and polyurethane. These intensity ratios reflect directly the higher number of C-R (R=amid) bonds per phenyl ring in the TDI polymers as compared to the MDI polymers.

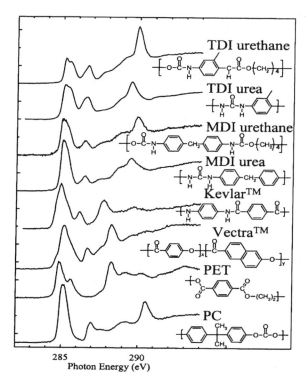

Figure 7. C 1s XANES spectra of various urethane compounds (and some other unsaturated polymers) (Figure provided by H. Ade)

A great number of polymeric systems have been analyzed by C-K XANES spectroscopy, for example, various polyurethane compounds [52,53] and reinforcing particles in polyurethanes [43], poly(phenylenevinylene) [54], polyacenes and poly-p-phenylenes [55], polymethyl-metacrylate (PMMA) [56], and plasma damaged polypropylene [57] and polycarbonate [58]. A study demonstrating once again the extreme sensitivity of XANES spectroscopy was carried out for various phthalate isomers by Urquhart et al.[59] The authors found characteristic differences in the C 1s (and also the O1s) spectra of 1,2-, 1,3-, and 1,4-substituted ester functionalities on a single benzene ring. These differences arise from the differences between the delocalization of the $\pi^*(C=O)$ orbitals of the methyl carboxylate groups and the $\pi^*(C=C)$ orbitals of the phenyl ring and they could, for example, be used for mapping phase segregation of mixed systems.

The sensitivity of C-K-XANES spectroscopy has also been used by many groups to investigate the chemisorption or physisorption of hydrocarbons on metal surfaces by carrying out measurement using the surface sensitive electron yield mode. Early results of these investigations are summarized in the book of J. Stöhr [33]. In a recent study Weiss et al.[60] have investigated the physisorption of saturated hydrocarbons on various metal surfaces. As compared to the corresponding gas phase spectra, the spectra of the molecules in direct contact with the metal – i.e. the first monolayer – show some significant differences: resonances that are assigned to Rydberg transitions are strongly quenched and new features arise that can be tentatively assigned to transitions into metal – molecule hybrid orbitals. A reliable assignment of the features observed in the here quoted study but also in many other C-K-XANES measurements of hydrocarbons is not always straight forward and in many cases rather sophisticated calculations are required. Most of those calculations are based on the so called $X\alpha$ multiple scattering method. An introduction into this method is given again in J. Stöhr's book.[33] A recent $X\alpha$ based theoretical analysis of the XANES spectrum of propane that is often used as a model for larger saturated hydrocarbons was carried out by Väterlein et al.[61] In another theoretical (and experimental) study Oji et al.[62] used ab initio molecular orbital calculations to analyze the spectra a several polycyclic aromatic hydrocarbons. They found that the spectra do not agree very well with the calculated density of unoccupied states. This is interpreted by the presence of a large core-hole effect and it explains the difficulties in obtaining reliable information about the density of unoccupied states for those aromatic systems directly from XANES spectra.

X-ray spectromicroscopy is perhaps the most important technique making use of the extreme chemical and functional group sensitivity of C-K XANES spectra. Here the differences in the energy position of "characteristic" absorption lines for various polymers are used to record XANES images in which the phases of the individual homopolymers can easily be determined down to the

100 nm scale. This type of investigation is especially important for polymer blends processed into thin films. For these systems a number of morphologies can evolve during spincasting and the following annealing processes. First experiments focusing on this time depend development were carried out by Slep et al.[63] for polystyrene (PS) and brominated polystyrene (PBrS) blends. The measurements showed that the morphology depends strongly on the blend composition. XANES – microscopy of annealed samples showed directly that with increasing concentration of PBrS, the morphology changes from droplets to surface holes in a continuous PBrS layer. Thin film blends of polysterene and poly(methyl–metacrylate) have also been studied with the same technique.[64]

Of course, also bulk polymer blends have been investigated by XANES microspectroscopy. Such studies have been carried out, for example, for blends composed of polypropylene and poly(styrene-r-acrylonitrile) [65], poly(methyl methacrylate) (PMMA) and poly(ethylene-alt-propylene) (PEP) [66] but also for blends that differ only in terms of their fraction of the same chemical moiety (here polystyrene) [67]. Figure 8 shows a "typical" image from the quoted PMMA/PEP study. The image is taken at an excitation energy of 288.0 eV where PEP has a characteristic absorption feature in its C-K-XANES spectrum and PMMA transmits more light. Thus, in the image PEP appears dark relative to PMMA. The picture shows clearly that PEP forms discrete domains within the continuous PMMA matrix.

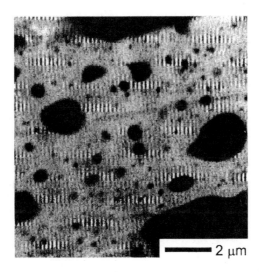

Figure 8. XANES spectromicroscopy image of a 75/25 PMMA/PEP blend taken at 288 eV, where PEP absorbs stronger than PMMA. Thus PEP appears darker than PMMA and it is obvious that PEP forms domains in the PMMA matrix (from the investigation described in Ref. 65) (Figure provided by H. Ade)

Beyond the here mentioned applications, there are a lot more areas in hydrocarbon research where XANES/NEXAFS microscopy has already been applied successfully, for example:

- Surfaces and thin films: polyimide films and surfaces, overcoats and lubricating layers on hard disks, thin film polymer bilayers, block copolymer thin films,
- Bulk morphology and composition of multicomponent, multiphasic polymers: engineered polyurethane polymers, multilayers, elastomer composites,
- Fibers.

This listing is taken from the already mentioned review article published by Ade and Urquhart [51] and references for research in all of these areas are given in this paper.

Besides the high intensity in the VUV and X-ray range, polarization properties are a major advantage of synchrotron radiation. In the electron plane the radiation is linearly polarized and above and below the plane there is elliptically/circularly polarized radiation. Making use of the linear polarization, XANES spectroscopy can be utilized to determine the orientation of specific chemical bonds (σ- or π-type bonds for example) as the transition probability of these transitions depends strongly on polarization, but also to determine the orientation of polymers at surfaces or on surfaces. This technique is called X-ray linear dichroism (XLD). XLD has been used, for example, to determine the orientation of small molecules adsorbed on surfaces,[33] the orientation of Langmuir Blodgett films,[68,69] and – as mentioned - also the orientation of surface polymer molecules. XLD measurements are normally carried out by taking angle dependant XANES spectra, i.e. the angle between the surface of interest and the electrical vector of the incoming radiation is varied. In this way the absolute orientation for planar and conjugated polymers can easily be determined as there is a clear σ/π separation for these molecules so that the orientation of the π* transition is well known. In aliphatic molecules the absolute orientation of the electronic moment has some ambiguity.[51] XLD experiments have been carried out, for example, for poly(tetrafluoroethylene) [70,71], pentacontane [72], and polyethylene terephthalate films [73].

5.   SULFUR – CROSSLINKS IN RUBBER

In the following section we will demonstrate some of the capabilities of XAFS emphasizing quantitative analyses by discussing one system in some greater detail. We have chosen rubber because of its tremendous importance in our daily life but also because we have worked in our group on the various aspects of this system for the last 10 years. In order to achieve a detailed understanding of the properties of a given rubber compound, it is crucial to gain information not only on the polymer strings, but also on the crosslinks as these determine – at least partly – the mechanical properties of the system. In most cases these crosslinks are formed using sulfur as crosslinking agent. Even though in principle XAFS analysis of the polymer backbones in such materials is possible, in practice excellent experimental techniques, which can be performed in a well equipped analytical laboratory, have been dominating the research performed on the polymer strings: at first IR spectroscopy,[74] (in spite of the inherent handicap of high absorption in many filled systems), and nowadays also ^{13}C-NMR,[75, 76] and proton-NMR [77] spectroscopy.

The situation encountered at the sulfur crosslinks is different. Using ^{13}C-NMR, it is possible to discriminate C-S and C-S-S structures, but this degree of site selectivity is not sufficient for the analysis of the crosslink distribution. In contrast to this, the S K-XANES spectra of sulfur in different chemical compounds which might be encountered in rubber and which are displayed in Figure 2b differ clearly from each other, allowing even the discrimination of sulfur chains with different chainlengths. Even bigger are the differences between the characteristic spectra of the different sulfur environments related to different stages of oxidation displayed in Figure 2a. Due to the additivity of XANES spectra, a weighted addition of such basis spectra can provide a detailed analysis of a complex spectrum, as shown in Figure 2a for the case of a spectrum of a rubber sample that has undergone thermooxidative aging.[86] At the same time and in contrast to IR spectroscopy, the penetration strength of X-rays resolves the problems connected with the investigation of highly absorbing samples, and the element selectivity inherent to the technique reduces the complexity of the encountered spectra. Thus, only destructive techniques, mostly wet chemical approaches can offer a comparable degree of precision in the description of crosslink distributions.

The general possibility to follow vulcanization and aging processes in sulfur cured rubber using S K-XANES has been demonstrated by many authors [79 - 82] and is directly recognized from Figure 9, which displays a series of spectra measured on samples extracted from a rubber batch subsequently during vulcanization at 130°C. Further information on the samples is found in Table 1.

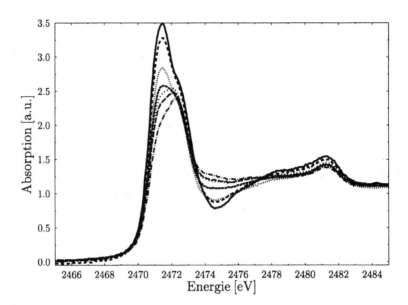

Figure 9. S-K-XANES spectra from rubber samples extracted from a rubber batch during vulcanization at 130⁰C: starting spectrum = solid line; final spectrum = dash - dot

Table 1. Vulcanization time and relative rheometer torque of the constituents of the db-series

Sample	db1089	db1090	db1091	db1093	db1094	db1095	db1096
T @130°C	10 min	20 min	22 min	29 min	40 min	50 min	100 min
% M_{max}	0	30	40	75	95	99	98

The interpretation of these spectra on the basis of a fingerprinting approach, using e.g. the references presented in Figure 2b, is rather straightforward: Clearly, two different types of transitions contribute to the white line, which are easily interpreted as S-S bond at about 2471.2 eV and S-C bond at about 2472.3 eV, respectively. Following the changes in the spectra as a function of time, it is evident that the contribution of S-S bonds is reduced, whereas the S-C contribution gains importance, which describes roughly what is known to happen during the process of vulcanization. Still, this is not equivalent to the precise extraction of a crosslink distribution, which presents a far more difficult task in spite of the additivity of XANES spectra which is the basis of the entire "blockbuilding" approach.[33]

A sample system which is well suited to demonstrate the pitfalls one may encounter when relying on the most simplified approaches to the analysis of XANES spectra are the *t*-butyl-ethinyl-n-sulfanes, whose S K-XANES spectra are displayed in Figure 10. As above, one might expect S-S and S-C bonds, located at typical energy positions. The spectra meet the first part of the expectation: in both di-and trisulfane, the whiteline is composed of a low-energy and a high-energy contribution, whereas in the monosulfane the low-energy contribution, which is assigned to the S-S bond is missing. Using Table 2 to compare the energy positions of the resonances in these spectra to the respective energy positions of the peaks of the corresponding mono- and disulfides displayed in Figure 2, first of all a considerable dependence of the respective energy position of these resonances on the moiety to which the sulfur atoms are bonding is observed. If one compares the energy position of the S-C peak in monosulfanes connected to *t*-butyl-ethinyl and octenyl moieties, respectively, a shift of 1.7 eV is observed. In fact, this sensitivity of the XANES spectra towards changes of the molecular environment is not an isolated observation. Detailed discussions of such effects can be found in Ref. 83 and 84 stressing that XANES spectra are clearly sensitive to changes in higher coordination shells.

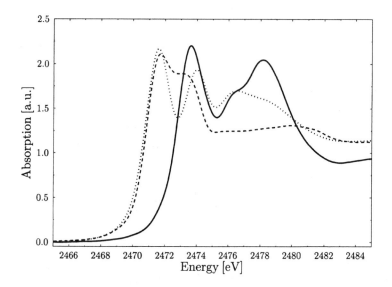

Figure 10. S-K-XANES spectra of t-butinylethynyl monosulfane (——), t-butinylethynyl disulfane (·····), t-butinylethynyl trisulfane (– – –).

Table 2. Energy positions of S-C and S-S resonance in the spectra of t-butyl-ethinyl- (t-but.) and di-octenyl- (di-oct.) n-sulfanes.

	S1, t-but.	S2, t-but.	S3, t-but.	S1, di-oct.	S2, di-oct.
S-C position	2473.6 eV	2474.0 eV	2473.3 eV	2472.3 eV	2472.5 eV
S-S position	----	2471.6 eV	4471.75 eV	----	2471.6 eV

In addition to that, the energy positions given in table 2 are not even "typical" for the S-S and S-C bonds within the series of t-butyl-ethinyl-n-sulfanes any more! Obviously, this observation cannot be related to changes in the local electron density around the absorbing atom. This value changes monotonously with increasing length of the sulfur chain whereas e.g. the position of the unoccupied sulfur- t-butyl-ethinyl orbital changes to a .4 eV higher energy position when comparing mono- to disulfide, but to an almost .8 eV lower energy position when comparing di- and trisulfide. Instead, the repulsive orbital-orbital interaction causes the corresponding increase of the splitting between the different molecular orbitals. Similar trends, although much smaller in magnitude can be found e.g. for the dioctenyl-n-sulfanes upon close inspection of Figure 2b.

The above example stresses the considerable sensitivity of XANES spectra to changes in the local environment of the absorbing atom, showing clearly that "local" in this context is far from restricted to the first coordination shell. On the one hand, it is only this degree of sensitivity which makes the extraction of a crosslink distribution from the spectrum of a rubber sample possible. After all, usually the white line of a rubber compound consists of several non-resolved contributions, and if the line shape of the different constituents was equal, this act would mean generation of information and no longer yield unique results. At the same time, one cannot escape to very large numbers of reference spectra for the basis sets for the linear space spanned by the spectra, covering every possible environment, because each increase of a basis set decreases its degree of linear independence and thus the amount of correlation within the basis as well as its ability to reproduce an arbitrary S K-spectrum. For example, it is not possible to perform a quantitative analysis of the above-mentioned db-series if one includes the spectra of the educts (i.e. the sulfur containing constituents of the batch) in the basis set, as in this case a combination of accelerator, sulfur and monosulfide consistently replaces all S2 bridges.

Consequently, the main problem when attempting a quantitative analysis in a complex system such as rubber is finding the correct minimal set of adequate basis spectra, which might vary for different types of polymer, but also for different rubber compounds and even different reaction conditions. For example, it is known that the degradation products of TBBS, a commonly used accelerator, can partly prevail in the late stages of vulcanization at low reaction temperatures if they have been added to the compound in excess, whereas they do not at high reaction temperatures. Some problems connected with the speciation and quantification of organically bound sulfur in fossil fuels are discussed in an article by Gorbaty and Kelemen [85].

Once a suitable and sufficiently reduced basis set has been found, the quantification of a crosslink distribution is possible.[19] Further necessary conditions for such a quantitative analysis are a reliable background removal in order to ensure correct normalization and, most important, a very stable energy calibration, as a change of the energy position of the white line of .1 eV(!) can induce significant changes in the extracted crosslink distribution. As a proof of principle, let us turn our attention to the results of a quantitative analysis of the "db-series". Figure 11a displays the development of the relevant contributions of the different types of sulfur crosslinks to the measured spectrum during the vulcanization process. In this case, the di-octenyl n-sulfanes shown in Figure 2b have been used as reference compounds for the basis set. If we restrict ourselves to this basis set, it turns out that an additional reference compound whose whiteline is located at higher energies than the one of the di-octenyl-monosulfane is required. As it is possible to determine the exact energy position of the white line for this "unknown" compound (HE-S1), the monosulfane spectrum shifted to higher energies has been used to simulate this reference compound, which can tentatively be related to a cyclic monosulfidic sulfur compound inducing an additional systematic error to the results of the analysis. For comparison, in Figure 11b the results of a previous study based on the results of wet-chemical analysis are displayed. Evidently, the quantitative descriptions yield comparable results, especially if one keeps in mind that compounding and vulcanization temperature of the two compounds was different. Based on the information obtained from the quantitative analysis of the S K-XANES spectra, it is now possible to understand details of the reaction process. For example, it is now possible to correlate the onset of the increase of rheometer torque to the first significant increase in the formation of disulfidic crosslinks. Furthermore, one observes in Figure 11a that the reduction of poylsufidic crosslinks is mainly correlated to the increase of S1 crosslinks and that a correlation between the decrease of disulfidic crosslinks and the increase of HE-S1 exists. Theoretical considerations have shown that from the thermodynamical point of view, the shortening of polysulfidic crosslinks to S1 bridges is highly preferred in contrast to a shortening of a disulfidic bridge. In contrast to this, the most important reaction for the decay of disulfidic crosslinks is known to be cyclization. Consequently, the quantitative analysis of the db-series yields not only qualitatively reasonable results, but allows detailed insight on a microscopic basis into the reactions occurring during vulcanization. Using similar techniques, only recently evidence on crosslink-filler interaction was extracted from S K-XANES spectra.[87]

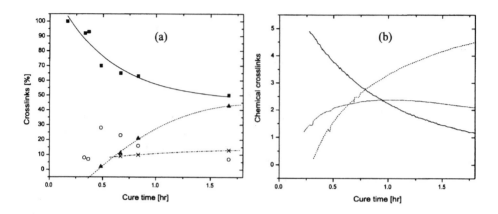

Figure 11. Development of the relevant sulfur compounds during vulcanization: (a) results of a quantitative S-K-XANES analysis: polysufides (■), disulfide (O), monosulfides (▲), HE-S1 (x); (b) results from a wet chemical analysis (redrawn from Ref. 110: polysulfides (solid line), disulfide (dots), monosulfide (short dashes)

6.   CATALYST FOR HYDROCARBON SYNTHESIS AND CATALYTIC REACTIONS

A broad overview of the various applications of XAFS spectroscopy for characterization of catalytic systems is given in a book edited by Y. Iwasawa [88]. In this book, a variety of catalysts are discussed, for example, metal oxide catalysts, supported metallic, bimetallic, and multi-metal systems, and sulfide catalysts. Also studies of the metal-support interface in supported metal catalysts and special techniques such as total reflection XAFS for the investigation of surfaces, diffraction anomalous fine structure (DAFS), and the combination of XAFS and XRD are discussed. Of special importance is the chapter on in situ XAFS measurements of catalysts where various types of reactor cells, for example sulfuration cells, high temperature and high pressure cells are discussed.[89]

Heterogeneous catalysis with a broad variety of catalysts is widely used in chemical industry. However, in fine chemicals industry just two systems dominate the applications: Raney nickel and the Sabather catalyst (supported Pd). Recent papers on the application of Pd catalysts for various catalytic reactions have focused on the influence of the support and/or the acidity on the activity [90, 91], on the influence of various "dopands" [92, 93] and on the preparation of bimetallic systems such as Pd/Pt[94] and Pd/Mn [95]. In some of these publications XAFS has been used to analyze at least some aspects of the system. For example, Lin et al.[91] determined via Pd-K EXAFS the increase of the average particle size from 6 Å to about 20 Å of Pd clusters on a styrene – divinylbenzene copolymer –

support during the ethyl acetate production from water – containing ethanol. Huang et al.[93] used Pd-L-III edge XANES spectroscopy to investigate the influence of Ag promotion on the d-band density of Pd. They found that catalysts with Ag have a less intense "white line" (for this system this is the transition from Pd 2p into the 3d band) which indicates that there is a charge transfer between Ag and Pd leading to an increased Pd-d-band electron density. The authors can correlate these observations also with the catalytic performance of the various samples. We have used the same technique for investigating the influence of various carbon supports on the electronic structure of Pd. Figure 12 shows the Pd-L-III spectra of three Pd samples supported on different types of carbon. There are clear differences not only in the intensity of the white line but also in the energy positions of this resonance. Also here the catalytic activity could be correlated with the differences in the XANES spectra.

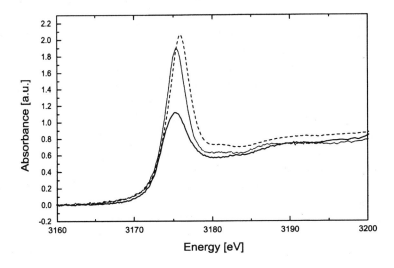

Figure 12. Pd-L-III XANES spectra of a Pd-catalyst supported on three different types of carbon.

In some catalytic reactions, for example, the combustion of carbon monoxide on transition metals, an oscillatory behavior in reaction rate as well as in reaction temperature is observed under certain conditions. This phenomenon is described is reviews by Schlüth et al.[96] and Imbihl and Ertl[97]. Ressler et al.[98] have studied these chemical oscillations during the oxidation of CO at atmospheric pressure on supported Pd catalyst in situ by XAFS spectroscopy in an energy dispersive mode. In contrast to the standard double crystal monochromators where spectra are recorded sequentially by a stepwise change of the Bragg angle, in the energy dispersive mode complete spectra are recorded in parallel. Here a bent monochromator crystal is used providing a linear variation of the Bragg angle and thus also of the reflected energy over its surface. The sample of interest is positioned in the polychromatic focus of the monochromator and spectra are recorded using a position sensitive detector.[99, 100] Applying this technique, XAFS spectra can be recorded in just several seconds so that a lot of chemical reactions (especially solid state reactions that are diffusion limited in their kinetics) can be observed in situ with a time resolution sufficient to observe also reaction intermediates. In their investigations Ressler et al. observed a periodic oxidation/reduction process accompanying the deactivation/ activation cycles of the catalyst and also an oscillation in the Pd coordination number (as deduced from the EXAFS analysis) and the Pd-Pd distance indicating an oscillation also of the oxygen surface coverage as well as the ratio of linearly and bridged bonded CO. Based on these observation it was possible to develop a model for the reaction mechanism of the observed chemical oscillations.

In the following, we will discuss in some more detail the characterization of Raney nickel by XAFS spectroscopy, as for this system the advantages of this technique are especially evident. Raney nickel catalyst is used widely in numerous industrial processes involving hydrogenation but also for anodes in liquid fuel cells.[101] In general, Raney nickel catalysts are produced by removing the aluminum from a Ni-Al alloy with an alkali solution. The residue is highly dispersed nickel with a sponge like structure and thus a very large specific surface area. In general, Raney nickel is so reactive that it has to be kept in an aqueous suspension as it is extremely pyrophoric in the dry state.

Because of its tremendous industrial importance, also in recent years several studies have been carried out to improve the performance of Raney nickel by modifying/improving the standard fabrication route. Lei et al.[102, 103], for example, prepared the Ni-Al alloys by melt – quenching claiming that this process results in a smaller grain size and a lower degree of segregation. These authors also apply H_2 – pretreatment to "activate" the rapidly solidified material. This treatment is claimed to give rise to changes of the morphology and the chemical composition of the samples and the authors report a significant increase of the Ni_2Al_3 phase as compared to $NiAl_3$ based on a quantitative phase analysis of their XRD measurements. The authors observe a high activity of the Ni_2Al_3 phase in the leaching process leading to a catalyst with high surface area and small pores.

"Although Raney catalysts have been used industrially for a very long time, there are still numerous gaps in our knowledge of the process involved in catalysis, e.g. of the hydrogen transfer and of aging and poisoning phenomena. If the findings obtained by chemical, kinetic, and physical methods are to be interpreted, the fullest possible morphological and analytical characterization of catalysts is required". These statements have been made in a paper by Birkenstock et al.[104] in 1985 and they are still true. Birkenstock et al. already emphasized the main problems that are connected with the characterization of Raney nickel – and also most other catalytic systems - as it is crucial to investigate the samples in the same states in which they are actually used in the corresponding catalytic process. In this respect, Raney nickel is a very instructive example for the advantages of using XAFS spectroscopy for the characterization of catalysts. As industrial Raney nickel is X-ray amorphous i.e. it has no long range order in its geometric structure, X-ray diffraction gives no detailed information about the geometric structure. More or less all other "standard" techniques that are applied for the characterization such as X-ray excited photoelectron spectroscopy (XPS/ESCA) or transmission electron microscopy (TEM) require high vacuum conditions around the sample so that no "in situ" experiments with wet samples are possible and thus, the samples have to be "dried" before measurements can be carried out. It is not at all clear whether this process does not change the structure/chemical composition of the samples. This could be especially crucial if XPS experiments are carried out that are surface sensitive. This problem had already been pointed out in an early XPS study by Holm and Storp [105]. Birkenstock et al. had at least tried to solve these problems partly as they developed various techniques that enable the Raney nickel samples to be transferred into the vacuum of the TEM and/or the XPS apparatus without undergoing changes. However, also in their investigation, the samples had to be dried before the actual measurements could be carried out. Figure 13 underlines the before discussed problems. This figure shows the Ni-K-XANES spectrum of a Raney nickel sample prepared on filter paper immediately after the transfer into the measuring chamber in comparison with the spectra of the same sample after drying under vacuum for about 10 and 30 minutes, respectively. Though the differences between the three spectra seem to be rather small they are significant and one should keep in mind these spectra have been taken in transmission mode, a mode that is not surface sensitive. However, it is very likely that the structural/chemical changes are more pronounced near the surface and if surface sensitive techniques (e.g. XPS) are applied for characterizing the catalysts, the wanted information about the "real" catalyst cannot be obtained.

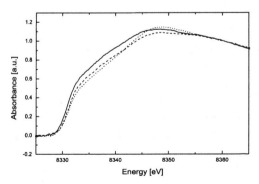

Figure 13. Ni-K-XANES spectra of a Raney nickel sample: immediately after the transfer into the experimental chamber (solid line), after 10 min (short dashes) and after 30 min (dots)

To our knowledge, four XAFS investigations have been carried out for Raney nickel samples. Nakabayashi et al.[101] have investigated sponge-type Raney that was used as anodes in liquid fuel cells. The Raney nickel electrodes were not prepared in the standard way but from substrates of spongy nickel plates coated with aluminum by use of a plasma flame gun. From the analysis of the Ni-K-EXAFS data the authors derive a coordination number of 6 in Raney nickel as compared to 12 for "ordinary" fcc - nickel. This is a strong indication that the "crystal structure" of Raney nickel contains a large number of crystal defects, i.e. there is a high degree of static disorder. The authors claim that there are just small differences between the XANES spectra of Ni-metal and Raney nickel. The main results obtained by Nakabayashi et al. from their analysis of the Ni-K EXAFS spectra of Raney nickel have been confirmed by Frenzel et al.[106]. These authors measured EXAFS spectra of Raney nickel and a metallic Ni foil in the temperature range between 20 and 320 K. Using a cumulant analysis technique that is more appropriate for systems with a high degree of disorder [107] and not the "standard" EXAFS analysis that is based on the assumption of a Gaussian distribution of the distances, Frenzel et al. determined the first shell Ni-Ni coordination number of the catalyst as 6 ± 1; they also found a significant asymmetry in the distributions of Ni-Ni distances. Both findings reflect the known sponge type structure with a high percentage of surface atoms.

In an attempt to correlate structural and chemisorptive properties of "pure" and "Cr-doped" Raney nickel, Hochard-Poncet et al.[108] have carried out a combined XAFS and XPS study. The authors observe an increased reactivity of their sample when Cr and NaOH are added simultaneously. This is explained by a purely geometric effect assuming that the simultaneous addition of Cr and NaOH induces a clustering of non- reactive oxides leading to an increase of the accessibility of the "active" nickel surface for the reactants (here hydrogen). From the Cr-XANES spectra the authors conclude that Cr has mainly valency +3 and from the analysis of the Cr-K-EXAFS spectra they deduce that Cr is not just surrounded by oxygen atoms, but that there are large clusters formed from Na^+, Al^{3+}, O^{2-}, and Cr^{3+} atoms.

Unfortunately Hochard-Poncet et al. have used an electron yield technique to measure the Ni K-EXAFS spectra of their samples (this is not absolutely clear from their description but it is very likely!). This technique is again surface sensitive (though not as much as XPS measurements) and thus requires "dry samples". Thus, the authors abandon one of the major advantages of XAFS spectroscopy that samples can be "really" measured in situ. From their XPS data (that are again surface sensitive and have been collected for dry samples) the authors calculate for all their samples (or at least for the surfaces of their samples!) a significant amount of nickel-oxide (up to 40%).

Rothe et al. [109] have investigated in some detail the activation process of Raney nickel and Fe-doped Raney nickel catalysts by measuring Ni-K-XANES (and for the doped system also Fe-K XANES) spectra of samples representing different "frozen" states of the leaching reaction. These samples have been prepared by neutralizing after different reaction times the alkaline solution used for leaching the aluminum from the alloys. Besides the "standard" Ni/Al alloy also a sample with 6 mass% Fe in the starting alloy was investigated. All spectra were measured in transmission mode. In all cases the valency of Ni – as derived from the first inflection point in the spectra - was "zero" so that there seems to be no net charge transfer between the various metals in the alloys. On the other hand, also significant differences in the fine structure of the spectra of metallic Ni and the Ni/Al alloy had been observed, pointing to a completely different local order around Ni atoms. The addition of Fe into the alloy causes slight but significant differences in the spectra. Similar conclusions (e.g. valency "zero" for the Fe atoms) could be drawn from the Fe-K-XANES spectra.

Some typical Ni-K-XANES spectra from samples where the strongly exothermal leaching reaction had been stopped after various periods of time are shown in Figure 14. Already after leaching times as short as 10 s drastic changes are observed in the rising edge of the absorption spectra, especially an increase of the intensity of the "shoulder" that is assigned to transitions into the 3d-band, reflecting a decrease of density of occupied electronic states at the Fermi level. The intensity of this feature increases for about 5 minutes before it starts decreasing. After about 60 minutes the spectrum approaches that of fcc – nickel. This is a clear indication that the Al in the samples is transferred almost instantaneously into a soluble state thus changing the electronic structure around the Ni – atoms. However, the transport of the aluminum out of the material and the re-arrangement of the geometric structure around the Ni-atoms are rather "slow" diffusion controlled processes. For the Ni/Fe/Al alloy the leaching reaction is less violent, changes in the intensity of the d-band transitions are less pronounced and the re-arrangement of the geometric structures starts earlier, but takes longer.

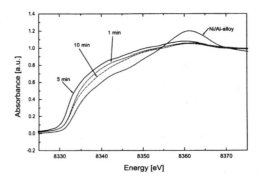

Figure 14. Ni-K-XANES spectra of a catalyst derived from Ni/Al alloy taken after various reaction times.

Figure 15 shows the Ni-K-XANES spectra of the "end products" of the leaching reaction compared once again with the spectrum of a fcc - Ni – foil. For both alloys the basic features of the electronic as well as the geometric structure of fcc Ni have developed. This can be concluded, for example, from the characteristic double peak of the fcc structure. The differences in the fine structure in the rising edge of the spectra reflect slight differences in the intensity of the transitions into the 3-d band. These could be due to a "particle size" effect and/or the interaction with some Al that still remains in the catalysts. In contrast to most XPS studies and some of the above-mentioned XAFS studies, Rothe et al. observe no significant content of Ni-oxides in their samples. This underlines once again the strength of XAFS spectroscopy, the crucial importance of "real in situ measurements" and the problems caused by, for example, drying catalytic samples and measuring surface sensitive spectra.

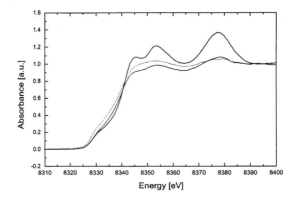

Figure 15. Ni-K-XANES spectra : fcc-Ni-foil (solid line), catalyst derived from Ni/Al alloy after a leaching time of 120 min (dots), catalyst derived from Ni/Fe/Al alloy after a leaching time of 120 min (short dashes).

7. ACHNOWLEDGEMENT

We would like to thank Prof. Dr. H. Ade for providing Figures 7 and 8. We have also profited a lot from his excellent review articles on microspectroscopy of hydrocarbons. Several former students of the Bonn synchrotron radiation group (Syli-group) have contributed with their results to this review. Their research was in most cases supported either by the German Federal Ministry of Education and Research (BMBF) or the "Deutsche Forschungsgemeinschaft" (DFG). This support is gratefully acknowledged.

8. REFERENCES

1. Fricke, H. *Phys. Rev.* **1920**, 16, 202.
2. Hertz, G. *Z. Phys.* **1920**, 3, 19.
3. de L. Kronig, R. *Z. Phys.* **1931**, 70, 317.
4. Sayers, D. E.; Stern, E. A.; Lytle, F. W. *Phys. Rev. Lett.* **1971**, 27, 1204.
5. Rehr, J. J.; Albers, R. C. *Reviews Modern Physics* **2000**, 72, 621.
6. *Synchrotron Radiation: Techniques and Applications*, Kunz, C. (Ed.), Topics Curr. Phys., Vol. 10, Springer: Berlin, 1979.
7. Sette, F.; Chen, C. T. *Rev. Sci. Instrum.* **1989**, 60, 1616.
8. Peterson, H. *Nucl. Instrum. Methods A* **1986**, 246, 260.
9. (a) Hitchcock, A. P.; *J. Electron. Spectrosc.* **1982**, 25, 245. (b) Hitchcock, A. P.; Mancini, D. C. *J. Electron. Spectrosc.* **1994**, 67, 1.
10. Oyanagi, H. "X-ray absorption fine structure", *in Applications of Synchrotron Radiation to Materials Analysis,* Saisho, H.; Gohshi, Y. (Eds), Elsevier: Amsterdam 1996.
11. *X-ray absorption: principles, applications, techniques of EXAFS, SEXAFS and XANES,* Koningsberger, D. C.; Prins, R. (Eds.), John Wiley & Sons: New York, 1988.
12. Teo, B.K. "EXAFS: Basic principles and data analysis", *Inorganic Chemistry Concepts,* Vol. 9, Springer: Berlin, 1986.
13. (a) "Proceedings of the Tenth International Conference on X-ray Absorption Fine Structure: XAFS X", Hasnain, S. S.; Helliwell, J. R.; Kamitsubo, H. (Eds.), *J. Synchrotron Radiation* **1999**, 6, 3; (b) "Proceedings of the Eleventh International Conference on X-ray Absorption Fine Structure: XAFS XI" Hasnain, S. S.; Kamitsubo, H.; Mills, D. M. (Eds.), *J. Synchrotron Radiation* **2001**, 8.
14. Bianconi, A. "XANES spectroscopy" in *X-ray absorption: principles, applications, techniques of EXAFS, SEXAFS and XANES,* Koningsberger, D. C.; Prins, R. (Eds.), John Wiley & Sons: New York, 1988; pp. 573.
15. Lengeler, B. *Phys. Bl,* **1990**. 46, 50.
16. Rehr, J. J.; Albers, R.C. *Phys. Rev. B* **1990**, 41, 8139.
17. (a) Munstre De Leon, J.; Rehr, J. J.; Zabinski, S. I.; Albers, R. C. *Phys. Rev. B* **1991**, B44, 4146; (b) Ankudinov, A. L.; Ravel, B.; Rehr, J. J.; Conradson, S. *Phys. Rev. B* **1998**, 58, 7565.
18. Natoli, C. R.; Misemer, D. K.; Doniach, S.; Kutzler, F. W. *Phys. Rev. A* **1980**, 22, 1104.
19. Modrow, H.; Hormes, J.; Visel, F.; Zimmer, R. *Rubber Chem. Technology* **2001**, 74, 281.
20. Vairavamurthy, A. *Spectrochimica Acta A* **1998**, 54, 2009.
21. Wong, J.; Lytle, F. W.; Messmer, R. P.; Maylotte, D. H. *Phys. Rev. B* **1984,** 30, 5596.
22. Hu, Z.; Kaindl, G.; Meyer, G. *J. Alloys and Compounds* **1998**, 274, 38.
23. Chetal, A. R; Sarode, P. R. *J. Phys. F* **1995**, 5, L217.
24. Srivastava, U. C. *Il Nuovo Cimento* **1972**, 11B, 68.

25. Hu, Z.; Mazundar, C.; Kaindl, G.; de Groot, F. M. F.; Warda, S.A.; Reinen, D. *Chem. Phys. Lett.* **1998**, 297, 321.
26. Dey, A. K.; Agarwal, B. K. *J. Chem. Phys.* **1973**, 59, 1397.
27. Salem, S. I.; Chang, C. N.; Lee, P. L.; Severson, V. *Journal of Physics C* **1978**.
28. Agarwal, B. K.; Verma, L. P. *Journal of Physics* **1970**, C3, 535.
29. Iwanowski, R. J.; Lawniczak-Jablonska, K. *Acta Physica Polonica* A **1997**, 97, 803.
30. Sarode, P. R.; Ramasesha, S.; Madhusudan, W. H.; Rao, C. N. R. *Journal of Physics C* **1979**, 12, 2439.
31. Kondawar, V. K.; Mande, C. *Journal of Physics C* **1976**, 9, 1351.
32. Kikuma, J.; Tonner, B. P. *J. El. Spectr. rel. phen.* **1996**, 82, 53.
33. **Stöhr, J.** "NEXAFS spectroscopy", *Springer Series in Surface Sciences Vol. 25*, Springer: Berlin, 1992.
34. Urquhart, S. G.; Ade, H.; Rafailovich, M.; Solkolov, J. S.; Zhang, Y. *Chem. Phys. Lett.* **2000**, 322, 412.
35. Ohara, H.; Yamamoto, Y.; Kajikawa, K.; Ishii, H.; Seki, K.; Ouchi, Y. *J. Synchrotron Rad.* **1999**, 6, 803.
36. Ade, H.; Zhang, X.; Cameron, S.; Costello, C.; Kirz, J.; Williams, S. *Science* **1992**, 258, 972.
37. Smith, A. P.; Ade, H. *Appl. Phys. Lett.* **1996**, 69, 3833.
38. Kirz, J.; Jacobsen, C.; Howells, M. *Quart. Rev. Biophysics* **1995**, 33, 33.
39. Hitchcock, A. *Amer. Lab.* **2001**, 33, 30.
40. Ade, H. in *Experimental Methods in Physical Sciences,* Vol. 32, Samson, J. A. R.; Ederer, D. L. (Eds.), Academic Press: New York, 1998; pp. 225.
41. Sayre, D.; Chapman, H. N. *Acta Crystallogr. A* **1995**, 51, 237.
42. Warwick, T.; Franck, K.; Kortwigh, J. B.; Meigs, G.; Moronne, M.; Myneni, S.; Rotenberg, E.; Seal, S.; Steele, W. F.; Ade, H.; Gracia, A.; Cerasari, S.; Denlinger, J.; Hayakawa, S.; Hitchcock, A. P.; Tyliszczak, T.; Rightor, E. G.; Shin, H.-J.; Tonner, B. *Rev. Sci. Instrum.* **1998**, 69, 2964.
43. Hitchcock, A. P.; Koprinarov, I.; Tyliszczak, T.; Rightor, E. G.; Mitchell, G. E.; Dineen, M. T.; Hayes, F.; Lidy, W.; Priester, R. D.; Urquhart, S. G.; Smith, A. P.; Ade, H. *Ultramicroscopy* **2001**, 88, 33.
44. Brion, C. E. *Comments At. Mol. Phys.* **1985**, 16, 249.
45. (a) Ade, H.; Smith, A. P.; Zhang, H.; Winn, B.; Kirz, J.; Rightor, E. G.; Hitchcock, A. P.; *J. Electron Spectrosc.* **1997**, 84, 53. (b) Urquhart, S. G.; Hitchcock, A. P.; Smith, A.P.; Ade, H.; Rightor, E.G. *J. Phys. Chem.* **1997**, B101, 2267.
46. Krause, M.; Olivier, J. H. *J. Phys. Chem. Ref. Data* **1979**, 8, 304.
47. Bressler, P. R.; Lübbe, M.; Zahn, D. R. T.; Braun, W. *J. Vac. Sci. Technol.* **1997**, A15, 2085.
48. Kikuma, J.; Tonner, B. P. *J. Electron Spectros. Relat. Phenom.* **1996**, 82, 53.
49. Unger, W. E. S.; Lippitz, A.; Wöll, C.; Heckmann, W. *Fresenius J. Anal. Chem.* **1997**, 358, 89.
50. Urquhart, S. G.; Hitchcock, A. P.; Smith, A. P.; Ade, H. A.; Lidy, W.; Rightor, E. G.; Mitchell, G. E. *J. Electron Spectrosc. Relat. Phenom.* **1999**, 100, 119.
51. Ade, H.; Urquhart, S. G. "NEXAFS Spectroscopy and Microscopy of Natural and Synthetic Polymers" in *Chemical Applications of Synchrotron Radiation*, Sham, T. K. (Ed.), World Scientific Publishing: Singapore, 2001.
52. Urquhart, S. G.; Ade, H. W.; Smith, A. P.; Hitchcock, A. P.; Rightor, E. G.; Lidy, W. *J. Phys. Chem. B* **1999**, 103, 4603.
53. Urquhart, S. G.; Hitchcock, A. P.; Priester, R. D.; Rightor, E. G. *J. Polym. Sci. B Polym. Phys.* 1995, 33, 1603.
54. Guo, J.-H.; Magnuson, M.; Sathe, C.; Nordgren, J.; Yang, L.; Luo, Y.; Agren, H.; Xing, K. Z.; Johansson, N.; Salaneckm, W. R.; Daik, R.; Feast, W. J. *J. Chem. Phys.* **1998**, 108, 5990.
55. Yokoyama, T.; Seki, K.; Morisada, I.; Edamatsu, K.; Ohta, T. *Physica Sripta* , **1990**, 41, 189.

56. Zhang, X.; Jacobsen, C.; Lindaas, S.; Williams, S. *J. Vac. Sci. Technol. B* **1995**, 13, 1477.
57. Gross, T.; Lippitz, A.; Unger, W. E. S.; Friedrich, J. F.; Wöll, C. *Polymer Communications* **1994**, 35, 5590.
58. Keil, M.; Rastomjee, C. S.; Rajagopal, A.; Sotobayashi, H.; Bradshaw, A. M; Lamont, C. L. A.; Gador, D.; Buchberger, C.; Fink, R.; Umbach, E. *Appl. Surf. Science* **1998**, 125, 273.
59. Urquhart, S. G.; Hitchcock, A. P.; Smith, A. P.; Ade, H.; Rightor, E. G. *J. Phys. Chem. B* **1997**, 101, 2267.
60. Weiss, K.; Weckesser, J.; Wöll, Ch. *J. Molecular Struct. (Theochem)* **1999**, 458, 143.
61. Väterlein, P.; Fink, R.; Umbach, E.; Wurth, W. *J. Chem. Phys.* **1998**, 108, 3313.
62. Oji, H.; Mitsumoto, R.; Ito, E.; Ishii, H.; Ouchi, Y.; Seki, K.; Yokoyama, T.; Ohta, T.; Kosugi, N. *J. Chem. Phys.* **1998**, 109, 10409.
63. Slep, D.; Asselta, J.; Rafailovich, M.; Sokolov, J.; Winesett, D. A.; Smith, A. P.; Ade, H.; Strzhemechny, Y.; Schwarz, S. A.; Sauer, B. B. *Langmuir* **1988**, 14, 4860.
64. Ade, H.; Winesett, D. A.; Smith, A. P.; Qu, S.; Ge, S.; Rafailovich, M. H.; Sokolov, J. *Europhys. Lett.* **1999**, 45, 526.
65. Ade, H.; Smith, A. P.; Cameron, S.; Cieslinski, R.; Mitchell, G.; Hsiao, B.; Rightor, B. *Polymer* **1995**, 36, 1843.
66. Smith, A. P.; Ade, H.; Smith, S. D.; Koch, C. C.; Spontak, R. J. *Macromolecules* **2001**, 34, 1536.
67. Smith, A. P.; Laurer, J. H.; Ade, H. W.; Smith, S. D.; Ashraf, A.; Spontak, R. J. *Macromolecules* **1997**, 30, 663.
68. Outka, D. A.; Stöhr, J.; Rabe, J. P.; Swalen, J.; Rotermund, H. H. *Phys. Rev. Lett.* **1987**, 59, 1321.
69. Hähner, G.; Kinzler, M.; Wöll, C.; Grunze, M.; Scheller, M. K.; Cederbaum, L. S. *Phys. Rev. Lett.* **1991**, 67, 851.
70. Ziegler, C.; Schedel-Niedrig, T.; Beamson, G.; Clark, D. T.; Salaneck, W. R.; Sotobayashi, H.; Bradshaw, A. M. *Langmuir* **1994**, 10, 4399.
71. Nagayama, K.; Sei, M.; Mitsumoto, R.; Ito, E.; Araki, T.; Ishii, H.; Ouchi, Y.; Seki, K.; Kondo, K. *J. Electron Spectrosc. Relat. Phenom.* **1996**, 78, 375.
72. Yamamoto, Y.; Mitsumoto, R.; Ito, E.; Araki, T.; Ouchi, Y.; Seki, K.; Takanishi, Y. *J. Electron Spectrosc. Relat. Phenom.* **1996**, 78, 367.
73. Okajima, T.; Teramoto, K.; Mitsumoto, R.; Oji, H.; Yamamoto, Y.; Mori, I.; Ishii, H.; Ouchi, Y.; Seki, K. *J. Phys. Chem. A* **1998**, 102, 7093.
74. Dikland, H. G.; Vanderdoes, L.; Bantjes, A. *Rubber Chemistry and Technology* **1993**, 66, 196.
75. Majumder, S.; Bhowmick, A. K. *Applied Polymer Science* **2000**, 77, 323.
76. Koenig, J. L. *Rubber Chemistry and Technology* **2000**, 73, 385.
77. Kumar, R. N.; Mehnert, R.; Scherzer, T.; Bauer, F. *Macromolecular Materials and Engineering* **2001**, 286, 598.
78. Hahn, J.; Palloch, P.; Thelen, N. *Rubber Chemistry and Technology* **2001**, 74, 28.
79. Chauvistre, R.; Hormes, J.; Brück, D.; Sommer, K.; Engels, H. W. *Kautschuk Gummi Kunststoffe* **1992**, 45, 808.
80. Chauvistre, R.; Hormes, J.; Sommer, K. *Kautschuk Gummi Kunststoffe* **1994**, 47, 808.
81. Paste, S.; Gotte, V.; Goulon-Ginet, C.; Rogalev, A.; Goulon, J.; Georget, P.; Marcilloux, J. *Journal de Physique IV* **1997**, 7, 665.
82. Dias, A. J.; McElrath, K. O.; Pfeiffer, D. G.; Sansone, M. Paper No.10, 153rd Spring Technical Meeting, Rubber Division, ACS, Indianapolis, 1998.
83. Chauvistre, R.; Hormes, J.; Hartmann, E.; Etzenbach, N.; Hosch, R.; Hahn, J. *Chem. Phys.* **1997**, 223, 293.
84. Flemming, B.; Modrow, H.; Hallmeier, K.-H.; Hormes, J.; Reinhold, J.; Szargan, R. *Chem. Phys.* **2001**, 270, 405.

85. Gorbaty, M. L.; Kelemen, S. R. *Fuel Processing Tech.* (Sp. Iss. SI) **2001**, 71, 71.
86. Modrow, H.; Zimmer, R.; Visel, F.; Hormes, J. *Kautschuk Gummi Kunststoffe* **2000,** 6, 308.
87. Brendebach, B.; Modrow, H. *Kautschuk Gummi Kunststoffe* 2002, **1**.
88. *X-ray absorption fine structure for catalysts and surfaces*, Iwasawa, Y. (Ed.) World Scientific: Singapore, 1996.
89. Bazin, D.; Dexpert, H.; Lynch, J. "In situ XAFS measurements of catalysts" in *X-ray Absorption Fine Structure for Catalysts and Surface*, Iwasawa, Y. (Ed.) World Scientific, 1996; p. 113
90. Zhang, J.-L.; Wang, X.-P.; Fang, K.-G.; Cai, T.-X.; Cheng, M.-J.; Bao, X.-H. *React. Kinet. Catal. Lett.* **2001**, 73, 13.
91. Lin, T.-B.; Chung, D.-L.; Chang, J. R. *Ind. Eng. Chem. Res.* **1999**, 38, 1271.
92. Shin, E. W.; Choi, C. H.; Chang, K. S.; Na, Y. H.; Moon, S. H. *Catalysis Today* **1998**, 44, 137.
93. Huang, D. C.; Chang, K. H.; Pong, W. F.; Tseng, P. K.; Hung, K. J.; Huang, W. F. *Catalysis Letters* **1998**, 53, 155.
94. Wang, Y.; Toshima, N. *J. Phys. Chem. B* **1997**, 101, 5301.
95. Renouprez, A. J.; Trillat, J. F.; Moraweck, B.; Massardier, J.; Bergeret, G. *J. Catalysis* **1998**, 179, 390.
96. Schlüth, F.; Henry, B. E.; Schmidt, L. *Adv. Catal.* **1993**, 39, 51.
97. Imbihl, R.; Ertl, G. *Chem. Rev.* **1995**, 95, 697.
98. Ressler, T.; Hagelstein, M.; Hatje, U.; Metz, W. *J. Phys. Chem.* **1997,** 101, 6680.
99. Hagelstein, M.; Cunis, S.; Frahm, R.; Nieman, W.; Rabe, P. *Phys. B* **1989**, 158, 324.
100. Blank, H.; Neff, B.; Steil, St.; Hormes, J. *Rev. Sci. Instr.* **1992**, 63, 1334.
101. Nakabayashi, I.; Nagao, E.; Miyata, K.; Moriga, T.; Ashida, T.; Tomida, T.; Hyland, M.; Metson, J. *J. Mater. Chem.* **1995**, 5, 737.
102. Lei, H.; Song, Z.; Bao, X.; Mu, X.; Zong, B.; Min, E. *Surf. Interface Anal.* **2001**, 32, 210.
103. Lei, H.; Song, Z.; Tan, D.; Bao, X.; Mu, X.; Zong, B.; Min, E. *Appl. Catalysis A* **2001**, 214, 69.
104. Birkenstock, U.; Holm, R.; Reinfandt, B.; Storp, S. *J. Catalysis* **1985**, 93, 55.
105. Holm, R.; Storp, S. *J. Electron Spectrosc. Relat. Phenom.* **1976**, 8, 139.
106. Frenzel, B.; Rothe, J.; Hormes, J.; Fornasini, P. *J. Phys. IV France* **1997**, 7, Colloque C2, C2-273.
107. Dalba, G.; Fornasini, P.; Grazioli, M.; Rocca, P. *Phys. Rev.* **1995**, 52, 11034.
108. Hochard-Poncet, F.; Delichère, P.; Moraweck, B.; Jobic, H.; Renouprez, A. J. *J. Chem. Soc. Faraday Trans.* **1995**, 91, 2891.
109. Rothe, J.; Hormes, J.; Schild, C.; Pennemann, B. *J. Catalysis* **2000**, 191, 294.
110. Loo, C. T. *Polymer* **1974**, 15, 357.

Index